D0204001

Scattering
Theory

Scattering Theory:

The Quantum Theory on Nonrelativistic Collisions

John R. Taylor, *1939–*
University of Colorado
Boulder, Colorado

John Wiley & Sons, Inc.
New York   London   Sydney   Toronto

Samford University Library

Copyright © 1972, by John Wiley & Sons, Inc.

All rights reserved. Published simultaneously in Canada.

No part of this book may be reproduced by any means,
nor transmitted, nor translated into a machine language
without the written permission of the publisher.

Library of Congress Catalog Card Number: 75-37938
ISBN 0-471-84900-6

Printed in the United States of America.

10 9 8 7 6 5 4 3 2 1

Annex
539.754

QC
173
.T33

To My Wife

75—06368

Preface

This book is intended for any student of physics who wants a thorough grounding in the quantum theory of nonrelativistic scattering. It is designed for a reader who is already familiar with the general principles of quantum mechanics (as taught, for example, in the first year graduate course at most American universities) and who has some small acquaintance with scattering theory, as would be acquired in the same first-year course. It is my hope to bring the student of atomic or nuclear physics to the point where he can begin reading the literature and tackling real problems, with a complete grasp of the underlying principles. For the student of high-energy physics I have tried to provide the necessary background for his later study of relativistic problems.

Concerning my choice of subject matter, the one feature that seems to need explanation here is the decision to exclude relativistic scattering theory. I made this decision for a variety of reasons: The theory of nonrelativistic collisions has wide application in the low-energy processes of atomic, nuclear, and particle physics; and is sufficiently extensive to make up a book by itself. The basic laws of nonrelativistic quantum mechanics are well understood and, in sharp contrast to the relativistic case, nonrelativistic scattering theory is a logically complete and selfcontained structure. More-over, there are many important features common to the nonrelativistic and relativistic theories. (For example, both are formulated in terms of the unitary scattering operator S.) This makes the former an excellent intro-duction to the latter, and wherever possible I have presented the material so as to emphasize its relevance in the relativistic, as well as the nonrelativistic, domains.

Concerning my treatment of the subject, two comments are in order. First, the reader will find that I have tried always to present a topic in terms of the simplest relevant example (the most complicated atomic process discussed in any detail is electron–hydrogen scattering; the polarization experiment chosen for discussion is the scattering of spin-half particles off a spinless target; and the first two thirds of the book deal exclusively with the simple case of single-channel scattering). I have done this because I believe that by far the best way to learn scattering theory is to become familiar with all of the basic concepts—the S operator, cross sections, the T matrix, and so on—in their simplest context. Once these concepts are really well understood, their extension to more general situations is usually a straightforward matter and can often be left to the reader. This policy means that the coverage of the book is less comprehensive and general than that of most books on the subject[1] [Mott and Massey (1933), Goldberger and Watson (1964), and Newton (1966)] and, hence, perhaps less useful as a reference for the active researcher. However, it is my hope that in this way it may prove more useful to the student struggling to learn the subject. Second, the reader will find a greater emphasis than is traditional on the time-dependent approach to the subject. Historically, scattering theory developed in the late twenties and thirties around the time-independent stationary scattering states. Only in the late fifties was this formalism properly justified with the development of a realistic time-dependent theory. Now, however, it is possible to develop scattering theory in a much more satisfactory way, beginning with the time-dependent formalism and using this to define all of the basic concepts, and only then introducing the time-independent theory as a tool for computation and for establishing certain general properties.

With the exception of a few chapters, the book is designed to be read systematically from the beginning to the end, and I hope that this is how the reader will choose to use it.[2] I also hope that the reader will try to do most, if not all, of the small number of problems at the end of each chapter. Most of these have been tested by three successive groups of students at the University of Colorado. They are intended to improve the reader's grasp of the material just covered and to introduce him to some important developments not treated in the text.

A large number of colleagues and friends have helped me in the writing of this book. Special thanks are due to Professor Thomas Jordan, Professor Michael Whippman, Mr. Rayner Rosich, and Mr. David Goodmanson, all of whom read large portions of the manuscript and made numerous helpful suggestions and criticisms. Also to Martin Hidalgo, Alan Hunt, Rayner Rosich and Robert Stolt who did the calculations behind several of the graphs and tables. I am grateful to Professor Paul Matthews for hospitality at Imperial College where I began serious work on the book; and to several colleagues at Imperial College and the University of Colorado—Kenneth

---

[1] All references are given by author's name and date and can be found on p. 463.
[2] The principal exceptions are Chapters 7, 14, 15, 20, and 21. The reader can omit or postpone some or all of the material in these chapters without seriously affecting his understanding of the subsequent material.

Barnes, Wesley Brittin, Chris Zafiratos and many others—for invaluable conversations and encouragement. Above all I want to thank my wife Debby. She not only bore with three years of authorship agonies in a most wifely way; she edited the whole manuscript and typed it twice.

JOHN R. TAYLOR

Contents

Introduction   **1**

1 Mathematical Preliminaries   **6**

1-a	The Hilbert Space of State Vectors	7
1-b	Subspaces	9
1-c	Operators and Inverses	10
1-d	Unitary Operators	13
1-e	Isometric Operators	14
1-f	Convergence of Vectors	16
1-g	Operator Limits	19

2 The Scattering Operator for a Single Particle   **21**

2-a	Classical Scattering	22
2-b	Quantum Scattering	25
2-c	The Asymptotic Condition	28
2-d	Orthogonality and Asymptotic Completeness	31
2-e	The Scattering Operator	34
2-f	Unitarity	36

3 Cross Sections in Terms of the S Matrix   **38**

3-a	Conservation of Energy	39
3-b	The On-Shell T Matrix and Scattering Amplitude	42
3-c	The Classical Cross Section	44
3-d	Definition of the Quantum Cross Section	46
3-e	Calculation of the Quantum Cross Section	49
3-f	The Optical Theorem	53

4   Scattering of Two Spinless Particles   56

4-a   Two-Particle Wave Functions   57
4-b   The Two-Particle S Operator   60
4-c   Conservation of Energy-Momentum and the T Matrix   62
4-d   Cross Sections in Various Frames   64
4-e   The Center-of-Mass Cross Section   66

5   Scattering of Two Particles with Spin   69

5-a   The Hilbert Space for Particles with Spin   70
5-b   The S Operator for Particles with Spin   72
5-c   The Amplitudes and Amplitude Matrix   73
5-d   Sums and Averages Over Spins   76
5-e   The In and Out Spinors   78

6   Invariance Principles and Conservation Laws   81

6-a   Translational Invariance and Conservation of Momentum   82
6-b   Rotational Invariance and Conservation of Angular Momentum   83
6-c   The Partial-Wave Series for Spinless Particles   85
6-d   Parity   90
6-e   Time Reversal   91
6-f   Invariance Principles for Particles with Spin;
        Momentum-Space Analysis   95
6-g   Invariance Principles for Particles with Spin;
        Angular-Momentum Analysis   103

7   More About Particles with Spin   109

7-a   Polarization and the Density Matrix   109
7-b   The In and Out Density Matrices   113
7-c   Polarization Experiments in (Spin $\frac{1}{2}$) — (Spin 0) Scattering   114
7-d   The Helicity Formalism   119
7-e   Some Useful Formulas   124

8   The Green's Operator and the T Operator   128

8-a	The Green's Operator	129
8-b	The T Operator	134
8-c	Relation to the Møller Operators	135
8-d	Relation to the Scattering Operator	138

9   The Born Series   143

9-a	The Born Series	144
9-b	The Born Approximation	147
9-c	The Yukawa Potential	150
9-d	Scattering of Electrons off Atoms	153
9-e	Interpretation of the Born Series in Terms of Feynman Diagrams	157

10   The Stationary Scattering States   164

10-a	Definition and Properties of the Stationary Scattering States	165
10-b	Equations for the Stationary Scattering Vectors	168
10-c	The Stationary Wave Functions	170
10-d	A Spatial Description of the Scattering Process	173

11   The Partial-Wave Stationary States   180

11-a	The Partial-Wave S Matrix	181
11-b	The Free Radial Wave Functions	182
11-c	The Partial-Wave Scattering States	185
11-d	The Partial-Wave Lippmann–Schwinger Equation	188
11-e	Properties of the Partial-Wave Amplitude	191
11-f	The Regular Solution	197
11-g	The Variable Phase Method	197
11-h	Iterative Solution for the Regular Wave Function	201
11-i	The Jost Function	204
11-j	The Partial Wave Born Series	207

12 Analytic Properties of the Partial-Wave Amplitude   212

12-a Analytic Functions of a Complex Variable   213
12-b Analytic Properties of the Regular Solution   215
12-c Analytic Properties of the Jost Function and S Matrix   217
12-d Bound States and Poles of the S Matrix   223
12-e Levinson's Theorem   226
12-f Threshold Behavior and Effective Range Formulas   229
12-g Zeros of the Jost Function at Threshold   232

13 Resonances   238

13-a Resonances and Poles of the S Matrix   240
13-b Bound States and Resonances   245
13-c Time Delay   249
13-d Decay of a Resonant State   253

14 Additional Topics in Single-Channel Scattering   259

14-a Coulomb Scattering   259
14-b Coulomb Plus Short-Range Potentials   266
14-c The Distorted-Wave Born Approximation   270
14-d Variational Methods   273
14-e The K Matrix   280

15 Dispersion Relations and Complex Angular Momenta   285

15-a Partial-Wave Dispersion Relations   287
15-b Forward Dispersion Relations   291
15-c Nonforward Dispersion Relations   294
15-d The Mandelstam Representation   297
15-e Complex Angular Momenta   302
15-f Regge Poles   306
15-g The Watson Transform   308

16 The Scattering Operator in Multichannel Scattering 315

16-a Channels 316
16-b Channel Hamiltonians and Asymptotic States 321
16-c Orthogonality and Asymptotic Completeness 326
16-d A Little More Mathematics 331
16-e The Scattering Operator 334

17 Cross Sections and Invariance Principles in Multichannel Scattering 338

17-a The Momentum-Space Basis Vectors 339
17-b Conservation of Energy and the On-Shell T Matrix 342
17-c Cross Sections 346
17-d Rotational Invariance 351
17-e Time-Reversal Invariance 353

18 Fundamentals of Time-Independent Multichannel Scattering 358

18-a The Stationary Scattering States 359
18-b The Lippmann–Schwinger Equations 361
18-c The T Operators 363
18-d The Born Approximation; Elastic Scattering 364
18-e The Born Approximation; Excitation 367

19 Properties of the Multichannel Stationary Wave Functions 372

19-a Asymptotic Form of the Stationary Wave Functions;
       Collisions Without Rearrangement 373
19-b Asymptotic Form of the Stationary Wave Functions;
       Rearrangement Collisions 378
19-c Expansion in Terms of Target States 380
19-d The Optical Potential 383

20 Analytic Properties and Multichannel Resonances   391

20-a Analytic Properties   392
20-b Proof of Analytic Properties   396
20-c Bound States   405
20-d Resonances   407
20-e Decay of a Multichannel Resonance   413

21 Two More Topics in Multichannel Scattering 418

21-a The Distorted-Wave Born Approximation   418
21-b Final-State Interactions   424

22 Identical Particles   434

22-a The Formalism of Identical Particles   435
22-b Scattering of Two Identical Particles   441
22-c Multichannel Scattering with Identical Particles   448
22-d Transition Probabilities and Cross Sections   450
22-e Electron–Hydrogen Scattering   454

References   463

Index   467

Scattering Theory

Introduction

The most important experimental technique in quantum physics is the scattering experiment. That this is so is clear from even the briefest review of modern physics. In atomic physics Rutherford's discovery of the nucleus was based on his study of α particles scattering off a gold foil; and the Franck–Hertz experiment established the existence of atomic energy levels by observation of electrons scattering off mercury vapor. In nuclear physics, the first clear evidence of nuclear structure came from Rutherford's observation of the scattering reaction:

$$\alpha + {}^{14}N \rightarrow p + {}^{17}O$$

And in the study of elementary particles, the scattering experiment not only provides the experimental data, but is also the principal means for the creation of the particles themselves, as for example, in the pion-production process

$$p + p \rightarrow p + p + \pi^0$$

The theoretical tool for the analysis of scattering experiments is scattering theory. In discussing scattering theory it is convenient to recognize various possible divisions of the subject. First, there are the nonrelativistic and relativistic theories, and as discussed in the preface, this book restricts itself to the former. Second, there are the single-channel and multichannel parts of the theory. And third, there are the time-dependent and time-independent parts. These divisions have determined the organization of this book.

Before describing our handling of single-channel and multichannel scattering, we must discuss briefly the definition of the two concepts. In most collisions there are many different sets of particles that can emerge in the final state. For example, when α particles are fired at nitrogen some of the different possible final configurations are

$$\alpha + {}^{14}N \rightarrow \alpha + {}^{14}N$$
$$p + {}^{17}O$$
$$\alpha + \alpha + {}^{10}B$$
$$\text{etc.}$$

Each of the possible final sets of particles is called a *channel;* therefore, a process of this type is called a *multichannel collision.* However, there are

1

certain simple processes in which there is just one channel. Two examples of such *single-channel* processes are the low-energy scattering of electrons off protons or of neutrons off α particles. In either of these processes the only possible outcome is elastic scattering,

$$e + p \rightarrow e + p$$

and

$$n + \alpha \rightarrow n + \alpha$$

The concept of a single-channel process is really an idealization. If, for example, we increase the energy of the n–α collision above about 20 MeV the neutron can knock the α particle apart; and in the e–p example there is always the possibility of producing low-energy photons, whatever the incident energy. Thus, neither example is truly a single-channel collision. Nonetheless, there are many processes (including these two examples) that can be well approximated as single-channel collisions, under the right conditions. And within the framework of nonrelativistic quantum mechanics, the scattering of a single particle off a fixed potential, and of two particles off one another, provide completely consistent models of single-channel systems.

The formalism of single-channel scattering is naturally much simpler than that of the general multichannel problem. At the same time the former includes almost all of the basic concepts needed for the latter. We shall therefore treat single-channel scattering in considerable detail before we go on to the multichannel problem. Specifically, after one chapter of mathematical preliminaries, Chapters 2–15 cover all aspects of single-channel scattering, and then Chapters 16–21 give a parallel treatment of the multichannel problem. (In Chapter 22, we consider the special problems of identical particles.)

Our other main division of scattering theory is into its time-dependent and time-independent parts. The first of these deals with the time-dependent wave function that describes the progress of a collision as it actually occurs. Well before the collision begins, and again after it is all over, the particles involved behave just like free particles, and, therefore, the corresponding wave functions behave like free wave functions. It proves possible to relate the essentially free wave function before the collision to that after the collision by a certain unitary operator, called the *scattering operator* S. In practice all measurements are made on the particles before and after the collision. (Even an extremely slow collision normally lasts much less than 10^{-10} sec.) It follows, therefore, that all experimentally relevant information (at least as regards scattering experiments) is contained in this one operator S. In particular, the experimentally measured *scattering cross sections* can all be expressed in terms of the matrix elements of S.

The time-independent formalism arises (at least in its simplest form) from an expansion of the actual time-dependent wave functions in terms of the so-called stationary scattering states, which are just the appropriate eigenfunctions of the Hamiltonian. The principal usefulness of this formalism is that it provides the means for the actual computation of the scattering operator (or the related scattering amplitude) and for establishing a number of its general properties.

A natural (though not the historical) order to develop scattering theory is to begin with the time-dependent formalism, to use this to define the S operator and scattering cross sections, and only then to develop the time-independent formalism as a computational technique. This is the order followed in this book, the content of which can be summarized as follows:

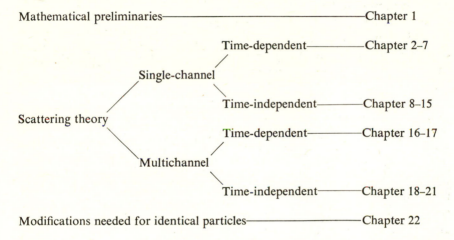

Mathematical preliminaries————————————————————Chapter 1

Scattering theory

Single-channel
- Time-dependent————————Chapter 2–7
- Time-independent————————Chapter 8–15

Multichannel
- Time-dependent————————Chapter 16–17
- Time-independent————————Chapter 18–21

Modifications needed for identical particles————————————Chapter 22

The order of presentation just outlined is by no means the order in which scattering theory was originally developed. Historically, the subject developed —in analogy with the theory of bound states—around the stationary scattering states; that is, in terms of the time-independent formalism. Only later was the time-dependent theory developed to provide a proper justification of the results already derived.

It is the traditional approach, beginning with the time-independent theory, that is presented in most elementary texts on quantum mechanics. Thus, in the scattering of a single particle off a fixed potential, the reader is undoubtedly familiar with the "scattering wave function" $\psi_{\mathbf{p}}^{+}(\mathbf{x})$, defined as the solution of the time-independent Schrödinger equation with the boundary conditions

$$\psi_{\mathbf{p}}^{+}(\mathbf{x}) \xrightarrow[r \to \infty]{} (2\pi)^{-\frac{3}{2}} \left\{ e^{i\mathbf{p} \cdot \mathbf{x}} + f(E, \theta) \frac{e^{ipr}}{r} \right\}$$

(Here, as throughout this book, we use units such that $\hbar = 1$.) This wave function is said to represent a steady incident beam of particles with momentum **p** plus a spherically spreading scattered wave with amplitude $f(E, \theta)$. This interpretation leads at once to the celebrated result for the differential cross section

$$\frac{d\sigma}{d\Omega} = \frac{\text{scattered flux/solid angle}}{\text{incident flux/area}} = |f(E, \theta)|^2$$

While this approach leads to the correct result (or, at least, what is the correct result under the right conditions) and is doubtless an acceptable first introduction to the subject, it is considerably less than satisfactory. A wave function depending on the one variable **x** should represent the state of one particle, not a beam of particles as claimed. And since $\psi_{\mathbf{p}}^+(\mathbf{x})$ is not normalizable, it cannot, in fact, represent a state at all. Furthermore, the function $\psi_{\mathbf{p}}^+(\mathbf{x})$ is an eigenfunction of the Hamiltonian and, therefore, corresponds to a steady–state situation—the complete opposite of the obviously time-dependent development of any real collision. Finally, the computation of the scattered and incident fluxes completely ignores the interference of the two waves.

Nonetheless, it is no accident that the traditional argument produces the right answer. In fact, all of the above objections can be removed and the desired conclusions justified if we build up a normalized, time-dependent wave packet by superposition of the wave functions $\psi_{\mathbf{p}}^+(\mathbf{x})$ with suitable momenta **p**. That is, the traditional time-independent approach can be justified by using it to construct a time-dependent formalism. This entirely legitimate procedure is in fact presented in a number of the more advanced texts on quantum mechanics. Nonetheless, in this book we shall follow the alternative and more natural procedure beginning with the realistic time-dependent formalism and introducing the time-independent theory only when it is needed as a means of computation. Probably the only regrettable consequence of this procedure is that the reader familiar with the traditional treatment in terms of the stationary states $\psi_{\mathbf{p}}^+(\mathbf{x})$ will have to wait until Chapter 10 before making contact with familiar ground.

We conclude with some brief comments on notation. For the vectors representing the states of a quantum-mechanical system we use Dirac's ket notation $|\psi\rangle$, $|\phi\rangle$, etc. Although this has certain undeniable disadvantages, it does often have the advantage of clarity. For example, the wave function $\psi_{\mathbf{p}}^+(\mathbf{x})$ discussed above is replaced by the compact $|\mathbf{p}+\rangle$. It also allows us to use the Greek letters $|\psi\rangle$, $|\phi\rangle$, ... exclusively for proper, normalizable vectors, since the improper vectors—the plane wave states $|\mathbf{p}\rangle$, the angular-momentum eigenstates $|E, l, m\rangle$, the scattering states $|\mathbf{p}+\rangle$—are labelled by the relevant (roman) eigenvalues.

As far as possible we label operators and matrices by capital letters and numbers by lower case. A typical operator is A and its eigenvalues a. The position and momentum operators of a single particle are $\mathbf{X}$ and $\mathbf{P}$, the corresponding eigenvalues $\mathbf{x}$ and $\mathbf{p}$. When we discuss two particles with position operators $\mathbf{X}_1$ and $\mathbf{X}_2$ and momenta $\mathbf{P}_1$ and $\mathbf{P}_2$, our rule forces us into the following unconventional notation: $\bar{\mathbf{X}}, \mathbf{X}$ = operators for the center-of-mass and relative positions; and $\bar{\mathbf{P}}, \mathbf{P}$ = operators for the total and relative momenta. These operators have corresponding eigenvalues $\bar{\mathbf{x}}, \mathbf{x}, \bar{\mathbf{p}}, \mathbf{p}$. Needless to say there are the inevitable exceptions to our general rule; these include lower case $\boldsymbol{\sigma}$ and ρ, which we use for the Pauli matrices and density matrix, capital E for the eigenvalues of energy, and several others.

With certain overworked letters we have recourse to various type faces. In particular, all unitary (and anti-unitary) operators are denoted by sans serif type. The scattering operator is S and its eigenvalues s, rotation operators are R, displacements D, parity P, time-reversal T, and so on.

One other convention concerning operators deserves mention. We shall be concerned both with problems in which momentum is conserved (e.g., scattering of two particles interacting with each other) and ones in which it is not (e.g., scattering of one particle off a fixed potential). It proves convenient to have different notations for the collision operators of these two types of system. Thus, the S operator for systems which do not conserve momentum we denote by S, and that for systems which do conserve momentum by the bold face **S**.

Sets of state vectors $|\psi\rangle$ we denote by script capital letters. Thus the Hilbert space of all state vectors (of a given system) we denote by $\mathscr{H}$ and the various subspaces of interest by $\mathscr{B}$, $\mathscr{D}$, $\mathscr{R}$, and $\mathscr{S}$. Finally, vectors in real three-dimensional space ($\mathbb{R}^3$) are denoted as usual by bold face. The magnitude of a general vector $\mathbf{a}$ is denoted by $a \equiv |\mathbf{a}|$ (in the case of the position vector $\mathbf{x}$ we use $r = |\mathbf{x}|$) and the unit vector in the direction of $\mathbf{a}$ is $\hat{\mathbf{a}} = \mathbf{a}/a$. The unit vectors pointing along the three coordinate axes are written $\hat{\mathbf{1}}, \hat{\mathbf{2}}, \hat{\mathbf{3}}$.

1 Mathematical Preliminaries

1-a   The Hilbert Space of State Vectors

1-b   Subspaces

1-c   Operators and Inverses

1-d   Unitary Operators

1-e   Isometric Operators

1-f   Convergence of Vectors

1-g   Operator Limits

One of the main objectives of this book is to show that the quantum theory of scattering need not be so difficult as is generally supposed. One of the principal means towards this end will be the use of a slightly more sophisticated mathematics than is used in many texts. For this reason the first chapter is devoted to a brief review of the mathematics that will be used in the remaining 21 chapters.[1]

The mathematics in question is not especially sophisticated by modern standards, and for most readers the bulk of this chapter will be no more than a review of familiar material. A few ideas, particularly the notions of an isometric operator and of convergence (both of central importance in scattering theory) may be new to many.

We shall begin the chapter with a quick survey of Hilbert spaces, eigenvector expansions, and subspaces. We shall then review some properties

---

[1] Much of the material in this chapter can be found in several texts on quantum mechanics (e.g., Messiah, 1961, Gottfried, 1966, or especially, see Jordan, 1969).

of operators, with particular reference to unitary and isometric operators. Finally we shall discuss the important notion of convergence of vectors and operators. We shall concentrate mainly on the formalism appropriate to the simplest of all systems, a single spinless particle moving in a fixed potential. The machinery for handling more complicated systems will be introduced later as it is needed.

The reader who is already conversant with all of this material may safely omit the whole of this Chapter.

1-a.   The Hilbert Space of State Vectors

The states of a single spinless particle are labelled by wave functions $\psi(\mathbf{x})$ satisfying

$$\int d^3x \, |\psi(\mathbf{x})|^2 < \infty \tag{1.1}$$

Each wave function $\psi(\mathbf{x})$ can be regarded as specifying the coordinates of an infinite-dimensional "*state vector*" $|\psi\rangle$. With any two of these vectors $|\psi\rangle$, $|\phi\rangle$ one can form linear combinations $a\,|\psi\rangle + b\,|\phi\rangle$ and the scalar product $\langle \psi \mid \phi \rangle = \int d^3x\psi(\mathbf{x})^*\phi(\mathbf{x})$. The *norm*, or length, of a vector is defined as $\|\psi\| = +\langle \psi \mid \psi \rangle^{1/2}$

With these definitions the set of all state vectors of a single particle forms a *linear vector space* of the particular variety known as a *Hilbert space*.[2] This is, in fact, the common feature of all quantum-mechanical systems. The precise nature of the wave functions depends on the character of the system under consideration, but in every case the wave functions define a Hilbert space, for which we use the general symbol $\mathscr{H}$. The Hilbert space appropriate to the single spinless particle discussed above is $\mathscr{H} = \mathscr{L}^2(\mathbb{R}^3)$, the space of all Lebesgue square-integrable[3] functions $\psi(\mathbf{x})$ of the variable $\mathbf{x}$ in real three-dimensional space $\mathbb{R}^3$. For a system of N distinct spinless particles it is $\mathscr{L}^2(\mathbb{R}^{3N})$, the space of all square-integrable functions $\psi(\mathbf{x}_1, \ldots, \mathbf{x}_N)$ of N positions $\mathbf{x}_1, \ldots, \mathbf{x}_N$. For a single particle of spin s (and no other degrees of freedom) it is the $(2s + 1)$-dimensional Hilbert space of all $(2s + 1)$-component spinors.[4]

---

[2] More precisely, a complex separable Hilbert space. For the record, we note that this is defined as a complex linear vector space which has the familiar scalar product $\langle \psi \mid \phi \rangle$, which possesses a countable orthonormal basis, and which has the convergence properties discussed in Section 1-f. For more details see Jordan (1969).

[3] A function is Lebesgue square-integrable if it satisfies (1.1) where the integral is a Lebesgue integral. The precise nature of the integral is a technical point that need not worry the reader who is unfamiliar with the Lebesgue theory of integration.

[4] In the mathematical literature the name Hilbert space is often reserved for infinite-dimensional spaces. In physics it is usual to allow both finite- and infinite-dimensional Hilbert spaces. We shall follow the physicists' practice.

An important property of any quantum-mechanical Hilbert space $\mathscr{H}$ is that it has a countable orthonormal basis; that is, there is a set of vectors $|1\rangle, |2\rangle, \ldots$ in $\mathscr{H}$ such that

$$\langle n' \mid n \rangle = \delta_{n'n} \tag{1.2}$$

and every $|\psi\rangle$ in $\mathscr{H}$ can be expanded as

$$|\psi\rangle = \sum_n \psi_n |n\rangle \tag{1.3}$$

The expansion coefficients ψ_n can be regarded as the coordinates of $|\psi\rangle$ in the coordinate system, or *representation*, defined by $|1\rangle, |2\rangle, \ldots$. According to (1.2) they are given by

$$\psi_n = \langle n \mid \psi \rangle \tag{1.4}$$

The observables of a quantum-mechanical system are given by Hermitian operators whose eigenvectors are (in a sense which we now discuss) bases of $\mathscr{H}$. There are some observables (e.g., the Hamiltonian of a simple harmonic oscillator) whose eigenvectors form an orthonormal basis of exactly the kind described above. Unfortunately, most observables do *not* have this property. In fact, many observables, such as the one-particle position and momentum operators $\mathbf{X}$ and $\mathbf{P}$ and the free Hamiltonian $H^0 = \mathbf{P}^2/2m$, have no eigenvectors at all. (No "proper" eigenvectors in $\mathscr{H}$, that is.) However, it was shown by von Neumann that, for the purposes of quantum mechanics, a suitable generalization of the basis of eigenvectors is provided by the *spectral decomposition*, and further, that every *self-adjoint* operator does have a spectral decomposition.[5] For this reason it is always assumed in quantum mechanics that the observables correspond to self-adjoint operators.

The formalism of the spectral decomposition, as developed by von Neumann and others, has not achieved wide usage among physicists, who prefer the formalism of Dirac. In the work of the latter, the observable, or self-adjoint operator, is treated *as if* its eigenvectors were a basis of $\mathscr{H}$, by the introduction of "improper vectors." [A "proper" vector is a vector in $\mathscr{H}$; that is, a normalizable vector, or a vector of finite length. Improper vectors have infinite length—see (1.5)—and do not belong to $\mathscr{H}$.] Thus, for example, even though the position operator $\mathbf{X}$ has no proper eigenvectors we

---

[5] The definition of a self-adjoint (or "hypermaximal") operator is quite technical and need not be given here. For a large class of operators self-adjointness is equivalent to the more familiar hermiticity (or "hermitian symmetry" or "symmetry"). In general, however, self-adjointness is the stronger property (i.e., self-adjointness implies hermiticity but not the converse). For an account of these points and of spectral decomposition see Jordan (1969).

introduce the improper eigenvectors $|\mathbf{x}\rangle$ satisfying $\mathbf{X}\,|\mathbf{x}\rangle = \mathbf{x}\,|\mathbf{x}\rangle$ and re-producing as closely as possible the properties (1.2) to (1.4) of an ortho-normal basis. The normalization condition is

$$\langle \mathbf{x}' \mid \mathbf{x}\rangle = \delta_3(\mathbf{x}' - \mathbf{x}) \qquad (1.5)$$

and the expansion of any proper $|\psi\rangle$ is

$$|\psi\rangle = \int d^3x\,\psi(\mathbf{x})\,|\mathbf{x}\rangle$$

The expansion coefficient $\psi(\mathbf{x})$ is just the familiar spatial wave function and, according to (1.5) is given by

$$\psi(\mathbf{x}) = \langle \mathbf{x} \mid \psi\rangle \qquad (1.6)$$

This clarifies the sense in which $\psi(\mathbf{x})$ can be regarded as the coordinate of $|\psi\rangle$ in a particular coordinate system in $\mathscr{H}$; namely the representation in which $\mathbf{X}$ is diagonal.

Similarly, we introduce momentum eigenvectors $|\mathbf{p}\rangle$, which we normalize so that $\langle \mathbf{p}' \mid \mathbf{p}\rangle = \delta_3(\mathbf{p}' - \mathbf{p})$. For any $|\psi\rangle$ the quantity $\langle \mathbf{p} \mid \psi\rangle$ is just the momentum-space wave function of $|\psi\rangle$, which we shall often write as $\psi(\mathbf{p})$ when there is no danger of confusion with the spatial wave function $\psi(\mathbf{x})$. With our normalizations the spatial wave function of $|\mathbf{p}\rangle$ is

$$\langle \mathbf{x} \mid \mathbf{p}\rangle = (2\pi)^{-3/2}e^{i\mathbf{p}\cdot\mathbf{x}}$$

In this book we shall use the improper vectors of Dirac. However, it cannot be overemphasized that *only the proper vectors (the vectors in $\mathscr{H}$) represent physically realizable states*. Improper vectors, such as $|\mathbf{p}\rangle$, do not represent physical states and have significance only as objects in terms of which the proper vectors can be expanded. This distinction is especially important in scattering theory where several results that must obviously be true for a physical state vector are nonetheless false for improper vectors. For example, the central result of scattering theory is that any vector representing the evolution of a collision process behaves just like a free-particle state vector long before and long after the collision takes place. This result is not true when applied to the improper scattering eigenstates.[6]

1-b. Subspaces

A *subspace* $\mathscr{S}$ of a Hilbert space $\mathscr{H}$ is any subset that itself has all the properties of a Hilbert space; in particular, if $|\psi\rangle$ and $|\phi\rangle$ are in $\mathscr{S}$ then so is

---

[6] In Chapter 10 we shall discuss at length the sense in which the so-called asymptotic condition applies to the improper eigenstates. Here we wish only to emphasize that it does not apply to improper vectors in the same way as it does to proper vectors and that there is no reason why it should.

$a \left| \psi \right\rangle + b \left| \phi \right\rangle$. Subspaces play an important role in quantum mechanics. For example, if the Hamiltonian H of some system has the eigenvalue E, then the set of all vectors representing states of energy E (that is, the set of all $\left| \psi \right\rangle$ satisfying $H \left| \psi \right\rangle = E \left| \psi \right\rangle$ for fixed E) is a subspace.

We shall say that a vector $\left| \phi \right\rangle$ is orthogonal to the subspace $\mathscr{S}$ if it is orthogonal to every vector in $\mathscr{S}$; that is, $\left\langle \phi \mid \psi \right\rangle = 0$ for all $\left| \psi \right\rangle$ in $\mathscr{S}$. (For example, in the real vector space $\mathbb{R}^3$, the unit vector $\hat{\mathbf{3}}$ pointing along the z axis is orthogonal to the subspace defined by the x–y plane.) It is easy to see that the set of all vectors orthogonal to a subspace $\mathscr{S}$ is itself a subspace, which we call the *orthogonal complement* $\mathscr{S}^{\perp}$ of $\mathscr{S}$;

$$\mathscr{S}^{\perp} = \{\left| \phi \right\rangle \text{ in } \mathscr{H}; \left| \phi \right\rangle \text{ orthogonal to } \mathscr{S}\}$$

(For example, the z axis is the orthogonal complement of the x–y plane in $\mathbb{R}^3$.) We write $\mathscr{H} = \mathscr{S} \oplus \mathscr{S}^{\perp}$ and, as can be easily checked, every $\left| \psi \right\rangle$ in $\mathscr{H}$ can be uniquely expressed as $\left| \psi \right\rangle = \left| \phi \right\rangle + \left| \chi \right\rangle$ where $\left| \phi \right\rangle$ is in $\mathscr{S}$ and $\left| \chi \right\rangle$ in $\mathscr{S}^{\perp}$.

Because the two subspaces $\mathscr{S}$ and $\mathscr{S}^{\perp}$ can themselves be split up in the same way, we introduce the more general notation

$$\mathscr{H} = \mathscr{S}_1 \oplus \cdots \oplus \mathscr{S}_n$$

to mean that the subspaces $\mathscr{S}_i$ are mutually orthogonal and that every $\left| \psi \right\rangle$ in $\mathscr{H}$ can be expressed as

$$\left| \psi \right\rangle = \left| \psi_1 \right\rangle + \cdots + \left| \psi_n \right\rangle$$

with each $\left| \psi_i \right\rangle$ in $\mathscr{S}_i$. Under these circumstances we say that $\mathscr{H}$ is the *direct sum* of $\mathscr{S}_1, \ldots, \mathscr{S}_n$.

An important example of a direct sum will appear in our discussion of single-particle scattering in Chapter 2. We shall find that the appropriate Hilbert space $\mathscr{H} = \mathscr{L}^2(\mathbb{R}^3)$ can be decomposed as

$$\mathscr{H} = \mathscr{R} \oplus \mathscr{B}$$

where $\mathscr{R}$ is the subspace of scattering states and $\mathscr{B}$ is the subspace spanned by the bound states (i.e., $\mathscr{B}$ is the subspace composed of arbitrary linear combinations of bound state vectors). This means that every scattering state is orthogonal to any bound state and that the most general state of the particle is a superposition of one scattering state vector in $\mathscr{R}$, and one state vector in $\mathscr{B}$.

1-c.   Operators and Inverses

The reader is certainly familiar with the concept of the linear operator, which can be defined as follows:

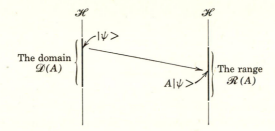

FIGURE 1.1.   The operator A maps certain vectors $|\psi\rangle$ onto their images $A\,|\psi\rangle$. Those vectors $|\psi\rangle$, for which $A\,|\psi\rangle$ is defined, comprise the *domain* of A; the set of image vectors $A\,|\psi\rangle$ is the *range* of A.

A linear operator A on a space $\mathcal{H}$ associates with each of certain vectors $|\psi\rangle$ in $\mathcal{H}$ a unique vector $A\,|\psi\rangle$ in such a way that $A(a\,|\psi\rangle + b\,|\phi\rangle) = aA\,|\psi\rangle + bA\,|\phi\rangle \cdots$ [any complex a and b].

It is often useful to visualize a linear operator A as a mapping of the vectors $|\psi\rangle$ in $\mathcal{H}$ onto their image vectors $|\psi'\rangle = A\,|\psi\rangle$, also in $\mathcal{H}$ (Fig. 1.1).

As indicated in Fig. 1.1, it is not necessarily true that an operator is defined for all $|\psi\rangle$ in $\mathcal{H}$. For example, in the space $\mathscr{L}^2(\mathbb{R}^1)$ of a spinless particle in one dimension, the position operator X is defined as the operator of multiplication by x on the spatial wave function: $X\,|\psi\rangle = |\psi'\rangle$ where $\psi'(x) = x\psi(x)$. Now the function $\psi'(x)$ defines a vector in $\mathscr{L}^2(\mathbb{R}^1)$ only if $\int dx\,|\psi'(x)|^2 < \infty$. Thus, the operator X is defined only on those vectors $|\psi\rangle$ in $\mathscr{L}^2(\mathbb{R}^1)$ with the additional property that $\int dx\,x^2\,|\psi(x)|^2 < \infty$; and, in fact, there is no useful way in which the definition can be extended to any other vectors. For this reason it is convenient to introduce the name *domain* $\mathscr{D}(A)$ of the operator A for the set of vectors on which A is defined. Obviously operators whose domain is the whole of $\mathcal{H}$ are easier to handle; however, it must be accepted that many important operators do not have this desirable property.[7]

It should be clear that in general not every vector $|\psi'\rangle$ in $\mathcal{H}$ will be the image under A of some $|\psi\rangle$. For this reason we introduce the name *range* $\mathscr{R}(A)$ of A for the set of image vectors onto which A maps.

In general, an operator can map two distinct vectors onto the same image

---

[7] An example of the kind of trouble that can occur when operators are not defined for all $|\psi\rangle$ is given by the definition of the sum $(A + B)$ of two operators: $(A + B)\,|\psi\rangle = A\,|\psi\rangle + B\,|\psi\rangle$. This definition makes sense only if $|\psi\rangle$ is in the domains of both A and B, and it is possible that there are no such vectors (except zero). Only if both A and B are defined on the whole of $\mathcal{H}$ can one be sure that no such problems arise. Here we shall not worry explicitly about these problems. We mention the point only to emphasize the sharp distinction between operators that are defined everywhere and those that are not.

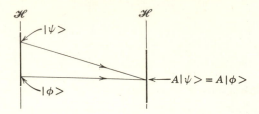

FIGURE 1.2.   If A maps two distinct vectors onto the same image vector, then A has no inverse.

vector; $|\psi\rangle \neq |\phi\rangle$, but $A|\psi\rangle = A|\phi\rangle = |\chi\rangle$, say. In this case there is obviously no unique vector from which $|\chi\rangle$ came under the action of A, and we say that A cannot be inverted (Fig. 1.2). On the other hand, it may happen for a given operator A that

$$\text{if}\quad |\psi\rangle \neq |\phi\rangle, \quad \text{then}\quad A|\psi\rangle \neq A|\phi\rangle \qquad (1.7)$$

in which case every $|\psi'\rangle$ in $\mathscr{R}(A)$ is the image of a unique $|\psi\rangle$ in $\mathscr{D}(A)$, $|\psi'\rangle = A|\psi\rangle$. In this case we define the inverse operator A^{-1} by the relation $A^{-1}|\psi'\rangle = |\psi\rangle$, [for $|\psi'\rangle$ in $\mathscr{R}(A)$]. With this definition it is easily checked that A^{-1} is a linear operator defined on $\mathscr{R}(A)$ and mapping $\mathscr{R}(A)$ back onto $\mathscr{D}(A)$ (Fig. 1.3).

Because A is linear we can rewrite the condition (1.7) for the existence of A^{-1} as

$$\text{if}\quad |\chi\rangle \neq 0, \quad \text{then}\quad A|\chi\rangle \neq 0 \qquad (1.8)$$

In words, the condition (1.7) means that A is a *one-to-one* mapping of $\mathscr{D}(A)$ onto $\mathscr{R}(A)$.[8]

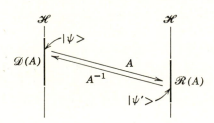

FIGURE 1.3.   The operator A maps $\mathscr{D}(A)$ onto $\mathscr{R}(A)$; the inverse A^{-1} maps $\mathscr{R}(A)$ back onto $\mathscr{D}(A)$.

[8] There is some confusion in the terminology of operator inverses. Our definition is that used by mathematicians and some physicists. However, some physicists reserve the term inverse for these cases where A^{-1} is defined on the whole of $\mathscr{H}$; that is, $\mathscr{R}(A) = \mathscr{H}$.

1-d.   Unitary Operators

A familiar example of an operator that does have an inverse is the unitary operator. A unitary operator can be defined as follows:

A unitary operator on $\mathscr{H}$ is a linear operator U that maps the whole of $\mathscr{H}$ onto the whole of $\mathscr{H}$ and preserves the norm. That is, $\mathscr{D}(U) = \mathscr{R}(U) = \mathscr{H}$ and $\|U\psi\| = \|\psi\|$ for all $|\psi\rangle$.

It is easily seen from the definition that a unitary operator has an inverse that is defined on the whole of $\mathscr{H}$. Since $\|U\psi\| = \|\psi\|$ the condition (1.8) is satisfied and U^{-1} exists; since $\mathscr{R}(U) = \mathscr{H}$ the inverse is defined everywhere on $\mathscr{H}$.

We have chosen this definition of unitary operators because it corresponds closely to their role in quantum mechanics. This can be illustrated by the example of the time evolution operator $U(t)$. The reader will recall that the time evolution of any system is determined (in the Schrödinger picture of quantum mechanics) by the Schrödinger equation:

$$i \frac{d}{dt} |\psi_t\rangle = H |\psi_t\rangle$$

For conservative systems (which we shall always be considering) the Hamiltonian H is independent of t and the general solution of the Schrödinger equation has the form:

$$|\psi_t\rangle = U(t) |\psi\rangle \equiv e^{-iHt} |\psi\rangle$$

It follows from a basic theorem on linear operators that, since H is self-adjoint, the evolution operator $U(t)$ is unitary. (Jordan, 1969, p. 52.) The evolution operator maps the state vector for time zero (that is, $|\psi\rangle$) onto the corresponding vector for time t.

Returning to our definition of a unitary operator we can interpret the unitarity of the evolution operator as follows (Fig. 1.4): The fact that $U(t)$ is defined on the whole of $\mathscr{H}$ means that for every state $|\psi\rangle$ at $t = 0$ there is a unique $|\psi'\rangle = U(t) |\psi\rangle$ into which is evolves. The fact that $U(t)$ has an

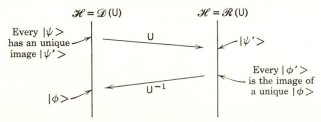

FIGURE 1.4.   If U is unitary, every normalized $|\psi\rangle$ has a unique normalized image $|\psi'\rangle$ and vice versa.

inverse defined everywhere on $\mathscr{H}$ means that for every state $|\phi'\rangle$ at t there is a state $|\phi\rangle = U(t)^{-1}|\phi'\rangle$ at $t = 0$ from which $|\phi'\rangle$ has evolved. Finally, the fact that $U(t)$ preserves the norm simply reflects the convention that all states are represented by normalized vectors in $\mathscr{H}$.

If we expand the condition $\|U\psi\| = \|\psi\|$ as $\langle\psi|\,U^\dagger U\,|\psi\rangle = \langle\psi\,|\,\psi\rangle$ and employ the famous trick of inserting first $|\psi\rangle = |\phi\rangle + |\chi\rangle$ and then $|\psi\rangle = |\phi\rangle + i\,|\chi\rangle$, we find that

$$\langle\phi|\,U^\dagger U\,|\chi\rangle = \langle\phi\,|\,\chi\rangle \qquad [\text{all } |\phi\rangle \text{ and } |\chi\rangle \text{ in } \mathscr{H}]$$

Thus, because U preserves the norm, it follows that

$$U^\dagger U = 1 \tag{1.9}$$

This result, together with the condition that $\mathscr{R}(U) = \mathscr{H}$, implies also that $UU^\dagger = 1$. We can see this by multiplying (1.9) by U to give $UU^\dagger U = U$ and hence

$$UU^\dagger(U\,|\psi\rangle) = U\,|\psi\rangle \qquad [\text{all } |\psi\rangle \text{ in } \mathscr{H}]$$

As $|\psi\rangle$ ranges over $\mathscr{H}$ so does $U\,|\psi\rangle$. [Here is where we use the condition $\mathscr{R}(U) = \mathscr{H}$.] We can therefore rewrite this equation as

$$UU^\dagger\,|\psi'\rangle = |\psi'\rangle \qquad [\text{all } |\psi'\rangle \text{ in } \mathscr{H}]$$

which is the desired result.

It is easy to see that the two conditions $U^\dagger U = UU^\dagger = 1$ are characteristic of a unitary operator and so can, if desired, be used as an alternative definition. As we shall see directly, both of the conditions are necessary; an operator can satisfy $\Omega^\dagger\Omega = 1$ but not be unitary.

1-e. Isometric Operators

More general than unitary operators are the so-called isometric operators. The importance of these is that the Møller wave operators $\Omega_\pm$, which will play a central role in our description of scattering, turn out to be isometric. The definition of an isometric operator is:

An isometric operator on $\mathscr{H}$ is a linear operator, Ω, which is defined on the whole of $\mathscr{H}$ and preserves the norm. That is, $\mathscr{D}(\Omega) = \mathscr{H}$ and $\|\Omega\psi\| = \|\psi\|$, for all $|\psi\rangle$.

This definition differs from that of a unitary operator only in that we do not require that Ω map $\mathscr{H}$ onto the whole of $\mathscr{H}$. In general $\mathscr{R}(\Omega) \neq \mathscr{H}$.

Obviously any unitary operator is isometric. On a *finite* dimensional space

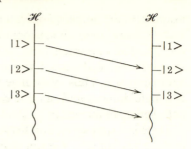

FIGURE 1.5. Schematic illustration of the isometric operator Ω defined in (1.10).

the converse is also true.[9] However, on an *infinite* dimensional space an operator can be isometric without being unitary, as the following example shows. Let $|1\rangle$, $|2\rangle$, ... be an orthonormal basis of an infinite dimensional Hilbert space $\mathcal{H}$ and define the linear operator Ω such that $\Omega\,|1\rangle = |2\rangle$, $\Omega\,|2\rangle = |3\rangle$, and in general,

$$\Omega\,|n\rangle = |n+1\rangle \tag{1.10}$$

It is clear that this defines Ω on all of $\mathcal{H}$ and that Ω preserves the norm; thus Ω is isometric. However, Ω does not map $\mathcal{H}$ onto the whole of $\mathcal{H}$ (It completely misses the vector $|1\rangle$—Fig. 1.5) and is therefore not unitary.

Because $\|\Omega\psi\| = \|\psi\|$, the vector $\Omega\,|\psi\rangle$ cannot vanish unless $|\psi\rangle$ does. Thus, Ω has an inverse Ω^{-1}, which is not in general defined on the whole of $\mathcal{H}$, however. This is clear in Fig. 1.5, where $\Omega^{-1}\,|2\rangle = |1\rangle$, $\Omega^{-1}\,|3\rangle = |2\rangle$, etc., but Ω^{-1} is simply not defined on the vector $|1\rangle$. Because Ω preserves the norm we can deduce, exactly as for unitary operators, that $\Omega^{\dagger}\Omega = 1$. On the other hand it is *not* generally true that $\Omega\Omega^{\dagger} = 1$, as we shall see.

The reader who has not met the isometric operator before may be feeling a little uneasy. He may feel that if Ω is a one-to-one linear map, then its range should be as big as its domain, namely the whole of $\mathcal{H}$. This is correct if $\mathcal{H}$ if finite-dimensional, but is false when $\mathcal{H}$ is infinite-dimensional, since an infinite set can be mapped one-to-one onto a proper subset of itself (as in Fig. 1.5).

Also, the reader may be tempted to feel that the condition $\Omega^{\dagger}\Omega = 1$ should automatically imply that $\Omega\Omega^{\dagger} = 1$ and, hence, that Ω is unitary. This second feeling is also based on experience with finite-dimensional spaces, where, as we have said, an isometric operator always is unitary. The difference between finite and infinite dimensional spaces can be seen if we reexamine the

[9] To prove this one has only to choose an orthonormal basis, in terms of which Ω is represented by an $n \times n$ matrix (Ω). We shall see that an isometric operator satisfies $\Omega^{\dagger}\Omega = 1$ and the matrix (Ω) therefore satisfies $(\Omega)^{\dagger}(\Omega) = 1$. This means that (Ω) is non-singular, has a matrix inverse $[(\Omega)^{\dagger}]$ and, hence, that $(\Omega)(\Omega)^{\dagger} = 1$. Thus, (Ω) is a unitary matrix and Ω a unitary operator.

example of an isometric operator shown in Fig. 1.5. Suppose, for example, that $\mathcal{H}$ were three dimensional and that we let Ω map $|1\rangle$ onto $|2\rangle$ and $|2\rangle$ on $|3\rangle$. The condition $\Omega^{\dagger}\Omega = 1$ means that orthogonal vectors must be mapped onto orthogonal vectors and we are therefore forced to map $|3\rangle$ back onto $|1\rangle$. Thus, the range has to be the whole of $\mathcal{H}$. Only when $\mathcal{H}$ is infinite-dimensional can one continue to map $|n\rangle$ onto $|n + 1\rangle$ *ad infinitum* and, hence, never return to $|1\rangle$.

There is a simple relation between the inverse Ω^{-1} of an isometric operator and the adjoint $\Omega^{\dagger}$. We can write $\Omega^{\dagger}\Omega = 1$ as

$$\Omega^{\dagger}(\Omega\,|\psi\rangle) = |\psi\rangle \qquad [\text{all } |\psi\rangle \text{ in } \mathcal{H}]$$

If we substitute $\Omega\,|\psi\rangle = |\phi\rangle$ we conclude that for any $|\phi\rangle$ in $\mathcal{R}(\Omega)$

$$\Omega^{\dagger}\,|\phi\rangle = \Omega^{-1}\,|\phi\rangle \qquad [\text{all } |\phi\rangle \text{ in } \mathcal{R}(\Omega)]$$

If, on the other hand, $|\phi\rangle$ is orthogonal to $\mathcal{R}(\Omega)$, we find that

$$\langle\phi|\,\Omega\,|\psi\rangle = 0 \qquad [\text{all } |\psi\rangle \text{ in } \mathcal{H}]$$

and, hence,

$$\langle\psi|\,\Omega^{\dagger}\,|\phi\rangle = 0 \qquad [\text{all } |\psi\rangle \text{ in } \mathcal{H}]$$

or $\Omega^{\dagger}\,|\phi\rangle = 0$. Thus,

$$\Omega^{\dagger} = \begin{cases} \Omega^{-1} & \text{on } \mathcal{R}(\Omega) \\ 0 & \text{on } \mathcal{R}(\Omega)^{\perp} \end{cases}$$

In the example of Fig. 1.5 this means that $\Omega^{\dagger}\,|1\rangle = 0$, which shows clearly that $\Omega\Omega^{\dagger} \neq 1$.

1-f. Convergence of Vectors

We now turn to the crucial question of convergence of vectors and operators. In scattering theory we shall be concerned with vectors $|\psi_t\rangle$ and operators A_t, depending on the continuous time variable t, and with their limits as $t \to \pm\infty$. We start our discussion with the vector $|\psi_t\rangle$ and make the following definition:

The vectors $|\psi_t\rangle$ converge to the limit vector $|\psi\rangle$ as $t \to \infty$ if and only if[10]

$$\|\psi_t - \psi\| \xrightarrow[t \to \infty]{} 0 \qquad (1.11)$$

With this definition the statement $|\psi_t\rangle \to |\psi\rangle$ simply means that $|\psi_t\rangle$ gets close to $|\psi\rangle$ (in the sense that the length of $|\psi_t\rangle - |\psi\rangle$ tends to zero).

Our definition of convergence can be interpreted as follows: If $|\psi_t\rangle \to |\psi\rangle$,

---

[10] The symbol $\|\psi - \phi\|$ means the norm of the vector $|\psi\rangle - |\phi\rangle$. It would have been more precise to use the monstrosity $\|\,|\psi\rangle - |\phi\rangle\,\|$, but there seems little danger of confusion in the omission of the ket symbols.

then as $t \to \infty$ the state represented by $|\psi_t\rangle$ becomes physically indistinguishable from that represented by $|\psi\rangle$. To understand the sense in which this is true, we note that the physical state represented by any $|\psi\rangle$ is completely identified if we measure the numbers $|\langle\phi \mid \psi\rangle|$ for all normalized $|\phi\rangle$. (The number $|\langle\phi \mid \psi\rangle|^2$ is just the "overlap probability" that a system known to be in the state $|\psi\rangle$ is observed in the state $|\phi\rangle$. It is a simple exercise to check that measurement of these numbers for all $|\phi\rangle$ determines $|\psi\rangle$ within the usual arbitrary phase factor.) Now from the Schwartz inequality[11] it follows that

$$
\begin{aligned}
|\langle\phi \mid \psi_t\rangle - \langle\phi \mid \psi\rangle| &= |\langle\phi| (|\psi_t\rangle - |\psi\rangle)| \\
&\leqslant \|\phi\| \cdot \|\psi_t - \psi\| \\
&= \|\psi_t - \psi\|
\end{aligned}
\tag{1.12}
$$

since $\|\phi\| = 1$. This number is independent of the particular $|\phi\rangle$ considered and tends to zero as $t \to \infty$. Thus by making t large enough we can arrange that the difference between $\langle\phi \mid \psi_t\rangle$ and $\langle\phi \mid \psi\rangle$, for *all* normalized $|\phi\rangle$, becomes smaller than any prescribed ϵ. It is in this sense that the states $|\psi_t\rangle$ and $|\psi\rangle$ become experimentally indistinguishable.

In scattering theory we shall be concerned with two normalized vectors $|\psi_t\rangle$ and $|\psi_t^0\rangle$, the first being the actual state of two (or more) colliding particles, the second describing some possible motion of the same particles in the absence of any forces. We shall prove that as $t \to \infty$, long after the collision, the difference $|\psi_t\rangle - |\psi_t^0\rangle$ tends to zero (with a similar result as $t \to -\infty$). We shall write this result as

$$
|\psi_t\rangle \xrightarrow[t \to \infty]{} |\psi_t^0\rangle
$$

It means that long after (and long before) the collision the actual motion of the particles is physically indistinguishable from that of two free particles.

In practice, the definition (1.11) does not provide a useful test for the convergence of a given $|\psi_t\rangle$. This is because it is impossible to determine whether $\|\psi_t - \psi\| \to 0$ without knowing $|\psi\rangle$ in advance. There is an analogous situation in the theory of real or complex numbers where a function a_t tends to a limit a if $|a_t - a| \to 0$, as $t \to \infty$. As the reader will probably recall, a simple test for convergence of a function is the so-called *Cauchy test*, according to which a_t has a limit if and only if $|a_t - a_{t'}| \to 0$, as t and $t' \to \infty$. It turns out that the analogous test works in a Hilbert space.[12] The vector $|\psi_t\rangle$ has a limit, if and only if $\|\psi_t - \psi_{t'}\| \to 0$, as t and $t' \to \infty$.

---

[11] $|\langle\phi \mid \chi\rangle| \leqslant \|\phi\| \cdot \|\chi\|$.

[12] One of the axioms that defines a Hilbert space is that a sequence satisfying the Cauchy criterion is convergent. Needless to say one does not get something for nothing simply by defining a Hilbert space as a space in which the Cauchy test works—one simple transfers the problem. In showing that $\mathscr{L}^2(\mathbb{R}^3)$, for example, is a Hilbert space, one has to show that for functions in $\mathscr{L}^2(\mathbb{R}^3)$ the Cauchy criterion does imply convergence.

The particular form in which we shall use this test is as follows: The vector $|\psi_t\rangle$ whose convergence we wish to establish will appear as an integral,

$$|\psi_t\rangle = \int_0^t d\tau\,|\phi_\tau\rangle \tag{1.13}$$

The Cauchy test tells us that this is convergent if:

$$\left\|\int_t^{t'} d\tau\,|\phi_\tau\rangle\right\| \to 0$$

as t and $t' \to \infty$. This is certainly satisfied if[13]

$$\int_t^{t'} d\tau\,\|\phi_\tau\| \to 0$$

as t and $t' \to \infty$. Now by the Cauchy test for real numbers this is true if and only if:

$$\int_0^\infty d\tau\,\|\phi_\tau\| < \infty \tag{1.14}$$

that is, this integral converges. Thus, to establish convergence of the vector integral (1.13), it is sufficient to show that the scalar integral (1.14) is convergent.

One final point concerning the convergence of vectors. The inequality (1.12) makes clear that if $|\psi_t\rangle \to |\psi\rangle$, then $\langle\phi\,|\,\psi_t\rangle \to \langle\phi\,|\,\psi\rangle$ for any fixed $|\phi\rangle$. In other words, if $|\psi_t\rangle$ converges, so does its component in any fixed direction. In an infinite-dimensional space the converse result is *false*.[14] Even if $\langle\phi\,|\,\psi_t\rangle$ has a limit for every fixed $|\phi\rangle$, the vector $|\psi_t\rangle$ may not have a limit. This can be easily understood with the help of an example. If $|\psi_t\rangle = U^0(t)\,|\psi\rangle \equiv e^{-iH^0t}\,|\psi\rangle$ where $H^0 = \mathbf{P}^2/2m$ is the Hamiltonian of a free particle, then $|\psi_t\rangle$ describes the motion of a freely moving wave packet. Its center moves through space with some mean velocity and it spreads. Thus, its wave function $\langle\mathbf{x}\,|\,\psi_t\rangle$ at any fixed point eventually tends to zero, and it follows that $\langle\phi\,|\,\psi_t\rangle$, the overlap between $|\psi_t\rangle$ and any fixed $|\phi\rangle$, tends to zero[15];

$$\langle\phi\,|\,\psi_t\rangle \to 0 \qquad \text{[any fixed } |\phi\rangle\text{]} \tag{1.15}$$

---

[13] It follows from the triangle inequality, $\|\psi + \phi\| \leqslant \|\psi\| + \|\phi\|$, that $\|\int d\tau \cdots\| \leqslant \int d\tau\,\|\cdots\|$.

[14] It is easily seen that in a finite dimensional space (e.g., $\mathbb{R}^3$) the converse is true.

[15] It is not hard to prove the result properly. For example, if one considers the case where $\phi(\mathbf{x})$ and $\psi(\mathbf{x})$ are both Gaussians, then the famous $t^{-3/2}$ behavior of $U^0(t)\,|\psi\rangle$ is explicitly known and it is easily shown that $\langle\phi\,|\,\psi_t\rangle$ goes to zero like $t^{-3/2}$. This establishes the result for Gaussians or finite linear combinations of Gaussians. One then has only to note that any function of $\mathscr{L}^2(\mathbb{R}^3)$ can be approximated arbitrarily well by sums of Gaussians to complete the proof.

Nonetheless, $\|\psi_t\| = \|\psi\| = 1$, which certainly does not tend to zero. In the literature, a vector $|\psi_t\rangle$ satisfying (1.15) is sometimes said to *converge weakly* to zero.[16] Thus we can say that if $|\psi_t\rangle$ converges, then it converges weakly, but that the converse is not true. In this book we shall not use the terminology of weak convergence explicitly. We shall, however, need to understand the difference between convergence of a vector $|\psi_t\rangle$ and convergence of its components $\langle\phi \mid \psi_t\rangle$, for all $|\phi\rangle$.

1-g.   Operator Limits

We shall say that an operator A_t has a limit A if for every $|\psi\rangle$ the vector $A_t |\psi\rangle$ has a limit,

$$A_t |\psi\rangle \xrightarrow[t\to\infty]{} |\phi\rangle \equiv A |\psi\rangle$$

where one can readily check that A defined in this way is a linear operator. In this case we write either $A_t \to A$ or $A = \lim A_t$.

Extreme care is needed in handling operator limits. Many results that seem obviously true are, in fact, false. For instance, if $A_t \to A$ it is not necessarily true that $A_t^\dagger \to A^\dagger$; and if also $B_t \to B$ it does not follow that $A_t B_t \to AB$. An important example of this kind of oddity will appear in Chapter 2 where we shall introduce the Møller wave operators $\Omega_\pm$ as the limits as $t \to \mp\infty$ of a certain unitary operator. Although one would naturally expect that the limit of a unitary operator must be unitary, this is actually not so. If

$$U_t |\psi\rangle \xrightarrow[t\to\infty]{} \Omega |\psi\rangle \qquad [\text{all } |\psi\rangle \text{ in } \mathscr{H}]$$

then $\|\Omega\psi\| = \lim \|U_t\psi\| = \|\psi\|$ and Ω is at least isometric, satisfying $\Omega^\dagger\Omega = 1$. (See Problem **1.3**.) However, even though $U_t U_t^\dagger = 1$ it is impossible to establish that $\Omega\Omega^\dagger = 1$ and, hence, that Ω is unitary. We shall see explicitly in Chapter 2 that in general Ω is not unitary and $\Omega\Omega^\dagger \neq 1$.

PROBLEMS

1.1.   The domain $\mathscr{D}(A)$ of a linear operator A is by definition a subspace [i.e., if $|\psi\rangle$ and $|\phi\rangle$ are in $\mathscr{D}(A)$ so is $a |\psi\rangle + b |\phi\rangle$ for any complex numbers a and b]. Show that the range $\mathscr{R}(A)$ is a subspace. (We have here glossed over the distinction between a closed subspace and an arbitrary subspace or linear manifold.)

---

[16] In which case, convergence (in our sense) is called *strong* convergence.

1.2.   Let $|\psi\rangle$ be any state of a single particle, and let $|\psi_\rho\rangle$ be the state obtained by a rigid translation of the particle through the vector ρ. This means that the spatial wave function of $|\psi_\rho\rangle$ is given by $\psi_\rho(\mathbf{x}) = \psi(\mathbf{x} - \rho)$, or the momentum–space wave function by $\bar{\psi}_\rho(\mathbf{p}) = \exp(-i\rho \cdot \mathbf{p})\bar{\psi}(\mathbf{p})$. Convince yourself of these statements and then prove that for any fixed $|\phi\rangle$ the overlap $\langle \phi \mid \psi_\rho \rangle$ goes to zero as $\rho \to \infty$, but that $|\psi_\rho\rangle$ does not tend to zero. Explain this result in words. (Your proof should be as rigorous as you can make it. It should certainly make clear *why* the result holds.)

1.3.   If $|\psi_t\rangle \to |\psi\rangle$ as $t \to \infty$ show that $\|\psi_t\| \to \|\psi\|$. Show that if U_t is a unitary operator that tends to the limit Ω (i.e., $U_t |\psi\rangle \to \Omega |\psi\rangle$ for all $|\psi\rangle$ in $\mathscr{H}$), then Ω is isometric.

The Scattering Operator for a Single Particle

2-a   **Classical Scattering**

2-b   **Quantum Scattering**

2-c   **The Asymptotic Condition**

2-d   **Orthogonality and Asymptotic Completeness**

2-e   **The Scattering Operator**

2-f   **Unitarity**

In this and the next chapter we shall discuss the simplest of all scattering processes, the elastic scattering of a spinless particle off a fixed target. Of course no real target is perfectly fixed. Nonetheless, our formalism does give an approximate description of an experiment where a single light particle scatters slowly off a heavy target (such as the scattering of a slow electron off a heavy atom). In addition, there is a close relationship between the scattering of one particle off a fixed potential and the scattering of two particles off one another (discussed in Chapter 4). However, our principal reason for studying this simplest of processes is that it provides an elementary introduction to most of the essential concepts encountered in all scattering problems—the scattering operator S, the Møller wave operators $\Omega_{\pm}$, the cross section, and the T operator. These concepts are so important that they deserve careful study with a minimum of inessential complications. Once they are thoroughly understood in the simplest case, their extension to more complicated situations will, for the most part, be quickly and easily accomplished.

As discussed in the Introduction, we shall base our scattering formalism on an analysis of the time-dependent state vector that describes the actual motion of the particle during the scattering process. In this chapter we shall see how the particle's essentially free motion long before collision can be directly related to the free motion long after collision by means of a unitary scattering operator S. In the next chapter we shall show how the experimentally measured quantity, the differential cross section, can be expressed in terms of the matrix elements of S.

2-a.   Classical Scattering

The time-dependent description of quantum scattering has a natural and instructive parallel in classical mechanics and we begin with a brief description of this classical theory. Figure 2.1 shows a typical classical scattering, which we may imagine to be the scattering of an electron by some fixed atom. The trajectory can be roughly divided into three parts: (1) the approach of the electron along an almost straight orbit until it reaches its region of interaction with the atom, (2) the possibly very complicated orbit during the inter-action, and (3) the departure of the electron along some other approximately straight orbit. Although these three divisions are only roughly defined, one point should be clear. The region of interaction is certainly no larger than a few atomic diameters and so is, in practice, completely unobservable. All that will be visible in the cloud chamber, bubble chamber, or whatever is used to observe the event, is a pair of straight tracks corresponding to the free motion before and after the collision. Therefore, in seeking a mathe-matical description of the scattering process, we shall try (as far as possible) to suppress the precise details of the orbit in the neighborhood of the target

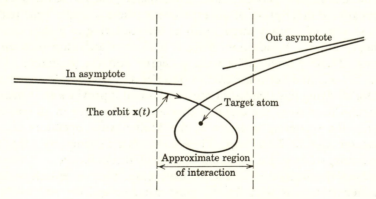

FIGURE 2.1.   Typical scattering orbit.

atom and concentrate on the relation between the asymptotically free incoming and outgoing trajectories.[1]

To make these ideas precise we must introduce some notation. We denote by $\mathbf{x}(t)$ the actual orbit of the scattered electron, obtained by solving Newton's equation $m\ddot{\mathbf{x}} = -\nabla V$ for the actual potential $V(\mathbf{x})$. The essentially free behavior of the electron before the collision means that as $t \to -\infty$ the orbit $\mathbf{x}(t)$ is asymptotic to some free orbit,

$$\mathbf{x}(t) \xrightarrow[t \to -\infty]{} \mathbf{x}_{in}(t) \equiv \mathbf{a}_{in} + \mathbf{v}_{in}t \qquad (2.1)$$

for some $\mathbf{a}_{in}$ and $\mathbf{v}_{in}$. [By $\mathbf{x}(t) \to \mathbf{y}(t)$ we mean $|\mathbf{x}(t) - \mathbf{y}(t)| \to 0$.] Similarly, after the collision,

$$\mathbf{x}(t) \xrightarrow[t \to +\infty]{} \mathbf{x}_{out}(t) \equiv \mathbf{a}_{out} + \mathbf{v}_{out}t \qquad (2.2)$$

The asymptotic orbits $\mathbf{x}_{in}(t)$ and $\mathbf{x}_{out}(t)$, which satisfy the equation of motion for a free particle, are called the *incoming* and *outgoing* (or "*in*" and "*out*") *asymptotes* of the actual scattering orbit $\mathbf{x}(t)$. *As far as observations are concerned, a scattering orbit is completely characterized once its two asymptotes are known;* and, if for any given incoming asymptote $\mathbf{x}_{in}(t)$ we could calculate the corresponding outgoing asymptote $\mathbf{x}_{out}(t)$, then the scattering problem would, for all practical purposes, be completely solved.

For the sake of comparison with the quantum mechanical case, we note some properties which one might expect of the correspondence between the in and out asymptotes. First, one might reasonably expect that the correspondence would be one-to-one; or, more precisely, that any six real numbers $(\mathbf{a}_{in}, \mathbf{v}_{in})$ should represent a possible in asymptote and should define a unique corresponding out asymptote given by six numbers $(\mathbf{a}_{out}, \mathbf{v}_{out})$, and vice versa. Second, although our principal interest is in the in and out asymptotes, we must recognize that the correspondence between them is defined by the actual orbit; that is, the correspondence has the form:

$$\mathbf{x}_{in}(t) \to \qquad \mathbf{x}(t) \qquad \to \mathbf{x}_{out}(t)$$

or

$$\text{in asymptote} \to \text{actual orbit} \to \text{out asymptote}$$

For every in or out asymptote one may reasonably expect there will be a corresponding orbit $\mathbf{x}(t)$. On the other hand, one would *not* expect every orbit $\mathbf{x}(t)$ to define in and out asymptotes, simply because a general potential will support some *bounded* orbits (corresponding to the bound states of

---

[1] Of course in classical mechanics one could discuss a process like the scattering of a comet by the sun in which the details of the orbit certainly are observable. Our interest however is in atomic and sub-atomic processes, where they are not.

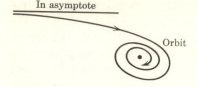

FIGURE 2.2. For sufficiently attractive potentials, a particle coming in from infinity may get caught in a spiral and never emerge.

quantum mechanics) and a particle in such an orbit will never escape from the potential and, hence, never behave freely. Thus, we must expect two kinds of orbit: scattering and bounded. The scattering orbits should have both in and out asymptotes,

in asymptotes → scattering orbits → out asymptotes

and the scattering orbits *together with the bounded orbits* should make up all possible orbits of the system.

Reasonable as all these properties may seem, one must recognize that they will certainly not hold for all potentials. If the potential does not fall off fast enough at infinity, then even as the particle moves far away it will not behave freely and the scattering orbits will have no asymptotes. (The Coulomb potential is a notorious example of a potential for which the scattering states never approach free asymptotes.) If the potential is too attractive at close range (e.g., $V = -1/r^3$), then a particle coming in from infinity may get caught in a spiralling orbit and never emerge; thus, an orbit with a perfectly good in asymptote may have no out asymptote at all, or vice versa (see Fig. 2.2).

While it is not our concern here to discuss in detail which potentials in the classical theory of scattering display these difficulties and which do not, it is important to recognize that such difficulties may occur. The same problems occur in quantum scattering theory; and they prove much less formidable if one recognizes in advance that they are to be expected, and that their origin can be easily (and even classically) understood. In particular, we must anticipate that only for some special class of well behaved potentials will our scattering theory go through smoothly without any difficulties. Fortunately, we shall find that this class includes almost all potentials of any interest.[2]

One final comment before we move on to quantum scattering: The condition of asymptotic free motion is expressed mathematically by the

[2] In the interests of accuracy we should admit that with respect to the *classical* theory of scattering we have implied an over-simplification. Even for "well-behaved" potentials there can be certain exceptional orbits that have an in asymptote but no out asymptote or vice versa. This does not happen in quantum scattering theory and so need not concern us here. See Newton (1966) Chapter 5.

limits (2.1) and (2.2), which we have claimed should hold as t tends to infinity. Of course this does not mean that one really has to wait an infinite time to observe the asymptotic free motion. Quite the contrary; even for a very slow projectile (e.g., a thermal neutron) incident on a large target (some big molecule) the total collision time will normally not exceed 10^{-10} sec; this means that if the collision is centered on $t = 0$ then for times before $t \approx -10^{-10}$ sec and after $t \approx +10^{-10}$ sec the motion is experimentally indistinguishable from free motion; that is, in practice t becomes "infinite" at $t \approx \pm 10^{-10}$ sec.

Nonetheless, the appropriate mathematical statement involves the limits $t \to \pm \infty$. The reason is that for any given orbit there is generally no finite time beyond which $\mathbf{x}(t)$ is *exactly* equal to $\mathbf{x}_{\mathrm{out}}(t)$. (To be definite we consider the out asymptotic motion.) Rather, our measuring devices have some minimum resolution, and there is a finite time beyond which the difference between $\mathbf{x}(t)$ and $\mathbf{x}_{\mathrm{out}}(t)$ is smaller than we can resolve. However, this is precisely what is meant by the mathematical limit (2.2): Given any $\epsilon > 0$ (the resolution) there is a time t_0 such that $\mathbf{x}(t) - \mathbf{x}_{\mathrm{out}}(t)$ is smaller than ϵ for any t later than t_0.

In our discussion of quantum scattering we shall establish limits that are analogous to (2.1) and (2.2). These limits are to be interpreted in exactly the same way—that for times more than some small amount (usually 10^{-10} sec or less) before or after the collision, the motion is experimentally indistinguishable from free motion.

2-b.   Quantum Scattering

The description of quantum scattering closely parallels the classical formalism outlined in Section 2-a. In lieu of the classical orbit $\mathbf{x}(t)$ satisfying Newton's equation, we now have a state vector $|\psi_t\rangle$ satisfying the time-dependent Schrödinger equation:

$$i \frac{d}{dt} |\psi_t\rangle = H |\psi_t\rangle$$

As discussed in Chapter 1, the general solution of this equation has the form $|\psi_t\rangle = U(t) |\psi\rangle \equiv e^{-iHt} |\psi\rangle$, where $U(t)$ is the so-called evolution operator, and $|\psi\rangle$ is any vector in the appropriate Hilbert space $\mathscr{H}$. We shall adopt the classical terminology and refer to the solution $U(t) |\psi\rangle$ as an *orbit*, although of course it is no longer an orbit in real space $\mathbb{R}^3$. Every orbit $U(t) |\psi\rangle$ can be uniquely identified by the fixed vector $|\psi\rangle$, which is just the state vector at the instant $t = 0$.

In this chapter we are considering a single spinless particle in a fixed potential. Thus $\mathscr{H}$ is the space $\mathscr{L}^2(\mathbb{R}^3)$, with wave functions $\psi(\mathbf{x}) \equiv \langle \mathbf{x} | \psi \rangle$

depending on a single variable **x**; and the Hamiltonian H has the form $H = H^0 + V$, where H^0 is the Hamiltonian of a free particle, $H^0 = \mathbf{P}^2/2m$, and V is the potential. We shall, for the moment, suppose that the potential is local; that is, that V is a function of the particle's position only. Assumptions about the function $V(\mathbf{x})$ will be discussed below.

Let us suppose that the orbit $U(t)\,|\psi\rangle$ describes the evolution of some scattering experiment. This means that when followed back to a time well before the collision $U(t)\,|\psi\rangle$ represents a wave packet that is localized far away from the scattering center and, therefore, behaves like a *free* wave packet. Now, the motion of a free particle is given by the free evolution operator $U^0(t) \equiv e^{-iH^0 t}$, and we therefore expect that as $t \to -\infty$,

$$U(t)\,|\psi\rangle \xrightarrow[t \to -\infty]{} U^0(t)\,|\psi_{\mathrm{in}}\rangle \qquad (2.3)$$

for some vector $|\psi_{\mathrm{in}}\rangle$. [As discussed in Section 1-f, the limit (2.3) simply means that the difference of the two vectors tends to zero and, hence, that the actual state $U(t)\,|\psi\rangle$ becomes experimentally indistinguishable from the freely evolving state $U^0(t)\,\psi_{\mathrm{in}}\rangle$.[3]] Similarly, after the collision the particle moves away again and we expect that

$$U(t)\,|\psi\rangle \xrightarrow[t \to +\infty]{} U^0(t)\,|\psi_{\mathrm{out}}\rangle \qquad (2.4)$$

for some $|\psi_{\mathrm{out}}\rangle$. These two limits are analogous to the classical limits (2.1) and (2.2) and in analogy with the classical terminology we shall call the asymptotic free orbits of (2.3) and (2.4) the *in and out asymptotes* of the actual orbit $U(t)\,|\psi\rangle$. The classical asymptotes could be identified by the fixed numbers $(\mathbf{a}_{\mathrm{in}}, \mathbf{v}_{\mathrm{in}})$ and $(\mathbf{a}_{\mathrm{out}}, \mathbf{v}_{\mathrm{out}})$. Similarly, it is convenient to label the quantum asymptotes (2.3) and (2.4) by the fixed vectors $|\psi_{\mathrm{in}}\rangle$ and $|\psi_{\mathrm{out}}\rangle$. That the limits (2.3) and (2.4) do hold (for the appropriate orbits) will be shown in Section 2-c.

As in the classical case we do not expect that every orbit $U(t)\,|\psi\rangle$ will have asymptotes. We expect rather that there will be certain scattering orbits that do have asymptotes, and that the scattering states together with the bound states will span the space $\mathscr{H}$ of *all* states. This result will be discussed in Section 2-d.

To conclude this section we discuss the conditions that the potential $V(\mathbf{x})$ must satisfy to give a physically reasonable scattering theory. As discussed in Section 2-a, the results of scattering theory will certainly not hold for all possible potentials. For example, if $V(\mathbf{x})$ does not fall off sufficiently fast as $\mathbf{x} \to \infty$, the particle will not behave like a free particle as it moves far away.

---

[3] Strictly speaking, we should only *expect* (2.3) to hold within a phase factor. However, we shall prove that it does hold exactly.

Unfortunately, the conditions under which some of the principal results have been proved are quite complicated; different proofs use different conditions. Further, a set of conditions that is both sufficient and necessary for all results is not known. We shall content ourselves here with stating a simple set of three conditions under which all of the relevant results can be proved. These conditions apply to a *spherical potential*, $V(r)$. The first condition constrains the behavior of $V(r)$ as $r \to \infty$, the second restricts its behavior at the origin, and the third constrains its behavior in between. The conditions are as follows [The notation $V(r) = O(r^p)$ means that $|V(r)| \leqslant c\,|r|^p$ for some constant c.][4]:

I $V(r) = O(r^{-3-\epsilon})$ as $r \to \infty$ (some $\epsilon > 0$)

II $V(r) = O(r^{-3/2+\epsilon})$ as $r \to 0$ (some $\epsilon > 0$)

III $V(r)$ is continuous for $0 < r < \infty$, except perhaps at a finite number of finite discontinuities.

These conditions can be paraphrased roughly as: I $V(r)$ falls off quicker than r^{-3} at infinity; II $V(r)$ is less singular than $r^{-3/2}$ at the origin; and III $V(r)$ is "reasonably smooth" in between.

In making the precise statements of Sections 2-c and 2-d we shall (for simplicity) suppose that these conditions hold. However it must be emphasized that the results can certainly be proved for more general potentials. In particular, it is certainly not necessary that the potential be spherically symmetric. For example, all results can be proved for the scattering by several fixed force centers (which would give a simple model for scattering of a particle off several targets) and for spin-dependent and other nonlocal potentials (see Chapter 5).

Nonetheless, the general features of our three conditions do give a reliable indication of the sort of potential for which a scattering theory can be constructed. The conditions include almost all potentials of general interest— the potential of an electron scattering off a rigid atom, the square well, the Yukawa potential, etc. Condition I excludes any potential (such as the Coulomb potential) that falls off more slowly than r^{-3} at infinity. In fact, none of the principal results of scattering theory do hold for the Coulomb potential.[5] Condition II excludes any potential more singular than $r^{-3/2}$ at the origin. Such potentials are sometimes called singular potentials. Repulsive singular potentials *can* be included in scattering theory, although they need

---

[4] I am much indebted to Dr. Walter Hunziker for help in assembling these conditions from their widely scattered sources. See Hack (1959), Ikebe (1960), Hunziker (1961), Faddeev (1965), and Hunziker (1968).

[5] For methods of handling the Coulomb potential, see Chapter 14.

very special treatment.[6] Attractive singular potentials suffer difficulties somewhat analogous to the classical spiralling orbits discussed in Section 2-a.

In summary, the principal results of scattering theory hold for a wide class of "reasonable" potentials, including those spherical potentials satisfying the simple conditions I–III above, but definitely excluding the Coulomb and attractive singular potentials. In the remainder of this chapter we shall assume (for simplicity) that V satisfies the conditions I–III.

2-c.   The Asymptotic Condition

We first establish that every vector in $\mathscr{H}$ (labelled $|\psi_{\text{in}}\rangle$ or $|\psi_{\text{out}}\rangle$ as appropriate) does represent the asymptote of some actual orbit; that is, for every vector $|\psi_{\text{in}}\rangle$ in $\mathscr{H}$ there is a solution $U(t)\,|\psi\rangle$ of the Schrödinger equation that is asymptotic to the free orbit $U^0(t)\,|\psi_{\text{in}}\rangle$ as $t \to -\infty$; and likewise for every $|\psi_{\text{out}}\rangle$ as $t \to +\infty$. This result is known as the asymptotic condition.

The Asymptotic Condition. If the potential V satisfies the conditions discussed in Section 2-b,[7] then for every $|\psi_{\text{in}}\rangle$ in $\mathscr{H}$ there is a $|\psi\rangle$ such that

$$U(t)\,|\psi\rangle - U^0(t)\,|\psi_{\text{in}}\rangle \xrightarrow[t \to -\infty]{} 0$$

and likewise for every $|\psi_{\text{out}}\rangle$ in $\mathscr{H}$ as $t \to +\infty$.

Proof: Multiplying by the unitary operator $U(t)^\dagger$, we can rewrite the desired result as:

$$|\psi\rangle - U(t)^\dagger U^0(t)\,|\psi_{\text{in}}\rangle \xrightarrow[t \to -\infty]{} 0 \tag{2.5}$$

In this form it is clear that all we have to prove is that for every $|\psi_{\text{in}}\rangle$, the vector $U(t)^\dagger U^0(t)\,|\psi_{\text{in}}\rangle$ has a limit. The trick is to write $U^\dagger U^0$ as the integral of its derivative:

$$\frac{d}{dt}\,U(t)^\dagger U^0(t) = ie^{iHt}(H - H^0)e^{-iH^0 t}$$

$$= iU(t)^\dagger V U^0(t)$$

This gives:

$$U(t)^\dagger U^0(t)\,|\psi_{\text{in}}\rangle = |\psi_{\text{in}}\rangle + i\int_0^t d\tau\, U(\tau)^\dagger V U^0(\tau)\,|\psi_{\text{in}}\rangle \tag{2.6}$$

---

[6] In discussions of the radial Schrödinger equation, the term singular potential is often used for a potential more singular than r^{-2} at $r = 0$. However, in a general three-dimensional analysis of the scattering process the difficulties start at $r^{-3/2}$. See Hunziker (1968) and references there cited.

[7] It will be clear that our proof actually holds provided only the integral $\int d^3x\,|V(\mathbf{x})|^2$ in (2.8) converges. In fact, the asymptotic condition can be proved under even weaker conditions. See Hunziker (1968).

which has the desired limit if and only if the integral on the right converges as $t \to -\infty$. As discussed in Section 1-f, a sufficient condition for this is that:

$$\int_{-\infty}^{0} d\tau \, \| U(\tau)^\dagger V U^0(\tau) \psi_{\text{in}} \| < \infty$$

or, since U is unitary,

$$\int_{-\infty}^{0} d\tau \, \| V U^0(\tau) \psi_{\text{in}} \| < \infty \qquad (2.7)$$

Let us restrict our attention for a moment to vectors with Gaussian wave functions,

$$\langle \mathbf{x} \mid \psi_{\text{in}} \rangle = e^{-(\mathbf{x}-\mathbf{a})^2/2\xi^2}$$

where the center $\mathbf{a}$ and width ξ are arbitrary. For these Gaussian wave functions the effect of the free evolution operator $U^0(\tau)$ in (2.7) is explicitly known[8]:

$$| \langle \mathbf{x} | \, U^0(\tau) \, | \psi_{\text{in}} \rangle |^2 = \left(1 + \frac{\tau^2}{m^2 \xi^4} \right)^{-3/2} \exp \left[- \frac{(\mathbf{x} - \mathbf{a})^2}{\xi^2 + \tau^2/m^2\xi^2} \right]$$

Thus, the norm appearing in (2.7) can be bounded as follows:

$$\| V U^0(\tau) \psi_{\text{in}} \|^2 = \int d^3x \, |V(\mathbf{x})|^2 \left(1 + \frac{\tau^2}{m^2\xi^4} \right)^{-3/2} \exp \left[- \frac{(\mathbf{x} - \mathbf{a})^2}{\xi^2 + \tau^2/m^2\xi^2} \right]$$

$$\leqslant \left[\int d^3x \, |V(\mathbf{x})|^2 \right] \left(1 + \frac{\tau^2}{m^2\xi^4} \right)^{-3/2} \qquad (2.8)$$

where the integral over $|V(\mathbf{x})|^2$ is certainly convergent under the conditions of the last section. Inserting this bound into (2.7) we find that:

$$\int_{-\infty}^{0} d\tau \, \| V U^0(\tau) \psi_{\text{in}} \| \leqslant \left[\int d^3x \, |V(\mathbf{x})|^2 \right]^{1/2} \int_{-\infty}^{0} d\tau \left(1 + \frac{\tau^2}{m^2\xi^4} \right)^{-3/4} < \infty$$

Thus, the well known $t^{-3/2}$ spreading of the Gaussian wave packet guarantees the convergence of the integral (2.6) and, hence, of the vector $U(t)^\dagger U^0(t) | \psi_{\text{in}} \rangle$ as required in (2.5).[9]

---

[8] This result is quoted in almost any text on quantum mechanics. See Messiah (1961), p. 75, problem 6.
[9] In general the integrand actually converges much more rapidly than $t^{-3/2}$. The point is that our proof has used only the spreading of the free wave packets. However, the general wave packet spreads *and moves*. Both the spreading and the movement contribute to the vanishing of $\| V U^0 \psi_{\text{in}} \|$ and the movement is usually the main contributor.

This completes the proof for any Gaussian wave function of any width and, hence, for any finite linear combinations of such Gaussians. The non-mathematical reader may choose to regard this as sufficient. For the mathematically inclined we leave it as an exercise to show that, because any vector in $\mathscr{H}$ can be arbitrarily well approximated by a finite sum of Gaussians, the result is in fact true for any $|\psi_{in}\rangle$ in $\mathscr{H}$ (see Problem **2.1**).

We can apply the same analysis as $t \to +\infty$.   Q.E.D.

The asymptotic condition guarantees that any $|\psi_{in}\rangle$ in $\mathscr{H}$ is in fact the in asymptote of some actual orbit $U(t)|\psi\rangle$. Furthermore, it is clear from the proof that the actual state $|\psi\rangle$ of the system at $t = 0$ is linearly related to the in asymptote $|\psi_{in}\rangle$; specifically,

$$|\psi\rangle = \lim_{t \to -\infty} U(t)^{\dagger} U^{0}(t) |\psi_{in}\rangle \equiv \Omega_{+} |\psi_{in}\rangle$$

Similarly the actual state $|\psi\rangle$ at $t = 0$ that will evolve into the out asymptote labelled by $|\psi_{out}\rangle$ is

$$|\psi\rangle = \lim_{t \to +\infty} U(t)^{\dagger} U^{0}(t) |\psi_{out}\rangle \equiv \Omega_{-} |\psi_{out}\rangle$$

The two operators $\Omega_{\pm}$, defined as the limits

$$\Omega_{\pm} = \lim_{t \to \mp\infty} U(t)^{\dagger} U^{0}(t)$$

are called the *Møller wave operators*.[10] They are the limits of a unitary operator and so, as discussed in Section 1-g, are isometric—a property we shall be using shortly. Their significance should be clear: Acting on any vector in $\mathscr{H}$, they give the actual state at $t = 0$ that would evolve from (or to) the asymptote represented by that vector. This is illustrated symbolically in Fig. 2.3.

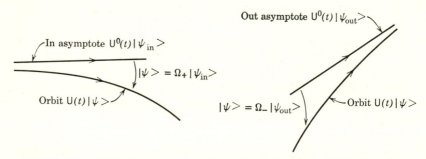

FIGURE 2.3.   Classical representation of the roles of the Møller operators.

[10] Notice the choice of the subscripts $\pm$ on $\Omega_{\pm}$, which are limits as $t \to \mp\infty$ of $U^{\dagger}U^{0}$. The reason for this apparently perverse choice will appear when we discuss the time independent scattering formalism.

In practice the accelerator and collimators produce some definite incoming wave packet, $|\psi_{\text{in}}\rangle = |\phi\rangle$; and the measuring apparatus is arranged to detect a definite out asymptote, $|\psi_{\text{out}}\rangle = |\chi\rangle$. For this reason it is usual and convenient to introduce the following additional notation:

$$\Omega_+ |\phi\rangle \equiv |\phi+\rangle \qquad [\text{any } |\phi\rangle]$$

and

$$\Omega_- |\chi\rangle \equiv |\chi-\rangle \qquad [\text{any } |\chi\rangle]$$

The vector $|\phi+\rangle$ represents the actual state of the system at $t = 0$, *if the in asymptote was* $|\psi_{\text{in}}\rangle = |\phi\rangle$; the vector $|\chi-\rangle$ represents the actual state at $t = 0$ *if the out asymptote were going to be* $|\psi_{\text{out}}\rangle = |\chi\rangle$.

The various notations can be summarized as follows:

$$\text{in asymptote} \xrightarrow{\;\;\Omega_+\;\;} \left\{ \begin{matrix} \text{actual state} \\ \text{at } t = 0 \end{matrix} \right\} \xleftarrow{\;\;\Omega_-\;\;} \text{out asymptote}$$

$$|\psi_{\text{in}}\rangle \longrightarrow \qquad |\psi\rangle \qquad \longleftarrow \qquad |\psi_{\text{out}}\rangle$$

$$|\phi\rangle \longrightarrow \qquad |\phi+\rangle$$

$$\qquad\qquad\qquad |\chi-\rangle \qquad \longleftarrow \qquad |\chi\rangle$$

In connection with this tabulation it may be well to remind the reader that throughout this chapter we are discussing the vectors that represent the evolution of an actual scattering event (either asymptotically or at all times). All vectors in this table and elsewhere are, therefore, normalized proper vectors in the space $\mathscr{H}$.

2-d. Orthogonality and Asymptotic Completeness

We have seen that every vector in $\mathscr{H}$ (denoted $|\psi_{\text{in}}\rangle$ or $|\psi_{\text{out}}\rangle$ as appropriate) labels the in or out asymptote of some actual orbit $U(t) |\psi\rangle$. We must now consider the converse question: Does every $|\psi\rangle$ in $\mathscr{H}$ define an orbit $U(t) |\psi\rangle$ that has in and out asymptotes? Just as in the classical case the answer to this question is, in general, no. The Hamiltonian $H = H^0 + V$ will usually have some bound states; and if $|\phi\rangle$ is a bound state, then the orbit $U(t) |\phi\rangle$ describes a stationary state in which the particle remains localized close to the potential and, hence, never behaves freely.

The situation is analogous to that of the classical problem and we can summarize the results one would expect to find as follows: Every orbit with an in asymptote should also have an out asymptote, and vice versa; and the set of all states with asymptotes (the scattering states) together with the bound states should span $\mathscr{H}$—the space of *all* states. We shall approach these results in two steps. We first prove the orthogonality theorem, which asserts that any bound state is orthogonal to all states with in or out asymptotes. To state this result we introduce some more notation. We denote by $\mathscr{B}$ the subspace spanned by the bound states. We next note that any state with an in asymptote is given by a vector of the form $|\psi\rangle = \Omega_+ |\psi_{in}\rangle$. These vectors make up the *range* of Ω_+. We denote this range by $\mathscr{R}_+$, which is therefore the set of all states that have in asymptotes. Similarly, $\mathscr{R}_-$ will denote the range of Ω_- and is the set of all states with out asymptotes. We can now state the orthogonality theorem.

Orthogonality Theorem. If V satisfies our usual assumptions,[11] then $\mathscr{R}_+ \perp \mathscr{B}$ and $\mathscr{R}_- \perp \mathscr{B}$

Proof: We consider the case of $\mathscr{R}_+$. We suppose that $|\psi\rangle$ is in $\mathscr{R}_+$ (that is, the orbit $U(t) |\psi\rangle$ has an in asymptote $U^0(t) |\psi_{in}\rangle$, with $|\psi\rangle = \Omega_+ |\psi_{in}\rangle$) and that $|\phi\rangle$ is a bound state, with $H |\phi\rangle = E |\phi\rangle$. We must prove that $\langle \phi \mid \psi \rangle = 0$.

The scalar product $\langle \phi \mid \psi \rangle$ can be evaluated at any time during the evolution of the corresponding orbits; that is:

$$\langle \phi \mid \psi \rangle = \langle \phi | U(t)^\dagger U(t) | \psi \rangle \qquad [\text{any } t] \qquad (2.9)$$

If we let t become large and negative then $U(t) |\psi\rangle$ represents a state in which the particle has moved far away, while the state $U(t) |\phi\rangle$ is always localized close to the potential. This means that the overlap of the two states tends to zero. But from (2.9) this overlap is independent of t, and hence is actually, always zero.

This argument can be written in detail, starting with (2.9), as

$$\langle \phi \mid \psi \rangle = e^{iEt}\langle \phi | U(t) | \psi \rangle = \lim e^{iEt}\langle \phi | U^0(t) |\psi_{in}\rangle = 0$$

The second equality follows since $U |\psi\rangle \to U^0 |\psi_{in}\rangle$ and the last equality from the result discussed in Section 1-f, (1.15), Q.E.D.

The orthogonality theorem is more easily proved than our second result, asymptotic completeness. This asserts that the set of all states with in asymptotes is precisely the same as that of all states with out asymptotes—that is, that $\mathscr{R}_+ = \mathscr{R}_-$; and also that the subspace $\mathscr{R}_+$ together with the subspace $\mathscr{B}$ of bound states spans the whole of $\mathscr{H}$. Because we know that $\mathscr{R}_+$ are orthogonal to $\mathscr{B}$, this means that $\mathscr{H}$ should be the *direct sum* of $\mathscr{R}_+$ and $\mathscr{B}$. If we write $\mathscr{H} = \mathscr{B} \oplus \mathscr{R}$, which simply defines $\mathscr{R}$ as the subspace of all vectors orthogonal to $\mathscr{B}$, we can give a simple statement of the desired result.

---

[11] It will be clear from the proof that the result is in fact true under any conditions for which the asymptotic condition holds.

Asymptotic Completeness.   A scattering theory will be called asymptotically complete if $\mathscr{R}_+ = \mathscr{R}_- = \mathscr{R}$ or, in words,

$$\begin{Bmatrix} \text{all states with} \\ \text{in asymptotes} \end{Bmatrix} = \begin{Bmatrix} \text{all states with} \\ \text{out asymptotes} \end{Bmatrix} = \begin{Bmatrix} \text{all states orthogonal} \\ \text{to the bound states} \end{Bmatrix}$$

The proof that for suitable potentials the scattering theory *is* asymptotically complete is very difficult and we refer the interested reader to the literature.[12] We shall simply accept that for a wide class of potentials, including those satisfying the conditions of Section 2-b, asymptotic completeness does hold.

With asymptotic completeness our description of the scattering process is almost complete. It can be summarized as follows: As far as the actual orbits of the system are concerned, the Hilbert space $\mathscr{H}$ is divided into two orthogonal parts; the subspace spanned by the bound states $\mathscr{B}$, and the subspace of scattering states $\mathscr{R}$. For every $|\psi\rangle$ in $\mathscr{R}$, the orbit $U(t)|\psi\rangle$ describes a scattering process with in and out asymptotes,

$$\begin{aligned} U(t)|\psi\rangle &\xrightarrow[t\to-\infty]{} U^0(t)|\psi_{\text{in}}\rangle \\ &\xrightarrow[t\to+\infty]{} U^0(t)|\psi_{\text{out}}\rangle \end{aligned}$$

Every $|\psi_{\text{in}}\rangle$ (or $|\psi_{\text{out}}\rangle$) in $\mathscr{H}$ labels the in (or out) asymptote of a unique actual orbit $U(t)|\psi\rangle$ and the Møller operators $\Omega_\pm$ map each $|\psi_{\text{in}}\rangle$ (or $|\psi_{\text{out}}\rangle$) in $\mathscr{H}$ onto the corresponding scattering state $|\psi\rangle$ in $\mathscr{R}$,

$$|\psi\rangle = \Omega_+ |\psi_{\text{in}}\rangle = \Omega_- |\psi_{\text{out}}\rangle \tag{2.10}$$

As we have already noted, the Møller operators are isometric. This means that for each normalized $|\psi_{\text{in}}\rangle$ or $|\psi_{\text{out}}\rangle$ in $\mathscr{H}$, there is a unique corresponding normalized $|\psi\rangle$ in $\mathscr{R}$; and conversely, for each normalized $|\psi\rangle$ in $\mathscr{R}$ there are unique normalized asymptotes $|\psi_{\text{in}}\rangle$ and $|\psi_{\text{out}}\rangle$. This situation is summarized schematically in Fig. 2.4.

Figure 2.4 serves to emphasize two points that may need clarification. First, Fig. 2.4 clearly indicates (and we have in fact proved) that *every* vector in $\mathscr{H}$ can represent an in (or out) asymptote. Now that we have had occasion to discuss bound states, the reader may well be asking himself how a bound state vector $|\phi\rangle$ can possibly represent a free asymptote. To answer this, one has only to recall the *significance* of the asymptotes. The statement

---

[12] The proofs use the time-independent formalism developed in Chapters 8 and 10. See Ikebe (1960) and Faddeev (1965).

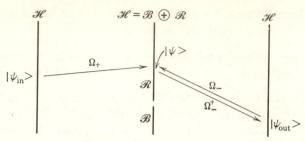

FIGURE 2.4.   The Møller operators $\Omega_\pm$ map the in and out asymptotes, represented by $|\psi_{\text{in}}\rangle$ and $|\psi_{\text{out}}\rangle$, onto the actual orbit labelled by $|\psi\rangle$.

that $|\phi\rangle$ represents a possible in asymptote (for example) means that if we multiply $|\phi\rangle$ by the *free* evolution operator $U^0(t)$ and take t large and negative, then $U^0(t) |\phi\rangle$ will look very like some actual scattering orbit. Now, the free evolution operator spreads *all* states; in particular, it will take no notice of the fact that $|\phi\rangle$ is an eigenstate of the full Hamiltonian $H^0 + V$. Thus, at the times in question (t large and negative) $U^0(t) |\phi\rangle$ does indeed represent the free motion of a particle far from the potential; and as such is a legitimate in asymptote. It is for this reason that all vectors in $\mathscr{H}$ can represent in or out asymptotes.

Second, because the Møller operators map $\mathscr{H}$ (representing the asymptotes) onto the subspace $\mathscr{R}$ (representing the scattering orbits), the Møller operators are isometric but, in general, not unitary. In the special case that H has no bound states the equation $\mathscr{H} = \mathscr{B} \oplus \mathscr{R}$ becomes $\mathscr{H} = \mathscr{R}$, and $\Omega_\pm$ are then unitary. However, in general they are not unitary, the number of bound states being the measure of their failure to be unitary. Perhaps it should be emphasized that this is in no sense surprising or distressing. All that is required is that $\Omega_\pm$ map the asymptotes one-to-one onto the scattering states, and for this it is sufficient that they be isometric.

2-e.   The Scattering Operator

So far we have expressed the actual scattering state at $t = 0$ in terms of either of its two asymptotes. Our ultimate goal is to express the out asymptote in terms of the in asymptote without reference to the experimentally uninteresting actual orbit, and this we can now do. Because Ω_- is isometric, the relation $|\psi\rangle = \Omega_- |\psi_{\text{out}}\rangle$ of (2.10) can be inverted. In fact, since $\Omega_-^\dagger \Omega_- = 1$, we simply multiply on the left by $\Omega_-^\dagger$ to give

$$|\psi_{\text{out}}\rangle = \Omega_-^\dagger |\psi\rangle = \Omega_-^\dagger \Omega_+ |\psi_{\text{in}}\rangle$$

If we define the *scattering operator* as[13]

$$S = \Omega_-^\dagger \Omega_+$$

this equation becomes

$$|\psi_{\text{out}}\rangle = S\,|\psi_{\text{in}}\rangle \tag{2.11}$$

which is the desired result.

The scattering operator S gives $|\psi_{\text{out}}\rangle$ directly in terms of $|\psi_{\text{in}}\rangle$; if a particle enters the collision with in asymptote $|\psi_{\text{in}}\rangle$, then it leaves with out asymptote $|\psi_{\text{out}}\rangle = S\,|\psi_{\text{in}}\rangle$. Since only the asymptotic free motion is observable in practice, the single operator S contains all information of experimental interest. If we know how to calculate S, then the scattering problem is solved. Needless to say, the problem of computing S will occupy several chapters of this book, starting with Chapter 8.

In terms of S we can calculate the experimentally relevant scattering probabilities. The particle emerges from the accelerator moving freely along the in asymptote $U^0(t)\,|\psi_{\text{in}}\rangle$ where $|\psi_{\text{in}}\rangle = |\phi\rangle$ (for instance) is characteristic of the accelerator. The experimental counter arrangement monitors for some definite out asymptote, given by $|\psi_{\text{out}}\rangle = |\chi\rangle$, say. The quantity of interest is therefore the probability that a particle that entered the collision with in asymptote $|\phi\rangle$ will be observed to emerge with out asymptote $|\chi\rangle$. To evaluate this probability we note that the actual state at $t = 0$, which will evolve from the in asymptote $|\phi\rangle$ is $|\phi+\rangle = \Omega_+\,|\phi\rangle$, while the actual state at $t = 0$, which would evolve into the out asymptote $|\chi\rangle$ is $|\chi-\rangle = \Omega_-\,|\chi\rangle$. Because the required probability amplitude is just the scalar product of the actual states at any given time ($t = 0$, for instance), we find that the probability is:

$$
\begin{aligned}
w(\chi \leftarrow \phi) &= |\langle \chi- \mid \phi+\rangle|^2 \\
&= |\langle \chi|\,\Omega_-^\dagger \Omega_+\,|\phi\rangle|^2 \\
&= |\langle \chi|\,S\,|\phi\rangle|^2
\end{aligned}
$$

The probability amplitude for the process $(\chi \leftarrow \phi)$ is just the S-matrix element $\langle \chi|\,S\,|\phi\rangle$.

In practice even the quantity $w(\chi \leftarrow \phi)$ is not directly observable; this is because one cannot actually produce or identify uniquely defined wave packets $|\phi\rangle$ and $|\chi\rangle$. However, as we shall show in Chapter 3, the quantity that *is* experimentally observable, the differential cross-section, can be expressed directly in terms of the matrix elements of S.

---

[13] We should mention that an alternative definition of the scattering operator $S = \Omega_+\Omega_-^\dagger$ is also found in the literature. The reasons for its existence are mainly historical and we shall have no occasion to use it.

2-f.   Unitarity

An important property of the S operator, which we can prove immediately, is that it is unitary. That this is so follows directly from asymptotic completeness and the definition $S = \Omega_-^\dagger \Omega_+$. The Møller operators Ω_+ and Ω_- are isometric operators mapping $\mathscr{H}$ onto the subspace $\mathscr{R}$ of scattering states. This means that Ω_+ is a linear, norm-preserving map of $\mathscr{H}$ onto $\mathscr{R}$, while $\Omega_-^\dagger$ is likewise linear and norm preserving (on $\mathscr{R}$), but maps $\mathscr{R}$ back onto $\mathscr{H}$ (see Fig. 2.4). It immediately follows that S is linear and norm preserving from $\mathscr{H}$ onto $\mathscr{H}$; that is, S is unitary.

One should not be deceived into thinking this result trivial by the simplicity of the argument. The hard work of the proof is, of course, in the proof of asymptotic completeness. In particular, an essential element of the present argument is that Ω_+ and Ω_- map $\mathscr{H}$ onto the *same range* $\mathscr{R}$—by no means a trivial property.

The significance of the unitarity of S can be easily seen. The in and out asymptotes of the system are labelled by the vectors of $\mathscr{H}$ and are related by the S operator, $|\psi_{out}\rangle = S\,|\psi_{in}\rangle$. The fact that S is unitary means that for every normalized $|\psi_{in}\rangle$ there is a unique normalized $|\psi_{out}\rangle$ and vice versa; and also (because S is linear) that the correspondence between $|\psi_{in}\rangle$ and $|\psi_{out}\rangle$ preserves superposition, that is, if $|\psi_{in}\rangle = a\,|\phi_{in}\rangle + b\,|\chi_{in}\rangle$, then $|\psi_{out}\rangle = a\,|\phi_{out}\rangle + b\,|\chi_{out}\rangle$.

We shall see later that the concept of the S operator generalizes to more complicated processes—the elastic scattering of two particles (Chapters 4 and 5), and the multiparticle processes involving arbitrary reactions (Chapter 16). In relativistic quantum field theory a corresponding S operator exists, and in the recent attempts to construct a relativistic scattering theory based directly on the properties of S (the so-called analytic S-matrix theory) the existence of S is taken as a fundamental postulate.[14] In all cases the significance of S is the same: S maps the in asymptote of any scattering orbit directly onto the corresponding out asymptote. In all cases one can expect that S should be unitary.[15]

----

[14] See, for example, Eden, et al. (1966).

[15] It may be worth commenting on a possible confusion concerning the status of the unitarity of S. Because of its simple interpretation, one often hears the claim that the unitarity of S is "obvious". But if S is "obviously" unitary then one could reasonably ask why people expend quite so much energy to prove it. This apparent conflict is, of course, only a question of point of view. If one chooses to *assume* that the S operator exists and that it has the two properties mentioned above, then indeed it is "obviously" unitary. This is the point of view of the analytic S-matrix theory and is certainly entirely reasonable. If, on the other hand, one wishes to prove that S is unitary within some prescribed dynamical theory (such as nonrelativistic Schrödinger theory with a definite potential) then the proof may be very difficult and, in the present case, it is.

The fact that S is unitary has been of the greatest importance in the recent history of scattering theory. Quite generally, it should be clear that any attempt to calculate S will be greatly facilitated by the knowledge that it belongs to the very restricted class of unitary operators. More specifically, unitarity has been an essential tool in the use of dispersion relations (see Chapter 15). We shall discuss a simple example of this kind of application in connection with the optical theorem in Chapter 3.

Our next task is to derive an expression for the quantity that is actually measured—the differential cross section—in terms of S. This we shall do in Chapter 4.

PROBLEMS

2.1. (A little mathematically oriented.)
(a) In Section 2-c the asymptotic condition was proved for a vector $|\psi_{\text{in}}\rangle$ with a Gaussian wave function. Show clearly that it is therefore true for any finite sum of such vectors.
(b) It is a fact that any vector in $\mathscr{L}^2(\mathbb{R}^3)$ can be approximated arbitrarily well by a finite sum of Gaussian vectors; that is, for any $|\psi\rangle$ and any $\epsilon > 0$ there is a finite sum of Gaussians $|\phi\rangle$ such that $\|\psi - \phi\| < \epsilon$. Use this fact to prove the asymptotic condition for any vector $|\psi_{\text{in}}\rangle$.

2.2. Prove the asymptotic condition for a one-particle potential operator of the form $V = |\zeta\rangle\langle\zeta|$, where $|\zeta\rangle$ is some fixed vector in the one-particle Hilbert space. [This is the so-called separable, or factorable, potential. It is our first example of a "nonlocal" potential. That is, it is not just a function of the position operator; or equivalently, the matrix element $\langle\mathbf{x}'|\,V\,|\mathbf{x}\rangle$ is not proportional to $\delta_3(\mathbf{x}' - \mathbf{x})$. In fact, $\langle\mathbf{x}'|\,V\,|\mathbf{x}\rangle = \zeta(\mathbf{x}')\zeta(\mathbf{x})^*$, which is the form in which a separable potential is often defined. The separable potential has been found to give excellent fits to the data in some low-energy nuclear problems, notably the three-nucleon problem. See Watson and Nuttall, 1968, pp. 75–79.] Assume that $|\zeta\rangle$ has a Gaussian wave function.

2.3. Because the Møller operators have the form $\Omega = \lim U(t)^\dagger U^0(t)$ where $U^\dagger U^0$ is unitary, it follows that $\Omega^\dagger\Omega = 1$ and the Møller operators are iso-metric. If we could take the adjoint of the above equation (i.e., if $\Omega^\dagger$ were $\lim U^{0\dagger}U$) it would follow that $\Omega\Omega^\dagger = 1$ and Ω would be unitary, which we know to be false. Show by example that, in fact, $U^{0\dagger}U$ does not have a limit for all vectors in $\mathscr{H}$. (Hint: We know that Ω fails to be unitary because of the bound states.)

3 Cross Sections in Terms of the S Matrix

3-a   **Conservation of Energy**

3-b   **The On-Shell T Matrix and Scattering Amplitude**

3-c   **The Classical Cross Section**

3-d   **Definition of the Quantum Cross Section**

3-e   **Calculation of the Quantum Cross Section**

3-f   **The Optical Theorem**

By introducing the S operator we have been able to relate the asymptotic free motion of a particle leaving a collision directly to its asymptotic free initial state. In particular, the probability that a particle that enters the collision with in asymptote $|\phi\rangle$ be observed to leave with out asymptote $|\chi\rangle$ is given in terms of the corresponding S-matrix element as

$$w(\chi \leftarrow \phi) = |\langle\chi| \, S \, |\phi\rangle|^2$$

Unfortunately, even though we have successfully eliminated all details of the unobservable actual orbit, we have still not arrived at a quantity that can be directly measured. This is because the wave packets $|\phi\rangle$ and $|\chi\rangle$ cannot in practice be uniquely identified (even though in principle they can). In practice, concerning the initial state $|\phi\rangle$ we know only that it is a wave packet whose position and momentum are both reasonably well defined. Concerning the final state, we measure only whether or not the outgoing direction of motion lies in some element of solid angle $d\Omega$. The nature of the measurement on the outgoing particle is easily allowed for; instead of

$w(\chi \leftarrow \phi)$ we have only to calculate the probability $w(d\Omega \leftarrow \phi)$ that the direction of motion of the out state lie inside the element of solid angle $d\Omega$. Our ignorance of the precise in asymptote $|\phi\rangle$ means that we must then average this probability over all relevant states $|\phi\rangle$. It is this averaging process that leads to the notion of the differential cross section, the principal subject of this chapter.

The reader familiar with the elementary discussions of cross sections may feel that the present analysis is unduly cumbersome. The reason for this is that the elementary treatments are always in terms of plane waves and, as most texts admit, can be properly justified only by building up the plane waves into suitable wave packets. The "building up of suitable wave packets" turns out to need some care and it is this operation that occupies the bulk of this chapter. We do not indulge in all this hard work from a whimsical desire for rigor. By defining the cross section properly in terms of wave packets, we shall gain considerable insight into its meaning; in particular, we shall see clearly what are its limits of usefulness and applicability.

Before we discuss the cross section it is convenient to establish two important properties of the S operator, conservation of energy and the decomposition of S in terms of the scattering amplitude. These two results will be discussed in Sections 3-a and 3-b, respectively. Then in Section 3-c we begin our discussion of the cross section with a brief description of a simple classical experiment. In Sections 3-d and 3-e we discuss the quantum cross section and derive its expression in terms of the scattering amplitude. In Section 3-f we prove the optical theorem—an important result that follows directly from the unitarity of S and our expression for the cross section in terms of the amplitude.

3-a.   Conservation of Energy

One of the most important properties of the S operator is that it conserves energy. This property is a little more subtle than one might expect. Because the Hamiltonian H is independent of time, energy is, of course, conserved; and the expectation value of H for any actual orbit is a constant. However, the S operator is a mapping of the asymptotic free orbits, which label the particle's state only when it is far away from the scatterer and does not feel the potential. As far as the asymptotic states are concerned, the actual energy is simply the kinetic energy and we should therefore expect to find that S commutes with the kinetic energy operator H^0, rather than H. This is what we shall now prove.

The essential step in the proof is the so-called *intertwining relation* for the Møller operators,

$$H\Omega_\pm = \Omega_\pm H^0 \qquad\qquad (3.1)$$

These important relations are proved by the following manipulations (which the mathematical reader will easily see to be rigorously justifiable):

$$
\begin{aligned}
e^{iH\tau}\Omega_\pm &= e^{iH\tau}[\lim e^{iHt}e^{-iH^0t}] \\
&= \lim[e^{iH(\tau+t)}e^{-iH^0t}] \\
&= [\lim e^{iH(\tau+t)}e^{-iH^0(\tau+t)}]e^{iH^0\tau} \\
&= \Omega_\pm e^{iH^0\tau}
\end{aligned}
$$

Differentiating with respect to τ and setting $\tau = 0$, we obtain the desired result.

Since the Møller operators are isometric and satisfy $\Omega^\dagger\Omega = 1$ we can rewrite the intertwining relations as

$$
\Omega_\pm^\dagger H \Omega_\pm = H^0 \tag{3.2}
$$

This lets us interpret $\Omega_\pm$ as operators whose action on the full Hamiltonian H is to give the free Hamiltonian H^0. This shows clearly, what we already know, that the Møller operators cannot in general be unitary. For, if they were, then (3.2) would imply that H and H^0 must have the same spectrum; since H^0 has no bound states it would follow that H could not. Thus, only when H has no bound states can $\Omega_\pm$ be unitary, exactly as we saw in Chapter 2.

We can now prove very simply that S commutes with H^0, by two applications of the intertwining relation. Thus

$$
SH^0 = \Omega_-^\dagger\Omega_+ H^0 = \Omega_-^\dagger H\Omega_+ = H^0\Omega_-^\dagger\Omega_+ = H^0S \tag{3.3}
$$

This result expresses the conservation of energy in a scattering experiment. The mean initial energy for the in state $|\psi_{in}\rangle$ is the expectation value

$$
\text{in energy} = \langle\psi_{in}| H^0 |\psi_{in}\rangle
$$

Similarly, the mean final energy for the corresponding out state $|\psi_{out}\rangle = S|\psi_{in}\rangle$ is:

$$
\text{out energy} = \langle\psi_{out}| H^0 |\psi_{out}\rangle = \langle\psi_{in}| S^\dagger H^0 S |\psi_{in}\rangle
$$

These two energies are equal since, according to (3.3), $H^0 = S^\dagger H^0 S$.

As every student of quantum mechanics knows, a convenient way to exploit the fact that an operator S commutes with an observable H^0 is to use a basis of eigenvectors of the observable. Of course, the free Hamiltonian H^0 has no proper eigenvectors. However, as discussed in Chapter 1, we shall follow the usual and convenient practice of expanding with "improper eigenvectors," which we treat (as far as possible) like ordinary vectors. A convenient choice for eigenvector of H^0 is the momentum eigenvector $|\mathbf{p}\rangle$, whose spatial wave function is the plane wave

$$\langle \mathbf{x} \mid \mathbf{p} \rangle = (2\pi)^{-3/2} e^{i\mathbf{p}\cdot\mathbf{x}}$$

with normalisation $\langle \mathbf{p}' \mid \mathbf{p} \rangle = \delta_3(\mathbf{p}' - \mathbf{p})$. This vector satisfies

$$H^0 |\mathbf{p}\rangle = \frac{\mathbf{p}^2}{2m} |\mathbf{p}\rangle \equiv E_p |\mathbf{p}\rangle$$

where we have introduced the notation:

$$\boxed{E_p = \frac{p^2}{2m}}$$

for the energy of a free particle of momentum $\mathbf{p}$.

We shall write the matrix elements of S in the momentum representation as $\langle \mathbf{p}'| S |\mathbf{p}\rangle$ and shall often refer to $\{\langle \mathbf{p}'| S |\mathbf{p}\rangle\}$ as "the S matrix." It is important to remember that just as the vector $|\mathbf{p}\rangle$ does not represent a physically realizable state, so $\langle \mathbf{p}'| S |\mathbf{p}\rangle$ is not the amplitude for any physically realizable process. The significance of the plane-wave "states" is that they form a convenient basis for the expansion of proper states,

$$|\psi\rangle = \int d^3p\ \psi(\mathbf{p}) |\mathbf{p}\rangle$$

In the same way the physical significance of the improper matrix elements $\langle \mathbf{p}'| S |\mathbf{p}\rangle$ is in the corresponding expansion of the proper matrix element $\langle \chi| S |\phi\rangle$; or equivalently of the out wave function $\psi_{\text{out}}(\mathbf{p})$ in terms of $\psi_{\text{in}}(\mathbf{p})$,

$$\psi_{\text{out}}(\mathbf{p}) = \int d^3p' \langle \mathbf{p}| S |\mathbf{p}'\rangle \psi_{\text{in}}(\mathbf{p}') \tag{3.4}$$

With these reservations, it is nonetheless convenient to visualize $\langle \mathbf{p}'| S |\mathbf{p}\rangle$ as the probability amplitude that an in state of momentum $\mathbf{p}$ lead to an out state of momentum $\mathbf{p}'$.

Since S commutes with H^0, its momentum-space matrix elements satisfy:

$$0 = \langle \mathbf{p}'| [H^0, S] |\mathbf{p}\rangle = (E_{p'} - E_p)\langle \mathbf{p}'| S |\mathbf{p}\rangle$$

This implies that $\langle \mathbf{p}'| S| \mathbf{p}\rangle$ is zero except when $E_{p'} = E_p$ and hence that it has the form:

$$\langle \mathbf{p}'| S |\mathbf{p}\rangle = \delta(E_{p'} - E_p) \times (\text{remainder}) \tag{3.5}$$

This is the momentum-space expression of energy conservation.

3-b.   The On-Shell *T* Matrix and Scattering Amplitude

To explore further the structure of the momentum-space *S* matrix it is convenient to introduce a second operator R defined by the relation S = 1 + R. Obviously R represents the difference between the actual value of S and its value in the absence of all interactions (namely S = 1). Since S commutes with H^0, so does R. Therefore, just like $\langle \mathbf{p'}| S |\mathbf{p} \rangle$ in (3.5), the matrix element $\langle \mathbf{p'}| R |\mathbf{p} \rangle$ contains a factor $\delta(E_{p'} - E_p)$. We write this as:

$$\langle \mathbf{p'}| R |\mathbf{p} \rangle = -2\pi i \, \delta(E_{p'} - E_p) \, t(\mathbf{p'} \leftarrow \mathbf{p}) \tag{3.6}$$

where the factor $-2\pi i$ is introduced for future convenience.[1] For the *S* matrix this gives:

$$\boxed{\langle \mathbf{p'}| S |\mathbf{p} \rangle = \delta_3(\mathbf{p'} - \mathbf{p}) - 2\pi i \, \delta(E_{p'} - E_p) \, t(\mathbf{p'} \leftarrow \mathbf{p})} \tag{3.7}$$

The significance of the two terms in this decomposition of the *S* matrix is easily understood. The first term, $\delta_3(\mathbf{p'} - \mathbf{p})$ is obviously the amplitude for the particle to pass the force center without being scattered. The second is therefore the amplitude that it actually is scattered. Now, when the particle is scattered its momentum changes, while its energy stays the same. Thus, the second (scattering) term in (3.7) should conserve energy, but not the individual components of momentum. This means that this term should contain an energy delta function, but no more delta functions. In other words, the factor $t(\mathbf{p'} \leftarrow \mathbf{p})$ is expected to be a smooth function of its arguments. It can be shown that for a large class of potentials $t(\mathbf{p'} \leftarrow \mathbf{p})$ is actually an analytic function of the relevant variables. Here we shall simply assume, as is entirely reasonable, that $t(\mathbf{p'} \leftarrow \mathbf{p})$ is at least a continuous function of its arguments.[2]

Because of the factor $\delta(E_{p'} - E_p)$ in (3.7), the quantity $t(\mathbf{p'} \leftarrow \mathbf{p})$ is defined only for $E_{p'} = E_p$; that is, in the space of the variables $\mathbf{p'}$ and $\mathbf{p}$ the function

---

[1] There is no universal agreement on this definition, and definitions differing by factors of 2 and π are to be found. The rationale for the present choice is that in Born approximation (Chapter 9) $t(\mathbf{p'} \leftarrow \mathbf{p})$ is exactly the same as $\langle \mathbf{p'}| V |\mathbf{p} \rangle$ (without any extra factors).

[2] It should be emphasized that this is really a quite profound result, which depends on the short range of the forces. If the forces were sufficiently long range (e.g., the Coulomb force) *all* particles would be scattered and $\langle \mathbf{p'}| S |\mathbf{p} \rangle$ would not have the structure (3.7). The result (3.7) [including the smoothness of $t(\mathbf{p'} \leftarrow \mathbf{p})$] is the simplest example of the so-called *cluster decomposition* of the *S* matrix, which has played an important role in relativistic scattering theory.

$t(\mathbf{p}' \leftarrow \mathbf{p})$ is defined only on the "shell" $\mathbf{p}'^2 = \mathbf{p}^2$. For this reason $t(\mathbf{p}' \leftarrow \mathbf{p})$ is called the T matrix *on the energy shell*, or just the *on-shell* T *matrix*.

Because the on-shell T matrix is defined only for $\mathbf{p}'^2 = \mathbf{p}^2$, it obviously does not define an operator T of which $t(\mathbf{p}' \leftarrow \mathbf{p})$ is the matrix element. (To define an operator one must give the matrix elements for *all* $\mathbf{p}'$ and $\mathbf{p}$.) However, we shall see in Chapter 8 that it is possible and convenient to define an operator T whose matrix elements $\langle \mathbf{p}' | T | \mathbf{p} \rangle$ coincide with $t(\mathbf{p}' \leftarrow \mathbf{p})$ when[3] $\mathbf{p}'^2 = \mathbf{p}^2$. The matrix $\langle \mathbf{p}' | T | \mathbf{p} \rangle$ is defined for all $\mathbf{p}'$ and $\mathbf{p}$ and is called the *off-shell T matrix*. As far as the observation of scattering experiments is concerned, only the *on-shell T matrix* is relevant, because knowledge of it alone determines the S matrix via (3.7). On the other hand, it turns out that the *off-shell T matrix* is a useful tool in calculations; in particular, it satisfies the important Lippmann–Schwinger equation, which we shall discuss in Chapters 8–10.[4]

The on-shell T matrix is closely related to the *scattering amplitude*. In fact the function:

$$\boxed{f(\mathbf{p}' \leftarrow \mathbf{p}) = -(2\pi)^2 m \, t(\mathbf{p}' \leftarrow \mathbf{p})} \qquad (3.8)$$

is precisely the amplitude of the elementary scattering theory where f is defined as the coefficient of the outgoing wave in the stationary scattering states, as discussed in the Introduction. Since we shall not be in a position to show this for a while, we shall adopt (3.8) as our *definition* of the scattering amplitude; in Chapter 10 we shall show that this definition coincides with the elementary one.

The reader may reasonably wonder why we bother to use both of the two functions $t(\mathbf{p}' \leftarrow \mathbf{p})$ and $f(\mathbf{p}' \leftarrow \mathbf{p})$ when they only differ by the trivial constant $-(2\pi)^2 m$. Of two reasons the first is tradition; in the literature of scattering theory both functions appear with more or less equal frequency. The second is that it is actually convenient to use both functions. In discussing the connection with the off-shell T matrix (as we shall in Chapter 8) it is convenient to use $t(\mathbf{p}' \leftarrow \mathbf{p})$. On the other hand, it is more convenient to use $f(\mathbf{p}' \leftarrow \mathbf{p})$ when discussing the cross section, which we shall show in Section 3-e to

---

[3] This is an oversimplification. What we shall actually define is an operator $T(z)$ depending on a complex variable z. If we set $z = E + i\epsilon$ (E real), and let $\epsilon \to 0$, then the matrix element $\langle \mathbf{p}' | T(E + i0) | \mathbf{p} \rangle$ for the special case $E_{p'} = E_p = E$ coincides with $t(\mathbf{p}' \leftarrow \mathbf{p})$.

[4] It should also be noted that we are at the moment discussing just the scattering of one particle off a fixed potential (which is closely related to scattering of two particles off one another, as we shall see in Chapter 4). When we go on to discuss processes involving three or more particles (e.g., neutron–deuteron scattering) we shall find that in various approximations the on-shell many-body T matrix can be written in terms of certain off-shell two-body matrices.

be just $d\sigma/d\Omega = |f(\mathbf{p}' \leftarrow \mathbf{p})|^2$. Needless to say it is not necessary to remember all basic formulas in terms of both functions. A simple procedure is to remember the decomposition (3.7) of the S matrix in terms of $t(\mathbf{p}' \leftarrow \mathbf{p})$ (all one has to worry about is the factor $-2\pi i$) and the definition (3.8) of $f(\mathbf{p}' \leftarrow \mathbf{p})$ in terms of $t(\mathbf{p}' \leftarrow \mathbf{p})$. From these one can, of course, derive the decomposition of the S matrix in terms of the amplitude,

$$\langle \mathbf{p}' | \, S \, | \mathbf{p} \rangle = \delta_3(\mathbf{p}' - \mathbf{p}) + \frac{i}{2\pi m} \, \delta(E_{p'} - E_p) f(\mathbf{p}' \leftarrow \mathbf{p}) \qquad (3.9)$$

but there is obviously no need to memorize this result.

3-c.   The Classical Cross Section

Having established conservation of energy and the decomposition of S in terms of the on-shell T matrix or amplitude, we can now return to the main topic of this chapter, the scattering cross section. We first discuss briefly the notion of the cross section in a very simple classical process, the scattering of a point particle by a fixed rigid body (Fig. 3.1). The target is placed in some suitable container and the projectile is fired in. We can measure the momentum of the projectile as it approaches the target ($\mathbf{p}_0$, say) and again as it leaves. On the other hand, we cannot measure the microscopic details of the event; in particular we cannot measure the impact parameter $\boldsymbol{\rho}$, which is defined as the perpendicular (vector) distance from a suitably chosen axis through the target to the line of the incident trajectory, as shown in Fig. 3.1. Our problem is to extract the maximum information about the target given these facts.

We first note that we can learn little from a single passage of the projectile. If it emerges with momentum different from $\mathbf{p}_0$ we know simply that it must have hit the target; if it emerges with its momentum unchanged we know that it has missed.

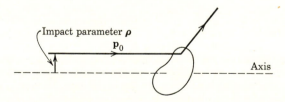

FIGURE 3.1.   Classical scattering of a point particle with initial momentum $\mathbf{p}_0$ and impact parameter $\boldsymbol{\rho}$, off a fixed rigid body.

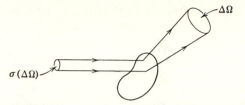

FIGURE 3.2.   The cross section $\sigma(\Delta\Omega)$ is the area of that part of the target which scatters into $\Delta\Omega$.

Let us suppose, however, that we can repeat the experiment many times, always with the same incident momentum, but with *random impact parameters*. If we do this often enough we can speak realistically of the number n_{inc} of projectiles incident per unit area perpendicular to $\mathbf{p}_0$. We can then assert that the number of particles that hit the target is just n_{inc} times the cross-sectional area σ of the target normal to $\mathbf{p}_0$. Because we can identify those particles that hit the target by the fact that they are scattered (i.e., change direction) we have the relation:

$$N_{sc} = n_{inc}\sigma \tag{3.10}$$

where N_{sc} denotes the total number of particles scattered. Since both N_{sc} and n_{inc} can be measured, this lets us find the cross section σ of the target.[5]

In fact, we can obtain much more information because we can count the number of scatterings in any given direction. If we denote by $N_{sc}(\Delta\Omega)$ the number of particles scattered into the solid angle $\Delta\Omega$, then

$$\boxed{N_{sc}(\Delta\Omega) = n_{inc}\sigma(\Delta\Omega)} \tag{3.11}$$

where $\sigma(\Delta\Omega)$ is just the cross section of that part of the target which scatters into $\Delta\Omega$ (Fig. 3.2) and can be measured using this relation. If the solid angle $\Delta\Omega$ is small ($d\Omega$), then $\sigma(d\Omega)$ is proportional to $d\Omega$ and it is usual to write:

$$\sigma(d\Omega) = \frac{d\sigma}{d\Omega}\, d\Omega \tag{3.12}$$

where $d\sigma/d\Omega$ is the so-called differential cross section and represents the most detailed information obtainable in this way.

---

[5] Some readers may feel suspicious of the fact that we discuss (and shall continue to discuss) cross sections in terms of the total number of particles scattered, rather than the scattered flux. One can, of course, use either, since the flux is just the total number divided by the total time. The preference for using flux is an historical accident related to the use of plane waves (for which only the flux is meaningful). When using wave packets, each of which describes one individual projectile, it is obviously more natural simply to count particles.

Before discussing the corresponding quantum problems, let us pause to make three comments. First, it is obviously essential that the incident particles be projected with properly random impact parameters so that we can safely speak of a uniform incident density n_{inc}. Second, it is (fortunately) not necessary that we send in particles uniformly over an infinite front; all we require is that they impinge uniformly over an area that is large compared to the target size (particles with sufficiently large impact parameters are not scattered, so do not contribute to N_{sc}). Third, the differential cross section in the *forward direction* is an undefined quantity because one cannot distinguish a particle that is scattered in the forward direction (whatever that means) from a particle that is not scattered at all.

3-d.   Definition of the Quantum Cross Section

In the quantum scattering problem the incident projectile approaches the target with some definite in asymptote $|\psi_{\text{in}}\rangle$, which we shall identify by its momentum-space wave function $\psi_{\text{in}}(\mathbf{p}) \equiv \langle \mathbf{p} \mid \psi_{\text{in}}\rangle$. The corresponding outgoing wave function $\psi_{\text{out}}(\mathbf{p}) \equiv \langle \mathbf{p} \mid \psi_{\text{out}}\rangle$ determines the probability that long after the collision the particle is found with momentum $\mathbf{p}$, $w(d^3p \leftarrow \psi_{\text{in}}) = d^3p|\psi_{\text{out}}(\mathbf{p})|^2$. The probability of the particle emerging with momentum anywhere in the element of solid angle $d\Omega$ about the direction $\hat{\mathbf{p}}$ is obtained by integrating over all $|\mathbf{p}|$,

$$ w(d\Omega \leftarrow \psi_{\text{in}}) = d\Omega \int_0^\infty p^2 \, dp \, |\psi_{\text{out}}(\mathbf{p})|^2 $$

where as usual $\mathbf{p} = p\hat{\mathbf{p}}$.[6]

Much as in the classical case, this result is of little value for a single event because we do not know the precise in state $|\psi_{\text{in}}\rangle$.[7] All we know is that the wave function $\psi_{\text{in}}(\mathbf{p})$ is well peaked about some value $\mathbf{p}_0$, and, just as in the classical case, we must average over a large number of experiments subject to this partial information.

An accelerator will certainly not produce precisely the same wave packet over and over again. However, we shall temporarily indulge in the fiction that we could arrange for the particles to be produced in wave packets that differ only by *random lateral displacements* perpendicular to $\mathbf{p}_0$. Imagine a succession of experiments with in states $|\psi_{\text{in}}\rangle = |\phi_\rho\rangle$ where $|\phi_\rho\rangle$ is the state

---

[6] The limits 0 and ∞ in this integral reflect the fact that one is not interested in the magnitude of $\mathbf{p}$, but only its direction. One can, of course, measure the magnitude but it is fixed by energy conservation and is therefore not an interesting quantity to observe.

[7] In addition, the quantity $w(d\Omega \leftarrow \psi_{\text{in}})$ is only a probability. This is an added complication of the quantum mechanical problem. Nonetheless, the essential feature of the present discussion is not the probabilistic nature of $w(d\Omega \leftarrow \psi_{\text{in}})$ but our ignorance of $|\psi_{\text{in}}\rangle$.

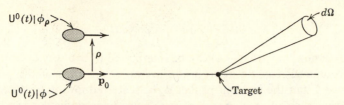

FIGURE 3.3.   Two incoming wave packets $|\phi\rangle$, $|\phi_\rho\rangle$ differ by rigid displacement through ρ perpendicular to their mean momentum.

obtained by rigid displacement of some definite state $|\phi\rangle$ through the distance ρ; that is, $|\phi_\rho\rangle$ is given by the wave function[8] $\phi_\rho(\mathbf{p}) = e^{-i\rho \cdot \mathbf{p}}\phi(\mathbf{p})$. We suppose that the vector $|\phi\rangle$ is the same in all cases but the displacements ρ are randomly distributed in the plane perpendicular to $\mathbf{p}_0$. This arrangement with random lateral displacements of the wave packet $\phi(\mathbf{p})$ corresponds to the random impact parameters of our classical discussion and we can refer to the displacement ρ as the impact parameter (Fig. 3.3).

If we repeat the experiment often enough, with random displacements $\rho_1, \rho_2, \ldots$, then the total number of observed scatterings into $d\Omega$ will be just the sum of the individual probabilities $w(d\Omega \leftarrow \phi_{\rho_i})$,

$$N_{sc}(d\Omega) = \sum_i w(d\Omega \leftarrow \phi_{\rho_i})$$

With enough events the sum can be replaced by an integral,

$$N_{sc}(d\Omega) = \int d^2\rho \, n_{inc} w(d\Omega \leftarrow \phi_\rho)$$

where n_{inc} is the incident density. (In practice this integral runs over an area that is finite but much larger than the target dimensions. However, because the probability of scattering is zero for sufficiently large ρ, the integral can be taken over the whole plane without changing its value.) Because the ρ_i are random, the density n_{inc} is uniform and can be taken outside the integral to give:

$$N_{sc}(d\Omega) = n_{inc} \int d^2\rho \, w(d\Omega \leftarrow \phi_\rho)$$

We see that the number of scatterings is proportional to n_{inc}, as one would expect, and comparison with the classical result (3.11) leads us to write this as:

$$N_{sc}(d\Omega) = n_{inc}\sigma(d\Omega \leftarrow \phi) \qquad (3.13)$$

[8] Recall that the operator of displacement through a vector ρ is $\exp(-i\rho \cdot \mathbf{P})$; its effect on the momentum-space wave function is just to multiply by $\exp(-i\rho \cdot \mathbf{p})$. See Problem **1.2** or Messiah (1961) p. 652.

where

$$\sigma(d\Omega \leftarrow \phi) = \int d^2\rho\, w(d\Omega \leftarrow \phi_\rho) \qquad (3.14)$$

This expression for $\sigma(d\Omega \leftarrow \phi)$ can be interpreted in two ways. Obviously we have introduced it as an appropriate average over all ρ of the probability $w(d\Omega \leftarrow \phi_\rho)$ that the incident packet ϕ_ρ be scattered into $d\Omega$. Alternatively, we can view it as an *area integral* over the plane perpendicular to $\mathbf{p}_0$, in which each element of area $d^2\rho$ is weighted by $w(d\Omega \leftarrow \phi_\rho)$. This view makes clear that $\sigma(d\Omega \leftarrow \phi)$ is just the *effective cross-sectional area* of the target potential for scattering of the wave packets ϕ_ρ (with ρ random) into $d\Omega$. In particular in the classical limit $w(d\Omega \leftarrow \phi_\rho)$ has the values 0 or 1 and the integral picks out precisely the actual area which scatters into $d\Omega$.

We shall find that, *provided the wave function is sufficiently peaked about its mean momentum* $\mathbf{p}_0$, the cross section $\sigma(d\Omega \leftarrow \phi)$ is independent of any of the details of $\phi(\mathbf{p})$ except for $\mathbf{p}_0$ itself. We can therefore write $\sigma(d\Omega \leftarrow \phi) = \sigma(d\Omega \leftarrow \mathbf{p}_0)$, for $\phi(\mathbf{p})$ sufficiently peaked about $\mathbf{p}_0$. This result allows us to remove the unrealistic assumption that the accelerator always produces the same wave packet $\phi(\mathbf{p})$ (apart from the random lateral displacements). If the accelerator produces wave packets of different shapes, as well as different impact parameters, then we must average over both of these variables. If, however, all packets are well peaked about the same momentum $\mathbf{p}_0$ the averaging over impact parameters produces a result that is independent of shape. The averaging over shapes then makes no further modification.

We should perhaps reemphasize how the cross section (3.14) is arrived at. The experiment consists of a sequence of independent collisions between a single particle and the single fixed scattering center, the incident wave packets being random with respect to impact parameter and shape. As we have said $\sigma(d\Omega \leftarrow \phi)$ is just a suitable average over impact parameters and shapes of the probability $w(d\Omega \leftarrow \phi)$. In practice this averaging is achieved in two ways: by using a beam of many particles and by having many scattering centers in a single target assembly. Provided that either the scattering centers are uniformly distributed in the target or the particles in the beam, our requirement of a uniform distribution of impact parameters is realized. However, to satisfy all conditions for our idealized definition of the cross section it is necessary that each incident particle scatter separately off one scattering center at most. For this to be so the beam must be sufficiently weak that the incident particles do not interact with one another; the target must be so thin that multiple scattering is negligible and the distribution of scatterers must be such as to avoid coherent scattering off two or more centers.

Whether these conditions for the applicability of (3.13) and (3.14) are actually realized is a question that must be examined separately for every experiment. As is well known, there are many experiments in which the conditions are very well fulfilled, and we shall, for the present, confine attention to these experiments.[9] We therefore proceed with the calculation of the cross section defined in (3.14).

3-e. Calculation of the Quantum Cross Section

At last we have all the results needed to express the observable cross section (3.14),

$$\sigma(d\Omega \leftarrow \phi) = \int d^2\rho \, w(d\Omega \leftarrow \phi_\rho) \tag{3.14}$$

in terms of the scattering amplitude. We have found that the probability $w(d\Omega \leftarrow \phi_\rho)$ in (3.14) is:

$$w(d\Omega \leftarrow \phi_\rho) = d\Omega \int_0^\infty p^2 \, dp \, |\psi_{\text{out}}(\mathbf{p})|^2 \tag{3.15}$$

where $\mathbf{p}$ points in the direction of observation and $\psi_{\text{out}}(\mathbf{p})$ is the out wave function corresponding to the in state $|\psi_{\text{in}}\rangle = |\phi_\rho\rangle$. According to (3.4),

$$\psi_{\text{out}}(\mathbf{p}) = \int d^3p' \langle \mathbf{p}| \, \mathsf{S} \, |\mathbf{p}'\rangle \psi_{\text{in}}(\mathbf{p}')$$

or, replacing $\langle \mathbf{p}| \, \mathsf{S} \, |\mathbf{p}'\rangle$ by its decomposition (3.9) in terms of the amplitude,

$$\psi_{\text{out}}(\mathbf{p}) = \psi_{\text{in}}(\mathbf{p}) + \frac{i}{2\pi m} \int d^3p' \, \delta(E_p - E_{p'}) f(\mathbf{p} \leftarrow \mathbf{p}') \psi_{\text{in}}(\mathbf{p}') \tag{3.16}$$

Here the first term is the unscattered wave and the second the scattered wave. In our case, $\psi_{\text{in}}(\mathbf{p}) = e^{-i\mathbf{p}\cdot\mathbf{\rho}}\phi(\mathbf{p})$.

We now make the essential restriction that we do not make our observations in the forward direction; that is, we require that the direction of observation $\hat{\mathbf{p}}$ avoid the small neighborhood of $\mathbf{p}_0$ where $\phi(\mathbf{p})$ is nonzero. With this proviso the first term in (3.16) is zero and we can write:

$$\psi_{\text{out}}(\mathbf{p}) = \frac{i}{2\pi m} \int d^3p' \, \delta(E_p - E_{p'}) f(\mathbf{p} \leftarrow \mathbf{p}') e^{-i\mathbf{p}\cdot\mathbf{\rho}'} \phi(\mathbf{p}') \tag{3.17}$$

provided $\mathbf{p}$ *is not in that neighborhood of* $\mathbf{p}_0$ *where the incident wave is nonzero.*

---

[9] A well-known example where the conditions are *not* fulfilled is the Davisson–Germer experiment. There the incident wave packet is comparable in size to the spacing of the scattering centers in the target and the particles scatter coherently off several centers.

Samford University Library

We are ready to take the plunge and substitute (3.17) into (3.15) and thence into (3.14). This gives,

$$\sigma(d\Omega \leftarrow \phi) = \frac{d\Omega}{(2\pi)^2 m^2} \int d^2\rho \int_0^\infty p^2\, dp$$

$$\times \int d^3p'\, \delta(E_p - E_{p'}) f(\mathbf{p} \leftarrow \mathbf{p'}) e^{-i\boldsymbol{\rho}\cdot\mathbf{p'}} \phi(\mathbf{p'})$$

$$\times \int d^3p''\, \delta(E_p - E_{p''}) f(\mathbf{p} \leftarrow \mathbf{p''})^* e^{i\boldsymbol{\rho}\cdot\mathbf{p''}} \phi(\mathbf{p''})^* \quad (3.18)$$

where the extra integral over $\mathbf{p}''$ comes from writing $|\psi_{\text{out}}(\mathbf{p})|^2$ as $\psi_{\text{out}}(\mathbf{p})$ times its complex conjugate.

This complicated expression simplifies quite easily. We first note that:

$$\int d^2\rho\, e^{i\boldsymbol{\rho}\cdot(\mathbf{p''}-\mathbf{p'})} = (2\pi)^2\, \delta_2(\mathbf{p'_\perp} - \mathbf{p''_\perp}) \quad (3.19)$$

where, since the integral over $\boldsymbol{\rho}$ is in the plane normal to $\mathbf{p}_0$, the two-dimensional delta function refers to the components of $\mathbf{p}'$ and $\mathbf{p}''$ in that plane. [For example if $\mathbf{p}_0$ points along the z axis, $\delta_2(\mathbf{p'_\perp} - \mathbf{p''_\perp})$ is just $\delta(p_1' - p_1'')$ $\delta(p_2' - p_2'')$.]

We next rewrite the second energy delta function in (3.18) as[10]:

$$\delta(E_{p'} - E_{p''}) = 2m\, \delta(\mathbf{p'}^2 - \mathbf{p''}^2) \quad (3.20)$$

The delta function (3.19) already requires that the components of $\mathbf{p}'$ and $\mathbf{p}''$ perpendicular to $\mathbf{p}_0$ are equal. We can therefore replace the argument of (3.20) by $(p_\parallel'^2 - p_\parallel''^2)$, where $p_\parallel'$ is the component of $\mathbf{p}'$ along $\mathbf{p}_0$. Therefore, the effect of the delta function (3.20) is to set $p_\parallel' = \pm p_\parallel''$. Provided the wave functions ϕ in (3.18) are sufficiently narrow, the points $p_\parallel' = -p_\parallel''$ do not contribute to the integral and the combined effect of the delta functions (3.19) and (3.20) is simply,

$$\delta_2(\mathbf{p'_\perp} - \mathbf{p''_\perp})\, \delta(E_{p'} - E_{p''}) = \frac{m}{p_\parallel'}\, \delta_3(\mathbf{p'} - \mathbf{p''})$$

When we use this delta function in (3.18) we obtain

$$\sigma(d\Omega \leftarrow \phi) = \frac{d\Omega}{m} \int_0^\infty p^2\, dp \int d^3p'\, \frac{1}{p_\parallel'}\, \delta(E_p - E_{p'})\, |f(\mathbf{p} \leftarrow \mathbf{p'})\phi(\mathbf{p'})|^2 \quad (3.21)$$

---

[10] Notice that we have taken advantage of the first δ function $\delta(E - E')$ to rewrite $\delta(E - E'')$ as $\delta(E' - E'')$. Recall also that $\delta(ax) = |a|^{-1}\, \delta(x)$. For this and other properties of the delta function see Problem **3.2**.

and, if we use the last delta function to do the integral over p, this becomes,

$$\sigma(d\Omega \leftarrow \phi) = d\Omega \int d^3p' \frac{p'}{p'_{\parallel}} |f(\mathbf{p} \leftarrow \mathbf{p}')\phi(\mathbf{p}')|^2 \qquad (3.22)$$

where $|\mathbf{p}| = |\mathbf{p}'|$.

We come now to the final and crucial step. If the region where $\phi(\mathbf{p}')$ is appreciably different from zero is so small that the variation of $f(\mathbf{p} \leftarrow \mathbf{p}')$ in this region is insignificant, then we can replace $f(\mathbf{p} \leftarrow \mathbf{p}')$ and $(p'/p'_{\parallel})$ by their values at $\mathbf{p}' = \mathbf{p}_0$. This gives:

$$\sigma(d\Omega \leftarrow \phi) = d\Omega |f(\mathbf{p} \leftarrow \mathbf{p}_0)|^2 \int d^3p' |\phi(\mathbf{p}')|^2$$

or

$$= d\Omega |f(\mathbf{p} \leftarrow \mathbf{p}_0)|^2 \qquad (3.23)$$

because the one remaining integral is just the normalisation integral of the wave function ϕ.

We can see clearly that, as anticipated, the result is independent of the shape of ϕ, provided only that $\phi(\mathbf{p})$ is sufficiently peaked about $\mathbf{p}_0$ and we are therefore justified in rewriting $\sigma(d\Omega \leftarrow \phi)$ as $\sigma(d\Omega \leftarrow \mathbf{p}_0)$. As in the classical case we define the differential cross section by the relation:

$$\sigma(d\Omega \leftarrow \mathbf{p}_0) = \frac{d\sigma}{d\Omega} (\mathbf{p} \leftarrow \mathbf{p}_0) \, d\Omega$$

where $\mathbf{p}$ points in the direction of observation around which $d\Omega$ is defined and $|\mathbf{p}| = |\mathbf{p}_0|$. With this definition our result becomes

$$\frac{d\sigma}{d\Omega} (\mathbf{p} \leftarrow \mathbf{p}_0) = |f(\mathbf{p} \leftarrow \mathbf{p}_0)|^2 \qquad (3.24)$$

which is the familiar result given in all elementary texts and quoted in the Introduction.[11]

The result (3.24) expresses the observable differential cross section in terms of the matrix elements of S, and is the central result of this chapter. Our chief remaining task for one-particle scattering is to set up means for the actual computation of S in terms of V. Before taking up this task in Chapter

[11] As emphasized earlier, we have not yet shown that the amplitude $f(\mathbf{p} \leftarrow \mathbf{p}_0)$ defined in (3.8) is in fact the same as that of the traditional approach in terms of the stationary scattering states. The result (3.24) can be regarded as confirming that they are the same within a phase. In Chapter 10 we shall show that they are exactly the same.

8, we shall describe in Chapters 4–7 how our formalism can be extended to cover two-particle scattering (including particles with spin) and discuss those general properties of the S operator and scattering amplitude that follow from the various possible invariance principles. In the remainder of this chapter we make some comments on the result (3.24) and its derivation, and then use the result to prove the optical theorem.

First, we can now see precisely what is meant by the statement that the wave packets $\phi(\mathbf{p}')$ must be "sufficiently peaked" in momentum. The requirement is simply that we can take the factor $f(\mathbf{p} \leftarrow \mathbf{p}')$ in (3.22) outside of the integral to give (3.23); that is $f(\mathbf{p} \leftarrow \mathbf{p}')$ must be essentially constant in the region where $\phi(\mathbf{p}')$ is appreciably different from zero. Since $f(\mathbf{p} \leftarrow \mathbf{p}')$ is continuous while $\phi(\mathbf{p}')$ can be chosen arbitrarily narrow, it follows that, *in principle*, this requirement can always be met. Whether it is met *in practice* is a question that must be decided separately for each experiment. It is possible that at certain energies $f(\mathbf{p} \leftarrow \mathbf{p}')$ may vary rapidly (a phenomenon usually associated with resonances) while the width of the incident packet may be large (due to intrinsic width at source, broadening caused by motion of either source or target, etc.); in this case our answer (3.24) may not be relevant. On the other hand, under normal circumstances the conditions for validity of (3.24) are usually met, as we shall discuss in detail in Chapter 10. Here we remark only that, as we shall see in Chapter 9, the amplitude $f(\mathbf{p} \leftarrow \mathbf{p}')$ is closely related to the Fourier transform of the potential $V(\mathbf{x})$. Thus, by the well known property of Fourier transforms, the condition that the *momentum-space* wave function be sharply peaked compared to $f(\mathbf{p} \leftarrow \mathbf{p}')$ can be expected to require that the *spatial* wave function be *broad* compared to $V(\mathbf{x})$. That is, the incident wave packet must be large compared to the size of the target potential. This condition, as well as other conditions discussed in Chapter 10, is usually satisfied.

Second, it should be clear that it was essential to our calculation that the direction of observation, defined by $d\Omega$, did not include the forward direction (i.e., $\mathbf{p} \neq \mathbf{p}_0$). This was necessary in order that the first term in (3.16), the unscattered wave, not contribute. It is easily seen that if $d\Omega$ did include the direction of $\mathbf{p}_0$, and hence the unscattered wave in (3.16) were present, then the integral over ρ in (3.18) would diverge. We conclude that, just as in the classical case, the differential cross section in the forward direction is a quantity without direct observational meaning.[12]

Third, since $d\sigma/d\Omega = |f|^2$, it is clear that in general only the modulus of f can be directly measured. We shall see later that there are various methods

---

[12] This is not to say that it cannot be given a mathematical meaning, which in fact it can. In most cases, it turns out, the amplitude is continuous in the neighborhood of the forward direction. The forward amplitude is therefore a well defined quantity and the forward cross section can be measured by extrapolation.

of measuring f itself, but that they usually involve at least partial knowledge of the underlying interactions.

Finally, some comment on alternative derivations of the result (3.24). Our attitude to any method using plane waves should already be clear. Such methods can never be regarded as physically satisfactory; but as an introductory route to the correct answer they obviously accomplish an important purpose. Another alternative approach uses the same initial wave packets as ours but analyzes the final state in terms of its spatial wave function. The reader will have noted that we have defined the cross section in terms of the probability that the final momentum lies in the cone defined by $d\Omega$ in *momentum* space. One could reasonably argue that for many experiments (e.g., one using counters) a more directly relevant definition would use the probability that as $t \to \infty$ the particle's position lies somewhere in the cone defined by $d\Omega$ in *coordinate* space. While these two points of view are in fact equivalent, a derivation based directly on the latter (and therefore using spatial wave functions) would obviously have some advantages. We shall outline such a derivation in Chapter 10. However, it is worthy of note that the momentum-space arguments of this chapter generalize immediately to relativistic scattering problems, which is not true of any method based on spatial wave functions (whose role in relativistic quantum mechanics is not at all clear).

3-f. The Optical Theorem

Having established the expression (3.24) for the cross section we can now prove the so-called optical theorem. This result is a direct consequence of the unitary of S and is really no more than the diagonal matrix element of the equation $S^\dagger S = 1$. Thus, if we insert $S = 1 + R$ into $S^\dagger S = 1$ we find that $R + R^\dagger = -R^\dagger R$. If we now take matrix elements and insert a complete set of states $|\mathbf{p}''\rangle$ on the right we find that,

$$\langle \mathbf{p}'| R |\mathbf{p}\rangle + \langle \mathbf{p}| R |\mathbf{p}'\rangle^* = -\int d^3 p'' \langle \mathbf{p}''| R |\mathbf{p}'\rangle^* \langle \mathbf{p}''| R |\mathbf{p}\rangle \qquad (3.25)$$

Now according to (3.6) and (3.8),

$$\langle \mathbf{p}'| R |\mathbf{p}\rangle = -2\pi i \, \delta(E_{p'} - E_p) \, t(\mathbf{p}' \leftarrow \mathbf{p}) = \frac{i}{2\pi m} \, \delta(E_{p'} - E_p) f(\mathbf{p}' \leftarrow \mathbf{p})$$

Inserting this into (3.25) we can factor out a common delta function to give:

$$f(\mathbf{p}' \leftarrow \mathbf{p}) - f(\mathbf{p} \leftarrow \mathbf{p}')^* = \frac{i}{2\pi m} \int d^3 p'' \, \delta(E_p - E_{p''}) f(\mathbf{p}'' \leftarrow \mathbf{p}')^* f(\mathbf{p}'' \leftarrow \mathbf{p})$$

$$(3.26)$$

where of course $E_{p'} = E_p$. Finally we can, if we wish, set $\mathbf{p}' = \mathbf{p}$ and use the delta function on the right to do the radial integration over p'' (see problem **3.2**). The result is,

$$\operatorname{Im} f(\mathbf{p} \leftarrow \mathbf{p}) = \frac{p}{4\pi} \int d\Omega_{\mathbf{p}''} \, |f(\mathbf{p}'' \leftarrow \mathbf{p})|^2 \qquad (3.27)$$

or

$$\boxed{\operatorname{Im} f(\mathbf{p} \leftarrow \mathbf{p}) = \frac{p}{4\pi}\, \sigma(\mathbf{p})} \qquad (3.28)$$

where $\sigma(\mathbf{p})$ denotes the *total* cross-section for scattering from the initial momentum $\mathbf{p}$,

$$\sigma(\mathbf{p}) = \int d\Omega_{\mathbf{p}'} \frac{d\sigma}{d\Omega} (\mathbf{p}' \leftarrow \mathbf{p})$$

In words, the optical theorem states that the imaginary part of the forward amplitude $f(\mathbf{p} \leftarrow \mathbf{p})$ is proportional to the total cross section $\sigma(\mathbf{p})$.

The optical theorem leads to a number of useful results. First, it shows clearly that in general the amplitude cannot be purely real and that it has a positive imaginary part near the forward direction—two surprisingly useful pieces of information. Second, we have already noted that measurement of $d\sigma/d\Omega$ in general determines only $|f|$; however, by exploiting the optical theorem one can measure $\operatorname{Im} f$ and hence $\operatorname{Re} f$ separately in the forward direction.[13]

A third and important application is in the use of dispersion relations. As we shall discuss in Chapter 15, dispersion relations express the real part of the amplitude as an integral over its imaginary part. If we exploit the optical theorem this means we can express the forward amplitude as an integral over the total cross-section and obtain a relation that can be directly checked with experiment. The importance of this sort of relation is that it is true independently of the precise details of the interactions (which are often unknown) and relates quantities that can be directly measured.

Finally, we shall see in Chapter 17 that the optical theorem generalizes to multichannel problems. In fact, precisely the relation (3.28) holds if the amplitude on the left is taken to be the forward *elastic* amplitude and the cross-section on the right is the total cross-section for elastic and inelastic scattering. There is every reason to think that the same relation holds in relativistic scattering since it depends only on the unitary of S. Now, it is an experimental fact that the total cross-sections for scattering of elementary

---

[13] Measurement of $\sigma(\mathbf{p})$ determines $\operatorname{Im} f(\mathbf{p} \leftarrow \mathbf{p})$. Measurement of $d\sigma/d\Omega(\mathbf{p} \leftarrow \mathbf{p})$ (by extrapolation from nonforward measurements) determines $|f(\mathbf{p} \leftarrow \mathbf{p})|$. One can then calculate $\operatorname{Re} f(\mathbf{p} \leftarrow \mathbf{p})$ within a sign.

particles appear to become constant at very high energies, and also that elastic scattering becomes sharply peaked around the forward direction. It follows from the optical theorem that Im $f(\mathbf{p} \leftarrow \mathbf{p})$ must grow like p as $p \to \infty$ and, hence, that the forward elastic peak of $d\sigma/d\Omega$ must rise at least like p^2—a prediction that is well confirmed by experiment.[14]

PROBLEMS

3.1. We have seen that S commutes with H^0 and have claimed that this implies conservation of energy, since,

$$\langle \psi_{\text{in}} | H^0 | \psi_{\text{in}} \rangle = \langle \psi_{\text{out}} | H^0 | \psi_{\text{out}} \rangle \tag{3.29}$$

when $|\psi_{\text{out}}\rangle = S |\psi_{\text{in}}\rangle$. To understand this one must check that these two expectation values are indeed the correct measures of the in and out energies. To this end, suppose that the potential V is time-dependent and, hence, that energy is *not* conserved. Convince yourself that provided V is suitably well behaved, the proof of Section 2-c that $U(t) |\psi\rangle \to U^0(t) |\psi_{\text{in/out}}\rangle$ as $t \to \mp \infty$ still holds. Then show that for any scattering orbit, the actual energy (i.e., the expectation value of H) tends to $\langle \psi_{\text{in}} | H^0 | \psi_{\text{in}} \rangle$ or $\langle \psi_{\text{out}} | H^0 | \psi_{\text{out}} \rangle$ as $t \to \mp \infty$. This means that, whether or not energy is conserved, these two expectation values are indeed the in and out energies, and that (3.29) expresses conservation of energy in collisions. [To do this really rigorously requires more care than it probably deserves; the main point is to understand the significance of (3.29).]

3.2. To refresh your memory about delta functions, show, by making suitable changes of variables, that

(a)
$$\delta(ax) = |a|^{-1} \delta(x)$$

(b)
$$\int_0^\infty dp \, g(p) \, \delta(E_p - E_{p'}) = \frac{m}{p'} \, g(p')$$

[The first relation was used in (3.20); the second in doing the radial part of the integral (3.26) for the optical theorem.]

---

[14] The high energy physicist would probably prefer to see this result stated in terms of $d\sigma/dt$, where t is the square of the momentum transfer $t = -(\mathbf{p}' - \mathbf{p})^2$. Since $d\sigma/dt \propto p^{-2} \, d\sigma/d\Omega$, the peak in $d\sigma/dt$ tends to a fixed height as $E \to \infty$.

4 Scattering of Two Spinless Particles

4-a   **Two-Particle Wave Functions**

4-b   **The Two-Particle S Operator**

4-c   **Conservation of Energy-Momentum and the T Matrix**

4-d   **Cross Sections in Various Frames**

4-e   **The Center-of-Mass Cross Section**

In Chapters 2 and 3 we have developed a rather complete description of the scattering of a single spinless particle in terms of the scattering operator S and amplitude f. We now go on to show that the formalism developed so far is already sufficient, with some simple extensions, to cover some more general and more interesting processes: the elastic scattering of two spinless particles in this chapter, and of two particles with spin in Chapter 5. For the moment we consider only the scattering of two *distinct* particles; the special problems associated with identical particles we defer to Chapter 22 (to which the reader can jump after Chapter 5, if he wishes).

The main purpose of this chapter is to establish the well known result that the elastic scattering of two particles, viewed in their center-of-mass frame of reference, is the "same thing" as the scattering of a single particle by a fixed potential. As an essential preliminary we first describe the mathematical relation of the Hilbert space of two-particle states to the two spaces of the individual one-particle systems.

4-a.   Two-Particle Wave Functions

The states of a system of two distinct spinless particles are represented by wave functions $\psi(\mathbf{x}_1, \mathbf{x}_2)$ depending on the positions $\mathbf{x}_1$ and $\mathbf{x}_2$ of the two particles. A special case of such functions is a product of the form $\psi(\mathbf{x}_1, \mathbf{x}_2) = \phi(\mathbf{x}_1)\chi(\mathbf{x}_2)$ whose corresponding state vector we shall denote by $|\psi\rangle = |\phi\rangle \otimes |\chi\rangle$. Here, of course, $|\psi\rangle$ is a vector in the two-particle Hilbert space $\mathcal{H}$, while $|\phi\rangle$ is in the one-particle space $\mathcal{H}_1$ of the first particle, and $|\chi\rangle$ is in the corresponding space $\mathcal{H}_2$ of the second. This product vector represents a two-particle state where the first particle is in the state $|\phi\rangle$ and the second is in the state $|\chi\rangle$. Two important properties of product vectors (both easily verified in terms of the corresponding wave functions) are: first, if $\{|n\rangle_1\}$ and $\{|m\rangle_2\}$ are orthonormal bases of the one-particle spaces $\mathcal{H}_1$ and $\mathcal{H}_2$, then the products,

$$|n\rangle_1 \otimes |m\rangle_2 \qquad [n, m = 1, 2, 3, \ldots]$$

form an orthonormal basis of the two-particle space $\mathcal{H}$. Second, the scalar products satisfy:

$$(\langle\phi'| \otimes \langle\chi'|)(|\phi\rangle \otimes |\chi\rangle) = \langle\phi' | \phi\rangle\langle\chi' | \chi\rangle$$

When a space $\mathcal{H}$ can be built up in this way from products of vectors in two smaller spaces $\mathcal{H}_1$ and $\mathcal{H}_2$, we say that $\mathcal{H}$ is the *tensor product* of $\mathcal{H}_1$ and $\mathcal{H}_2$ and we write[1] $\mathcal{H} = \mathcal{H}_1 \otimes \mathcal{H}_2$. It is important to remember that this equation does *not* mean that every vector in $\mathcal{H}$ is a product vector; it means only that $\mathcal{H}$ is *spanned* by product vectors; that is, every vector can be expressed as a *sum* of product vectors. [This is very clear in terms of wave functions. Obviously not every function $\psi(\mathbf{x}_1, \mathbf{x}_2)$ is a product function.]

The tensor product occurs whenever a system has two or more independent degrees of freedom. Another example is the Hilbert space of a single particle with spin s, for which

$$\mathcal{H} = \mathcal{H}_{\text{space}} \otimes \mathcal{H}_{\text{spin}}$$

where $\mathcal{H}_{\text{space}}$ is just $\mathcal{L}^2(\mathbb{R}^3)$, the space of ordinary wave functions, while $\mathcal{H}_{\text{spin}}$ is the $(2s + 1)$-dimensional spin space. This means that the vectors in $\mathcal{H}$ are given by functions $\psi_m(\mathbf{x})$ of the coordinate $\mathbf{x}$ and an index m (such as the eigenvalue of S_3) labelling some basis in $\mathcal{H}_{\text{spin}}$. In this case one usually groups the functions together to form $(2s + 1)$-component spinor wave functions.

---

[1] We omit the formal definition of the tensor product, which is in any case quite complicated and unilluminating. For our purposes it is sufficient always to regard $\mathcal{H}_1 \otimes \mathcal{H}_2$ as defined by wave functions that can be built up from products of wave functions, one from $\mathcal{H}_1$ and the other from $\mathcal{H}_2$.

We can regard the product $\mathcal{H}_1 \otimes \mathcal{H}_2$ as a *factoring* of $\mathcal{H}$, and it is an important fact that a given space can be factored in many different ways. For example, every wave function $\psi(\mathbf{x}_1, \mathbf{x}_2)$ can be rewritten as a function of the familiar center-of-mass (CM) and relative coordinates, which we denote by:

$$\bar{\mathbf{x}} = \frac{m_1 \mathbf{x}_1 + m_2 \mathbf{x}_2}{m_1 + m_2}$$

$$\mathbf{x} = \mathbf{x}_1 - \mathbf{x}_2$$

Obviously the space of functions $\psi(\bar{\mathbf{x}}, \mathbf{x})$ can be spanned by product functions of the form $\phi(\bar{\mathbf{x}}) \chi(\mathbf{x})$ and it can therefore be written as

$$\mathcal{H} = \mathcal{H}_1 \otimes \mathcal{H}_2 = \mathcal{H}_{\text{cm}} \otimes \mathcal{H}_{\text{rel}}$$

where $\mathcal{H}_{\text{cm}}$ and $\mathcal{H}_{\text{rel}}$ are the spaces of wave functions of the CM coordinate $\bar{\mathbf{x}}$ and relative coordinate $\mathbf{x}$ respectively. We shall see that for many purposes this second factoring of $\mathcal{H}$ is more convenient than the first.

Just as certain vectors in $\mathcal{H} = \mathcal{H}_1 \otimes \mathcal{H}_2$ are product vectors so certain operators are product operators. In fact, if A and B are any operators acting on $\mathcal{H}_1$ and $\mathcal{H}_2$ respectively, we can define an operator $(A \otimes B)$ acting on $\mathcal{H}$ by the relation:

$$(A \otimes B)(|\phi\rangle \otimes |\chi\rangle) = A \,|\phi\rangle \otimes B \,|\chi\rangle \tag{4.1}$$

In particular, the basic dynamical variables of a two-particle system are all operators of the form $A \otimes 1$ or $1 \otimes B$. For instance, the momentum operator for the first particle on the two-particle space $\mathcal{H}$ is defined in terms of the corresponding operator on $\mathcal{H}_1$ as

$$\mathbf{P}_1(\text{on } \mathcal{H}) = \mathbf{P}_1(\text{on } \mathcal{H}_1) \otimes 1(\text{on } \mathcal{H}_2)$$

that is, acting on a two-particle vector $|\phi\rangle \otimes |\chi\rangle$, it replaces $|\phi\rangle$ by $\mathbf{P}_1 |\phi\rangle$ but leaves $|\chi\rangle$ unchanged. [With respect to the spatial wave function $\psi(\mathbf{x}_1, \mathbf{x}_2)$ this definition means simply that $\mathbf{P}_1$ is given by the familiar $-i\nabla_1$.] Similarly, the operator for the momentum of the second particle is:

$$\mathbf{P}_2(\text{on } \mathcal{H}) = 1(\text{on } \mathcal{H}_1) \otimes \mathbf{P}_2(\text{on } \mathcal{H}_2)$$

As one would expect (since they refer to different degrees of freedom), any two operators $A \otimes 1$ and $1 \otimes B$ commute, as can be directly checked using (4.1). (A specific and familiar example is given by the momentum operators $-i\nabla_1$ and $-i\nabla_2$.)

In practice it is usually unnecessary to distinguish $A \otimes 1$ from A and we shall write just A for either; by the same token $1 \otimes B$ will be abbreviated to B and hence $A \otimes B$, which is the same as $(A \otimes 1)(1 \otimes B)$, to AB. However, when we wish to emphasize the product structure of an operator we shall revert to the tensor-product notation.

In the light of these ideas we can now consider the two-particle Hamiltonian and the corresponding evolution operator. The Hamiltonian is:

$$H = \frac{\mathbf{P}_1^2}{2m_1} + \frac{\mathbf{P}_2^2}{2m_2} + V \equiv H^0 + V \tag{4.2}$$

where (since we shall assume that the interaction is local and translationally invariant) V is a function of the relative coordinate $\mathbf{x}_1 - \mathbf{x}_2 = \mathbf{x}$ only. As is well known it is convenient to reexpress H in terms of the operators for the total and relative momenta

$$\bar{\mathbf{P}} = \mathbf{P}_1 + \mathbf{P}_2$$

and

$$\mathbf{P} = \frac{m_2 \mathbf{P}_1 - m_1 \mathbf{P}_2}{m_1 + m_2}$$

which, are, of course, the momenta conjugate to the CM and relative position operators

$$\bar{\mathbf{X}} = \frac{m_1 \mathbf{X}_1 + m_2 \mathbf{X}_2}{m_1 + m_2}$$

and

$$\mathbf{X} = \mathbf{X}_1 - \mathbf{X}_2$$

Of these four operators, $\bar{\mathbf{P}}$ and $\bar{\mathbf{X}}$ act only on $\mathscr{H}_{\text{cm}}$, while $\mathbf{P}$ and $\mathbf{X}$ act only on $\mathscr{H}_{\text{rel}}$. In terms of these operators,[2]

$$H = \frac{\bar{\mathbf{P}}^2}{2M} + \left[\frac{\mathbf{P}^2}{2m} + V(\mathbf{x}) \right]$$

$$\equiv H_{\text{cm}} + H_{\text{rel}} \tag{4.3}$$

where M and m are the total and reduced masses,

$$M = m_1 + m_2 \quad \text{and} \quad m = \frac{m_1 m_2}{m_1 + m_2}$$

In the form of (4.3), H is obviously the sum of two terms, one of which acts only on $\mathscr{H}_{\text{cm}}$, the other on $\mathscr{H}_{\text{rel}}$. These two terms commute (as they must,

[2] We should really write the *operator* V as a function $V(\mathbf{X})$ of the *operator* $\mathbf{X}$. However, this is defined as the operator that multiplies the spatial wave function by the (number) function $V(\mathbf{x})$; since this is how it is usually thought of, we avoid the pedantic $V(\mathbf{X})$ in favor of the more natural looking $V(\mathbf{x})$.

since they act on different spaces) and so the evolution operator can be factored as follows:[3]

$$\mathsf{U}(t) \equiv e^{-iHt} = e^{-i(H_{\mathrm{cm}}+H_{\mathrm{rel}})t}$$

$$= e^{-iH_{\mathrm{cm}}t}e^{-iH_{\mathrm{rel}}t}$$

$$= e^{-iH_{\mathrm{cm}}t} \otimes e^{-iH_{\mathrm{rel}}t} \tag{4.4}$$

Here the last expression serves to emphasize that $\mathsf{U}(t)$ is the product of two evolution operators, one for $\mathscr{H}_{\mathrm{cm}}$ and the other for $\mathscr{H}_{\mathrm{rel}}$. This result means that the motions of the center of mass and the relative coordinate are independent. In particular, since H_{cm} is just $\bar{\mathbf{P}}^2/2M$, the center of mass moves like a free particle of mass M. Since H_{rel} has the same form as the Hamiltonian of a single particle with the reduced mass m and in the fixed potential V, the evolution in $\mathscr{H}_{\mathrm{rel}}$ will exactly resemble the evolution discussed in Chapters 2 and 3. This result is the crucial simplifying feature in two-particle scattering and is the origin of the celebrated connection between this problem and the scattering of one particle off a fixed potential.[4]

Finally, we note that the free evolution operator can also be factored in the same way,

$$\mathsf{U}^0(t) \equiv e^{-iH^0 t} = e^{-iH_{\mathrm{cm}}t} \otimes e^{-iH^0_{\mathrm{rel}}t} \tag{4.5}$$

Of course $U^0(t)$ can also be factored as:

$$\mathsf{U}^0(t) = \exp\left(-i\,\frac{\mathbf{P}_1^2}{2m_1}\,t\right) \otimes \exp\left(-i\,\frac{\mathbf{P}_2^2}{2m_2}\,t\right)$$

corresponding to the factoring of $\mathscr{H}$ as $\mathscr{H}_1 \otimes \mathscr{H}_2$. This second result expresses the obvious fact that two noninteracting particles move independently. The same result obviously does not hold for the full evolution operator $\mathsf{U}(t)$.

4-b.   The Two-Particle S Operator

At first sight the problem of two-particle scattering appears very different from that of a single particle in a fixed potential. [For example, the two

---

[3] We recall that $\exp(A + B)$ is *not* equal to $\exp(A)\exp(B)$ unless A and B commute.

[4] By simply *assuming* that H has the form (4.2) we have obscured the fact that this elegant result depends only on Galilean invariance. The argument is this: Obviously any H can be put in the form (4.3) if we allow V to be an arbitrary operator. Translational invariance requires that V commute with $\bar{\mathbf{P}}$, while invariance under Galilean boosts requires that V commute with $\bar{\mathbf{X}}$ (see Jordan, 1969, p. 124). This means that V acts only on $\mathscr{H}_{\mathrm{rel}}$ and the factoring (4.4) follows. The connection between (4.4) and Galilean invariance is particularly noteworthy since there is no factoring analogous to (4.4) in relativistic quantum mechanics.

problems are formulated on completely different Hilbert spaces $\mathscr{L}^2(\mathbb{R}^6)$ and $\mathscr{L}^2(\mathbb{R}^3)$, respectively.] However, with the results just established we can now demonstrate the sense in which the two problems are equivalent.

The general orbit of our two-particle system is $U(t)\,|\psi\rangle$ where $|\psi\rangle$ is any vector in the two-particle space $\mathscr{H}$. Just as in the one-particle case, we expect there to be certain scattering orbits for which the two particles move far apart as $t \to \mp\infty$ and $U(t)\,|\psi\rangle$ behaves like some freely evolving

$$U^0(t)\,|\psi_{\text{in/out}}\rangle$$

As in the one-particle case, this expectation is correct and can be precisely stated in the form of two results: the asymptotic condition and asymptotic completeness.

The asymptotic condition states that every vector $|\psi_{\text{in}}\rangle$ in $\mathscr{H}$ is the in asymptote of some actual orbit $U(t)\,|\psi\rangle$,

$$U(t)\,|\psi\rangle \xrightarrow[t\to-\infty]{} U^0(t)\,|\psi_{\text{in}}\rangle$$

(similarly for the out asymptotes). To prove this we must (as before) prove that for every vector $|\psi_{\text{in}}\rangle$ the vector $U(t)^\dagger U^0(t)\,|\psi_{\text{in}}\rangle$ has a limit, or what is the same thing, that the operator $U(t)^\dagger U^0(t)$ converges. To this end we substitute the expressions (4.4) for U and (4.5) for U^0 into $U^\dagger U^0$. We note that both U and U^0 contain the same term $\exp(-iH_{\text{cm}}t)$ as their first factor. In the product $U^\dagger U^0$ these two terms cancel and we find the simple result:

$$U(t)^\dagger U^0(t) = 1_{\text{cm}} \otimes (e^{iH_{\text{rel}}t}e^{-iH^0_{\text{rel}}t})$$

where 1_{cm} denotes the unit operator on $\mathscr{H}_{\text{cm}}$.

This operator has a limit if and only if the second factor does. Moreover, the mathematical structure of this second factor is precisely that of the corresponding operator for a one-particle system with Hamiltonian

$$H_{\text{rel}} = \frac{\mathbf{P}^2}{2m} + V(\mathbf{x}) \equiv H^0_{\text{rel}} + V(\mathbf{x})$$

Thus, if we make the same assumptions on $V(\mathbf{x})$ as we made in Chapter 2, the second factor does have a limit, which we call Ω_+, and the desired result follows. The actual orbit whose in asymptote was $|\psi_{\text{in}}\rangle$ is given by,

$$|\psi\rangle = \lim_{t\to-\infty} U(t)^\dagger U^0(t)\,|\psi_{\text{in}}\rangle$$

$$= (1_{\text{cm}} \otimes \Omega_+)\,|\psi_{\text{in}}\rangle \equiv \Omega_+\,|\psi_{\text{in}}\rangle \tag{4.6}$$

The corresponding result obviously holds for any $|\psi_{\text{out}}\rangle$, and our proof of the asymptotic condition is complete.

We have introduced the notation $\mathbf{\Omega}_\pm$ to denote the two-particle Møller operators acting on the two-particle space $\mathcal{H} = \mathscr{L}^2(\mathbb{R}^6)$. They have the simple form:

$$\mathbf{\Omega}_\pm = 1_{cm} \otimes \Omega_\pm \tag{4.7}$$

where

$$\Omega_\pm = \lim_{t \to \mp\infty} e^{iH_{rel}t}e^{-iH^0_{rel}t}$$

The operators $\Omega_\pm$ act on $\mathcal{H}_{rel} = \mathscr{L}^2(\mathbb{R}^3)$ and have precisely the structure of the Møller operators for a single particle in a fixed potential. The simple form of $\Omega_\pm$ is, of course, a direct consequence of the factorization (4.4) of the evolution operator. In particular the factor 1_{cm} in $\mathbf{\Omega}_\pm$ reflects the fact that the center of mass moves like a free particle and is not scattered.

The proof of asymptotic completeness now carries over directly from the corresponding one-particle results. Under the conditions on V given in Chapter 2 we have the results that: First, the orbits with in asymptotes are precisely the same as those with out asymptotes, (i.e., Ω_+ and Ω_- map $\mathcal{H}$ onto the same range $\mathcal{R}$), and second, the direct sum of $\mathcal{R}$, the space of scattering states, and $\mathcal{B}$, the space spanned by the bound states, is the whole space $\mathcal{H}$.

It follows at once that the operator

$$\mathbf{S} = \mathbf{\Omega}^\dagger_- \mathbf{\Omega}_+$$

is a unitary operator mapping any $|\psi_{in}\rangle$ in $\mathcal{H}$ onto the corresponding $|\psi_{out}\rangle = \mathbf{S}\,|\psi_{in}\rangle$. According to (4.7) it has the simple structure:

$$\boxed{\mathbf{S} = 1_{cm} \otimes S}$$

where $S = \Omega^\dagger_- \Omega_+$ acts on $\mathcal{H}_{rel}$ and is precisely the one-particle S operator computed from the Hamiltonian H_{rel}.

4-c.   Conservation of Energy-Momentum and the T Matrix

From the expression $\mathbf{S} = 1_{cm} \otimes S$ it is immediately clear that $\mathbf{S}$ commutes with $\bar{\mathbf{P}}$ (which acts only on $\mathcal{H}_{cm}$) and, hence, that total momentum is conserved.[5] Just as in the one-particle case $\mathbf{S}$ commutes with H^0 and energy is conserved. From conservation of energy and momentum it follows that the matrix elements,

$$\langle \mathbf{p}'_1, \mathbf{p}'_2|\,\mathbf{S}\,|\mathbf{p}_1, \mathbf{p}_2\rangle$$

---

[5] This result can of course be traced back to the translational invariance of the system as we shall discuss in Chapter 6.

contain the factors

$$\delta(E_1' + E_2' - E_1 - E_2)\, \delta_3(\mathbf{p}_1' + \mathbf{p}_2' - \mathbf{p}_1 - \mathbf{p}_2)$$

where $E_1 = \mathbf{p}_1^2/2m_1$ and so on.

As before it is convenient to decompose this matrix element in terms of an on-shell T matrix. To this end we first note that the eigenvector $|\mathbf{p}_1, \mathbf{p}_2\rangle$ of $\mathbf{P}_1$ and $\mathbf{P}_2$ is also an eigenvector of the total and relative momenta $\bar{\mathbf{P}}$ and $\mathbf{P}$. We can, without danger of confusion, write it as:

$$|\mathbf{p}_1, \mathbf{p}_2\rangle \equiv |\bar{\mathbf{p}}, \mathbf{p}\rangle \equiv |\bar{\mathbf{p}}\rangle \otimes |\mathbf{p}\rangle$$

where, of course, $\bar{\mathbf{p}} = \mathbf{p}_1 + \mathbf{p}_2$ and $\mathbf{p} = (m_2\mathbf{p}_1 - m_1\mathbf{p}_2)/(m_1 + m_2)$. (This corresponds to the identities

$$e^{i(\mathbf{p}_1 \cdot \mathbf{x}_1 + \mathbf{p}_2 \cdot \mathbf{x}_2)} \equiv e^{i(\bar{\mathbf{p}} \cdot \bar{\mathbf{x}} + \mathbf{p} \cdot \mathbf{x})} \equiv e^{i\bar{\mathbf{p}} \cdot \bar{\mathbf{x}}} \, e^{i\mathbf{p} \cdot \mathbf{x}}$$

for the wave functions.) If we then write $\mathbf{S}$ as $1_{\mathrm{cm}} \otimes S$ it is immediately clear that

$$\langle \mathbf{p}_1', \mathbf{p}_2' | \, \mathbf{S} \, | \mathbf{p}_1, \mathbf{p}_2\rangle = \langle \bar{\mathbf{p}}, \mathbf{p}' | \, (1_{\mathrm{cm}} \otimes S) \, | \bar{\mathbf{p}}, \mathbf{p}\rangle$$
$$= \delta_3(\bar{\mathbf{p}}' - \bar{\mathbf{p}})\langle \mathbf{p}' | \, S \, | \mathbf{p}\rangle$$

Because the operator S has precisely the structure of a one-particle S operator, its matrix elements have the familiar decomposition

$$\langle \mathbf{p}' | \, S \, | \mathbf{p}\rangle = \delta_3(\mathbf{p}' - \mathbf{p}) - 2\pi i \, \delta(E_{p'} - E_p)\, t(\mathbf{p}' \leftarrow \mathbf{p}) \qquad (4.8)$$

Combining these last two equations we get the desired result,

$$\boxed{\begin{aligned} \langle \mathbf{p}_1', \mathbf{p}_2' | \, \mathbf{S} \, | \mathbf{p}_1, \mathbf{p}_2\rangle &= \delta_3(\mathbf{p}_1' - \mathbf{p}_1)\, \delta_3(\mathbf{p}_2' - \mathbf{p}_2) \\ &\quad - 2\pi i \, \delta(\textstyle\sum E_i' - \sum E_i) \\ &\quad \times \delta_3(\textstyle\sum \mathbf{p}_i' - \sum \mathbf{p}_i)\, t(\mathbf{p}' \leftarrow \mathbf{p}) \end{aligned}}$$

$$(4.9)$$

The first of these two terms is the amplitude that each particle passes through unscattered; the second is the amplitude that the two particles actually scatter. The latter conserves total energy and total momentum but not, of course, the individual components of the relative momentum.

Just as in one-particle scattering it is convenient to define a *scattering amplitude*,

$$f(\mathbf{p}' \leftarrow \mathbf{p}) = -(2\pi)^2 m \, t(\mathbf{p}' \leftarrow \mathbf{p})$$

where in this case m denotes the *reduced* mass of the two particles.

At this point some comment on our notation is in order. The reader will have noted that we have used bold-face type for the operators $\mathbf{\Omega}_\pm$ and $\mathbf{S}$ of the translationally invariant two-particle problem, and ordinary type for the corresponding operators $\Omega_\pm$ and S describing the relative motion. The operators $\mathbf{\Omega}_{\pm}$ and $\mathbf{S}$ act on the full space $\mathscr{H} = \mathscr{L}^2(\mathbb{R}^6)$ and their matrix elements contain a momentum conserving $\delta_3(\bar{\mathbf{p}}' - \bar{\mathbf{p}})$. The operators $\Omega_{\pm}$ and S act on $\mathscr{H}_{\mathrm{rel}} = \mathscr{L}^2(\mathbb{R}^3)$ and have precisely the structure of the corresponding one-particle operators.

Throughout this book we shall use bold-face type for the collision operators $\mathbf{\Omega}_{\pm}$ and $\mathbf{S}$ of systems with translational invariance. These operators conserve total momentum and their matrix elements contain the factor $\delta_3(\bar{\mathbf{p}}' - \bar{\mathbf{p}})$. Ordinary type $\Omega_{\pm}$ and S will refer *either* to the relative motion of a translationally invariant system ($\mathbf{S} = 1_{\mathrm{cm}} \otimes \mathrm{S}$, etc.) *or* to the motion of a system with fixed potentials. In this way we guarantee that $\Omega_{\pm}$ and S always have the same structure; in particular, their matrix elements do *not* contain an overall momentum-conserving delta function. In either case the on-shell T matrix $t(\mathbf{p}' \leftarrow \mathbf{p})$ is defined by (4.8) (or its multichannel analogue) and contains no delta functions.

4-d.   Cross Sections in Various Frames

The cross sections for two-particle collisions can be defined in various frames of reference. In most experiments the target particle is initially at rest, and the frame defined by this condition is therefore called the *laboratory (lab) frame of reference*. For theoretical purposes the most convenient frame is that in which the center of mass is at rest (or more precisely the expectation value of the total momentum is zero); this frame is called the *center-of-mass (CM) frame*. In some situations (e.g., molecular collisions in a gas) one has occasion to discuss frames in which both particles move with arbitrary momenta.

We shall find that the cross section can be defined so that the number of particles scattered into any element $\Delta\Omega$ of solid angle is given by the familiar formula (3.11) *in all frames of reference;* that is:

$$N_{\mathrm{sc}}(\Delta\Omega \leftarrow \mathbf{p}_1, \mathbf{p}_2) = n_{\mathrm{inc}}(\mathbf{p}_1, \mathbf{p}_2)\sigma(\Delta\Omega \leftarrow \mathbf{p}_1, \mathbf{p}_2) \qquad (4.10)$$

where the initial momenta $\mathbf{p}_1$ and $\mathbf{p}_2$ of the two particles are inserted to identify the frame in which all quantities are measured.

We first consider a given collision in one definite frame, its CM frame for example. If we repeat the collision many times in the same way as in Chapter 3, then the number of scatterings into any $\Delta\Omega$ will be proportional to the incident density n_{inc}. (In the classical case this is obvious; for the quantum case we shall prove it in Section 4-e.) Thus, in this frame we can use (4.10) to *define* the cross section $\sigma(\Delta\Omega \leftarrow \mathbf{p}_1, \mathbf{p}_2)$.

Having defined the CM cross section so that the relation $N_{\mathrm{sc}} = n_{\mathrm{inc}}\sigma$ is true in the CM frame, let us consider the same sequence of collisions as viewed by an observer in some other frame of reference. To determine what this new observer sees, we must consider how the quantities involved in (4.10) transform from one frame to another. The momenta that the CM observer $\mathcal{O}$ calls $\mathbf{p}_1$ and $\mathbf{p}_2$ (with $\mathbf{p}_2 = -\mathbf{p}_1$, of course) will be labelled $\mathbf{p}_1'$ and $\mathbf{p}_2'$ by the second observer $\mathcal{O}'$. Similarly, the element of solid angle called $\Delta\Omega$ by $\mathcal{O}$ is called $\Delta\Omega'$ by $\mathcal{O}'$. (We are using $\Delta\Omega$ to identify both the magnitude of solid angle and the direction about which it is defined. In general, both the magnitude and direction will be different as measured by the two observers.) However, the *actual number of counts registered in any actual solid angle* must, of course, be the same as measured by either observer. This means that,

$$N_{sc}(\Delta\Omega' \leftarrow \mathbf{p}_1', \mathbf{p}_2') = N_{sc}(\Delta\Omega \leftarrow \mathbf{p}_1, \mathbf{p}_2)$$

for any two observers $\mathcal{O}$ and $\mathcal{O}'$; that is, N_{sc} is *invariant*.

The transformation of the incident density n_{inc} is a little more subtle. We suppose first that the incident momenta $\mathbf{p}_1'$ and $\mathbf{p}_2'$ in the second frame are collinear (as they are in the lab frame, for instance). In this case, it is easy to see that the total number of incident particles to cross unit area perpendicular to the incident momentum is the same, as counted in either frame. That is, the incident densities as seen by $\mathcal{O}$ and $\mathcal{O}'$ are the same,

$$n_{inc}(\mathbf{p}_1', \mathbf{p}_2') = n_{inc}(\mathbf{p}_1, \mathbf{p}_2) \tag{4.11}$$

If we consider a frame in which the two particles approach obliquely, it is not immediately clear what should be meant by the incident density n_{inc} in this new frame. However, a moment's reflection shows that the natural definition of n_{inc} is the incident density *perpendicular to the relative motion* of projectile and target. With this definition $n_{inc}(\mathbf{p}_1', \mathbf{p}_2')$ is exactly the same as the original $n_{inc}(\mathbf{p}_1, \mathbf{p}_2)$. That is, with the proper definitions, (4.11) holds for any $\mathcal{O}$ and $\mathcal{O}'$.

The situation is now this: In the CM frame we have defined a cross section so that the number of scatterings is given by the relation $N_{sc} = n_{inc}\sigma$; and the quantities N_{sc} and n_{inc} are invariant as measured in all frames of reference. It immediately follows that we can define the cross section in any frame such that $N_{sc} = n_{inc}\sigma$ and that, with this definition, the cross section has the same value in all frames:

$$\sigma(\Delta\Omega' \leftarrow \mathbf{p}_1', \mathbf{p}_2') = \sigma(\Delta\Omega \leftarrow \mathbf{p}_1, \mathbf{p}_2)$$

that is, $\sigma(\Delta\Omega \leftarrow \mathbf{p}_1, \mathbf{p}_2)$ is an invariant.

The differential cross section is defined by considering a small solid angle $d\Omega$ and setting

$$\sigma(d\Omega \leftarrow \mathbf{p}_1, \mathbf{p}_2) = \left(\frac{d\sigma}{d\Omega}\right) d\Omega$$

It follows that the differential cross section is not invariant since $d\Omega$ is not. Obviously, in fact,

$$\boxed{\left(\frac{d\sigma}{d\Omega}\right)' = \left(\frac{d\sigma}{d\Omega}\right)\frac{d\Omega}{d\Omega'}} \tag{4.12}$$

The problem of computing the differential cross section in one frame given its value in another is therefore a purely kinematic problem of computing $d\Omega/d\Omega'$. For example, the cross section of experimental interest is usually the lab cross section while, as we shall see, the most convenient cross section to compute is that in the CM frame. It is a simple exercise to check that these are related by

$$\left(\frac{d\sigma}{d\Omega}\right)_{\text{lab}} = \left(\frac{d\sigma}{d\Omega}\right)_{\text{cm}} \frac{(1 + 2\lambda \cos\theta_{\text{cm}} + \lambda^2)^{3/2}}{|1 + \lambda \cos\theta_{\text{cm}}|} \tag{4.13}$$

where $\lambda = m_1/m_2$ is the ratio of projectile to target mass and θ_{cm} is the CM scattering angle.

In conclusion, we remark that nothing we have said is changed if we consider observers related by a Lorentz transformation. Thus, our conclusions— that, with the appropriate definitions, $\sigma(\Delta\Omega \leftarrow \mathbf{p}_1, \mathbf{p}_2)$ is invariant, while $d\sigma/d\Omega$ transforms as in (4.12)—are also correct in relativistic scattering.

4-e.   The Center-of-Mass Cross Section

The computation of the cross section in terms of the on-shell T matrix or the amplitude is similar to the corresponding one-particle calculation of Chapter 3 and need not be given in complete detail. It is most easily performed in the CM frame for the following simple reason: In the CM frame the total momentum is strongly peaked about zero and measurement of $\mathbf{p}_1$ is therefore the same thing as measurement of the relative momentum $\mathbf{p}$. (If $\bar{\mathbf{p}} = 0$ then $\mathbf{p}_1 = -\mathbf{p}_2 = \mathbf{p}$.) Now, the cross section is related to the probability that after the collision $\mathbf{p}_1$ lie in some $d\Omega$; on the other hand, since $\mathbf{S}$ has the form $1_{\text{cm}} \otimes \mathsf{S}$ it is the probability that $\mathbf{p}$ lie in some $d\Omega$ which is easily calculated. In the CM frame these two quantities are the same.

The probability that long after the collision the relative momentum $\mathbf{p}$ lie in some $d\Omega$ is just

$$w(d\Omega \leftarrow \psi_{\text{in}}) = d\Omega \int d^3\bar{p} \int_0^\infty p^2 \, dp \, |\psi_{\text{out}}(\bar{\mathbf{p}}, \mathbf{p})|^2 \tag{4.14}$$

(Note that we integrate over all $\bar{\mathbf{p}}$ since the total final momentum is not interesting to measure.) Using the decomposition (4.9) of the S matrix we can write the function ψ_{out} in terms of ψ_{in} as

$$\psi_{\text{out}}(\bar{\mathbf{p}}, \mathbf{p}) = \psi_{\text{in}}(\bar{\mathbf{p}}, \mathbf{p}) + \frac{i}{2\pi m} \int d^3p' \, \delta(E_p - E_{p'}) f(\mathbf{p} \leftarrow \mathbf{p}') \psi_{\text{in}}(\bar{\mathbf{p}}, \mathbf{p}') \quad (4.15)$$

(Note the trivial dependence on the total momentum $\bar{\mathbf{p}}$, arising from conservation of momentum.)

The in state for a typical collision process viewed in the CM frame has a wave function of the form $\phi_1(\mathbf{p}_1) \phi_2(\mathbf{p}_2)$. The function $\phi_1(\mathbf{p}_1)$ is the wave function of the projectile as it emerges from the accelerator or collimators; $\phi_2(\mathbf{p}_2)$ is that of the target. Both functions are well peaked in momentum, ϕ_1 about some $\mathbf{p}_0$ and ϕ_2 about $-\mathbf{p}_0$.

To define the cross section we repeat the experiment many times with the projectile displaced randomly in the plane perpendicular to $\mathbf{p}_0$. That is, we take

$$\psi_{\text{in}}(\bar{\mathbf{p}}, \mathbf{p}) = \phi_\rho(\bar{\mathbf{p}}, \mathbf{p}) \equiv e^{-i\rho \cdot \mathbf{p}_1} \phi_1(\mathbf{p}_1) \phi_2(\mathbf{p}_2) \quad (4.16)$$

with ρ taking on random values $\rho_i (i = 1, 2, \ldots)$ in the plane perpendicular to $\mathbf{p}_0$. After many such experiments the number of particles scattered into $d\Omega$ will be, as before,

$$N_{\text{sc}}(d\Omega \leftarrow \phi) = \sum_i w(d\Omega \leftarrow \phi_{\rho i})$$

$$= n_{\text{inc}} \int d^2\rho \, w(d\Omega \leftarrow \phi_\rho)$$

As anticipated, this is proportional to n_{inc} and can be written as

$$N_{\text{sc}}(d\Omega \leftarrow \phi) = n_{\text{inc}} \sigma(d\Omega \leftarrow \phi),$$

where

$$\sigma(d\Omega \leftarrow \phi) = \int d^2\rho \, w(d\Omega \leftarrow \phi_\rho)$$

Substitution of (4.14)–(4.16) into this expression, followed by a computation almost identical to that of Section 3-e yields the answer (which the reader should check; see Problem **4.2**):

$$\sigma(d\Omega \leftarrow \phi) = d\Omega \, |f(\mathbf{p} \leftarrow \mathbf{p}_0)|^2 \quad (4.17)$$

(where $\mathbf{p}$ points in the direction of $d\Omega$) provided only that the initial wave functions $\phi_1(\mathbf{p}_1)$ and $\phi_2(\mathbf{p}_2)$ are sufficiently peaked about $\mathbf{p}_0$ and $-\mathbf{p}_0$, and the solid angle $d\Omega$ does not include the forward direction $\mathbf{p}_0$.

As in Chapter 3 this answer is independent of the precise shape of the initial wave function ϕ and is proportional to $d\Omega$. We can therefore write it as:

$$\sigma(d\Omega \leftarrow \mathbf{p}_0) = \frac{d\sigma}{d\Omega} (\mathbf{p} \leftarrow \mathbf{p}_0) \, d\Omega$$

[In the CM frame we omit the second $\mathbf{p}_0$ from $\sigma(d\Omega \leftarrow \mathbf{p}_0, -\mathbf{p}_0)$.] and we find for the CM differential cross section

$$\frac{d\sigma}{d\Omega}(\mathbf{p} \leftarrow \mathbf{p}_0) = |f(\mathbf{p} \leftarrow \mathbf{p}_0)|^2 \qquad (4.18)$$

$\mathbf{p}_0$ and $\mathbf{p}$ being the initial and final momenta of the projectile measured in the CM frame.

This result completes our demonstration that two-particle scattering when viewed in its CM frame is equivalent to the problem of one particle that has the reduced mass m and scatters off the fixed potential V. In summary, the two-particle scattering operator has the form $\mathbf{S} = 1_{\text{cm}} \otimes S$, where S, the scattering operator of the relative motion, is precisely the S operator of the equivalent one-particle problem. The CM differential cross section is given by the usual one-particle formula (4.18). And the cross section in any frame can be computed from that of the CM frame using the kinematic relation (4.12). The two-particle problem is therefore reduced to that of computing the amplitude f of the corresponding one-particle problem defined by the relative motion.

PROBLEMS

4.1. Derive the expression for the lab differential cross section in terms of the CM differential cross section (see 4.13).

4.2. (a) Derive in detail the expression for the CM differential cross section in terms of the amplitude; i.e., fill in the details of the calculation leading from (4.14) to (4.18). [The main differences between this calculation and that of Section 3-e are the extra variable $\bar{\mathbf{p}}$ and the integral $\int d^3\bar{p}$ (these should carry straight through your calculation and drop out in a normalization integral at the end). Also the random displacements of particle 1 give the factor $\exp(-i\boldsymbol{\rho} \cdot \mathbf{p}_1)$ in (4.16) whereas it would be nicer to have a factor $\exp(-i\boldsymbol{\rho} \cdot \mathbf{p})$; this is easily achieved by expressing $\mathbf{p}_1$ in terms of $\bar{\mathbf{p}}$ and $\mathbf{p}$.]
(b) Show clearly that one would get the same answer if one treated particle 1 as "target" and particle 2 as "projectile;" i.e., made the random displacements on particle 2. (In practice, cross sections are usually achieved in both ways at once. There are many particles in the beam, giving an average over the impact parameter of particle 1, and there are many particles in the target assembly, giving an average over that of particle 2.)

5 Scattering of Two Particles with Spin

5-a   The Hilbert Space for Particles with Spin

5-b   The S Operator for Particles with Spin

5-c   The Amplitudes and Amplitude Matrix

5-d   Sums and Averages Over Spins

5-e   The In and Out Spinors

In this chapter we shall extend our scattering formalism to include particles with spin—an extension whose importance should be clear if one recalls that more than half of all particles, nuclei, and atoms do have nonzero spin.[1]

In moving from spinless particles to particles with spin we must anticipate two obvious complications. First, the Hilbert space for particles with spin is more complicated, because it must describe spin as well as spatial degrees of freedom. Second, the Hamiltonian may be spin-dependent and contain terms such as the spin-orbit interaction of an electron with a nucleus or the tensor interaction between two nucleons. Despite these apparent complications we shall see that the scattering formalism for particles with spin can

---

[1] Nonetheless, there is some feeling (at least partially justified) that the complications due to spin are inessential to the mainstream of scattering theory. The reader who feels this can omit or postpone some or all of the material related to spin, without seriously affecting his understanding of what follows (although this procedure is not particularly recommended because some of the best illustrations of general principles come from the spin formalism). The material in question is covered in Chapter 5, the last two sections of Chapter 6, and Chapter 7.

be set up in exactly the same way as that for spinless particles for the following simple reason. The essential feature that allows one to establish a scattering theory is that before and after their collisions the particles become spatially separated and cease to interact. This is as true for particles with spin as for those without, and is true whether or not the interactions are spin-dependent.

The scattering theory that we now construct differs from the corresponding spinless theory in just one important respect. In place of the single amplitude and single differential cross section of the spinless case, is a large number of amplitudes and cross sections, one each for every possible choice of initial and final spin states.

5-a. The Hilbert Space for Particles with Spin

The Hilbert space appropriate to a single particle of spin s is the tensor product,

$$\mathscr{H} = \mathscr{H}_{\text{space}} \otimes \mathscr{H}_{\text{spin}}$$

where $\mathscr{H}_{\text{space}}$ is the space $\mathscr{L}^2(\mathbb{R}^3)$ of ordinary wave functions and $\mathscr{H}_{\text{spin}}$ is the $(2s + 1)$-dimensional spin space. As a basis for $\mathscr{H}_{\text{spin}}$ it is usual to use the eigenvectors $|m\rangle$ of the third component of the spin operator,

$$S_3 |m\rangle = m |m\rangle$$

the eigenvalue m running from $-s$ to s in integer steps. An arbitrary vector $|\chi\rangle$ in $\mathscr{H}_{\text{spin}}$ can be expanded as

$$|\chi\rangle = \sum_{m=-s}^{s} \chi_m |m\rangle \tag{5.1}$$

In connection with this expansion, it may be worth emphasizing that although there are just $(2s + 1)$ basic states $|m\rangle$, a particle with spin nonetheless has *infinitely many* different spin states, corresponding to the infinite number of possible combinations (5.1).

The spin state $|\chi\rangle$ in (5.1) is uniquely identified by the coefficients χ_m, which can conveniently be grouped into a $(2s + 1)$-component spinor

$$\chi = \begin{bmatrix} \chi_s \\ \cdot \\ \cdot \\ \cdot \\ \chi_{-s} \end{bmatrix}$$

In fact, it is often convenient in discussing spin to overlook the distinction between the abstract vector $|\chi\rangle$ and its representative spinor χ, and to regard the two as one and the same thing.[2] In this case the operators on $\mathcal{H}_{\text{spin}}$ are regarded as $(2s + 1)$-dimensional square matrices. For example, we can write the spin operator for a particle of spin half as $\mathbf{S} = \boldsymbol{\sigma}/2$, where σ_1, σ_2, σ_3 are the usual Pauli spin matrices.

A basis for the space $\mathcal{H}$ can be constructed from any bases of $\mathcal{H}_{\text{space}}$ and $\mathcal{H}_{\text{spin}}$. For example, one convenient basis is given by the eigenvectors of $\mathbf{P}$ and S_3,

$$|\mathbf{p}, m\rangle = |\mathbf{p}\rangle \otimes |m\rangle$$

which are products of the momentum eigenvectors $|\mathbf{p}\rangle$ in $\mathcal{H}_{\text{space}}$ and the S_3 eigenvectors $|m\rangle$ in $\mathcal{H}_{\text{spin}}$.

The Hilbert space for two distinct particles with spins s_1 and s_2 is of course the product $\mathcal{H} = \mathcal{H}_1 \otimes \mathcal{H}_2$ of the two one-particle spaces, each of which is itself a product of the type just described. The spatial wave functions have the form $\psi_{m_1 m_2}(\mathbf{x}_1, \mathbf{x}_2)$ and just as in the spinless case it is convenient to rewrite these as functions $\psi_{m_1 m_2}(\bar{\mathbf{x}}, \mathbf{x})$ of the CM and relative positions, $\bar{\mathbf{x}}$ and $\mathbf{x}$. Because these can clearly be spanned by products of the form $\phi(\bar{\mathbf{x}})\chi_{m_1 m_2}(\mathbf{x})$, we can regard the space $\mathcal{H}$ as $\mathcal{H} = \mathcal{H}_{\text{cm}} \otimes \mathcal{H}_{\text{rel}}$ where $\mathcal{H}_{\text{cm}}$ describes the motion of the CM position only, while $\mathcal{H}_{\text{rel}}$ describes the relative motion, *including both spins*. The space $\mathcal{H}_{\text{rel}}$ can itself be regarded as a product of one space for the relative coordinate $\mathbf{x}$ and another for both spins.

As a basis for the spin space we can use either the eigenvectors $|m_1, m_2\rangle$ of the two z components, or the eigenvectors $|s, m\rangle$ of the total spin and its z component. The relation between these is:

$$|s, m\rangle = \sum_{m_1, m_2} |m_1, m_2\rangle \langle s_1 s_2 m_1 m_2 \,|\, sm\rangle$$

where $\langle s_1 s_2 m_1 m_2 \,|\, sm\rangle$ is the usual Clebsch–Gordan coefficient.[3] When we do not wish to commit ourselves to a particular basis we shall use the notation $|\xi\rangle$ to label any convenient choice. In practice, ξ usually stands for either (m_1, m_2) or (s, m), and in any case is a label taking on $(2s_1 + 1)(2s_2 + 1)$

---

[2] Of course one can always take the view that the state vector *is* the wave function (or, in the present case, the spinor). However, it is usually preferable to distinguish between the two and to regard the state vector as being *represented* by the wave function. Nonetheless, in discussing spin it is sometimes convenient to let the distinction become a little blurred.

[3] For the Clebsch–Gordan coefficients (also known as Wigner, or vector-coupling coefficients) we shall follow the popular Condon and Shortley phase conventions. (See Condon and Shortley, 1935, or Messiah, 1961.)

distinct values. The general spin state of the two particles can be expanded as:

$$|\chi\rangle = \sum_{\xi} \chi_{\xi} |\xi\rangle$$

and is completely identified by the numbers χ_{ξ}, which can be grouped into a column spinor χ of $(2s_1 + 1)(2s_2 + 1)$ components.

For any basis $\{|\xi\rangle\}$ of the spin space there are several corresponding bases of the complete space $\mathcal{H}$ and the space of the relative motion $\mathcal{H}_{rel}$. The most important basis of $\mathcal{H}$ consists of the momentum eigenvectors, which we write (without serious danger of confusion) in either of the forms:

$$|\mathbf{p}_1, \mathbf{p}_2, \xi\rangle \equiv |\bar{\mathbf{p}}, \mathbf{p}, \xi\rangle$$

where $\bar{\mathbf{p}}$ and $\mathbf{p}$ are the total and relative momenta as usual. The corresponding basis vectors of $\mathcal{H}_{rel}$ are just $|\mathbf{p}, \xi\rangle$ in terms of which we can write $|\bar{\mathbf{p}}, \mathbf{p}, \xi\rangle$ as $|\bar{\mathbf{p}}\rangle \otimes |\mathbf{p}, \xi\rangle$.

5-b.   The S Operator for Particles with Spin

Apart from the fact that the Hilbert space is a little more complicated than before, we can now set up a scattering formalism exactly as for the spinless case. The orbits have the usual form $U(t) |\psi\rangle$ where $|\psi\rangle$ is any vector in the space $\mathcal{H}$ just described. The evolution operator $U(t)$ is determined by the Hamiltonian

$$H = H^0 + V$$

where H^0 is the same as for the spinless case,

$$H^0 = \frac{\mathbf{P}_1^2}{2m_1} + \frac{\mathbf{P}_2^2}{2m_2} = \frac{\bar{\mathbf{P}}^2}{2M} + \frac{\mathbf{P}^2}{2m}$$

Typical examples of the interaction V would be a nucleon–nucleon interaction of the form:

$$V = V_1(r) + \mathbf{S}_1 \cdot \mathbf{S}_2 \, V_2(r) + \mathbf{S}_1 \cdot \mathbf{x} \, \mathbf{S}_2 \cdot \mathbf{x} \, V_3(r) \tag{5.2}$$

(Remember that $\mathbf{x} = \mathbf{x}_1 - \mathbf{x}_2$ and $r = |\mathbf{x}|$.) or the spin-orbit interaction of an electron in an atom or nucleon in a nucleus,

$$V = V_1(r) + \mathbf{L} \cdot \mathbf{S} \, V_2(r) \tag{5.3}$$

In all cases, whether or not V depends on the spins, one expects that V will go to zero as the two particles move apart. Thus, with the potentials (5.2) and (5.3) we expect that the coefficients $V_i(r)$ will go to zero suitably rapidly as $r \to \infty$.

Using the same arguments as in the spinless case we can now prove the asymptotic condition, which asserts that every $|\psi_{in}\rangle$ in $\mathcal{H}$ labels the in asymptote of some actual orbit $U(t)\,|\psi\rangle$,

$$U(t)\,|\psi\rangle \xrightarrow[t\to-\infty]{} U^0(t)\,|\psi_{\text{in}}\rangle$$

Exactly as before, this is true provided the vector

$$U(t)^\dagger U^0(t)\,|\psi_{\text{in}}\rangle = |\psi_{\text{in}}\rangle + i\int_0^t d\tau\, U(\tau)^\dagger V U^0(\tau)\,|\psi_{\text{in}}\rangle$$

converges. And exactly as before it is sufficient to prove that the integral

$$\int_{-\infty}^0 d\tau\, \|U(\tau)^\dagger V U^0(\tau)\psi_{\text{in}}\| = \int_{-\infty}^0 d\tau\, \|V U^0(\tau)\psi_{\text{in}}\|$$

is convergent. The evolution operator $U^0(\tau)$ (which is precisely the same as in the spinless case) conveniently spreads the two particles apart as τ becomes large. Thus, provided V is suitably short range the integral does converge and the asymptotic condition holds exactly as before.

Even if it were possible to precisely enumerate the class of interactions V for which the asymptotic condition and asymptotic completeness both hold, it would probably not be very interesting to do so. We content ourselves here with the assertion that both results surely hold for all "reasonable" potentials and shall confine attention to these from now on. (As usual this excludes the Coulomb potential.) Thus the Møller operators $\boldsymbol{\Omega}_\pm$ exist as the limits of $U(t)^\dagger U^0(t)$ and map each in or out asymptote $|\psi_{\text{in}}\rangle$ or $|\psi_{\text{out}}\rangle$ onto the corresponding actual state $|\psi\rangle$ at $t = 0$. The operator $\mathbf{S} = \boldsymbol{\Omega}_-^\dagger \boldsymbol{\Omega}_+$ is unitary and maps each $|\psi_{\text{in}}\rangle$ directly onto the corresponding $|\psi_{\text{out}}\rangle$.

Exactly as in the spinless case the S operator has the structure

$$\mathbf{S} = 1_{\text{cm}} \otimes S \tag{5.4}$$

where, of course, 1_{cm} refers to the motion of the CM position only, while S acts on the space of the relative motion, including both spins. This result can be regarded as reducing the problem of two particles with spin to an equivalent *quasi-one-particle* problem, namely the scattering of a single particle with spin s_1 off a fixed target of spin s_2.

5-c.   The Amplitudes and Amplitude Matrix

For the same reasons as before **S** conserves energy and momentum and its matrix elements can be decomposed into two terms,

$$\begin{aligned}
\langle \mathbf{p}_1', \mathbf{p}_2', \xi' | \mathbf{S} | \mathbf{p}_1, \mathbf{p}_2, \xi \rangle = {} & \delta_3(\mathbf{p}_1' - \mathbf{p}_1)\,\delta_3(\mathbf{p}_2' - \mathbf{p}_2)\delta_{\xi'\xi} \\
& - 2\pi i\,\delta(\textstyle\sum E_i' - \sum E_i) \\
& \times \delta_3(\textstyle\sum \mathbf{p}_i' - \sum \mathbf{p}_i) t(\mathbf{p}', \xi' \leftarrow \mathbf{p}, \xi)
\end{aligned} \tag{5.5}$$

[Compare (4.9) for the spinless case.] In (5.5) the first term is the amplitude for no scattering and leaves both momenta and spins unchanged. The second conserves energy and total momentum but can in general connect states of different relative momenta and different spins. Just as before, the on-shell T matrix is related directly to the operator S of the relative motion; specifically, if we insert (5.4) into (5.5) and factor out the total-momentum delta function, we find:

$$\langle \mathbf{p}', \xi' | \, \mathsf{S} \, | \mathbf{p}, \xi \rangle = \delta_3(\mathbf{p}' - \mathbf{p})\delta_{\xi'\xi} - 2\pi i \, \delta(E_{p'} - E_p) \, t(\mathbf{p}', \xi' \leftarrow \mathbf{p}, \xi)$$

The on-shell T matrix is proportional to the scattering amplitude, which we define as usual to be:

$$f(\mathbf{p}', \xi' \leftarrow \mathbf{p}, \xi) = -(2\pi)^2 m \, t(\mathbf{p}', \xi' \leftarrow \mathbf{p}, \xi)$$

Exactly as in the spinless case we can now calculate the CM cross sections in terms of the amplitude. However, in place of the single differential cross section of the spinless case, we now find an infinite number of differential cross sections because the particles can enter the collision in any spin state $|\chi\rangle$ and one can (in principle, at least) measure the number of particles emerging into $d\Omega$ with any given spin state $|\chi'\rangle$. We consider first the case where the particles enter in one of the basic spin states $|\xi\rangle$ and we count the number emerging into $d\Omega$ in the basic spin state $|\xi'\rangle$. [Recall that ξ is just the label for any convenient basis in the spin space; e.g., $\xi = (m_1, m_2)$.] For this case a calculation identical to that in Section 4-e gives:

$$\frac{d\sigma}{d\Omega}(\mathbf{p}', \xi' \leftarrow \mathbf{p}, \xi) = |f(\mathbf{p}', \xi' \leftarrow \mathbf{p}, \xi)|^2$$

which is the CM differential cross section for observation of the final particles in the direction of $\mathbf{p}'$ with spins given by $|\xi'\rangle$ if the initial particles had relative momentum $\mathbf{p}$ and spins $|\xi\rangle$.

The cross section for observing an arbitary final spin state $|\chi'\rangle$ coming from any initial spin state $|\chi\rangle$ can be evaluated in the same way. If

$$|\chi\rangle = \sum_{\xi} \chi_{\xi} \, |\xi\rangle$$

and similarly $|\chi'\rangle$, we can calculate the relevant S-matrix element from (5.5) and, hence, the corresponding CM cross section, which is easily seen to be (see Problem **5.1**.)

$$\frac{d\sigma}{d\Omega}(\mathbf{p}', \chi' \leftarrow \mathbf{p}, \chi) = \left| \sum_{\xi', \xi} \chi_{\xi'}'^{*} f(\mathbf{p}', \xi' \leftarrow \mathbf{p}, \xi) \chi_{\xi} \right|^2$$

This result expresses the cross section for arbitrary spins ($|\chi'\rangle \leftarrow |\chi\rangle$) in terms of the amplitudes for the $[(2s_1 + 1)(2s_2 + 1)]^2$ basic processes ($|\xi'\rangle \leftarrow |\xi\rangle$).

Its form suggests that we rewrite the basic amplitudes as

$$f(\mathbf{p}', \xi' \leftarrow \mathbf{p}, \xi) = f_{\xi'\xi}(\mathbf{p}' \leftarrow \mathbf{p})$$

and then regard them as the elements of an *amplitude matrix*,

$$F(\mathbf{p}' \leftarrow \mathbf{p}) = \{f_{\xi'\xi}(\mathbf{p}' \leftarrow \mathbf{p})\}$$

[This matrix is often called $M(\mathbf{p}', \mathbf{p})$ in the literature.] Our result can then be written in the compact form[4]

$$\frac{d\sigma}{d\Omega}(\mathbf{p}', \chi' \leftarrow \mathbf{p}, \chi) = |\chi'^{\dagger}F(\mathbf{p}' \leftarrow \mathbf{p})\chi|^2 \tag{5.6}$$

from which it is clear that all information relevant to the scattering of the two particles is contained in the matrix $F(\mathbf{p}' \leftarrow \mathbf{p})$, just as, in the spinless case, all information was contained in the single amplitude $f(\mathbf{p}' \leftarrow \mathbf{p})$.

It may help the reader to become familiar with the idea of the amplitude matrix to focus attention on some specific example, the simplest of which is the case of a spin-half particle scattering off a spinless target. This example includes such important processes as the scattering of electrons off a spin-zero atom, of nucleons off a spinless nucleus, and a number of elementary particle processes, of which the most important is pion–nucleon scattering. Because one particle is spinless, the spin space of the whole system is just the two-dimensional spin space of the spin-half projectile. We use the usual S_3 basis with basic vectors $|+\rangle$ and $|-\rangle$ corresponding to the eigenvalues $m = \pm\tfrac{1}{2}$. According to (5.6), the scattering is determined by a (2×2) amplitude matrix:

$$F(\mathbf{p}' \leftarrow \mathbf{p}) = \begin{pmatrix} f_{++}(\mathbf{p}' \leftarrow \mathbf{p}) & f_{+-}(\mathbf{p}' \leftarrow \mathbf{p}) \\ f_{-+}(\mathbf{p}' \leftarrow \mathbf{p}) & f_{--}(\mathbf{p}' \leftarrow \mathbf{p}) \end{pmatrix} \tag{5.7}$$

The element $f_{m'm}(\mathbf{p}' \leftarrow \mathbf{p})$ is the amplitude for an initial particle with momentum $\mathbf{p}$ and z component of spin m to be scattered into the direction $\mathbf{p}'$ and observed with z component of spin m'. (For obvious reasons f_{+-} and f_{-+} are referred to as *spin-flip amplitudes* and f_{++} and f_{--} as *spin-nonflip amplitudes*.) The most general initial and final spin states are given by two-component spinors χ and χ' and, according to (5.6), the amplitude for observing the corresponding process $(\mathbf{p}', \chi' \leftarrow \mathbf{p}, \chi)$ is just the number $\chi'^{\dagger}F(\mathbf{p}' \leftarrow \mathbf{p})\chi$.

---

[4] I have chosen to use the notation $\chi'^{\dagger}F\chi$, rather than $\langle\chi'|F|\chi\rangle$ to emphasize the view of F as a matrix and χ and χ' as column spinors. This result includes as a special case the cross section $|f_{\xi'\xi}|^2$ for the basic process $(\mathbf{p}', \xi' \leftarrow \mathbf{p}, \xi)$, since the spinor for the basic state $|\xi\rangle$ consists of a 1 in the ξth place and zeros elsewhere.

5-d.   Sums and Averages Over Spins

In principle, the result of the last section contains all the information we need in the elastic scattering of two particles with spin. If we can calculate the amplitude matrix $F(\mathbf{p}' \leftarrow \mathbf{p})$ (by any of the methods to be developed in Chapters 8–14 for example) then we can predict the cross sections for all possible spin states χ and χ'. Alternatively, if we can observe the cross sections for sufficiently many different χ and χ', then we can *measure* the amplitude matrix within an undetermined overall phase.[5]

In practice, the situation is not so simple. In most real experiments the incident particles do not all occupy a single definite initial spin state χ and the scattered particles are counted by detectors that cannot discriminate between the different possible final spin states χ'. These complications are most easily handled in general using the density matrix of quantum statistical mechanics (discussed in Chapter 7). However, in certain simple (and frequently occurring) situations it is just as convenient to use the more elementary method of averaging over initial and summing over final spin states, as we now describe.

We begin by considering (what is in fact the most common experimental situation) the case that the particles of the incident beam and target are *completely unpolarized;* that is, instead of all occupying some definite spin state, the spins of the particles are oriented completely randomly. We shall show in Chapter 7 that a beam or target in which the particles are oriented at random is the same thing as one in which the particles are equally distributed among the $(2s + 1)$ states of any conveniently chosen orthonormal basis. (For example, a beam of electrons randomly oriented is indistinguishable from a beam of electrons 50% of which are spin up and 50% spin down.) For the present we shall simply take as our definition of an unpolarized beam or target one in which the $(2s + 1)$ states $|m\rangle$, $m = -s, \ldots, s$, are occupied by equal numbers of particles.

We consider an experiment in which an unpolarized beam of spin s_1 impinges with relative momentum $\mathbf{p}$ on an unpolarized target of spin s_2. We imagine that we count the number of particles emerging with momentum in $d\Omega$ about $\mathbf{p}'$ and *with some definite basic spin state* $|\xi'\rangle = |m'_1, m'_2\rangle$. Because the beam particles are equally distributed among the $(2s_1 + 1)$ basic states $|m_1\rangle$, and similarly the target particles among their $(2s_2 + 1)$ states

---

[5] See Problem 5.3. Just as in the spinless case the overall phase *can* be measured but not by simple measurement of $d\sigma/d\Omega$ in a given direction. (For example, measurement of the total cross section allows one to determine the forward amplitude completely, using the optical theorem. If one assumes that at low energies only a few partial waves contribute, then fitting $d\sigma/d\Omega$ at all angles usually allows one to determine the amplitude with, at worst, a finite number of ambiguities.)

$|m_2\rangle$, the proportion of collisions for which the initial spin state is any definite $|\xi\rangle = |m_1, m_2\rangle$ is exactly $1/(2s_1 + 1)(2s_2 + 1)$. Thus, when the total incident density is n_{inc}, the incident density for collisions in which the initial spin state is $|\xi\rangle$ is just $n_{inc}/(2s_1 + 1)(2s_2 + 1)$. The contribution of *these* collisions to our scattered counts is therefore:

$$\frac{n_{inc}}{(2s_1 + 1)(2s_2 + 1)} \frac{d\sigma}{d\Omega} (\mathbf{p}', \xi' \leftarrow \mathbf{p}, \xi)\, d\Omega \qquad (5.8)$$

Since the total number of counts is the sum of the contributions from the various initial states, the total count is obtained by summing the expression (5.8) over all ξ, that is, the total count can be written as:

$$n_{inc} \frac{d\sigma}{d\Omega} (\mathbf{p}', \xi' \leftarrow \mathbf{p})\, d\Omega$$

where

$$\frac{d\sigma}{d\Omega} (\mathbf{p}', \xi' \leftarrow \mathbf{p}) = \frac{1}{(2s_1 + 1)(2s_2 + 1)} \sum_{\xi} \frac{d\sigma}{d\Omega} (\mathbf{p}', \xi' \leftarrow \mathbf{p}, \xi) \qquad \begin{bmatrix} \text{in spins} \\ \text{unpolarized} \end{bmatrix}$$

This last is seen to be the effective cross section for scattering from initial momentum $\mathbf{p}$ to final momentum $\mathbf{p}'$ and final spin state $|\xi'\rangle$ if the *initial beam and target are unpolarized;* it is seen to be the *average over the initial spins* ξ of the cross section for the basic process $(\mathbf{p}', \xi' \leftarrow \mathbf{p}, \xi)$.

We next consider the complementary situation in which the initial particles *are polarized* (they are all in some definite spin state) but we use spin-insensitive detectors that count all emerging particles irrespective of their spin state. If the initial spin state is one of the basic states $|\xi\rangle$ and the incident density n_{inc}, then the number of particles emerging into solid angle $d\Omega$ about $\mathbf{p}'$ with any definite spin state $|\xi'\rangle$ is of course

$$n_{inc} \frac{d\sigma}{d\Omega} (\mathbf{p}', \xi' \leftarrow \mathbf{p}, \xi)\, d\Omega$$

Because our counters accept all particles irrespective of their spin state the total count is obtained by summing this expression over all ξ'. That is, the total count can be written as:

$$n_{inc} \frac{d\sigma}{d\Omega} (\mathbf{p}' \leftarrow \mathbf{p}, \xi)\, d\Omega$$

where

$$\frac{d\sigma}{d\Omega} (\mathbf{p}' \leftarrow \mathbf{p}, \xi) = \sum_{\xi'} \frac{d\sigma}{d\Omega} (\mathbf{p}', \xi' \leftarrow \mathbf{p}, \xi) \qquad \begin{bmatrix} \text{out spins} \\ \text{not monitored} \end{bmatrix}$$

is the effective cross section appropriate to the initial spin state $|\xi\rangle$ if *we use spin-insensitive detectors for the final particles.* That is, the cross section appropriate to spin-insensitive counters is obtained from the basic cross section $(\mathbf{p'}, \xi' \leftarrow \mathbf{p}, \xi)$ by *summing over the final spins* ξ'. Obviously it is not essential that our incident particles occupy one of the basic spin states $|\xi\rangle$; if they occupy instead an arbitrary state given by the spinor χ, then we can define a corresponding cross section for the process $(\mathbf{p'} \leftarrow \mathbf{p}, \chi)$, which is obtained by summing that for the process $(\mathbf{p'}, \xi' \leftarrow \mathbf{p}, \chi)$ over all ξ'.

In practice, the most common type of experiment uses initial particles that are completely unpolarized *and* spin-insensitive counters. In this case we can immediately combine the above discussions to give the effective cross section:

$$\frac{d\sigma}{d\Omega}(\mathbf{p'} \leftarrow \mathbf{p}) = \frac{1}{(2s_1 + 1)(2s_2 + 1)} \sum_{\xi} \sum_{\xi'} \frac{d\sigma}{d\Omega}(\mathbf{p'}, \xi' \leftarrow \mathbf{p}, \xi)$$

$$\begin{bmatrix} \text{in spins unpolarized} \\ \text{out spins not monitored} \end{bmatrix} \quad (5.9)$$

This cross section is generally referred to as the *unpolarized cross section.* It is obtained from the basic cross section $(\mathbf{p'}, \xi' \leftarrow \mathbf{p}, \xi)$ by averaging over the initial spins ξ and summing over the final spins ξ'.

If we consider the example of (spin $\frac{1}{2}$) $-$ (spin 0) scattering discussed above then the various cross sections can all be expressed in terms of the four basic amplitudes $f_{++}, f_{+-}, f_{-+}, f_{--}$. Thus, for example, the unpolarized cross section (5.9) is just

$$\frac{d\sigma}{d\Omega}(\mathbf{p'} \leftarrow \mathbf{p}) = \tfrac{1}{2}\{|f_{++}|^2 + |f_{+-}|^2 + |f_{-+}|^2 + |f_{--}|^2\} \quad (5.10)$$

This makes clear the important fact that as long as we have only unpolarized beams and spin-insensitive methods of counting, the only relevant quantity, and the only measurable quantity, is this particular combination of all four amplitudes. To obtain more complete information on the amplitudes (individual magnitudes or relative phases) we must clearly use a polarized beam or some spin-sensitive method of detection. We shall return to these questions in Chapter 7.

5-e.   The In and Out Spinors

With the results of the last two sections we can give a useful alternative interpretation of the amplitude matrix $F(\mathbf{p'} \leftarrow \mathbf{p})$. To this end we consider particles incident in some definite spin state, which we now label by the normalized spinor χ^{in}; and in terms of χ^{in} we define a second spinor,

$$\boxed{\chi^{out} = F(\mathbf{p}' \leftarrow \mathbf{p})\chi^{in}} \qquad (5.11)$$

What we shall show is that χ^{out} is precisely the actual (unnormalized) spinor of those particles which emerge with momentum $\mathbf{p}'$ if the initial momentum and spins were $\mathbf{p}$ and χ^{in}. To see this we focus attention on those particles emerging with momentum $\mathbf{p}'$. The result (5.6) for the cross section $(\mathbf{p}', \chi' \leftarrow \mathbf{p}, \chi^{in})$ can be rewritten as:

$$\frac{d\sigma}{d\Omega} (\mathbf{p}', \chi' \leftarrow \mathbf{p}, \chi^{in}) = |\chi'^{\dagger} F(\mathbf{p}' \leftarrow \mathbf{p})\chi^{in}|^2 = |\chi'^{\dagger}\chi^{out}|^2$$

which shows that the probability of those particles emerging in the direction $\mathbf{p}'$ being found with spins χ' is proportional to $|\chi'^{\dagger}\chi^{out}|^2$, for any χ'. But according to the elementary principles of quantum mechanics this means simply that χ^{out} *is* the actual spin state of these particles.[6]

The spinor χ^{out} is not normalized; on the contrary,

$$\|\chi^{out}\|^2 = \sum_{\xi'} |\chi_{\xi'}^{out}|^2 = \sum_{\xi'} \left| \sum_{\xi} f_{\xi'\xi}(\mathbf{p}' \leftarrow \mathbf{p})\chi_{\xi}^{in} \right|^2$$

$$= \sum_{\xi'} \frac{d\sigma}{d\Omega} (\mathbf{p}', \xi' \leftarrow \mathbf{p}, \chi^{in})$$

This sum will be recognized as the cross section $(\mathbf{p}' \leftarrow \mathbf{p}, \chi^{in})$ that is measured if we use spin-insensitive counters, which accept all particles irrespective of their spin state. Thus, the result is simply that

$$\|\chi^{out}\|^2 = \frac{d\sigma}{d\Omega} (\mathbf{p}' \leftarrow \mathbf{p}, \chi^{in}) \qquad \text{[out spins not monitored]} \qquad (5.12)$$

To conclude, we summarize the principal results of this chapter. In Section 5-c we showed that if the two particles in a collision have spins s_1 and s_2, then in place of the single amplitude $f(\mathbf{p}' \leftarrow \mathbf{p})$ of the spinless case, one has an amplitude matrix $F(\mathbf{p}' \leftarrow \mathbf{p})$ made up of the $[(2s_1 + 1)(2s_2 + 1)]^2$ basic amplitudes $f_{\xi'\xi}(\mathbf{p}' \leftarrow \mathbf{p})$. The cross section for any process $(\mathbf{p}', \chi' \leftarrow \mathbf{p}, \chi)$ involving arbitrary initial and final spin states χ and χ' is just $|\chi'^{\dagger}F\chi|^2$, which includes as a special case the cross section $|f_{\xi'\xi}|^2$ for the basic process $(\mathbf{p}', \xi' \leftarrow \mathbf{p}, \xi)$. In Section 5-d we gave the cross sections appropriate to

---

[6] The resemblance between the equation (5.11) $\chi^{out} = F\chi^{in}$ and the familiar result $|\psi_{out}\rangle = S|\psi_{in}\rangle$ is, needless to say, no accident. In fact the matrix $F(\mathbf{p}' \leftarrow \mathbf{p})$ is, apart from assorted factors of $2\pi i$ and delta functions, the restriction of the S operator to the subspaces defined by the momenta $\mathbf{p}$ and $\mathbf{p}'$. Thus, it is entirely legitimate to think of (5.11) as an expression of the basic result $|\psi_{out}\rangle = S|\psi_{in}\rangle$. The reason that we have not derived the results of this section from this point of view is that it is hard to keep track of the normalization of χ^{out} this way. (Note well that χ^{out} is *not* normalized when χ^{in} is.)

experiments using unpolarized beams or detectors that cannot distinguish between the various spin states. Finally, in this section, we have shown that if the particles enter with relative momentum $\mathbf{p}$ and spin state χ^{in} and if they are observed to emerge with momentum $\mathbf{p}'$, then their actual spin state is given by $\chi^{\text{out}} = F(\mathbf{p}' \leftarrow \mathbf{p})\chi^{\text{in}}$, where χ^{out} is normalized such that $\|\chi^{\text{out}}\|^2$ is just the cross section (5.12) for their emerging with momentum $\mathbf{p}'$ irrespective of spin state.

PROBLEMS

5.1. Using the decomposition (5.5), write down the decomposition of an arbitrary S-matrix element $\langle \mathbf{p}_1', \mathbf{p}_2', \chi' | \mathbf{S} | \mathbf{p}_1, \mathbf{p}_2, \chi \rangle$ in terms of $\chi'^\dagger F(\mathbf{p}' \leftarrow \mathbf{p})\chi$. From this derive the cross section (5.6) for the process $(\mathbf{p}', \chi' \leftarrow \mathbf{p}, \chi)$ in terms of the amplitude matrix $F(\mathbf{p}' \leftarrow \mathbf{p})$. (Obviously there is no point in reproducing those messy details that you have already given for the spinless case. The point is to understand how the spins affect the calculation.)

5.2. (a) Suppose that the interaction of two particles with spin is spin-*independent*. Show that the amplitude matrix has the form $F(\mathbf{p}' \leftarrow \mathbf{p}) = f(\mathbf{p}' \leftarrow \mathbf{p})I$, where I is the unit spin matrix and f is the amplitude one would obtain for two spinless particles with the same Hamiltonian. (This means one can simply ignore the spins when calculating the amplitude.)
(b) Show that the particles emerge from any collision with the same spins as those with which they entered, and that an experimenter without spin-sensitive counters can detect no evidence that the particles have spin (in this experiment). (This last is in sharp contrast to the situation when the interactions depend on the spin. In this case certain initial spin states can lead to asymmetries in the outgoing distribution that can be detected even without spin-sensitive detectors.)

5.3. Consider the scattering of a spin-half projectile off a spinless target. Show clearly that by measuring the differential cross section $(\mathbf{p}', \chi' \leftarrow \mathbf{p}, \chi)$ for sufficiently many different spin states χ and χ' (and given $\mathbf{p}$ and $\mathbf{p}'$) it is possible to measure all four elements of the amplitude matrix $F(\mathbf{p}' \leftarrow \mathbf{p})$, apart from a single overall phase. Show that in general 10 measurements are needed (assuming all 4 elements of F are independent) and describe some suitable sequence of measurements.

5.4. (a) State and prove the optical theorem for particles with spin. (Review the proof of Section 3-f for spinless particles, and remember to consider an arbitrary initial spin state χ.)
(b) For (spin $\frac{1}{2}$) $-$ (spin 0) scattering show that the optical theorem lets one measure $\operatorname{Im} f_{++}$, $\operatorname{Im} f_{--}$, and two other quantities involving f_{+-} and f_{-+} all in the forward direction).

Invariance Principles and Conservation Laws

6

6-a   **Translational Invariance and Conservation of Momentum**

6-b   **Rotational Invariance and Conservation of Angular Momentum**

6-c   **The Partial-Wave Series for Spinless Particles**

6-d   **Parity**

6-e   **Time Reversal**

6-f   **Invariance Principles for Particles with Spin; Momentum-Space Analysis**

6-g   **Invariance Principles for Particles with Spin; Angular-Momentum Analysis**

In this chapter we shall discuss the application of invariance principles to the scattering theory of the preceding chapters. We shall find that invariance of the system under any of the possible symmetry operations (rotations, parity, time reversal, etc.) implies severe restrictions on the possible form of the scattering amplitude. For example, rotational invariance implies that the amplitude $f(\mathbf{p}' \leftarrow \mathbf{p})$ for two spinless particles depends only on the magnitude of $\mathbf{p}$ and the angle between $\mathbf{p}$ and $\mathbf{p}'$; and that $f(\mathbf{p}' \leftarrow \mathbf{p})$ can be expanded in the famous partial-wave series.

The results of this chapter are of great practical value since they give one information on the form of the amplitude even before one attempts to calculate it. But perhaps their greatest importance lies in their complete generality. They depend only on the assumed invariance principles and are quite independent of the precise form of the Hamiltonian (and indeed of the very existence of a Hamiltonian). Thus, all of the results of this chapter carry over to any theory where there is an S operator and where the relevant invariance principles apply. In particular, they have played an especially important role in relativistic scattering theory, where there is good evidence in favor of the various symmetries (at least for some systems), but where we do not know the precise nature of the underlying interactions.

6-a.   Translational Invariance and Conservation of Momentum

We begin our discussion of invariance principles by re-deriving the conservation of momentum directly from translational invariance. This derivation will be the prototype for our later discussion of all other invariance principles.

The effect of a rigid translation through a vector $\mathbf{a}$ on any system is given by the unitary *translation, or displacement, operator*

$$D(\mathbf{a}) = e^{-i\mathbf{a}\cdot\bar{\mathbf{P}}} \tag{6.1}$$

where $\bar{\mathbf{P}}$ is the total momentum operator of the system. This means simply that if the system occupies any state $|\psi\rangle$ and is then rigidly displaced through $\mathbf{a}$, then the resulting state is[1] $D(\mathbf{a})|\psi\rangle$. The dynamics are said to be *translationally invariant* if the Hamiltonian is unchanged by any displacement. If this is the case, then $D(\mathbf{a})^\dagger H D(\mathbf{a}) = H$ and H commutes with the displacement operators.

We now return to our two-particle system with Hamiltonian $H = H^0 + V$, and note that H^0 automatically commutes with $D(\mathbf{a})$. Thus, if the system is translationally invariant the translation operators commute with both H and H^0, and from this it follows that they commute with the Møller operators $\Omega_\pm$,

$$D(\mathbf{a})\Omega_\pm = D(\mathbf{a})[\lim\, e^{iHt}e^{-iH^0t}]$$
$$= [\lim\, e^{iHt}e^{-iH^0t}]D(\mathbf{a}) = \Omega_\pm D(\mathbf{a})$$

This implies that the displacement operators commute with $\mathbf{S} = \Omega_-^\dagger\Omega_+$ and, hence, that $\mathbf{S} = D^\dagger S D$. Taking matrix elements we obtain $\langle\chi|\mathbf{S}|\phi\rangle = \langle\chi_D|\mathbf{S}|\phi_D\rangle$ where $|\phi_D\rangle$ denotes the translated state $D|\phi\rangle$. This result means that, as one would expect, translational invariance implies equality of the

---

[1] For the case of a single particle this is just the result used in Section 3-d when displacing the incident wave packet through ρ. For the general case, see Messiah (1961) p. 652, for example. It may help the reader to recall that for a single particle (6.1) reduces to $D(\mathbf{a}) = \exp(-\mathbf{a}\cdot\nabla)$ in the coordinate representation; this obviously generates the Taylor series for $\psi(\mathbf{x} - \mathbf{a})$ in terms of $\psi(\mathbf{x})$.

probabilities for any process $(\chi \leftarrow \phi)$ and for the translated process $(\chi_D \leftarrow \phi_D)$.

To establish conservation of momentum we have only to return to the expression $D(\mathbf{a}) = \exp(-i\mathbf{a} \cdot \bar{\mathbf{P}})$. Since S commutes with $D(\mathbf{a})$ for any $\mathbf{a}$, it must commute with $\bar{\mathbf{P}}$, the total momentum; $[\bar{\mathbf{P}}, S] = 0$. In particular, if we take momentum-space matrix elements of this equation we find that

$$(\bar{\mathbf{p}}' - \bar{\mathbf{p}})\langle \cdots' | S | \cdots \rangle = 0$$

Thus, the momentum-space matrix element of S is zero unless the initial and final total momenta are equal—that is, total momentum is conserved—and the S matrix contains the familiar factor $\delta_3(\bar{\mathbf{p}}' - \bar{\mathbf{p}})$.

6-b.   Rotational Invariance and Conservation of Angular Momentum

The effect of any rotation on a quantum mechanical system is given by a unitary rotation operator R. If we parametrize the rotation through an angle α clockwise about the direction $\hat{\mathbf{u}}$ by the vector $\boldsymbol{\alpha} = \alpha\hat{\mathbf{u}}$, then the corresponding rotation operator is[2]

$$R(\boldsymbol{\alpha}) = e^{-i\boldsymbol{\alpha} \cdot \bar{\mathbf{J}}} \tag{6.2}$$

where $\bar{\mathbf{J}}$ is the total angular momentum operator. As with translational invariance, the dynamics are said to be rotationally invariant if the Hamiltonian H commutes with all rotation operators $R(\boldsymbol{\alpha})$. In most (though not all) problems of physical interest the dynamics *are* rotationally invariant.[3]

If we consider again our two-particle system and assume rotational invariance, then we can show (using the same arguments as above) that $R(\boldsymbol{\alpha})$ commutes with $\Omega_{\pm}$ and, hence, with S. Because this is true for any $\boldsymbol{\alpha}$, it follows that S commutes with $\bar{\mathbf{J}}$ and the total angular momentum is conserved.

We know, of course, that the two-particle scattering operator factors as $S = 1_{cm} \otimes S$ and it will be no surprise that the interesting consequences of rotational invariance can be found by considering just the operator S of the

---

[2] The reader may be more familiar with the parametrization in terms of Euler angles, for which

$$R(\varphi, \theta, \psi) = \exp(-i\varphi \bar{J}_3)\exp(-i\theta \bar{J}_2)\exp(-i\psi \bar{J}_3)$$

(or some slight variant, depending on one's definition of the Euler angles). Nonetheless for our purposes the parametrization (6.2) is more compact and convenient.

[3] Actually it is one of the most strongly held beliefs in modern physics that *all* isolated systems are invariant under rotations of the whole system. However, it is often convenient to regard part of a system as "external" and *fixed*, in which case rotational invariance (under rotations of the rest of the system) does not hold. For example, the scattering of an electron off a fixed crystal lattice is not rotationally invariant.

relative motion. To see this we have only to recall the elementary result that the total angular momentum is the sum of the angular momentum of the center of mass plus that of the internal motion. For two particles this simply means that

$$\bar{\mathbf{J}} = \mathbf{J}_1 + \mathbf{J}_2$$
$$= \bar{\mathbf{X}} \times \bar{\mathbf{P}} + \mathbf{X} \times \mathbf{P} + \mathbf{S}_1 + \mathbf{S}_2$$
$$= \bar{\mathbf{X}} \times \bar{\mathbf{P}} + \mathbf{J}$$

($\mathbf{S}_1$ and $\mathbf{S}_2$ being the two spin operators). Here the operator $\bar{\mathbf{X}} \times \bar{\mathbf{P}}$ is the angular momentum of the center of mass and acts only on the space $\mathcal{H}_{cm}$ of the CM motion; while $\mathbf{J}$ is the internal angular momentum and acts only on the relative space $\mathcal{H}_{rel}$. Clearly $\mathbf{S} = 1_{cm} \otimes S$ commutes with the total angular momentum $\bar{\mathbf{J}}$ if and only if S commutes with the internal angular momentum $\mathbf{J}$.[4] There is therefore no point in discussing the two-particle operator $\mathbf{S}$ any further and we can confine attention to the operator S of the relative motion. Because this latter is the same as the S operator of the equivalent one-particle problem, we shall for simplicity present much of the remaining discussion using the terminology of one-particle scattering.

To conclude this section we return to the rotation operators $R(\boldsymbol{\alpha})$ and consider the special case where both particles are spinless. In this case, the condition that H commute with all rotations requires simply that V be spherically symmetric; that is, a function of r only, $V(\mathbf{x}) = V(r)$. If V is spherically symmetric, then S commutes with all rotations R and $S = R^\dagger S R$ (S being either the S operator of a single particle in a fixed potential or that of the relative motion of two particles). Taking momentum-space matrix .elements of this equation we find that:

$$\langle \mathbf{p}' | \, S \, | \mathbf{p} \rangle = \langle \mathbf{p}'_R | \, S \, | \mathbf{p}_R \rangle$$

where $\mathbf{p}_R$ denotes the momentum obtained from $\mathbf{p}$ by the rotation R. Substituting into the definitions (3.7) and (3.8) of the scattering amplitude, we find the same result for $f(\mathbf{p}' \leftarrow \mathbf{p})$,

$$\boxed{f(\mathbf{p}' \leftarrow \mathbf{p}) = f(\mathbf{p}'_R \leftarrow \mathbf{p}_R)} \cdots \text{[R invariance]} \qquad (6.3)$$

(Here, as elsewhere, we indicate in brackets the conditions under which the result holds; in this case rotational invariance.) This means, much as one would expect, that rotational invariance implies equality of the amplitudes for any process $(\mathbf{p}' \leftarrow \mathbf{p})$ and the rotated process $(\mathbf{p}'_R \leftarrow \mathbf{p}_R)$.

---

[4] Since the factoring $\mathbf{S} = 1_{cm} \otimes S$ is in reality a consequence of Galilean invariance (See last footnote of Section 4-a), so is the present result.

The result (6.3) means that the amplitude $f(\mathbf{p}' \leftarrow \mathbf{p})$, which is *a priori* a function of five variables, $\mathbf{p}$ and the direction of $\mathbf{p}'$ (recall that $|\mathbf{p}| = |\mathbf{p}'|$), is in fact only a function of two variables, which we take to be the energy E_p and the scattering angle θ between $\mathbf{p}$ and $\mathbf{p}'$. To prove this we must show that if $(\mathbf{p}' \leftarrow \mathbf{p})$ and $(\tilde{\mathbf{p}}' \leftarrow \tilde{\mathbf{p}})$ are any two processes that have the same energies and same scattering angles, then the corresponding amplitudes are equal:

$$f(\mathbf{p}' \leftarrow \mathbf{p}) = f(\tilde{\mathbf{p}}' \leftarrow \tilde{\mathbf{p}}) \tag{6.4}$$

According to (6.3) this is certainly so if we can find a rotation R that simultaneously carries $\mathbf{p}$ to $\tilde{\mathbf{p}}$ and $\mathbf{p}'$ to $\tilde{\mathbf{p}}'$. Because this is clearly possible,[5] (6.4) is true and $f(\mathbf{p}' \leftarrow \mathbf{p})$ is indeed a function of E_p and θ only. Thus for the scattering of two spinless particles interacting via a spherically symmetric potential, we can write

$$\boxed{f(\mathbf{p}' \leftarrow \mathbf{p}) = f(E_p, \theta)} \cdots \text{[R invariance]} \tag{6.5}$$

In practice this means there is no loss of generality in supposing that the initial momentum $\mathbf{p}$ lies along the z axis; and further, that with $\mathbf{p}$ along the z axis and $\mathbf{p}'$ in the direction (θ, φ) the amplitude depends on the energy and θ, but not on the azimuth φ.

We shall see later that the results (6.3) and (6.5), which apply to spinless particles, are, in a sense, deceptively simple. The corresponding result for particles with spin is rather more complicated.[6]

6-c.   The Partial-Wave Series for Spinless Particles

In this section we shall explore in greater depth the consequences of rotational invariance for spinless particles. In particular, we shall establish that rotational invariance implies the well-known partial-wave series.

We have just seen that rotational invariance means that S commutes with **J**, which for spinless particles is simply **L** (the orbital angular momentum of the single particle, or the relative orbital momentum of two particles). We

---

[5] For instance, let R_1 be any rotation that carries $\mathbf{p}$ onto $\tilde{\mathbf{p}}$. (This exists since $|\mathbf{p}| = |\tilde{\mathbf{p}}|$.) After performing R_1 we note that the initial momenta coincide, while the final momenta differ by at most a rotation R_2 about the initial direction (since the two scattering angles are the same.) Since this R_2 does not change the initial momenta the combination $R = R_2 R_1$ carries $\mathbf{p}$ and $\mathbf{p}'$ simultaneously onto $\tilde{\mathbf{p}}$ and $\tilde{\mathbf{p}}'$.

[6] The point is that from rotational invariance all that is immediately obvious is that the *probabilities* for the processes $(\mathbf{p}' \leftarrow \mathbf{p})$ and $(\mathbf{p}'_R \leftarrow \mathbf{p}_R)$ must be equal, and that (6.3) must hold *within a phase*. That (6.3) actually holds exactly is a simplifying feature of the spinless case, and when the particles have spin we shall have to worry about additional phase factors; see (6.36).

already know that S commutes with H^0. Now, for a single spinless particle (or equivalently for the relative motion of two spinless particles), the three operators H^0, $\mathbf{L}^2$, L_3 form a complete set of commuting observables,[7] and it follows that in the representation defined by these observables S is diagonal. We shall denote the "spherical wave" basis vectors of this representation by $|E, l, m\rangle$ where E, $l(l+1)$, m are the eigenvalues of H^0, $\mathbf{L}^2$, L_3 respectively. The corresponding spatial wave functions are:

$$\langle \mathbf{x} \mid E, l, m\rangle = i^l \left(\frac{2m}{\pi p}\right)^{\!\!1\!/\!2} \frac{1}{r} \hat{j}_l(pr) Y_l^m(\hat{\mathbf{x}}) \qquad [p \equiv (2mE)^{1\!/\!2}] \qquad (6.6)$$

Here we have introduced the Riccati–Bessel function $\hat{j}_l(z) = z j_l(z)$, where $j_l(z)$ is the usual spherical Bessel function, and $Y_l^m(\hat{\mathbf{x}})$ is the spherical harmonic, whose argument $\hat{\mathbf{x}}$ denotes the polar angles (θ, φ) of $\mathbf{x}$.[8] The normalisation is:

$$\langle E', l', m' \mid E, l, m\rangle = \delta(E' - E)\delta_{l'l}\delta_{m'm} \qquad (6.7)$$

It should be remembered that (as the δ function in this normalization shows) the $|E, l, m\rangle$ are *improper* vectors. Like the plane waves $|\mathbf{p}\rangle$ they do not represent physically realizable states; their significance is rather as a basis in terms of which the real physical states can be expanded.

Because S commutes with H^0 and $\mathbf{L}$ the S matrix in the angular-momentum representation is diagonal; that is, it has the form:

$$\langle E', l', m'| \, \mathsf{S} \, |E, l, m\rangle = \delta(E' - E)\delta_{l'l}\delta_{m'm}s_l(E) \qquad (6.8)$$

That the number $s_l(E)$ is actually independent of m (as our notation implies) is an immediate consequence of the Wigner–Eckart theorem, but can be proved directly by noting that since S commutes with $\mathbf{L}$ it commutes with the raising and lowering operators $L_\pm$. Thus, $L_- S L_+ = S L_- L_+$. Taking matrix elements of this equation we find, after a little algebra, that,

$$\langle \ldots, m+1 \, |\mathsf{S}| \ldots, m+1\rangle = \langle \ldots, m \, |\mathsf{S}| \ldots, m\rangle$$

that is, the matrix element is independent of m.

The number $s_l(E)$ on the right hand side of (6.8) is just the eigenvalue of S belonging to the eigenvector $|E, l, m\rangle$. Since S is unitary, each of its eigenvalues has modulus 1 and can be written as the exponent of a purely

---

[7] That is, there is an orthonormal basis each vector of which is a simultaneous eigenvector of all three operators and is uniquely identified by its corresponding eigenvalues.

[8] We follow the conventions of Messiah (1961) (Appendix B) for Bessel functions and spherical harmonics. The *phase* of the wave function (6.6) is purely a matter of convention. The factor i^l is inserted for convenience in discussing time-reversal invariance.

imaginary number. We can therefore rewrite (6.8) as:

$$\langle E', l', m'| \; S \; |E, l, m\rangle = \delta(E' - E)\delta_{l'l}\delta_{m'm}e^{2i\delta_l(E)} \qquad (6.9)$$

where the factor of 2 is inserted so that the real number $\delta_l(E)$ is just the conventional *phase shift*, as we shall show in Chapter 11. For the moment we shall adopt this equation as our definition of the phase shift $\delta_l(E)$. It should be noted that this defines $\delta_l(E)$ only to within addition of an arbitrary multiple of π.

The angular momentum basis $\{|E, l, m\rangle\}$ and the momentum basis $\{|\mathbf{p}\rangle\}$ are the only two bases of any great importance for the expansion of the asymptotic free states. (When the particles have spin there are, of course, several possibilities within each category, corresponding to the various possible choices of spin basis.) The importance of the angular-momentum, or "partial-wave," basis is that it diagonalizes S. This means, as we have just seen, that properties such as unitarity have an especially simple form in this representation. The importance of the momentum basis is that the amplitude $f(\mathbf{p}' \leftarrow \mathbf{p})$ defined in this basis is directly related to the observable cross section, as we have already seen.

One can pass from one representation to the other using the appropriate transformation matrix; in this case [9]

$$\langle \mathbf{p} \; | \; E, l, m\rangle = (mp)^{-\frac{1}{2}} \, \delta(E_p - E)Y_l^m(\hat{\mathbf{p}}) \qquad (6.10)$$

If, for example, we wish to express $\langle \mathbf{p}'| \, S \, |\mathbf{p}\rangle$ [or rather the amplitude $f(\mathbf{p}' \leftarrow \mathbf{p})$] in terms of the partial-wave matrix elements, we proceed in the standard manner as follows: [Since f is proportional to the matrix elements of $(S - 1)$, we work with $(S - 1)$ rather than S.] First,

$$\langle \mathbf{p}'| \, (S - 1) \, |\mathbf{p}\rangle = \frac{i}{2\pi m} \, \delta(E_{p'} - E_p)f(\mathbf{p}' \leftarrow \mathbf{p}) \qquad (6.11)$$

(This is simply the familiar decomposition of the S matrix written in terms of f.) To express this in terms of the partial-wave matrix elements we insert a complete set of states $|E, l, m\rangle$ into the left hand side to give:

$$= \int dE \sum_{l,m} \langle \mathbf{p}'| \, (S - 1) \, |E, l, m\rangle\langle E, l, m \; | \; \mathbf{p}\rangle$$

Now, $|E, l, m\rangle$ is an eigenvector of $(S - 1)$, which can therefore be replaced by the corresponding eigenvalue $(s_l - 1)$. Each of the remaining two

---

[9] This follows from the well known expansion of a plane wave in terms of spherical harmonics and Bessel functions, (11.12).

factors is simply the transformation matrix (6.10) and some simple algebra reduces them to:

$$= \frac{1}{mp} \delta(E_{p'} - E_p) \sum_{l,m} Y_l^m(\hat{\mathbf{p}}')[s_l(E_p) - 1]Y_l^m(\hat{\mathbf{p}})^* \qquad (6.12)$$

Comparing the two expressions (6.11) and (6.12) we obtain the desired answer,

$$f(\mathbf{p}' \leftarrow \mathbf{p}) = \frac{2\pi}{ip} \sum_{l,m} Y_l^m(\hat{\mathbf{p}}')[s_l(E_p) - 1]Y_l^m(\hat{\mathbf{p}})^*$$

By choosing $\mathbf{p}$ along the z axis, in which case only $Y_l^0(\hat{\mathbf{p}})$ is nonzero, we can rewrite this as:[10]

$$f(E_p, \theta) \equiv f(\mathbf{p}' \leftarrow \mathbf{p}) = \frac{1}{2ip} \sum_{l=0}^{\infty} (2l + 1)[s_l(E_p) - 1]P_l(\cos\theta) \qquad (6.13)$$

where

$$P_l(\cos\theta) = \left(\frac{4\pi}{2l + 1}\right)^{1/2} Y_l^0(\theta, \varphi)$$

is the Legendre polynomial and as usual θ is the angle between $\mathbf{p}'$ and $\mathbf{p}$. It is usual to introduce a *partial-wave amplitude*, defined as:

$$f_l(E) = \frac{s_l(E) - 1}{2ip} = \frac{e^{2i\delta_l(E)} - 1}{2ip} = \frac{e^{i\delta_l}\sin\delta_l}{p} \qquad (6.14)$$

in terms of which (6.13) becomes:

$$f(E, \theta) = \sum_{l=0}^{\infty} (2l + 1)f_l(E)P_l(\cos\theta) \qquad (6.15)$$

This is the so-called *partial-wave expansion* for the "full" amplitude $f(E, \theta)$ in terms of the partial-wave amplitude $f_l(E)$. It can of course be inverted. In fact, using the well-known orthogonality of the Legendre polynomials we find:

$$f_l(E) = \tfrac{1}{2}\int_{-1}^{1} d(\cos\theta)f(E, \theta)P_l(\cos\theta) \qquad (6.16)$$

[10] This follows equivalently from the spherical harmonic addition theorem

$$\sum_m Y_l^m(\hat{\mathbf{p}}')Y_l^m(\hat{\mathbf{p}})^* = [(2l + 1)/4\pi]P_l(\cos\theta)$$

(see Messiah, 1961, p. 496).

These two results make clear that knowledge of the partial-wave amplitude $f_l(E)$ [or equivalently the phase shifts $\delta_l(E)$] for all l implies knowledge of the full amplitude $f(E, \theta)$ and vice versa.[11]

The partial-wave series (6.15) for $f(E, \theta)$ gives a double series for the differential cross section $d\sigma/d\Omega = |f|^2$ as a sum of products of Legendre polynomials. If this is integrated over all angles the orthogonality of the Legendre polynomials cancels all cross terms and the total cross section is therefore:

$$\sigma(p) = \sum_l \sigma_l(p)$$

where

$$\sigma_l(p) = 4\pi(2l + 1)\,|f_l(p)|^2 = 4\pi(2l + 1)\,\frac{\sin^2 \delta_l}{p^2}$$

Since $|\sin \delta| \leqslant 1$, the maximum contribution of any partial wave to the total cross section is given by:

$$\sigma_l(p) \leqslant 4\pi\,\frac{2l + 1}{p^2}$$

This inequality is often referred to as the *unitarity bound* because it arises from the reality of δ_l, which in turn reflects the unitarity of S. A partial cross section attains its unitarity bound if and only if the phase shift is an odd multiple of $\pi/2$—a situation often associated with a resonance (see Chapter 13). Notice that the unitarity bound on the partial cross sections tells us nothing about the total cross section because insertion into the series $\sigma = \sum \sigma_l$ leads to the trivial result $\sigma(p) \leqslant \infty$.

Probably the principal importance of the partial-wave series is that at low energies only a small number of the phase shifts $\delta_l(E)$ are nonzero (as we shall see in Chapter 11). In this case the infinite series (6.13) reduces to a finite sum and gives a useful parametrization of $f(E, \theta)$ in terms of a small number of real parameters. This parametrization is especially attractive since the reality of the phase shifts ensures that it is automatically consistent with the unitarity of S. In addition it supplies a means of measuring the *phase*, as well as the magnitude, of the amplitude f. For example (to mention the simplest case) it is found that at very low energies differential cross sections become isotropic (in the CM frame). This indicates that at these energies f is independent of θ, and only the $l = 0$ phase shift is nonzero. The

---

[11] Notice that the possibility of expanding f as a series of Legendre polynomials was already clear from the result (6.5) that f is a function of E and θ only. However, the calculation leading from (6.10) to (6.15) was necessary to establish the relation (6.14) of the expansion coefficients $f_l(E)$ to the partial-wave S-matrix elements (or phase shifts).

partial wave series (6.13) then collapses to:

$$f(E, \theta) = \frac{1}{2ip} \, (e^{2i\delta_0(E)} - 1) = \frac{1}{p} \, e^{i\delta_0(E)} \sin \delta_0(E)$$

In this case measurement of $d\sigma/d\Omega = |f|^2$ determines $\sin \delta_0$ within a sign, and we can measure f itself with just one ambiguity.

6-d. Parity

In addition to rotational invariance, many systems satisfy invariance under parity. It is, of course, well known that there are processes (notably those involving the weak interactions) that violate invariance under parity. Nonetheless, there is an enormous class of processes (those involving the electromagnetic and strong interactions) that are apparently consistent with parity invariance. In this short section we explore the consequences of parity invariance for spinless particles. We return to the more interesting case of particles with spin in Section 6-f.

Parity is defined as a reversal of the three spatial axes. More precisely, the parity operator P for any system is defined as that operator (unique up to an overall phase factor) that changes the signs of all positions and momenta but leaves all angular momenta (and the nature of the system itself) unchanged. Here the system of interest contains two spinless particles, for which (as was the case with rotations) we need only consider the space of the relative motion. For this the definition of P implies that

$$\mathsf{P} \, |\mathbf{x}\rangle = \eta \, |-\mathbf{x}\rangle \quad \text{and} \quad \mathsf{P} \, |\mathbf{p}\rangle = \eta \, |-\mathbf{p}\rangle \tag{6.17}$$

where η is an undetermined number of modulus one. Since P in any case contains an arbitrary overall phase, we can redefine it so that[12]

$$\mathsf{P} \, |\mathbf{x}\rangle = |-\mathbf{x}\rangle \quad \text{and} \quad \mathsf{P} \, |\mathbf{p}\rangle = |-\mathbf{p}\rangle. \tag{6.18}$$

For the wave functions this means that if $|\psi_\mathsf{P}\rangle$ denotes $\mathsf{P} \, |\psi\rangle$, then $\psi_\mathsf{P}(\mathbf{x}) = \psi(-\mathbf{x})$; and likewise for the momentum-space wave function.

---

[12] It should perhaps be emphasized that we are here discussing systems in which particles cannot be created or destroyed. The situation when particles *can* be created or destroyed (as occurs typically in relativistic quantum mechanics) is more complicated. In the present case we can consistently confine attention to a system of fixed constituents and (6.17) holds with a *single* fixed number η, which can be absorbed into P. In the relativistic case, transitions occur between systems with different numbers and types of particles. The result (6.17) still holds, but the numbers η may be different for different types of particles. Since redefinition of P changes all of the factors η by the same amount, they cannot (in general) all be removed. In this case the ratio of the numbers η for two different particles (the intrinsic relative parity) is physically significant.

The dynamics are invariant under parity if and only if P commutes with H; or equivalently, $V(\mathbf{x}) = V(-\mathbf{x})$. [Note that this condition is automatically satisfied if $V(\mathbf{x})$ is spherically symmetric.] It follows exactly as in our discussion of rotations that invariance under parity implies that P commutes with S, and, hence, that $S = P^{\dagger}SP$. If $|\phi\rangle$ and $|\phi'\rangle$ are two states of definite parity,

$$P\,|\phi\rangle = p\,|\phi\rangle \quad \text{and} \quad P\,|\phi'\rangle = p'\,|\phi'\rangle$$

then $\langle\phi'|\,S\,|\phi\rangle = 0$ unless $p = p'$; that is, parity is conserved. More generally, for any initial and final states $\langle\phi'|\,S\,|\phi\rangle = \langle\phi'_P|\,S\,|\phi_P\rangle$. In particular,

$$\langle\mathbf{p}'|\,S\,|\mathbf{p}\rangle = \langle-\mathbf{p}'|\,S\,|-\mathbf{p}\rangle$$

and, hence, for the amplitude we have the very natural result

$$f(\mathbf{p}' \leftarrow \mathbf{p}) = f(-\mathbf{p}' \leftarrow -\mathbf{p}) \qquad \text{[P invariance]} \qquad (6.19)$$

This result clearly shows that invariance under parity is automatic if our system (of two spinless particles) is rotationally invariant. This is because a rotation through π about a direction perpendicular to $\mathbf{p}$ and $\mathbf{p}'$ carries $\mathbf{p}$ to $-\mathbf{p}$ and $\mathbf{p}'$ to $-\mathbf{p}'$; thus rotational invariance implies (6.19). However, as we shall see later, rotational invariance and invariance under parity are in general quite independent.

6-e. Time Reversal

So far we have discussed three types of symmetry operations—displacements, rotations, and parity—all of which are represented by unitary operators. We must now discuss a symmetry that is *not* given by a unitary operator: time reversal. To understand time reversal we must go back briefly to discuss some general properties of symmetry operators.

Wigner's Theorem and Anti-Unitary Operators. The effect of an invariance transformation on the state vectors of any quantum mechanical system is determined by a fundamental theorem of Wigner.[13] Wigner's theorem establishes that any symmetry is given by a *unitary or anti-unitary* operator on $\mathscr{H}$. More precisely, if a symmetry carries each state $|\psi\rangle$ into a transformed state $|\psi'\rangle$, then *either* it is possible to adjust the arbitrary phase in $|\psi'\rangle$ such that

$$|\psi'\rangle = U\,|\psi\rangle \qquad \text{[all } |\psi\rangle\text{]}$$

---

[13] The theorem was originally given by Wigner in the early thirties (See Wigner, 1959, Appendix to Chapter 20). It is discussed by Messiah (1961) Theorem III, p. 633, and by Gottfried (1967) p. 226.

where U is unitary, *or* it is possible to adjust the phases so that

$$|\psi'\rangle = W\,|\psi\rangle \qquad [\text{all } |\psi\rangle]$$

where W is an *anti-unitary* operator (which we shall define shortly). These two possibilities are exclusive; for a given symmetry just one alternative holds. Whichever alternative applies, the operator is uniquely determined apart from the inevitably arbitrary overall phase.

Associated with Wigner's theorem are various tests that determine whether a given symmetry operator is unitary or anti-unitary. Application of these tests shows that displacements, rotations and parity are given by unitary operators (results we have already used) but that time reversal is anti-unitary.[14] To discuss time-reversal invariance we must briefly review the properties of anti-unitary operators, which are defined as follows:

An anti-unitary operator W is a one-to-one map of $\mathcal{H}$ onto $\mathcal{H}$ that is norm-preserving and *anti-linear;* that is,

$$W(a\,|\psi\rangle + b\,|\phi\rangle) = a^*W\,|\psi\rangle + b^*W\,|\phi\rangle.$$

This differs from the definition of a unitary operator only in that W is anti-linear.

Some of the properties of anti-linear operators are very straightforward. For instance, it is easy to see that the product of two anti-linear operators is itself linear, while the product of one linear and one anti-linear operator is anti-linear.

The only difficulty with anti-linear operators arises in connection with the Dirac notation for matrix elements. This notation is designed so that for a *linear* operator A, the quantity $\langle\phi|\,A\,|\psi\rangle$ can be viewed in two ways:

$$\langle\phi|\,A\,|\psi\rangle = \langle\phi|\,(A\,|\psi\rangle) = ((\langle\phi|\,A)\,|\psi\rangle$$

where the bra $\langle\phi|\,A$ by definition corresponds to the ket $A^\dagger\,|\phi\rangle$. We can emphasize this point by rewriting the identity using the mathematicians' notation $\langle\ ,\ \rangle$ for the scalar product

$$\langle\phi|\,A\,|\psi\rangle = \langle\phi, A\psi\rangle = \langle A^\dagger\phi, \psi\rangle \tag{6.20}$$

Unfortunately if A is anti-linear these alternative views of $\langle\phi|\,A\,|\psi\rangle$ are no longer equivalent. This is because there cannot exist an operator $A^\dagger$ satisfying

---

[14] This can be shown in various ways. For example, one can consider the basic commutation relation $[X_i, P_j] = i\delta_{ij}$. By definition the time-reversal operator T has the effect $T^\dagger X T = X$ and $T^\dagger P T = -P$. Thus, if we sandwich the basic commutation relation between $T^\dagger$ and T the left hand side changes sign. It follows that the right side must also change sign; that is, $T^\dagger i T = -i$. This is obviously impossible if T is unitary, but *is* satisfied if T is anti-unitary (as we shall see shortly). An alternative and more general test uses the fact that the spectrum of H is bounded from below but not from above (see Jordan, 1969, p. 126).

the second equality of (6.20). (This is easy to understand. The quantity $\langle \phi, A\psi \rangle$ is anti-linear in $|\psi\rangle$, whereas $\langle A^\dagger \phi, \psi \rangle$ would have to be linear in $|\psi\rangle$ whatever $A^\dagger$ is.) In place of the definition:

$$\langle \phi, A\psi \rangle = \langle A^\dagger \phi, \psi \rangle \qquad [A \text{ linear}]$$

for $A^\dagger$, one defines the adjoint of an anti-linear operator by the relation,

$$\langle \phi, A\psi \rangle = \langle A^\dagger \phi, \psi \rangle^* \qquad [A \text{ anti-linear}] \qquad (6.21)$$

(It is easy to check that this does define an operator $A^\dagger$ and that $A^\dagger$ is anti-linear.) If one wishes to apply Dirac's notation to anti-linear operators one must distinguish between the brackets $\langle \phi | (A |\psi\rangle)$ and $(\langle \phi | A) |\psi\rangle$. An alternative, which we shall follow, is always to revert to the more precise mathematicians' notation $\langle \phi, A\psi \rangle$ whenever one has occasion to discuss anti-linear operators.

Returning to the anti-unitary operator W it is easy to use the definition (6.21) of W† together with the condition that W preserves the norm and maps $\mathscr{H}$ onto the whole of $\mathscr{H}$ to show that:

$$\mathsf{W}^\dagger \mathsf{W} = \mathsf{W} \mathsf{W}^\dagger = 1$$

The only difference between this and the corresponding equation $\mathsf{U}^\dagger \mathsf{U} = \mathsf{U} \mathsf{U}^\dagger = 1$ is that the individual operators W and W† are anti-linear. Their product is of course linear.

Time Reversal. Having established the main properties of anti-unitary operators in general we can now return to the example of interest, the time-reversal operator T. This operator is defined to change the sign of the momenta and spins of all particles but to leave their positions unchanged. Wigner's theorem implies that T must be anti-unitary and, for a single spinless particle, this implies that (after suitable adjustment of the arbitrary overall phase of T) T $|\mathbf{x}\rangle = |\mathbf{x}\rangle$ and T $|\mathbf{p}\rangle = |-\mathbf{p}\rangle$. If we expand any $|\psi\rangle$ as

$$|\psi\rangle = \int d^3x\, \psi(\mathbf{x}) \,|\mathbf{x}\rangle \qquad (6.22)$$

and remember that T is anti-linear, we find for T $|\psi\rangle$ that

$$|\psi_T\rangle \equiv \mathsf{T} \,|\psi\rangle = \mathsf{T} \int d^3x\, \psi(\mathbf{x}) \,|\mathbf{x}\rangle$$

$$= \int d^3x\, \psi(\mathbf{x})^* \mathsf{T} \,|\mathbf{x}\rangle$$

$$= \int d^3x\, \psi(\mathbf{x})^* \,|\mathbf{x}\rangle \qquad (6.23)$$

Comparing (6.22) for $|\psi\rangle$ and (6.23) for $|\psi_T\rangle$ we see that the effect of T (for a spinless particle) is simply to replace the spatial wave function by its complex conjugate,

$$\psi_T(\mathbf{x}) = \psi(\mathbf{x})^*$$

It should be noted that the operation of complex conjugation of the wave function in one representation is not generally the same as the corresponding operation in another representation. For example, it is easily seen that in momentum space

$$\psi_T(\mathbf{p}) = \psi(-\mathbf{p})^*$$

that is, complex conjugation of $\psi(\mathbf{x})$ is equivalent to complex conjugation of the momentum-space wave function *plus* a simultaneous change in the sign of $\mathbf{p}$.

Invariance under time reversal means that T commutes with H (and automatically with H^0), $TH = HT$. (For a spinless particle in a local potential, this requires simply that V be real, which it has to be anyway in order that H be Hermitian. Thus, in the present case T invariance is automatic.[15]) Because T is anti-unitary, this means that:

$$Te^{iHt} = e^{-iHt}T$$

(Note well the change of sign, which arises because $Ti = -iT$.) It follows that

$$T\Omega_{\pm} = T\left[\lim_{t\to\mp\infty} e^{iHt}e^{-iH^0t}\right]$$

$$= \left[\lim_{t\to\mp\infty} e^{-iHt}e^{iH^0t}\right]T$$

$$= \Omega_{\mp}T \tag{6.24}$$

or since $T^\dagger T = 1$,

$$\Omega_{\pm} = T^\dagger\Omega_{\mp}T$$

Thus, the effect of T on the Møller operators is to interchange Ω_+ and Ω_-.

From (6.24) it follows that:

$$TS = T\Omega_-^\dagger\Omega_+ = \Omega_+^\dagger T\Omega_+ = \Omega_+^\dagger\Omega_- T = S^\dagger T \tag{6.25}$$

or

$$S = T^\dagger S^\dagger T$$

---

[15] This is because we are confining attention to local potentials. A simple example of an interaction that is not time-reversal invariant is the interaction of a charged particle with a fixed, external, magnetic field. Complex potentials, which are used as models for the elastic part of multichannel scattering, also violate T invariance but this is because they are (in this respect at least) an oversimplified model. It is not because the actual multichannel situation necessarily violates T invariance.

that is, T changes S to its inverse $S^\dagger$. Taking matrix elements of this equation (and reverting to the mathematicians' notation to avoid trouble with the anti-linear operators) we find[16]:

$$\langle \chi, S\phi \rangle = \langle \chi, T^\dagger S^\dagger T\phi \rangle$$
$$= \langle T\chi, S^\dagger T\phi \rangle^*$$
$$= \langle \chi_T, S^\dagger \phi_T \rangle^*$$

or

$$\boxed{\langle \chi| S |\phi \rangle = \langle \phi_T| S |\chi_T \rangle} \cdots \text{[T invariance]} \tag{6.26}$$

This result shows that, as one would expect, time reversal invariance implies that the probability $w(\chi \leftarrow \phi)$ is the same as the probability $w(\phi_T \leftarrow \chi_T)$ for the process in which initial and final states are time reversed and their roles exchanged.

In particular, for one-particle scattering, the result (6.26)—when written in the momentum representation—gives

$$\langle \mathbf{p}'| S |\mathbf{p} \rangle = \langle -\mathbf{p}| S |-\mathbf{p}' \rangle \tag{6.27}$$

or equivalently,

$$\boxed{f(\mathbf{p}' \leftarrow \mathbf{p}) = f(-\mathbf{p} \leftarrow -\mathbf{p}')} \cdots \text{[T invariance]} \tag{6.28}$$

for the amplitude. Just as with parity conservation, this result follows automatically from rotational invariance (as the reader can check). This is because the particles are spinless and we shall see that, in general, T invariance is quite independent of rotational invariance.

6-f.   Invariance Principles for Particles with Spin; Momentum-Space Analysis[17]

We now wish to make an analysis for particles with spin similar to that given in the previous five sections for the spinless case. The basic arguments are exactly as before. For example, if the dynamics are rotationally invariant, then S commutes with the rotation operators and, hence:

$$\langle \chi| S |\phi \rangle = \langle \chi_R| S |\phi_R \rangle \tag{6.29}$$

---

[16] In deriving the second line remember that for an antilinear operator $\langle \chi, T^\dagger \psi \rangle = \langle T\chi, \psi \rangle^*$ as opposed to the usual result $\langle \chi, A^\dagger \psi \rangle = \langle A\chi, \psi \rangle$ for linear operators.

[17] The reader who regards the complications of spin as uninteresting or inessential can safely omit the last two sections of this chapter.

(As usual it is sufficient to consider the relative motion.) However, the situation is more complicated when we try to compute the rotated state $|\phi_R\rangle$ in terms of $|\phi\rangle$. For example, in the spinless case the momentum eigenstates rotate according to the transparent rule: $R\,|\mathbf{p}\rangle = |\mathbf{p}_R\rangle$; that is, the effect of rotating the basis vector of momentum $\mathbf{p}$ is to produce precisely the basis vector with rotated momentum $\mathbf{p}_R$. However, if we consider the momentum eigenstate $|\mathbf{p}, \xi\rangle$ for the relative motion of two particles with spin, then the rotation R will carry $\mathbf{p}$ to $\mathbf{p}_R$ and *simultaneously rotate the spin state*. Thus, for a general basis vector $|\mathbf{p}, \xi\rangle$ the rotated state $R\,|\mathbf{p}, \xi\rangle$ will be a certain linear combination of the $(2s_1 + 1)(2s_2 + 1)$ basis vectors $|\mathbf{p}_R, \xi'\rangle$, and the result (6.29) is generally rather complicated when expressed in terms of the basic amplitudes $f_{\xi'\xi}(\mathbf{p}' \leftarrow \mathbf{p})$.

Our task in the remainder of this chapter is to find practical and simple ways of expressing the constraints imposed on the amplitudes by invariance under the symmetries R, P, and T. Just as in the spinless case there are basically two approaches, one in terms of the momentum representation, and the other in terms of the angular-momentum (or partial-wave) representation.

Action of the Symmetry Operators.   Our first task is to determine the action of the various symmetry operators on the basic spin states, and we naturally begin by considering just the $(2s_1 + 1)(2s_2 + 1)$-dimensional spin space by itself. The effect of any rotation R on an arbitrary spin state $|\chi\rangle$ we denote by:

$$|\chi\rangle \xrightarrow{\;R\;} |\chi_R\rangle = R_s\,|\chi\rangle$$

where we have temporarily introduced the notation R_s to denote the unitary rotation operator on the spin space alone,

$$R_s(\boldsymbol{\alpha}) = e^{-i\boldsymbol{\alpha}\cdot\mathbf{S}}$$

(S being the total spin operator). If we consider a basis $\{|\xi\rangle\}$ of the spin space, then the effect of R_s on the basis vectors can be written as:

$$R_s\,|\xi\rangle = \sum_{\xi'} |\xi'\rangle\langle\xi'|\,R_s\,|\xi\rangle \qquad (6.30)$$

which simply says (as we already knew) that, with respect to any basis, R_s is given by a unitary *matrix* with elements $\langle\xi'|\,R_s\,|\xi\rangle$.

For future reference, it should be noted that for rotations about the z axis and for states that are eigenstates of S_3 (for example, $|s, m\rangle$) we have the simple result:

$$R_s(\alpha\hat{\mathbf{3}})\,|s, m\rangle = e^{-i\alpha S_3}\,|s, m\rangle = e^{-i\alpha m}\,|s, m\rangle \qquad (6.31)$$

That is, a rotation about the z axis simply changes the phase of the eigenstates of S_3. A general rotation will of course mix the various eigenstates, as in (6.30).

The action of a rotation R on the complete state vector (of the relative motion) can be immediately determined. If we consider, for example, the momentum-eigenstate $|\mathbf{p}, \xi\rangle = |\mathbf{p}\rangle \otimes |\xi\rangle$ then R sends $|\mathbf{p}\rangle$ to $|\mathbf{p_R}\rangle$, while it transforms $|\xi\rangle$ as in (6.30). Thus,

$$R\,|\mathbf{p}, \xi\rangle = R(|\mathbf{p}\rangle \otimes |\xi\rangle) = |\mathbf{p_R}\rangle \otimes \sum_{\xi'} |\xi'\rangle\langle\xi'|\, R_s\, |\xi\rangle$$

$$= \sum_{\xi'} |\mathbf{p_R}, \xi'\rangle\langle\xi'|\, R_s\, |\xi\rangle$$

For the general rotation this is, as anticipated, a linear combination of the vectors $|\mathbf{p_R}, \xi'\rangle$.

The effects of the discrete operators P and T are easily determined. Since P commutes with all angular momenta it leaves all spin states unchanged, and since P reverses all momenta, it follows that

$$P\,|\mathbf{p}, \xi\rangle = |-\mathbf{p}, \xi\rangle \tag{6.32}$$

The time-reversal operator reverses all angular momenta. With the usual conventions it can be shown that its effect on the eigenstates $|s, m\rangle$ of total spin (for example) is[18]:

$$T\,|s, m\rangle = (-)^{s-m}\,|s, -m\rangle \tag{6.33}$$

and hence,

$$T\,|\mathbf{p}, s, m\rangle = (-)^{s-m}\,|-\mathbf{p}, s, -m\rangle \tag{6.34}$$

Rotational Invariance. Now, if we consider the scattering of two particles and assume rotational invariance, we note that since $S = R^\dagger S R$,

$$\langle\mathbf{p}', \xi'|\, S\, |\mathbf{p}, \xi\rangle = \langle\mathbf{p}', \xi'|\, R^\dagger S R\, |\mathbf{p}, \xi\rangle \qquad [\text{any R}] \tag{6.35}$$

Because one can always choose a rotation R that brings any given $\mathbf{p}$ onto the z axis, this result lets us express the amplitudes for arbitrary $\mathbf{p}$ in terms of those for which the initial relative momentum lies along the z axis. In other words, we lose no essential generality if we assume that $\mathbf{p}$ is parallel to $\hat{\mathbf{3}}$.

Having fixed $\mathbf{p}$ along $\hat{\mathbf{3}}$ we can write the amplitude matrix as

$$F(\mathbf{p}' \leftarrow \mathbf{p}) = F(E_p, \theta, \varphi) \qquad [\hat{\mathbf{p}} = \hat{\mathbf{3}} \text{ and } \hat{\mathbf{p}}' = (\theta, \varphi)] \tag{6.36}$$

We now restrict attention to the case where the basic spin states $|\xi\rangle$ are eigenstates of S_3. Either $\xi = (s, m)$ or $\xi = (m_1, m_2)$ will do, and for definiteness we take $\xi = (s, m)$. In this case we can use rotational invariance about the z axis to determine each amplitude's dependence on the azimuth

[18] It should be emphasized that the phase factor here is substantially a matter of convention, both because T has an arbitrary overall phase and because the relative phase of the different vectors $|s, m\rangle$ is itself a matter of convention. For further discussion of (6.33) see Problem **6.3**.

φ. To understand how this works we note that a rotation about the z axis leaves the spin states (s, m) physically unchanged; this was the content of (6.31). Thus, if we consider a process $(\mathbf{p}', s', m' \leftarrow \mathbf{p}, s, m)$, (with $\mathbf{p}$ along the z axis) and if we make a rotation about the z axis, then the only thing that is changed by the rotation is the azimuth φ of the final momentum. Because the system is rotationally invariant, this means that the cross section for $(\mathbf{p}', s', m' \leftarrow \mathbf{p}, s, m)$ must be independent of φ. It follows that the corresponding amplitude has at most a φ-dependent phase factor, which we can determine as follows:

The final momentum $\mathbf{p}'$ with direction (θ, φ) can be rotated onto $\mathbf{p}''$ with direction $(\theta, 0)$ by a rotation through $-\varphi$ about $\hat{\mathbf{3}}$. Specifically, from (6.31):

$$R(-\varphi\hat{\mathbf{3}}) |\mathbf{p}', s', m'\rangle = e^{im'\varphi} |\mathbf{p}'', s', m'\rangle$$

while, since $\mathbf{p}$ lies along the z axis,

$$R(-\varphi\hat{\mathbf{3}}) |\mathbf{p}, s, m\rangle = e^{im\varphi} |\mathbf{p}, s, m\rangle$$

Substituting these back into (6.35) we find:

$$\langle \mathbf{p}', s', m'| S |\mathbf{p}, s, m\rangle = e^{i(m-m')\varphi}\langle \mathbf{p}'', s', m'| S |\mathbf{p}, s, m\rangle$$

and the corresponding relation for the amplitudes

$$f_{s'm',sm}(E, \theta, \varphi) = e^{i(m-m')\varphi}f_{s'm',sm}(E, \theta, 0) \qquad (6.36)$$

Thus, rotational invariance completely fixes the dependence of the elements of $F(E, \theta, \varphi)$ on φ. The problem of computing F is reduced to that of computing $[(2s_1 + 1)(2s_2 + 1)]^2$ functions of E and θ only.

We can illustrate this result with the important example of (spin $\frac{1}{2}$) − (spin 0) scattering discussed in Chapter 5. In this case the spin space is the two-dimensional space of the spin-half projectile with basis vectors $|+\rangle$ and $|-\rangle$, and the amplitude matrix $F(\mathbf{p}' \leftarrow \mathbf{p})$ is a 2×2 matrix with elements $f_{++}, f_{+-}, f_{-+}, f_{--}$. If we fix $\mathbf{p}$ along the z axis we can write $F(\mathbf{p}' \leftarrow \mathbf{p})$ as $F(E, \theta, \varphi)$, and then the result (6.36) shows that the most general form of F compatible with rotational invariance is:

$$F(E, \theta, \varphi) = \begin{pmatrix} a(E, \theta) & b(E, \theta)e^{-i\varphi} \\ c(E, \theta)e^{i\varphi} & d(E, \theta) \end{pmatrix} \qquad \text{[R invariance]} \qquad (6.37)$$

That is, the scattering is completely determined by four complex functions a, b, c, d of E and θ. Rotational invariance has nothing further to tell us concerning these functions.

Parity.   Let us next suppose that our system is invariant under rotations *and parity* (which we believe to be true of any process for which weak interactions are insignificant), and for simplicity let us continue to discuss the

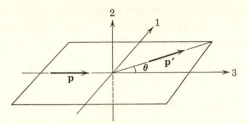

FIGURE 6.1. If initial and final momenta p and p′ lie in the x–z plane, rotation through π about the y axis reverses their signs. The combined effect of this rotation and parity leaves both momenta unchanged.

scattering of a spin-half projectile off a spinless particle. For this case it turns out that only two of the four functions a, b, c, d in (6.37) are actually independent. To see this we consider first a process $(\mathbf{p}', m' \leftarrow \mathbf{p}, m)$ with $\mathbf{p}$ along $\hat{3}$, as before, and $\mathbf{p}'$ lying in the half plane $\varphi = 0$ (Fig. 6.1). The effect of parity alone is to replace $\mathbf{p}$ and $\mathbf{p}'$ by $-\mathbf{p}$ and $-\mathbf{p}'$, which is inconvenient in view of our agreement to keep the initial momentum along $\hat{3}$. To get around this, we combine parity with a rotation through π about the y axis, which swings both momenta back to their original directions. The combined effect of P and $R(\pi\hat{2})$ is equivalent to a reflection through the x–z plane, which we denote by $M_2 = R(\pi\hat{2})P$ (M stands for "mirror"). Since P leaves the spin invariant while $R(\pi\hat{2})$ reverses its third component, invariance under M_2 should relate the two processes $(\mathbf{p}', m' \leftarrow \mathbf{p}, m)$ and $(\mathbf{p}', -m' \leftarrow \mathbf{p}, -m)$. Thus, of the four *a priori* independent processes $(\mathbf{p}', \pm \leftarrow \mathbf{p}, \pm)$, only two should in fact be independent. To check this, we proceed as above starting with the equation:

$$S = M_2^\dagger S M_2 \tag{6.38}$$

and taking matrix elements.

To find the effect of the reflection M_2 on the states $|\mathbf{p}, m\rangle$, we need to know the effect of the rotation $R(\pi\hat{2})$ on the spin state $|m\rangle$. As already discussed, this changes m to $-m$, and, in fact, it can be shown that

$$R(\pi\hat{2}) |m\rangle = (-)^{s-m} |-m\rangle \tag{6.39}$$

and, hence, that $R(\pi\hat{2}) |\mathbf{p}, m\rangle = (-)^{s-m} |-\mathbf{p}, -m\rangle$ (for $\mathbf{p}$ in the x–z plane). (See Problem **6.3**.) Combining this with the result (6.32) for the parity operator, we find that

$$M_2 |\mathbf{p}, m\rangle = (-)^{s-m} |\mathbf{p}, -m\rangle$$

(once again for any $\mathbf{p}$ in the x–z plane). Thus, taking matrix elements of (6.38) we find that

$$\langle \mathbf{p}', m'| \, S \, |\mathbf{p}, m\rangle = (-)^{m'-m} \langle \mathbf{p}', -m'| \, S \, |\mathbf{p}, -m\rangle$$

The corresponding relations for the amplitudes are $f_{++} = f_{--}$ and $f_{+-} = -f_{-+}$ or, in terms of the functions a, b, c, d of (6.37), $a = d$ and $b = -c$. Inserting these relations into (6.37) we find for the most general form of F compatible with invariance under *rotations and parity*:

$$F(E, \theta, \varphi) = \begin{pmatrix} a(E, \theta) & b(E, \theta)e^{-i\varphi} \\ -b(E, \theta)e^{i\varphi} & a(E, \theta) \end{pmatrix} \qquad \text{[R and P invariance]} \quad (6.40)$$

In this case the amplitude matrix is determined by just two complex functions of E and θ.[19]

It is clear from (6.40) that the cross section for a process involving definite spin states *can* depend on the azimuth φ. If, for instance, we consider an incident beam that is transversely polarized in the x direction, with incident spinor

$$\chi = \frac{1}{\sqrt{2}} \begin{pmatrix} 1 \\ 1 \end{pmatrix}$$

then the cross section for the particles to emerge with polarization unchanged is:

$$\frac{d\sigma}{d\Omega} (\mathbf{p}', \chi \leftarrow \mathbf{p}, \chi) = |\chi^\dagger F \chi|^2 = |a|^2 - 2(\text{Im } ab^*)\sin \varphi + |b|^2 \sin^2 \varphi$$

This result is in sharp contrast to that for spinless particles, for which rotational invariance implies a cross section independent of φ. The difference is, of course, that any transverse polarization of the particles with spin defines a preferred azimuthal direction—something which cannot occur in the spinless case. On the other hand, it should be clear that the *unpolarized* cross section for particles with spin has to be independent of φ, since in this case there can be no preferred azimuth. This is borne out by (6.40), from which we find

$$\frac{d\sigma}{d\Omega} (\mathbf{p}' \leftarrow \mathbf{p}) = \tfrac{1}{2}\{|f_{++}|^2 + |f_{+-}|^2 + |f_{-+}|^2 + |f_{--}|^2\}$$

$$= |a(E, \theta)|^2 + |b(E, \theta)|^2 \qquad \begin{bmatrix} \text{in spins unpolarized;} \\ \text{out spins not monitored} \end{bmatrix}$$

An Alternative Approach.   The approach we have used so far does not lend itself conveniently to a discussion of T invariance. This is because T relates the processes $(\mathbf{p}', \chi' \leftarrow \mathbf{p}, \chi)$ and $(-\mathbf{p}, \chi_T \leftarrow -\mathbf{p}', \chi_T')$ (where χ_T denotes the spinor obtained from χ by time reversal). If we wish to relate the latter process to one in which the initial momentum is along the z axis we must

---

[19] There are several alternative notations for the functions a and b. The most popular is g and $-ih$; Goldberger and Watson (1964) use f and $g \sin \theta$.

now make a rotation, and the required rotation depends on $\mathbf{p}'$. Rather than pursue this problem, we turn to an alternative approach. For simplicity, we continue to discuss the special case of (spin $\frac{1}{2}$) — (spin 0) scattering.

First, we note that some simple algebra allows us to rewrite (6.40) as

$$F(\mathbf{p} \leftarrow \mathbf{p}) = a\begin{pmatrix} 1 & 0 \\ 0 & 1 \end{pmatrix} + ib\begin{pmatrix} 0 & -\sin\varphi - i\cos\varphi \\ -\sin\varphi + i\cos\varphi & 0 \end{pmatrix}$$

or

$$\boxed{F(\mathbf{p}' \leftarrow \mathbf{p}) = a(E_p, \theta)I + ib(E_p, \theta)\hat{\mathbf{n}} \cdot \boldsymbol{\sigma}} \quad \cdots \text{ [R and P invariance]} \quad (6.41)$$

where I denotes the unit (2×2) matrix, $\hat{\mathbf{n}}$ is the unit vector in the direction of $\mathbf{n} = \mathbf{p} \times \mathbf{p}'$ [that is, $\hat{\mathbf{n}}$ is the unit normal to the scattering plane, $\hat{\mathbf{n}} = (-\sin\varphi, \cos\varphi, 0)$] and $\boldsymbol{\sigma}$ denotes the three Pauli matrices,

$$\sigma_1 = \begin{pmatrix} 0 & 1 \\ 1 & 0 \end{pmatrix}, \qquad \sigma_2 = \begin{pmatrix} 0 & -i \\ i & 0 \end{pmatrix}, \qquad \sigma_3 = \begin{pmatrix} 1 & 0 \\ 0 & -1 \end{pmatrix}$$

This form for F suggests the following alternative argument. Any (2×2) matrix can be expanded as a linear combination of the three Pauli matrices and the unit matrix. Thus, we can certainly write (for any $\mathbf{p}'$ and $\mathbf{p}$)

$$F(\mathbf{p}' \leftarrow \mathbf{p}) = a(\mathbf{p}', \mathbf{p})I + i\boldsymbol{\beta}(\mathbf{p}', \mathbf{p}) \cdot \boldsymbol{\sigma} \qquad (6.42)$$

for some functions $\alpha(\mathbf{p}', \mathbf{p})$ and $\beta_i(\mathbf{p}', \mathbf{p})$, $(i = 1, 2, 3)$. (In the general case we can expand F in terms of $[(2s_1 + 1)(2s_2 + 1)]^2$ suitably chosen matrices. Obviously the simple properties of the familiar Pauli matrices make the spin-half case especially easy.) To determine the nature of the expansion coefficients α and β we must of course consider how F behaves under rotations, which we do now. As usual we denote the effects of a rotation R by

$$\mathbf{p} \xrightarrow{\text{R}} \mathbf{p}_R \quad \text{and} \quad \chi \xrightarrow{\text{R}} \chi_R = \text{R}_s\chi$$

where, since it is now convenient to work with spinors, R_s denotes the $(2s_1 + 1)(2s_2 + 1)$-dimensional rotation matrix on the spin space. (For a moment we can let the spins be arbitrary.) Now, as we saw in Chapter 5, the amplitude for the general process $(\mathbf{p}', \chi' \leftarrow \mathbf{p}, \chi)$ is $\chi'^\dagger F(\mathbf{p}' \leftarrow \mathbf{p})\chi$ and rotational invariance therefore implies that

$$\chi'^\dagger F(\mathbf{p}' \leftarrow \mathbf{p})\chi = \chi_R'^\dagger F(\mathbf{p}_R' \leftarrow \mathbf{p}_R)\chi_R$$

for any χ and χ'. It follows that

$$\boxed{F(\mathbf{p}' \leftarrow \mathbf{p}) = \text{R}_s^\dagger F(\mathbf{p}_R' \leftarrow \mathbf{p}_R)\text{R}_s} \quad \cdots \text{ [R invariance]} \qquad (6.43)$$

which is just the extension to particles with spin of the result $f(\mathbf{p}' \leftarrow \mathbf{p}) = f(\mathbf{p}'_R \leftarrow \mathbf{p}_R)$ of the spinless case.

Substituting the expansion (6.42) into this result we find that

$$\alpha(\mathbf{p}', \mathbf{p})I + i\boldsymbol{\beta}(\mathbf{p}', \mathbf{p}) \cdot \boldsymbol{\sigma} = \alpha(\mathbf{p}'_R, \mathbf{p}_R)I + i\boldsymbol{\beta}(\mathbf{p}'_R, \mathbf{p}_R) \cdot (R_s^\dagger \boldsymbol{\sigma} R_s)$$

Under the transformation $\boldsymbol{\sigma} \to R_s^\dagger \boldsymbol{\sigma} R_s$, the three Pauli matrices transform among themselves like a rotational vector; that is, $R_s^\dagger \sigma_i R_s = O_{ij}(R)\sigma_j$ where O denotes the (3×3) real orthogonal rotation matrix. Thus, comparison of coefficients in the above identity shows that $\alpha(\mathbf{p}', \mathbf{p}) = \alpha(\mathbf{p}'_R, \mathbf{p}_R)$; that is, α is an invariant function of $\mathbf{p}'$ and $\mathbf{p}$; while $\boldsymbol{\beta}(\mathbf{p}'_R, \mathbf{p}_R)$ must be related to $\boldsymbol{\beta}(\mathbf{p}', \mathbf{p})$ like a rotational vector; that is, $\boldsymbol{\beta}$ is a covariant vector function of $\mathbf{p}'$ and $\mathbf{p}$.

We already know that a rotationally invariant function of $\mathbf{p}'$ and $\mathbf{p}$ is really a function of E_p and θ only—see (6.5); thus, we can immediately write $\alpha(\mathbf{p}', \mathbf{p}) = \alpha(E_p, \theta)$, as expected. To find the general form of the vector function $\boldsymbol{\beta}$ we note that the three vectors $\mathbf{p} \times \mathbf{p}' = \mathbf{n}$, $\mathbf{p}' + \mathbf{p}$, and $\mathbf{p}' - \mathbf{p}$ form an orthogonal set, and can therefore be used to expand any vector. Thus, we can write $\boldsymbol{\beta}$ as:

$$\boldsymbol{\beta}(\mathbf{p}', \mathbf{p}) = b\hat{\mathbf{n}} + c(\mathbf{p}' + \mathbf{p}) + d(\mathbf{p}' - \mathbf{p}) \tag{6.44}$$

where b, c, and d may, of course, depend on $\mathbf{p}'$ and $\mathbf{p}$. Now, since $\boldsymbol{\beta}$ is a rotational vector function of $\mathbf{p}'$ and $\mathbf{p}$, a rotation of $\mathbf{p}'$ and $\mathbf{p}$ must produce the same effect as a rotation of $\boldsymbol{\beta}$ itself. This implies that b, c, and d are invariant, and hence functions of E_p and θ only. Thus, returning to (6.42), we find the most general form of F consistent with rotational invariance is

$$F(\mathbf{p}' \leftarrow \mathbf{p}) = aI + i[b\hat{\mathbf{n}} + c(\mathbf{p}' + \mathbf{p}) + d(\mathbf{p}' - \mathbf{p})] \cdot \boldsymbol{\sigma}$$

$$\text{[R invariance]} \tag{6.45}$$

where the four coefficients a, b, c, d are arbitrary functions of E_p and θ [and are certain linear combinations of the four functions a, b, c, d used in (6.37)].

If parity is good, then an argument similar to that preceding (6.43) shows that

$$F(\mathbf{p}' \leftarrow \mathbf{p}) = P_s^\dagger F(-\mathbf{p}' \leftarrow -\mathbf{p})P_s$$

and, hence (since P leaves $\boldsymbol{\sigma}$ unchanged),

$$\boldsymbol{\beta}(\mathbf{p}', \mathbf{p}) = \boldsymbol{\beta}(-\mathbf{p}', -\mathbf{p})$$

Thus, if parity is good the coefficients c and d in (6.44) must be zero, and (6.45) reduces to the expected result (6.41).

Finally, it is easily seen that T invariance implies

$$\boldsymbol{\beta}(\mathbf{p}', \mathbf{p}) = -\boldsymbol{\beta}(-\mathbf{p}, -\mathbf{p}')$$

(see Problem **6.4** and remember that T reverses $\boldsymbol{\sigma}$). This requires that the coefficient d in (6.44) vanish; and the most general form for F compatible with invariance under rotations and T is therefore

$$F(\mathbf{p}' \leftarrow \mathbf{p}) = aI + i[b\hat{\mathbf{n}} + c(\mathbf{p}' + \mathbf{p})] \cdot \boldsymbol{\sigma} \qquad \text{[R and T invariance]} \quad (6.46)$$

Comparison of this result with (6.41) makes clear that, for (spin ½) − (spin 0) scattering, invariance under rotations and P automatically entails invariance under T; but that the converse (i.e., R and T → P) is false. We shall see later that this is a special case and that in general T and P are quite independent.

The general case of two particles with arbitrary spins can be analyzed similarly, but is naturally much more complicated (see Problem **6.5**).

6-g.   Invariance Principles for Particles with Spin; Angular-Momentum Analysis

Just as in the scattering of spinless particles, an alternative way to exploit rotational invariance is to evaluate the S matrix in an angular-momentum basis. In the present case, rotational invariance implies that the scattering operator S of the relative motion commutes with the angular momentum $\mathbf{J} = \mathbf{L} + \mathbf{S}_1 + \mathbf{S}_2$. To exploit this fact we use a basis of eigenvectors of $\mathbf{J}^2$ and J_3. This is constructed in the standard way. We start with the eigenvectors $|s, m\rangle$ of the total spin and form the product of these with the orbital eigenvectors $|E, l, m\rangle$ to give vectors

$$|E, l, s, m_l, m_s\rangle = |E, l, m_l\rangle \otimes |s, m_s\rangle$$

We then "couple" l and s to give

$$|E, l, s, j, m\rangle = \sum_{m_l, m_s} |E, l, s, m_l, m_s\rangle\langle lsm_lm_s \,|\, jm\rangle \qquad (6.47)$$

where $\langle lsm_lm_s \,|\, jm\rangle$ is the appropriate Clebsch–Gordan coefficient.

Because S commutes with $\mathbf{J}$, its matrix elements in this basis have the form:

$$\langle E', l', s', j', m'|\, S\, |E, l, s, j, m\rangle = \delta(E' - E)\delta_{j'j}\delta_{m'm}s^{j}_{l's', ls}(E)$$

$$\text{[R invariance]} \qquad (6.48)$$

This is the analogue of the result given in (6.9) for the spinless case, and for the same reasons as mentioned there, the numbers $s^{j}_{l's', ls}(E)$ are independent of m. The result means that for a rotationally invariant system S is diagonal with respect to E, j, m and for given values of E and j is determined by an m-independent matrix $S^{j}(E)$, made up of the numbers $s^{j}_{l's', ls}(E)$. The columns of the matrix $S^j(E)$ are labelled by the initial pair (l, s), the rows by the final pair (l', s'); the dimension of the matrix is just the number of different values of l and s that can contribute to a given j in (6.47).

Just as in the spinless case, the unitarity of S takes an especially simple form in this representation. In fact, as can be easily checked, it implies simply that each matrix $S^j(E)$ is a unitary matrix. In the special case where $s_1 = s_2 = 0$, the matrix is one-dimensional and reduces to the familiar form $\exp [2i\delta_l(E)]$.

The implications of P and T invariance are easily explored. The effect of P on the orbital eigenstates is well known to be

$$P \,|E, l, m\rangle = (-)^l \,|E, l, m\rangle$$

[This follows from the identity $Y_l^m(-\hat{\mathbf{x}}) = (-)^l Y_l^m(\hat{\mathbf{x}})$.] Since P leaves the spins unchanged, it follows that

$$P \,|E, l, s, j, m\rangle = (-)^l \,|E, l, s, j, m\rangle$$

that is the angular-momentum basis vectors are eigenvectors of P with parity $(-1)^l$. If the dynamics are P invariant and S therefore commutes with P, then S cannot link states of different parity. Thus, those matrix elements for which $l' - l$ is odd must vanish. In particular, if we assume invariance under *parity and rotations*, we have simply:

$$s^j_{l's', ls}(E) = 0, \quad \text{for } (l' - l)\text{ odd} \qquad \text{[R and P invariance]}$$

which approximately halves the number of elements of each matrix $S^j(E)$.

The anti-linear time-reversal operator T has the effect

$$T \,|E, l, s, j, m\rangle = (-)^{j-m} \,|E, l, s, j, -m\rangle$$

(see Problem **6.7**). Since T invariance implies (exactly as before) that $\langle \chi| \,S\, |\phi\rangle = \langle \phi_T| \,S\, |\chi_T\rangle$ we see that the matrix elements for the processes

$$(E', l', s', j', m' \leftarrow E, l, s, j, m) \quad \text{and} \quad (E, l, s, j, -m \leftarrow E', l', s', j', -m')$$

are equal within a factor $(-)^{j-j'-m+m'}$. If we also assume rotational invariance, then the matrix elements are nonzero only when $(E', j', m') = (E, j, m)$ and are independent of m. In terms of the matrix $S^j(E)$ the result then reduces to

$$s^j_{l's', ls}(E) = s^j_{ls, l's'}(E) \qquad \text{[R and T invariance]}$$

that is, the matrix $S^j(E)$ is symmetric.

In summary, rotational invariance implies that the scattering for given E and j is determined by the matrix $S^j(E)$. Invariance under rotations and P implies that those elements of S^j with $l' - l$ odd are zero. Invariance under rotations and T implies that S^j is symmetric.

We can again illustrate our results with the simple case of (spin $\frac{1}{2}$) − (spin 0) scattering. Since in this case only one particle has spin, we have only to couple the one spin, $s = \frac{1}{2}$, to l to give j. The possible values of l are the half-odd-integers $\frac{1}{2}, \frac{3}{2}, \ldots$, each arising from two possible j values, $l = j \pm \frac{1}{2}$. Thus, rotational invariance implies that for each j the scattering is determined by a (2×2) unitary matrix $S^j(E)$. Since the two l values, $l = j \pm \frac{1}{2}$, differ by one, invariance under P means that both off-diagonal elements of S^j are zero. That is, S^j is a diagonal unitary matrix and, hence, has the form:

$$
S^j = \begin{pmatrix} e^{2i\delta^j_+} & 0 \\ 0 & e^{2i\delta^j_-} \end{pmatrix} \begin{matrix} \leftarrow l = j + \frac{1}{2} \\ \leftarrow l = j - \frac{1}{2} \end{matrix} \qquad \text{[R and P invariance]} \qquad (6.49)
$$

Since a diagonal matrix is certainly symmetric, we see again that in this case invariance under rotations and P implies invariance under T. Obviously this is so only because S^j is two-dimensional, and in general P and T are quite independent.

The relation between the momentum and angular-momentum bases can be calculated in the standard way, though it is in general rather complicated. For the example of (spin $\frac{1}{2}$) − (spin 0) scattering a calculation exactly parallel to that of Section 6-c gives the amplitudes in terms of the angular-momentum matrix elements. For the case of maximum symmetry (rotations and P) the result for the spin-nonflip amplitude is

$$
a(E, \theta) = \sum_l [(l + 1)f_l^+(E) + lf_l^-(E)]P_l(\cos \theta) \qquad (6.50)
$$

while the spin-flip amplitude is:

$$
b(E, \theta) = \sum_l [f_l^+(E) - f_l^-(E)]\sin \theta P_l'(\cos \theta) \qquad (6.51)
$$

In these equations we have introduced the partial-wave amplitudes

$$
f_l^j(E) = \frac{e^{2i\delta_l^j(E)} - 1}{2ip} \qquad [p \equiv (2mE)^{\frac{1}{2}}]
$$

with $f_l^\pm \equiv f_l^{l\pm\frac{1}{2}}$, and $P_l'(z)$ is the derivative of the Legendre polynomial $P_l(z)$ (see Problem **6.9**).

Just as in the spinless case, these partial-wave series are especially useful at low energies where usually only the phase shifts δ^j of low l are nonzero. For example, in pion–nucleon scattering the data up to about 200 MeV can be fitted well if one assumes that all phase shifts with $l \geqslant 2$ are zero, which gives an expansion of a and b in terms of just three real parameters δ_0^+, δ_1^+, and δ_1^-.

PROBLEMS

6.1.   (a) Using the definition of an anti-unitary operator W, show that

$$\langle W\phi, W\psi \rangle = \langle \phi, \psi \rangle^*$$

(in contrast to the result $\langle U\phi, U\psi \rangle = \langle \phi, \psi \rangle$ for unitary operators).
(b) Show that $W^\dagger W = 1$ and, hence, that $WW^\dagger = 1$. (Review the proof that $UU^\dagger = 1$ in Section 1-d.)
(c) Show that the operator defined by complex conjugation of the spatial wave function is anti-unitary.

6.2.   Consider scattering of two spinless particles, whose Hamiltonian is *not* necessarily invariant under rotations.
(a) Show that invariance under PT (total inversion) implies that the momentum-space S matrix is symmetric.
(b) What does invariance under PT imply for the S matrix in the angular-momentum representation, $|E, l, m\rangle$? To answer this you will need to show from the definition (6.6) of $|E, l, m\rangle$ that

$$T |E, l, m\rangle = (-)^{l-m} |E, l, -m\rangle \qquad (6.52)$$

Remember that $Y_l^{m*} = (-)^m Y_l^{-m}$.
(c) If the S operator is represented by a symmetric matrix in one representation, is the same necessarily true in every other representation? Discuss.

6.3.   To get some understanding of the irritating phases that crop up under the action of the various symmetry operators, consider a single spin-half particle. In this case our conventions mean simply that the spin is given by the usual Pauli matrices.
(a) Verify that $R(\pi\hat{2}) |m\rangle = (-)^{s-m} |-m\rangle$ as in (6.39).
(b) Obviously $T |\pm\rangle = \eta_\pm |\mp\rangle$ for some phase factors $\eta_\pm$. Use the fact that T reverses *any* spin state (e.g., a state in which the spin points along the positive x axis is reversed so that the spin points along the negative x axis) to prove that $\eta_+ = -\eta_-$. Show that by adjustment of the overall phase of T one can arrange that $T |m\rangle = (-)^{s-m} |-m\rangle$ as in (6.33).
(c) Do the results of (a) and (b) mean that $R(\pi\hat{2})$ and T are really the same operator?

6.4.   Time-reversal invariance implies the equality of the S-matrix elements, and hence amplitudes, for the processes

$$(\mathbf{p}', \chi' \leftarrow \mathbf{p}, \chi) \quad \text{and} \quad (-\mathbf{p}, \chi_T \leftarrow -\mathbf{p}', \chi_T')$$

where $\chi_T = T_s \chi$ is the time-reversed spinor obtained from χ
(a) Deduce that $F(\mathbf{p}' \leftarrow \mathbf{p}) = T_s^\dagger F(-\mathbf{p} \leftarrow -\mathbf{p}')^\dagger T_s$.

(b) Show that for (spin $\frac{1}{2}$) — (spin 0) scattering the coefficient $\boldsymbol{\beta}(\mathbf{p'}, \mathbf{p})$ of (6.42) satisfies $\boldsymbol{\beta}(\mathbf{p'}, \mathbf{p}) = -\boldsymbol{\beta}(-\mathbf{p}, -\mathbf{p'})$ if T invariance holds; and deduce the form (6.46) for the amplitude matrix. (Remember that T_s is anti-unitary and satisfies $\mathsf{T}_s^\dagger \boldsymbol{\sigma} \mathsf{T}_s = -\boldsymbol{\sigma}$. It may help to write $\chi'^\dagger F\chi$ as $\langle \chi', F\chi \rangle$.)

6.5. In (spin $\frac{1}{2}$) — (spin 0) scattering the 2×2 amplitude matrix F was expanded in terms of the four independent matrices I and $\boldsymbol{\sigma}$. For a rotationally invariant system this led to (6.45), with four independent scalar amplitudes. Carry out the corresponding analysis for the (spin $\frac{1}{2}$) — (spin $\frac{1}{2}$) case (e.g., n–p scattering) along the following lines[20]:

The amplitude matrix F must first be expanded in terms of 16 independent matrices. The most convenient choice for these is given by products of the matrices I^1, $\boldsymbol{\sigma}^1$, and I^2, $\boldsymbol{\sigma}^2$ of particles 1 and 2 respectively. This gives

$$F = \alpha I + \beta_i \sigma_i^1 + \gamma_i \sigma_i^2 + \epsilon_{ij} \sigma_i^1 \sigma_j^2$$

(where repeated indices are summed). Show that R invariance implies that α is an invariant scalar function of $\mathbf{p'}$ and $\mathbf{p}$, that β_i and γ_i define vector functions, and ϵ_{ij} a tensor. Use the fact that the three vectors $\mathbf{n} = \mathbf{p} \times \mathbf{p'}$, $\mathbf{q} = \mathbf{p'} - \mathbf{p}$ and $\mathbf{k} = \mathbf{p'} + \mathbf{p}$ span $\mathbb{R}^3$ to express β_i, γ_i and ϵ_{ij} in terms of products of $\mathbf{n}$, $\mathbf{q}$, and $\mathbf{k}$ with scalar coefficients. Show that R invariance alone leaves 16 independent amplitudes, R and P leave 8, R and T 10, and R, P, and T 6. [If the particles are identical (e.g., p–p scattering) this last number drops to five as we shall discuss in Chapter 22.]

6.6. Show that unitarity of the operator S implies that the matrix $S^j(E)$ defined by (6.48) is unitary. (Any proof of more than a few lines is too long.)

6.7. Show that, with our phase conventions, the effect of time reversal on the angular-momentum eigenstates is

$$\mathsf{T} |E, l, s, j, m\rangle = (-)^{j-m} |E, l, s, j, -m\rangle$$

The transformation of the orbital state $|E, l, m_l\rangle$ is given in (6.52) and that of the spin state $|s, m_s\rangle$ in (6.33). The states of interest can be expressed by means of the Clebsch–Gordan series (6.47), and then all you need to know is that the Clebsch–Gordan coefficients (in the Condon and Shortley conventions) are real and satisfy

$$\langle l, s, m_l, m_s \,|\, j, m\rangle = (-)^{j-l-s}\langle l, s, -m_l, -m_s \,|\, j, -m\rangle$$

6.8. Assuming rotational invariance discuss the form of the matrix $S^j(E)$ for (spin $\frac{1}{2}$) — (spin $\frac{1}{2}$) scattering, explaining the labelling of the rows and

---

[20] The classical references for this problem are Wolfenstein (1956) and MacGregor, Moravcsik, and Stapp (1960). The scalar amplitudes that you will derive are called the Wolfenstein parameters.

columns. Explore the additional consequences of invariance under P, T, and P and T. In particular confirm the results of Problem **6.5**.

6.9. Use the methods of Section 6-c to derive the partial-wave series (6.50) and (6.51) for the amplitudes of (spin ½) − (spin 0) scattering under the assumption of R and P invariance. You will obviously need the Clebsch–Gordan coefficients $\langle lsm_l m_s \,|\, jm \rangle$ for coupling l and $s = \frac{1}{2}$ to give j. These are:

$$
\begin{array}{ccc}
 & m_s = +\tfrac{1}{2} & m_s = -\tfrac{1}{2} \\[2mm]
j = l + \tfrac{1}{2} & \left(\dfrac{l + m_l + 1}{2l + 1}\right)^{\!\frac{1}{2}} & \left(\dfrac{l - m_l + 1}{2l + 1}\right)^{\!\frac{1}{2}} \\[4mm]
j = l - \tfrac{1}{2} & -\left(\dfrac{l - m_l}{2l + 1}\right)^{\!\frac{1}{2}} & \left(\dfrac{l + m_l}{2l + 1}\right)^{\!\frac{1}{2}}
\end{array}
$$

7

More about Particles with Spin

7-a   **Polarization and the Density Matrix**

7-b   **The In and Out Density Matrices**

7-c   **Polarization Experiments in (Spin 1/2) − (Spin 0) Scattering**

7-d   **The Helicity Formalism**

7-e   **Some Useful Formulas**

In this chapter[1] we discuss two further topics in the scattering of particles with spin. In Sections 7-a to 7-c we describe the formalism of the spin density matrix, which provides the most satisfactory general method for handling experiments involving unpolarized beams and targets. Then in Sections 7-d and 7-e we discuss the uses of the helicity quantum number, which is defined as the component of a particle's spin in the direction of its motion; we shall see that, because the helicity (unlike the z component of spin) is a rotational invariant, there are considerable advantages to working with a basis of helicity eigenstates instead of the more familiar eigenstates of S_3.

7-a.   Polarization and the Density Matrix

In Chapter 5 we discussed the measurement of cross sections for particles that are completely unpolarized, and we saw that if one wishes to get complete information on the scattering amplitudes it is essential that one do experiments with particles that are polarized. Unfortunately, it is very hard to

---

[1] This whole chapter can be omitted without any serious effect on later chapters

attain a state of complete polarization, and in most experiments the best one can hope for is to have target and beam in known states of *partial polarization*. For example, if one tries to orient the particles (of spin s and magnetic moment μ) in a target by putting them in a magnetic field B, then at any temperature T the particles will distribute themselves among the $(2s + 1)$ states $|m\rangle$ in accordance with the Boltzmann distribution factor $\exp(m\mu B/kT)$. By strengthening B or lowering T, one can increase the proportion of particles occupying the single lowest state; nonetheless, *all* $(2s + 1)$ states will always be occupied by at least some particles. Such situations can still be handled by making suitable averages over spin states (much as discussed in Section 5-d), but in general this can become a very clumsy procedure. By far the best approach to the general problem is to use the machinery of the quantum-mechanical density matrix, as we now describe.

The reader will recall that a quantum-mechanical system that is known to be in a state given by some vector $|\psi\rangle$ is said to occupy the *pure state* $|\psi\rangle$. If, on the other hand, we know only that the system is in one of various states $|\psi^i\rangle$ with probabilities w_i, then we say that the system is in a *mixed state*, which can be conveniently characterized by the self-adjoint *density operator*

$$\rho = \sum_i w_i \, |\psi^i\rangle\langle\psi^i| \tag{7.1}$$

where the weights w_i satisfy $\sum w_i = 1$ and the vectors $|\psi^i\rangle$ are assumed to be normalized (although not necessarily orthogonal). Since the mean result of many measurements of an observable A on the state $|\psi^i\rangle$ is $\langle\psi^i| A |\psi^i\rangle$, it follows that the corresponding result for the mixed state with weights w_i is $\sum w_i\langle\psi^i| A |\psi^i\rangle$ which, with some simple algebra, reduces to the well-known expression:

$$\boxed{\langle A\rangle_\rho = \mathrm{tr}\, A\rho} \tag{7.2}$$

Here tr stands for the trace; that is,

$$\mathrm{tr}\, B = \sum_n \langle n| \, B \, |n\rangle$$

for any orthonormal basis $\{|n\rangle\}$. In the special case where the system is actually in a pure state $|\psi\rangle$,

$$\rho = |\psi\rangle\langle\psi| \qquad \text{[pure state]}$$

and, of course,

$$\langle A\rangle_\rho = \mathrm{tr}\, A\rho = \langle\psi| A |\psi\rangle$$

It is easily checked that the necessary and sufficient condition that a density operator ρ represent a pure state is that it satisfy $\rho^2 = \rho$. Whether the state of a system is pure or mixed it is uniquely characterized by its density operator

ρ. If two systems have the same ρ then all possible measurements on the systems yield the same results, and conversely.[2]

It is easily shown that the density operator defined by (7.1) above satisfies tr $\rho = 1$. However, it is often convenient to relax this normalization condition and to allow the operator ρ of (7.1) to be multiplied by an arbitrary constant; in which case $\langle A \rangle_\rho$ is given by

$$\langle A \rangle_\rho = \frac{\operatorname{tr} A\rho}{\operatorname{tr} \rho}$$

If we consider for our system a particle of spin s (and temporarily ignore its spatial degrees of freedom) then with respect to any orthonormal basis ρ becomes a $(2s + 1)$-dimensional matrix, the *density matrix*. If the particle is in a pure state $|\chi\rangle$ then $\rho = |\chi\rangle\langle\chi|$, which is therefore the density matrix appropriate to a beam or target which is completely polarized. At the opposite extreme is the density matrix for a particle that is completely un-polarized. In Chapter 5 we adopted as our definition of an unpolarized assembly of particles one in which each of the $(2s + 1)$ states of any ortho-normal basis is equally populated. The density matrix describing this situation is just

$$\rho = \sum_m \frac{1}{2s + 1} |m\rangle\langle m| = \frac{1}{2s + 1} I \tag{7.3}$$

where I is the $(2s + 1)$-dimensional unit matrix. That is, the density matrix for an unpolarized system is a multiple of the unit matrix—in agreement with the general result that a density matrix $\rho = cI$ corresponds to complete ignorance of the state of the system.

It is satisfactory that the result (7.3) is independent of what orthonormal basis is used, since we would certainly wish our definition of an unpolarized system to be basis-independent. (This means, for example, that an un-polarized beam of electrons can be regarded as being 50% spin up and 50% spin down, or alternatively as 50% spin right and 50% spin left; either view gives the same density matrix and therefore corresponds to the same experi-mental situation.) Nonetheless, the definition of an unpolarized system as one with the $(2s + 1)$ states of some basis equally occupied is hardly a natural definition. For any typical experimental system (a beam of particles emerging from a cyclotron, a gas of target atoms, etc.) a much more natural definition would be that an unpolarized system is one whose spins are randomly oriented. We can now show that these two definitions are actually equivalent. The characteristic feature of a system of randomly oriented spins is that the measurement of any spin observable should be unaffected by a rotation of all

---

[2] We ignore the slight complications introduced by super-selection rules.

spins. This means that the corresponding density matrix satisfies $\rho = R^{st}\rho R^s$ where R^s is the $(2s + 1)$-dimensional rotation matrix for any rotation. Now, it is a well-known result from the theory of the rotation group (Schur's lemma) that the only $(2s + 1)$-dimensional matrices that commute with all of the rotation matrices are multiples of the unit matrix. Thus, $\rho = cI$ and, taking into account the normalization, we see that ρ has precisely the form (7.3). That is, our two definitions of an unpolarized system lead to the same density matrix and are therefore (as anticipated in Section 5-d) completely equivalent.

To summarize, the density matrix for a particle of spin s is a $(2s + 1)$-dimensional Hermitian matrix ρ, usually normalized to satisfy tr $\rho = 1$. The density matrix appropriate to a completely *polarized* particle is characterized by the property $\rho^2 = \rho$; that for a completely *unpolarized* particle has the unique form $\rho = (2s + 1)^{-1}I$ and any ρ which has neither of these properties corresponds to a state of *partial polarization*.

For the case of a spin-half particle the density-matrix formalism is especially simple. The density matrix is a (2×2) Hermitian matrix and so can be expanded in terms of the three Pauli matrices and the unit matrix,

$$\rho = \lambda(1 + \boldsymbol{\pi} \cdot \boldsymbol{\sigma}) \tag{7.4}$$

where λ and $\boldsymbol{\pi}$ are real. Since the Pauli matrices are traceless, tr $\rho = 2\lambda$ and if we want ρ to be normalized then $\lambda = \frac{1}{2}$. The real vector $\boldsymbol{\pi}$ (which is often denoted by **P**) is called the *polarization vector* and has the following properties (which the reader should check):

(1) The vector $\boldsymbol{\pi}$ is twice the mean value of the spin, $\boldsymbol{\pi} = \langle \boldsymbol{\sigma} \rangle_\rho$, and hence $0 \leqslant |\boldsymbol{\pi}| \leqslant 1$.
(2) If $|\boldsymbol{\pi}| = 1$, then ρ represents a completely polarized pure state, which is an eigenstate of the component of spin along $\boldsymbol{\pi}$.
(3) If $\boldsymbol{\pi} = 0$, the particle is completely unpolarized.
(4) If $0 < |\boldsymbol{\pi}| < 1$, the particle is partially polarized.

In view of these simple properties it is often convenient to identify the spin state of a spin-half particle by the polarization vector $\boldsymbol{\pi}$ in preference to either a spinor or a density matrix.

The extension of the density-matrix formalism to the case of two particles with spins s_1 and s_2 is entirely straightforward. Since the only example of the formalism that we shall consider in detail is that of (spin $\frac{1}{2}$) $-$ (spin 0) scattering, we content ourselves with pointing out that in the general case ρ is obviously a matrix of dimension $(2s_1 + 1)(2s_2 + 1)$.

So far we have ignored the particles' spatial degrees of freedom. If the particles' spatial motion is given by a pure state $|\phi\rangle$ (in $\mathscr{H}_{\text{space}}$) while their spins are in a mixed state given by ρ (acting on $\mathscr{H}_{\text{spin}}$), then we can of course set up an appropriate total density operator, namely $(|\phi\rangle\langle\phi|) \otimes \rho$. However, in practice it is just as convenient to use a hybrid identification in terms of the *vector* $|\phi\rangle$ and the *matrix* ρ. In scattering theory the actual initial state is usually mixed with respect to both spatial and spin degrees of freedom. However, we have already set up a suitable process for averaging over the initial spatial wave packets, and we know that in the evaluation of cross sections we need only consider momentum eigenstates. These we shall identify by the vector $|\mathbf{p}_1, \mathbf{p}_2\rangle$ together with the appropriate spin density matrix ρ.

7-b.   The In and Out Density Matrices

We are now ready to return to the scattering problem. We saw in Section 5-e that if the particles enter a collision with relative momentum $\mathbf{p}$ and pure spin state χ_{in}, and if they are observed to emerge in the direction $\mathbf{p}'$, then the outgoing spin state is given by the spinor

$$\chi_{\text{out}} = F(\mathbf{p}' \leftarrow \mathbf{p})\chi_{\text{in}}$$

If $\|\chi_{\text{in}}\| = 1$, then χ_{out} is normalized to the probability that the particles emerge in the direction $\mathbf{p}'$ (irrespective of spin). Thus, if instead of the pure state χ_{in} the initial spin is a mixture given by the density matrix[3]

$$\rho_{\text{in}} = \sum_i w_i |\chi_{\text{in}}^i\rangle \langle \chi_{\text{in}}^i| \tag{7.5}$$

then the corresponding outgoing density matrix for those particles emerging in the direction $\mathbf{p}'$ is

$$\rho_{\text{out}} = \sum_i w_i |\chi_{\text{out}}^i\rangle \langle \chi_{\text{out}}^i|$$
$$= \sum_i w_i F |\chi_{\text{in}}^i\rangle \langle \chi_{\text{in}}^i| F^\dagger$$

or

$$\boxed{\rho_{\text{out}} = F(\mathbf{p}' \leftarrow \mathbf{p})\rho_{\text{in}}F(\mathbf{p}' \leftarrow \mathbf{p})^\dagger} \tag{7.6}$$

Since the spinors χ_{out} are not normalized, neither is ρ_{out}. In fact tr ρ_{out} has a simple interpretation. We have seen that $\|\chi_{\text{out}}\|^2$ is just the cross section that particles entering with relative momentum $\mathbf{p}$ and with spin χ_{in} be observed to emerge with relative momentum $\mathbf{p}'$ and with any spin. This was the

---

[3] The purist may reasonably object that if I wish to consider ρ as a matrix, then I should use the form $\chi\chi^\dagger$ rather than Dirac's $|\chi\rangle \langle\chi|$. In this case, it seems to me the clarity of Dirac's notation outweighs this slight inconsistency.

content of (5.12), which we rewrite as

$$\frac{d\sigma}{d\Omega}(\mathbf{p}' \leftarrow \mathbf{p}, \chi_{\text{in}}) = \|\chi_{\text{out}}\|^2 = \sum_{\xi} \langle \xi \mid \chi_{\text{out}} \rangle \langle \chi_{\text{out}} \mid \xi \rangle$$

$$= \text{tr} \, |\chi_{\text{out}}\rangle\langle\chi_{\text{out}}| \qquad \text{[out spins not monitored]}$$

The corresponding cross section for the mixed initial state given by (7.5) is obtained by averaging this expression over initial spins with the weights w_i,

$$\frac{d\sigma}{d\Omega}(\mathbf{p}' \leftarrow \mathbf{p}, \rho_{\text{in}}) = \sum_i w_i \frac{d\sigma}{d\Omega}(\mathbf{p}' \leftarrow \mathbf{p}, \chi_{\text{in}}^i)$$

$$= \sum_i w_i \, \text{tr} \, |\chi_{\text{out}}^i\rangle\langle\chi_{\text{out}}^i|$$

or

$$\boxed{\frac{d\sigma}{d\Omega}(\mathbf{p}' \leftarrow \mathbf{p}, \rho_{\text{in}}) = \text{tr} \, \rho_{\text{out}}} \cdots \text{[out spins not monitored]} \qquad (7.7)$$

To summarize, the spin density matrix $\rho_{\text{out}} = F\rho_{\text{in}}F^\dagger$ contains all information relevant to an experiment with initial relative momentum $\mathbf{p}$ and initial spins given by ρ_{in}. Its normalization tr ρ_{out} determines the total number of particles emerging in the direction $\mathbf{p}'$ irrespective of spin. And the result of any spin measurement made on these emerging particles is given by $\langle A \rangle_{\text{out}} = \text{tr}(A\rho_{\text{out}})/\text{tr} \, \rho_{\text{out}}$ for any spin operator A.

In the next section we shall apply these results to the case of (spin ½) − (spin 0) scattering. For this case we already know the general expressions for ρ_{in} in terms of the polarization vector $\boldsymbol{\pi}_{\text{in}}$ and for the amplitude F [(7.4) and (6.41), respectively]. We can therefore write down the corresponding form of ρ_{out} and, hence, express the cross section and final polarization $\boldsymbol{\pi}_{\text{out}}$ directly in terms of the initial polarization $\boldsymbol{\pi}_{\text{in}}$. We shall discuss two experimental applications. In the first the scattering amplitudes are known and measurement of the cross section is used to determine the original polarization $\boldsymbol{\pi}_{\text{in}}$. In the second the amplitudes are unknown and we discuss the fairly elaborate sequence of experiments needed to measure them.

7-c.   Polarization Experiments in (Spin ½) − (Spin 0) Scattering

In the scattering of spin-half particles off a spinless target the most general form for the amplitude matrix F consistent with invariance under rotations and parity was shown in (6.41) to be[4]

$$F(\mathbf{p}' \leftarrow \mathbf{p}) = a(\theta)I + ib(\theta)\hat{\mathbf{n}} \cdot \boldsymbol{\sigma}$$

---

[4] Throughout this section we suppress the variable E in the functions a(E, θ), etc.

where $\hat{\mathbf{n}}$ is the unit vector normal to the scattering plane ($\mathbf{n} = \mathbf{p} \times \mathbf{p}'$). The initial spin state (pure or mixed) can be labelled by the polarization vector $\boldsymbol{\pi}_{in}$, in terms of which the initial density matrix is

$$\rho_{in} = \tfrac{1}{2}[1 + \boldsymbol{\pi}_{in} \cdot \boldsymbol{\sigma}]$$

According to (7.6) the spin density matrix for those particles emerging in the direction $\mathbf{p}'$ is therefore

$$\rho_{out} = F(\mathbf{p}' \leftarrow \mathbf{p})\rho_{in}F(\mathbf{p}' \leftarrow \mathbf{p})^{\dagger}$$
$$= \tfrac{1}{2}(a + ib\hat{\mathbf{n}} \cdot \boldsymbol{\sigma})(1 + \boldsymbol{\pi}_{in} \cdot \boldsymbol{\sigma})(a^* - ib^*\hat{\mathbf{n}} \cdot \boldsymbol{\sigma}) \quad (7.8)$$

With a few lines of tedious algebra and use of the identity

$$\mathbf{u} \cdot \boldsymbol{\sigma} \, \mathbf{v} \cdot \boldsymbol{\sigma} = \mathbf{u} \cdot \mathbf{v} + i\mathbf{u} \times \mathbf{v} \cdot \boldsymbol{\sigma}$$

this expression can be multiplied out to give

$$\rho_{out} = \tfrac{1}{2}(|a|^2 + |b|^2 + 2 \operatorname{Im} ab^*\hat{\mathbf{n}} \cdot \boldsymbol{\pi}_{in})[1 + \boldsymbol{\pi}_{out} \cdot \boldsymbol{\sigma}] \quad (7.9)$$

where for the moment we shall not trouble with the precise form of $\boldsymbol{\pi}_{out}$.

According to the results of the previous section, the cross section for scattering into the direction $\mathbf{p}'$ with any final spin is just tr ρ_{out}. Since the Pauli matrices are traceless, this is precisely the quantity in parentheses($\cdots$) in (7.9), which we rewrite as

$$\frac{d\sigma}{d\Omega}(\mathbf{p}' \leftarrow \mathbf{p}, \boldsymbol{\pi}_{in})$$

$$= [|a(\theta)|^2 + |b(\theta)|^2][1 + v(\theta)\hat{\mathbf{n}} \cdot \boldsymbol{\pi}_{in}] \quad \text{[out spin not monitored]} \quad (7.10)$$

where we have introduced the function $v(\theta)$ [often denoted $P(\theta)$ or $S(\theta)$]

$$v(\theta) = \frac{2 \operatorname{Im} ab^*}{|a|^2 + |b|^2}$$

For the special case of an unpolarized initial beam ($\boldsymbol{\pi}_{in} = 0$) this cross section reduces to

$$\frac{d\sigma}{d\Omega}(\mathbf{p}' \leftarrow \mathbf{p}, \boldsymbol{\pi}_{in} = 0) = |a(\theta)|^2 + |b(\theta)|^2 \quad \begin{bmatrix} \text{in spin unpolarized} \\ \text{out spin not monitored} \end{bmatrix}$$

which is just the unpolarized cross section derived in Section 6-f by the elementary procedure of averaging over initial and summing over final spins.

If the incident beam *is* polarized ($\boldsymbol{\pi}_{in} \neq 0$) the term in (7.10)

$$v(\theta)\hat{\mathbf{n}} \cdot \boldsymbol{\pi}_{in} \propto v(\theta) |\boldsymbol{\pi}_{in}| \cos \varphi$$

gives the cross section a $\cos \varphi$ dependence on the azimuth φ.[5] If the

---

[5] In making this statement we assume the following geometry: $\mathbf{p}$ lies along $\hat{\mathbf{3}}$ as usual and $\boldsymbol{\pi}_{in}$ lies in the y–z plane with polar angle α. Then $\hat{\mathbf{n}} \cdot \boldsymbol{\pi}_{in} = |\boldsymbol{\pi}_{in}| \sin \alpha \cos \varphi$.

amplitudes a and b (and hence v) are known, then measurement of this cross section lets one determine $\hat{\mathbf{n}} \cdot \boldsymbol{\pi}_{\text{in}}$ and, hence $\boldsymbol{\pi}_{\text{in}}$. If, on the other hand, the initial polarization $\boldsymbol{\pi}_{\text{in}}$ is known, then the measurement lets one determine v, which together with $(|a|^2 + |b|^2)$ gives two of the three data needed to determine a and b (apart from their overall phase).

As a first example of the application of these ideas we consider the measurement of the polarization of electrons produced in β decay by observing their scattering off a charged nucleus.[6] Such electrons are usually relativistic, and indeed the spin-dependent effects that we wish to discuss only show up at relativistic energies. However, the polarization analyses are essentially the same whether the energies are relativistic or not.

Since the interaction of electrons with nuclei is known (the Coulomb interaction), the corresponding amplitudes can be explictly calculated. (For the relativistic case this was first done by Mott and the process is therefore known as Mott scattering.) The functions a, b and hence v are therefore known. This means that by measuring the cross section (7.10) we should be able to determine $\hat{\mathbf{n}} \cdot \boldsymbol{\pi}_{\text{in}}$ and, hence, $\boldsymbol{\pi}_{\text{in}}$.

It turns out that there is an immediate difficulty because the electrons produced in β decay are longitudinally polarized; that is, they have $\boldsymbol{\pi}$ parallel to $\mathbf{p}$. This means that, whatever the scattering direction, $\hat{\mathbf{n}} \cdot \boldsymbol{\pi}_{\text{in}}$ is zero and the desired effect vanishes. This difficulty can be overcome by bending the electron beam through 90 deg in an electric field. This does not affect the spins[7] and so produces a transversely polarized beam, which we take to have momentum along $\hat{\mathbf{3}}$ and polarization along $\hat{\mathbf{2}}$, as shown in Fig. 7.1. A piece of gold foil is placed in the path of the electrons and their scattering observed by two identical counters. These counters are located in the x–z plane at equal distances from the target and at equal angles to the left and right of the incident direction. The number of counts in either counter is proportional to $d\sigma/d\Omega$ and, from (7.10) is

$$N = n \, d\Omega(|a|^2 + |b|^2)(1 + v\hat{\mathbf{n}} \cdot \boldsymbol{\pi}_{\text{in}})$$

(Here n is the total number of incident electrons times the number of target nuclei per unit area of foil, and $d\Omega$ is the solid angle subtended by either counter.) Since the two counters are situated at the same polar angle θ they differ only in that the normals $\hat{\mathbf{n}}_L$ and $\hat{\mathbf{n}}_R$ point in opposite directions, $\hat{\mathbf{n}}_L = -\hat{\mathbf{n}}_R = \hat{\mathbf{2}}$. (Remember that $\mathbf{n} = \mathbf{p} \times \mathbf{p}'$.) Thus, the last term in the

---

[6] The experiment described here in much simplified form was performed by Greenberg et al. (1960). The interested reader can find more details and references in the original account.

[7] Actually, allowance has to be made for the fact that electrons moving rapidly through an electric field experience a magnetic field. Their spins therefore precess a little.

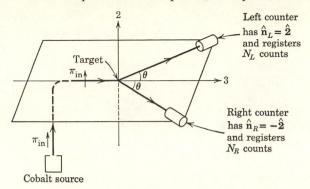

FIGURE 7.1.   A cobalt source produces longitudinally polarized electrons by β decay.   Electrons are bent through 90° by an electric field to give a transversely polarized beam.   Identical counters at equal angles to incident beam measure left-right asymmetry $\epsilon = (N_L - N_R)/(N_L + N_R)$.

expression for N has opposite signs in N_L and N_R.   In particular, if we form the asymmetry ratio $\epsilon = (N_L - N_R)/(N_L + N_R)$ all other factors cancel and

$$\epsilon = \frac{N_L - N_R}{N_L + N_R} = \nu(\theta)\hat{2} \cdot \boldsymbol{\pi}_{\text{in}} \tag{7.11}$$

Since $\nu(\theta)$ is known and since we have arranged that $\boldsymbol{\pi}_{\text{in}}$ is parallel to $\hat{2}$, measurement of ϵ allows a determination of the initial polarization of the electrons.[8]

In this example the scattering of electrons off a nucleus, for which the amplitude F is known, was used to measure the initial polarization of the electrons.   A more complicated situation occurs when the amplitude is *unknown* (as, for example, in nuclear collisions) and one wishes to use the data to measure it.   As an example of this we consider the scattering of protons off a spin zero nucleus such as carbon.

In this second example the first difficulty to be faced is that a beam of protons emerging from a cyclotron is unpolarized.   We have already pointed out that under these circumstances (and in the absence of spin sensitive counters) a single scattering experiment can tell us only the unpolarized cross section $|a|^2 + |b|^2$.   To obtain more information we need a way of polarizing the initially unpolarized beam.

The answer to this problem depends on the fact that, even if the beam is initially unpolarized, after a collision it *is* polarized.   To see this we need the

---

[8] Since the measured number ϵ is proportional to $\nu(\theta) |\boldsymbol{\pi}_{\text{in}}|$ the function $\nu(\theta)$ is often called the *polarization analyzing power* of the target.

expression for the final polarization $\boldsymbol{\pi}_{\mathrm{out}}$ that we omitted from (7.9). For the special case that $\boldsymbol{\pi}_{\mathrm{in}} = 0$ this is easily calculated, since then $\rho_{\mathrm{in}} = \frac{1}{2}I$ and so,

$$\rho_{\mathrm{out}} = F\rho_{\mathrm{in}}F^{\dagger} = \frac{1}{2}(a + ib\hat{\mathbf{n}}\cdot\boldsymbol{\sigma})(a^* - ib^*\hat{\mathbf{n}}\cdot\boldsymbol{\sigma})$$
$$= \frac{1}{2}(|a|^2 + |b|^2)[1 + \nu(\theta)\hat{\mathbf{n}}\cdot\boldsymbol{\sigma}] \qquad [\boldsymbol{\pi}_{\mathrm{in}} = 0]$$

where, as before, $\nu = (2\,\mathrm{Im}\,ab^*)/(|a|^2 + |b|^2)$. From this we see that the unpolarized cross section tr ρ_{out} is $(|a|^2 + |b|^2)$ (as we already knew) and the final polarization is

$$\boldsymbol{\pi}_{\mathrm{out}} = \nu(\theta)\hat{\mathbf{n}} \qquad [\boldsymbol{\pi}_{\mathrm{in}} = 0]$$

Thus, by scattering a beam of unpolarized protons we can produce protons with polarization $\nu(\theta)\hat{\mathbf{n}}$.[9] These can then be used in a second scattering experiment, and for this second experiment the incident protons *are* polarized.

A typical double scattering experiment is shown schematically in Fig. 7.2.[10] The beam of unpolarized protons scatters through angle θ_1 off a carbon target and emerges with polarization $\nu(\theta_1)\hat{\mathbf{2}}$. (We choose the scattering plane normal to $\hat{\mathbf{2}}$.) This polarization is then detected by a second scattering using exactly the same technique as described above in connection with electrons. Identical counters are situated at equal angles θ_2 to left and right of the second target; and applying the relation (7.11), $\epsilon = \nu(\theta)\hat{\mathbf{2}}\cdot\boldsymbol{\pi}_{\mathrm{in}}$, to this second collision (for which $\boldsymbol{\pi}_{\mathrm{in}} = \nu(\theta_1)\hat{\mathbf{2}}$) we obtain for the measureable asymmetry

$$\epsilon = \frac{N_L - N_R}{N_L + N_R} = \nu(\theta_2)\nu(\theta_1) \tag{7.12}$$

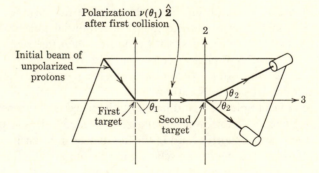

FIGURE 7.2.   **Unpolarized protons scatter off the carbon target producing protons of polarization $\nu(\theta_1)\hat{\mathbf{2}}$. These polarized protons then impinge on the second carbon target used to analyze polarization by measuring resultant left-right asymmetry. The experiment is performed as shown with both scatterings lying in the same plane.**

[9] For this reason $\nu(\theta)$ is sometimes called the *polarizing power* of the target.
[10] This is a much simplified account of the experiments of Chamberlain, et al. (1956).

In particular by choosing $\theta_2 = \theta_1 = \theta$, say, we can measure $v(\theta)$ within a sign as $v(\theta) = \pm \epsilon^{\frac{1}{2}}.$[11] Knowing $v(\theta_1)$ for one fixed θ_1, we have then only to vary θ_2 and use (7.12) to measure v for all angles.

We see that starting with an unpolarized beam and using single and double scattering experiments we can determine the quantities $(|a|^2 + |b|^2)$ and (Im ab^*). One more quantity is needed to determine the amplitudes a and b separately (apart from the elusive overall phase). It is quite easy to see that nothing further can be learned from the double scattering experiment. (For instance changing the plane of the second scattering will produce no new information.) One must therefore move on to a triple scattering experiment which can be considered as follows. Scattering 1 polarizes the protons, scattering 2 (which has a polarized incident beam) is regarded as the actual experiment of interest, and scattering 3 is used to analyze the final polarization produced in scattering 2. A somewhat tedious calculation shows that by measuring the left–right asymmetry of the final collision, one can determine the relative phase of the two functions $a \pm ib$. Knowledge of this angle, together with the values of $(|a|^2 + |b|^2)$ and Im ab^* obtained from single and double scattering, is sufficient to determine a and b separately (apart from their overall phase) and the results can then be compared with the calculated amplitudes derived from any proposed form of the underlying interactions.

7-d.   The Helicity Formalism

In the final two sections we give a brief introduction to the use of helicity states in scattering theory. We shall assume a slightly greater knowledge of the rotation group than is used elsewhere in the book; the reader unacquainted with the necessary properties can either find them in the standard texts (Rose, 1957, or Edmonds, 1957) or can simply omit these sections.

The helicity λ of a particle is defined as the component of the spin in the direction of the particle's momentum. The advantages of a basis of one-particle states labelled by $\mathbf{p}$ and λ over one labelled by $\mathbf{p}$ and m (the eigenvalue of S_3) are several. As we have already mentioned, the helicity λ, unlike

---

[11] The ambiguous sign in this determination of $v(\theta)$ arises because we have used the same scattering process to produce the polarization and to measure it. This means that $v(\theta)$ appears twice in the asymmetry (7.12) and its sign cannot be determined. One way to get around this is to replace the second target by one for which the amplitudes (and, in particular, the quantity v) are already known. For example, if we replace the second target by helium then the result (7.12) for the asymmetry becomes

$$\epsilon = v_{\text{He}}(\theta_2) v_{\text{C}}(\theta_1)$$

where the function v_{He} for proton–helium scattering is known. This obviously allows one to determine the proton–carbon function v_{C} completely.

m, is unchanged by rotations, which considerably simplifies the analysis of rotational invariance. Since the helicity operator $\mathbf{S} \cdot \hat{\mathbf{P}}$ can be rewritten as $\mathbf{J} \cdot \hat{\mathbf{P}}$ ($\mathbf{L} \cdot \hat{\mathbf{P}}$ being zero) it can be easily defined in the relativistic context, where the separate definitions of $\mathbf{L}$ and $\mathbf{S}$ are much more awkward. And in the case of massless particles there are only two spin states ($\lambda = \pm s$) of which only one need occur (as with the neutrino), and helicity is then the *only* natural labelling of the states. Thus, it is really in the relativistic domain that the helicity formalism has its greatest advantages, and the primary purpose of the present account is to pave the way for the reader's later study of the relativistic case.

For each value of a particle's momentum $\mathbf{p}$ there are $2s + 1$ eigenvalues of the helicity operator $\mathbf{S} \cdot \hat{\mathbf{p}}$, namely $\lambda = -s, \ldots, +s$, and it is clear that the space of all one-particle states can be spanned by momentum and helicity eigenvectors $|\mathbf{p}, \lambda\rangle$ just as well as by the momentum and S_3 eigenvectors $|\mathbf{p}, m\rangle$. In setting up the helicity basis the only ambiguity is the inevitable ambiguity in the choice of the phase of the vectors $|\mathbf{p}, \lambda\rangle$. This is usually chosen as follows: Consider first states whose momentum $\mathbf{p}$ lies along the positive z axis, $\hat{\mathbf{p}} = \hat{\mathbf{3}}$; for these states the helicity coincides with S_3, and so we can choose the states $|\mathbf{p}, \lambda\rangle$ to be precisely the usual S_3 eigenstates with the usual phase conventions

$$|\mathbf{p}, \lambda\rangle = |\mathbf{p}\rangle \otimes |S_3 = \lambda\rangle \qquad [\hat{\mathbf{p}} = \hat{\mathbf{3}}]$$

Next we note that an arbitrary momentum $\mathbf{p}$ with direction (θ, φ) can always be obtained from one in the z direction by a rotation. In fact this can be done in several ways, of which the simplest is just to rotate through θ about an axis perpendicular to $\hat{\mathbf{3}}$ and $\mathbf{p}$ (Fig. 7.3.) It is convenient to parametrize this rotation in terms of Euler angles, which we define such that

$$R(\varphi, \theta, \psi) = e^{-i\varphi J_3} e^{-i\theta J_2} e^{-i\psi J_3} \qquad (7.13)$$

(we use the so-called "active" definition, in which one rotates the system, not the axes). In terms of this definition the rotation just described is seen to be $R(\varphi, \theta, -\varphi)$ (Fig. 7.3). Since helicity is unchanged by rotations we can now define the state $|\mathbf{p}, \lambda\rangle$ as that state obtained by performing $R(\varphi, \theta, -\varphi)$ on the z-axis state $|p\hat{\mathbf{3}}, \lambda\rangle$.[12]

$$|\mathbf{p}, \lambda\rangle = R(\varphi, \theta, -\varphi)|p\hat{\mathbf{3}}, \lambda\rangle \qquad [\hat{\mathbf{p}} = (\theta, \varphi)] \qquad (7.14)$$

---

[12] It should perhaps be emphasized that the first rotation angle, $-\varphi$, is purely a matter of convention. In fact $R(\varphi, \theta, \psi)$ would carry $p\hat{\mathbf{3}}$ onto $\mathbf{p}$ with *any* value of ψ; use of a different ψ would just give $|\mathbf{p}, \lambda\rangle$ a different phase. The choice $\psi = -\varphi$ is the most common, and also most convenient, convention.

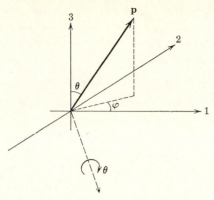

FIGURE 7.3. The momentum **p** can be obtained from $p\hat{3}$ by rotation through θ about an axis perpendicular to **p** and $\hat{3}$. This is equivalent to three successive rotations: $-\varphi$ about $\hat{3}$, θ about $\hat{2}$, and φ about $\hat{3}$. This rotation has the Euler parameters $(\varphi, \theta, -\varphi)$.

This procedure provides a well-defined orthonormal basis for the states of a single particle, satisfying

$$\langle \mathbf{p}', \lambda' \mid \mathbf{p}, \lambda \rangle = \delta_3(\mathbf{p}' - \mathbf{p})\delta_{\lambda'\lambda}$$

From this we could immediately construct a two-particle basis out of the products

$$|\mathbf{p}_1, \lambda_1\rangle \otimes |\mathbf{p}_2, \lambda_2\rangle \qquad (7.15)$$

However, our principal interest is in the relative motion, or equivalently in states of the CM frame, for which $\mathbf{p}_1 = -\mathbf{p}_2$; for such states there is a more convenient choice of basis. The point is that the states (7.15) are defined by two separate rotations (one for each particle) while for the special case $\mathbf{p}_1 = -\mathbf{p}_2$ we can define an equivalent state with a single rotation. Thus, we first define a state of the relative motion with the relative momentum along $\hat{3}$,

$$|\mathbf{p}, \lambda_1, \lambda_2\rangle = |\mathbf{p}\rangle \otimes |S_3^{(1)} = \lambda_1\rangle \otimes |S_3^{(2)} = -\lambda_2\rangle \qquad [\hat{\mathbf{p}} = \hat{3}] \qquad (7.16)$$

In the CM frame this represents a state in which particle 1 moves up the z axis with momentum **p** while particle 2 moves *down* the z axis with momentum $-\mathbf{p}$. The minus sign in the last factor of (7.16) is because, with its momentum along $-\hat{3}$, particle 2 has helicity $-S_3^{(2)}$. The general state (of the relative motion) with arbitrary relative momentum **p** is now obtained by the appropriate rotation as

$$|\mathbf{p}, \lambda_1, \lambda_2\rangle = R(\varphi, \theta, -\varphi)|p\hat{3}, \lambda_1, \lambda_2\rangle \qquad [\hat{\mathbf{p}} = (\theta, \varphi)] \qquad (7.17)$$

where R is now the two-particle rotation operator generated by the total angular momentum of the two particles.

The vectors (7.17) define states (of the relative motion) in which the particles have helicities λ_1 and λ_2 and relative momentum $\mathbf{p}$ (i.e., in the CM frame the two particles have equal and opposite momenta, $\mathbf{p}_1 = \mathbf{p}$ and $\mathbf{p}_2 = -\mathbf{p}$). They are normalized so that

$$\langle \mathbf{p}', \lambda_1', \lambda_2' \mid \mathbf{p}, \lambda_1, \lambda_2 \rangle = \delta_3(\mathbf{p}' - \mathbf{p})\delta_{\lambda_1'\lambda_1}\delta_{\lambda_2'\lambda_2}$$

and provide an orthonormal basis for the relative motion alternative to the basis $|\mathbf{p}, m_1, m_2\rangle$ used earlier. The S matrix in the helicity basis can of course be decomposed as

$$\langle \mathbf{p}', \lambda_1', \lambda_2'| \, S \, |\mathbf{p}, \lambda_1, \lambda_2 \rangle = \delta_3(\mathbf{p}' - \mathbf{p})\delta_{\lambda_1'\lambda_1}\delta_{\lambda_2'\lambda_2}$$
$$+ \frac{i}{2\pi m}\,\delta(E_{p'} - E_p)f(\mathbf{p}', \lambda_1', \lambda_2' \leftarrow \mathbf{p}, \lambda_1, \lambda_2) \quad (7.18)$$

and the CM cross section for a process involving definite helicities is just

$$\frac{d\sigma}{d\Omega}\,(\mathbf{p}', \lambda_1', \lambda_2' \leftarrow \mathbf{p}, \lambda_1, \lambda_2) = |f(\mathbf{p}', \lambda_1', \lambda_2' \leftarrow \mathbf{p}, \lambda_1, \lambda_2)|^2$$

Let us now suppose that our system is rotationally invariant. To exploit this property we naturally turn to a basis of angular-momentum eigenvectors. It is here that the advantages of using the helicity basis appear. Because the helicity operators $\mathbf{S}^{(1)} \cdot \hat{\mathbf{P}}$ and $-\mathbf{S}^{(2)} \cdot \hat{\mathbf{P}}$ are scalars, they commute with $\mathbf{J}$ and we can therefore choose a complete set of commuting observables including both $\mathbf{J}^2$, J_3 and the two helicities, Specifically, we can choose a basis (of the relative motion) $\{|E, j, m, \lambda_1, \lambda_2\rangle\}$ and the passage between this basis and the momentum basis $|\mathbf{p}, \lambda_1, \lambda_2\rangle$ can be made without the use of any Clebsch–Gordan coefficients to couple spins and orbital angular momenta.

At this point we shall appeal to a standard result from the theory of the rotation group and immediately write down the angular-momentum states in terms of the momentum states as

$$|E, j, m, \lambda_1, \lambda_2\rangle = \left[\frac{mp(2j+1)}{4\pi}\right]^{\frac{1}{2}}\int d\Omega_\mathbf{p} R_{m\lambda}^j(\varphi, \theta, -\varphi)^* \,|\mathbf{p}, \lambda_1, \lambda_2\rangle \quad (7.19)$$

where the $R_{m\lambda}^j(\varphi, \theta, \psi)$ are the elements of the $(2j + 1)$-dimensional rotation matrix for Euler angles (φ, θ, ψ) (see Section 7-e), where $\lambda \equiv \lambda_1 - \lambda_2$, and where $\mathbf{p}$ has direction (θ, φ) and fixed magnitude $(2mE)^{\frac{1}{2}}$. The vector (7.19) is an eigenvector of the energy, total angular momentum and the two helicities. It is normalized such that[13]

$$\langle E', j', m', \lambda_1', \lambda_2' \mid E, j, m, \lambda_1, \lambda_2 \rangle = \delta(E' - E)\delta_{j'j}\delta_{m'm}\delta_{\lambda_1'\lambda_1}\delta_{\lambda_2'\lambda_2} \quad (7.20)$$

---

[13] That (7.19) is an eigenvector of the energy and helicities follows from that fact it is a superposition of vectors all of which are energy and helicity eigenvectors with the same eigenvalues. That it is an eigenvector of $\mathbf{J}^2$ and J_3 follows from (7.26) below; its normalization from (7.27).

Using the definition (7.19), we can write down the transformation matrix that takes one from the states $|\mathbf{p}, \mu_1, \mu_2\rangle$ of momentum $\mathbf{p}$ and helicities μ_1, μ_2 to the state $|E, j, m, \lambda_1, \lambda_2\rangle$,

$$\langle \mathbf{p}, \mu_1, \mu_2 \mid E, j, m, \lambda_1, \lambda_2\rangle$$

$$= (mp)^{-\frac{1}{2}}\delta(E_p - E)\delta_{\mu_1\lambda_1}\delta_{\mu_2\lambda_2}\left(\frac{2j+1}{4\pi}\right)^{\frac{1}{2}}R^j_{m\lambda}(\varphi, \theta, -\varphi)^* \quad (7.21)$$

In particular, for spinless particles, $\lambda = 0$ and j is just the orbital momentum $j = l$. In this case the factor

$$\left(\frac{2j+1}{4\pi}\right)^{\frac{1}{2}}R^j_{m\lambda}(\varphi, \theta, -\varphi)^*$$

is precisely the spherical harmonic $Y^m_l(\theta, \varphi)$ and the result (7.21) reduces to the result (6.10) for the matrix $\langle \mathbf{p} \mid E, l, m\rangle$ of two spinless particles.

Equipped with the transformation matrix between our two bases we are now ready to exploit rotational invariance. In the angular-momentum basis the S matrix is diagonal with respect to E, j, and m, and so has the form:

$$\langle E', j', m', \lambda_1', \lambda_2'| \; S \; |E, j, m, \lambda_1, \lambda_2\rangle = \delta(E' - E)\delta_{j'j}\delta_{m'm}s^j_{\lambda_1'\lambda_2',\lambda_1\lambda_2}(E)$$

that is, the scattering for given E, j, m is determined by a unitary matrix $S^j(E)$ whose rows and columns are labelled by the final and initial helicities (λ_1', λ_2') and (λ_1, λ_2). As usual we introduce a partial wave amplitude,

$$f^j_{\lambda_1'\lambda_2',\lambda_1\lambda_2}(E) = \frac{s^j_{\lambda_1'\lambda_2',\lambda_1\lambda_2}(E) - 1}{2ip}$$

and we can now express the full amplitude in terms of the partial-wave amplitudes by exactly the procedure used in Section 6-c for the spinless case. Thus,

$$\langle \mathbf{p}', \lambda_1', \lambda_2'| \; (S - 1) \; |\mathbf{p}, \lambda_1, \lambda_2\rangle$$

$$= \frac{i}{2\pi m}\delta(E_{p'} - E_p)f(\mathbf{p}', \lambda_1', \lambda_2' \leftarrow \mathbf{p}, \lambda_1, \lambda_2) \quad (7.22)$$

Inserting complete sets of states $|E, j, m, \mu_1, \mu_2\rangle$ fore and aft of the factor $(S - 1)$ on the left, we can immediately replace the angular-momentum matrix element of $(S - 1)$ by $2ip$ times the partial-wave amplitude (times the appropriate delta functions). Just as in the spinless case, the sum on the left reduces to a sum over the numbers $f^j_{\lambda_1'\lambda_2',\lambda_1\lambda_2}$ times two factors, each of which is just the transformation matrix (7.21). Some simple algebra reduces this to

$$= \frac{2i}{m}\delta(E_{p'} - E_p)\sum_{j\,m}\frac{2j+1}{4\pi}R^j_{m\lambda'}(\varphi', \theta', -\varphi')^*f^j_{\lambda_1'\lambda_2',\lambda_1\lambda_2}(E_p)R^j_{m\lambda}(\varphi, \theta, -\varphi)$$

$$(7.23)$$

We now choose the initial momentum **p** along the z axis (and then use θ, φ for the direction of **p′**) in which case the last factor of (7.23) is just $\delta_{m\lambda}$. Comparing (7.22) and (7.23), we obtain the desired answer

$$f_{\lambda_1'\lambda_2',\lambda_1\lambda_2}(E, \theta, \varphi) = \sum_j (2j + 1)f^j_{\lambda_1'\lambda_2',\lambda_1\lambda_2}(E)\mathsf{R}^j_{\lambda\lambda'}(\varphi, \theta, -\varphi)^* \qquad (7.24)$$

where we have rewritten the amplitude $f(\mathbf{p}', \lambda_1', \lambda_2' \leftarrow \mathbf{p}, \lambda_1, \lambda_2)$ as a function of (E, θ, φ) for the case that the initial momentum lies along the z axis. Since the functions $\mathsf{R}^j_{\lambda\lambda'}(\varphi, \theta, -\varphi)$ are tabulated [see (7.25) and the remarks following] this series is no more complicated than the familiar partial-wave series (6.15) of the spinless case, $f = \sum (2l + 1)f_l P_l$. In particular for the special case that the particles *are* spinless one has only to recall that $\mathsf{R}^l_{00}(\varphi, \theta, -\varphi) = P_l(\cos \theta)$ to see that the result (7.24) *is* the partial-wave series (6.15).

As the reader who did Problem **6.9** will certainly recognize, the derivation of the helicity partial-wave series (7.24) *for arbitrary spins* is considerably easier than the simplest special case [(spin ½) − (spin 0) scattering] in the S_3 representation. When one recalls that the problem in the S_3 representation becomes progressively harder as the spins increase, the great superiority of the helicity formalism (for this particular problem at least) should be quite apparent.

For further discussion of the helicity formalism the reader can do no better than to read the original article of Jacob and Wick (1959). A few simple additional topics are taken up in the problems.

7-e.   Some Useful Formulas

Here we collect those formulas involving rotation matrices that are needed in Section 7-d and in the problems for this chapter. More complete collections can be found in Messiah (1961) Appendix C, Gottfried (1966) Sections 32 and 34, Edmonds (1957) Chapter IV, and Rose (1957) Chapter 4.

The rotation matrix $\mathsf{R}^j(\varphi, \theta, \psi)$, often denoted D^j, is defined by the relation

$$\mathsf{R}^j_{mm'}(\varphi, \theta, \psi) = \langle j, m| \, \mathsf{R}(\varphi, \theta, \psi) \, |j, m'\rangle$$

where $\mathsf{R}(\varphi, \theta, \psi)$ is the unitary rotation operator (7.13) for Euler angles φ, θ, ψ. Its dependence on φ and ψ is trivial and one can write:

$$\mathsf{R}^j_{mm'}(\varphi, \theta, \psi) = e^{-i(\varphi m + \psi m')}r^j_{mm'}(\theta) \qquad (7.25)$$

where

$$r^j_{mm'}(\theta) = \langle j, m| \, e^{-i\theta J_2} \, |j, m'\rangle$$

Since, with our conventions, J_2 is given by a purely imaginary matrix, the matrix $r^j(\theta)$ (often denoted d^j) *is real.* Explicit expressions for $r^j(\theta)$ can be given, and its values are tabulated (see any of the above references). In working with the helicity states in practice (see problems **7.4–7.7**) it is naturally more convenient to use r^j rather than the full matrix R^j.

To prove that the states $|E, j, m, \lambda_1, \lambda_2\rangle$ of (7.19) are eigenstates of $\mathbf{J}^2$ and J_3 it is convenient to rewrite (7.19) as

$$|E, j, m, \lambda_1, \lambda_2\rangle = \text{const} \int_0^{4\pi} d\varphi \int_0^{\pi} \sin\theta \, d\theta \int_0^{2\pi} d\psi R^j_{m\lambda}(\varphi, \theta, \psi)^*$$
$$\times \; \mathrm{R}(\varphi, \theta, \psi) \, |p\hat{3}, \lambda_1, \lambda_2\rangle \quad (7.26)$$

This integral is the invariant integral of the rotation group, and it follows immediately that $|E, j, m, \lambda_1, \lambda_2\rangle$ transforms under rotations like a state of angular momentum j, m; that is, that it *is* a state of angular momentum j, m.

The normalization of the states $|E, j, m, \lambda_1, \lambda_2\rangle$ follows from the group orthogonality relation

$$\int_0^{4\pi} d\varphi \int_0^{\pi} \sin\theta \, d\theta \int_0^{2\pi} d\psi R^j_{m\lambda}(\varphi, \theta, \psi)^* R^{j'}_{m'\lambda'}(\varphi, \theta, \psi) = \frac{16\pi^2}{2j+1} \delta_{jj'} \delta_{mm'} \delta_{\lambda\lambda'}$$

This is equivalent to

$$\int_0^{\pi} \sin\theta \, d\theta \, r^j_{m\lambda}(\theta)^* r^{j'}_{m\lambda}(\theta) = \frac{2}{2j+1} \delta_{jj'} \quad (7.27)$$

from which the normalization (7.20) follows easily (see Problem **7.4**).

In order to relate the helicity and angular-momentum eigenvectors $|E, j, m, \lambda_1, \lambda_2\rangle$ to the more familiar l, s eigenvectors one needs to decompose the product of two matrices $r^j(\theta)$ and $r^s(\theta)$ using the Clebsch–Gordan series

$$r^j_{mm'}(\theta) r^s_{m_s m_s'}(\theta) = \sum_l \langle jsmm_s | lm_l\rangle \langle jsm'm_s' | lm_l'\rangle r^l_{m_l m_l'}(\theta)$$

Using standard identities for the Clebsch–Gordan coefficients,[14] this can be rewritten as:

$$r^j_{mm'}(\theta) r^s_{m_s m_s'}(\theta) = \sum_l \frac{2l+1}{2j+1} \langle lsm_l m_s | jm\rangle \langle lsm_l'm_s' | jm'\rangle r^l_{m_l m_l'}(\theta) \quad (7.28)$$

where, of course, $m_l \equiv m - m_s$ and $m_l' \equiv m' - m_s'$. This is the form in which the decomposition is most conveniently applied in this context (see Problem **7.6b**).

---

[14] See Messiah (1961) p. 1056, Eq. (C.13). One also needs the relation [*ibid.* Eq. (C.65)]

$$r^j_{mm'} = (-)^{m-m'} r^j_{-m,-m'}$$

PROBLEMS

7.1.   An electron has spin density matrix $\rho = \frac{1}{2}(1 + \boldsymbol{\pi} \cdot \boldsymbol{\sigma})$. Show that $\langle s \rangle_\rho = \frac{1}{2}\boldsymbol{\pi}$, that $|\boldsymbol{\pi}| \leqslant 1$, and that $|\boldsymbol{\pi}| = 1$ if and only if the electron is in a pure state, which is an eigenstate of the component of spin in the direction of $\boldsymbol{\pi}$.

7.2.   The cross section for a process involving definite spin states is *not* necessarily rotationally invariant even if the interactions are. (The point is that the spins can define preferred directions.) However, the unpolarized cross section *is* invariant (provided the interactions are); that is,

$$\frac{d\sigma}{d\Omega}(\mathbf{p}' \leftarrow \mathbf{p}) = \frac{d\sigma}{d\Omega}(\mathbf{p}_R' \leftarrow \mathbf{p}_R) \qquad \begin{bmatrix} \text{in spins unpolarized;} \\ \text{out spins not monitored} \end{bmatrix}$$

Prove this by writing the unpolarized cross section as tr $\rho_\text{out} = \text{tr}\, F\rho_\text{in}F^\dagger$ with $\rho_\text{in} = (2s + 1)^{-1}I$, and noting that $F(\mathbf{p}_R' \leftarrow \mathbf{p}_R) = RF(\mathbf{p}' \leftarrow \mathbf{p})R^\dagger$ where R is the unitary rotation matrix.

7.3.   Calculate the final polarization $\boldsymbol{\pi}_\text{out}$ for (spin $\frac{1}{2}$) $-$ (spin 0) scattering in terms of the amplitudes a and b and the initial polarization $\boldsymbol{\pi}_\text{in}$; i.e., calculate $\boldsymbol{\pi}_\text{out}$ in (7.9). Check that your answer reduces to $\boldsymbol{\pi}_\text{out} = v\hat{\mathbf{n}}$ when $\boldsymbol{\pi}_\text{in} = 0$.

7.4.   Verify the normalization (7.20) of the states $|E, j, m, \lambda_1, \lambda_2\rangle$ defined in (7.19) [use the orthogonality relation (7.27).]

7.5.   The parity operator reverses helicities, since it reverses momenta but leaves all angular momenta unchanged. For the simple case of (spin s) $-$ (spin 0) scattering this means that P $|\mathbf{p}, \lambda\rangle$ is proportional to $|-\mathbf{p}, -\lambda\rangle$, and that P $|E, j, m, \lambda\rangle$ is proportional to $|E, j, m, -\lambda\rangle$. In fact, it can be shown (as the reader who is used to handling rotation matrices could check) that

$$\text{P }|E, j, m, \lambda\rangle = (-)^{j-s} |E, j, m, -\lambda\rangle$$

Use this result and the fact that S is unitary to explore the structure of the helicity matrix $S^j(E)$ in (spin $\frac{1}{2}$) $-$ (spin 0) scattering of a parity invariant system. Show that S^j is a (2×2) matrix with elements of the form:

$$s^j_{++} = s^j_{--} = \frac{1}{2}(e^{2i\delta^j_-} + e^{2i\delta^j_+})$$

and

$$s^j_{+-} = s^j_{-+} = \frac{1}{2}(e^{2i\delta^j_-} - e^{2i\delta^j_+})$$

for certain real numbers $\delta^j_\pm$ (depending on E of course).

7.6. (For the reader with some knowledge of rotation matrices.) It is naturally possible to express the helicity eigenvectors in terms of the more familiar S_3 eigenvectors and vice versa. Consider the scattering of a spin s particle off a spinless target and do the following:

(a) Show from its definition (7.14) that

$$|\mathbf{p}, \lambda\rangle = \sum_{m_s} |\mathbf{p}, m_s\rangle e^{i(\lambda - m_s)\varphi} r^s_{m_s\lambda}(\theta)$$

where $r^s(\theta)$ is defined in (7.25).

(b) With the help of this result, show from its definition (7.19) that

$$|E, j, m, \lambda\rangle = \sum_l \left(\frac{2l+1}{2j+1}\right)^{\frac{1}{2}} |E, l, j, m\rangle \langle ls0\lambda \mid j\lambda\rangle$$

[For this you need to combine two rotation matrices r^j and r^s using (7.28), to give an r^l. The integration over angles then gives the state $|E, l, m_l, m_s\rangle$.]

(c) Use the result of (b) in (spin ½) − (spin 0) scattering to identify the numbers $\delta^j_\pm$ of problem **7.5** with the phase shifts of (6.49). (Note well that S^j in Problem **7.5** is the helicity matrix with rows and columns labelled by $\lambda = \pm½$; whereas S^j in Section 6-g relates to the basis $|E, l, j, m\rangle$ and has rows and columns labelled by $l = j \pm ½$.)

7.7. Prove that time-reversal invariance (plus rotational invariance) implies that the helicity matrix $S^j(E)$ is symmetric. [All you need to establish is the effect of T on the eigenvectors of angular momentum and helicity. If you consider for simplicity (spin s) − (spin 0) scattering, this can be read off from Problem **7.6b** plus the known effect of T on $|E, l, j, m\rangle$—see Problem **6.7.**]

8 The Green's Operator and the T Operator

8-a The Green's Operator

8-b The T Operator

8-c Relation to the Møller Operators

8-d Relation to the Scattering Operator

In the preceding chapters we have set up a time-dependent description of collisions in which two particles scatter elastically—the so-called single-channel processes. The collisions were first described in terms of the scattering operator S. The matrix elements of S were then decomposed in terms of the scattering amplitude, and the differential cross section was expressed in terms of the amplitude. Finally, we explored a number of general properties of the amplitude that follow from invariance principles, and we discussed some problems in the measurement of cross sections for particles with spin.

Our principal remaining problem in one-channel scattering is to set up methods for the actual computation of the amplitude in terms of a given interaction. All such methods fall within the so-called time-*independent* scattering theory, which is therefore the subject of the next several chapters. This formalism is built around two operators and the so-called stationary scattering states. The two operators are the Green's operator $G(z)$ and the T operator $T(z)$, which are the subjects of Chapters 8 and 9. The stationary scattering states, denoted $|\mathbf{p}\pm\rangle$, correspond to the elementary "scattering wave functions" often denoted $\psi_{\mathbf{p}}^{\pm}(\mathbf{x})$, and are the subjects of Chapters 10–12.

In this chapter we introduce the operators $G(z)$ and $T(z)$ and establish their role in scattering theory. In Sections 8-a and 8-b we define the two operators and discuss their mathematical properties. In Sections 8-c and 8-d we show how the collision operators $\Omega_{\pm}$ and S can be expressed in terms of $G(z)$ and $T(z)$. In particular, we shall find that the on-shell T matrix is given in terms of the operator $T(z)$ by the important identity

$$t(\mathbf{p'} \leftarrow \mathbf{p}) = \lim_{\epsilon \to 0} \langle \mathbf{p'}| \, T(E_p + i\epsilon) \, |\mathbf{p}\rangle \tag{8.1}$$

Perhaps it should be emphasized that our order or presentation is chosen to make the nature of the operators $G(z)$ and $T(z)$ as clear as possible (and later the stationary scattering states $|\mathbf{p}\pm\rangle$); it in no sense follows the historical development of time-independent scattering theory. In particular, we shall introduce the operator $T(z)$, without any attempt at motivation, by the definition (8.8); next we shall deduce various mathematical properties of $T(z)$, and only then "discover" the connection (8.1) between the matrix elements of $T(z)$ and the on-shell T matrix. As we shall discuss later, this is certainly not the route by which $T(z)$ was originally introduced into scattering theory.

For simplicity's sake we will deal mainly with the scattering of a single particle off a fixed potential (which is, of course, equivalent to the scattering of two particles off one another), and most of the discussion will be about particles of spin zero. Most (though not all) results generalize in a completely straightforward way.

8-a. The Green's Operator

In the mathematical study of self-adjoint operators perhaps the most important single tool is the so-called *Green's operator* or *resolvent*[1] $G(z)$. In the scattering of a single particle the self-adjoint operators of interest are $H^0 = \mathbf{P}^2/2m$ and $H = H^0 + V$, and the corresponding Green's operators are defined to be

$$\boxed{\begin{aligned} G^0(z) &\equiv (z - H^0)^{-1} \\ G(z) &\equiv (z - H)^{-1} \end{aligned}}$$

both for any z, real or complex, for which the inverse exists.

To understand the name "Green's operator" we assume for a moment that these inverses do exist and are defined everywhere on the Hilbert space $\mathcal{H}$. (We shall see that for any z with Im $z \neq 0$ this is certainly true.) In this case it is clear from the definition that, for example,

$$(z - H^0)G^0(z) = 1$$

---

[1] The name resolvent is more popular in the mathematical literature. However, it should be noted that this name is sometimes used in a slightly different sense.

We now take coordinate-space matrix elements of this equation. In the coordinate representation, H^0 is represented by $-\nabla^2/2m$; that is

$$\langle \mathbf{x} |\, H^0\, |\psi\rangle = -\frac{\nabla^2}{2m}\, \langle \mathbf{x} \mid \psi\rangle$$

and the equation therefore becomes

$$\left(\frac{\nabla^2}{2m} + z\right)\langle \mathbf{x} |\, G^0(z)\, |\mathbf{x}'\rangle = \delta_3(\mathbf{x} - \mathbf{x}')$$

Thus, the coordinate-space matrix element of the free *Green's operator* $G^0(z)$ is the *Green's function* for the differential operator $[(\nabla^2/2m) + z]$. In exactly the same way the matrix element of $G(z)$ is the Green's function for the operator $[(\nabla^2/2m) - V(\mathbf{x}) + z]$.

The Green's operators do not exist for all values of z. For example, if E_n is any proper eigenvalue of H, then there is a vector (the corresponding eigenvector $|n\rangle$) for which

$$(E_n - H)\, |n\rangle = 0$$

According to our discussion in Section 1-c, this means that there is no inverse $(E_n - H)^{-1}$. That is, the operator $G(z)$ is undefined for z equal to any eigenvalue E_n.

To see in more detail what happens to $G(z)$ as z approaches an eigenvalue, let us suppose for a moment that H has a purely discrete and nondegenerate spectrum. (Hamiltonians with purely discrete spectra have no place in scattering theory. However, for a moment we can consider as an example the Hamiltonian of a one-dimensional simple harmonic oscillator.) If $\{|n\rangle\}$ denotes the orthonormal basis of eigenvectors with energies E_n, then $1 = \sum |n\rangle\langle n|$ and

$$G(z) = (z - H)^{-1}1 = \sum_n \frac{|n\rangle \langle n|}{z - E_n} \tag{8.2}$$

It is easy to see that for any z not equal to an eigenvalue, this operator is well defined on any $|\psi\rangle$ in $\mathscr{H}$. [That is, $G(z)\, |\psi\rangle$ is a well-defined vector, itself in $\mathscr{H}$.] We can go further if we introduce the definition:

An operator $G(z)$ is said to be an *analytic (operator) function of the complex variable z*, if the number $\langle \chi |\, G(z)\, |\psi\rangle$ is analytic for all $|\chi\rangle$ and $|\psi\rangle$.[2]

Since the matrix element

$$\langle \chi |\, G(z)\, |\psi\rangle = \sum_n \frac{\langle \chi \mid n\rangle \langle n \mid \psi\rangle}{z - E_n}$$

---

[2] There are several different definitions of an analytic operator. Fortunately, for a bounded operator (which the Green's operator is) they are all equivalent.

is obviously analytic as long as $z \neq E_1, E_2, \ldots$[3] we see that the operator $G(z)$ is an analytic function of z for all $z \neq E_1, E_2, \ldots$. Furthermore, at $z = E_n$ the matrix element has a simple pole with residue $\langle \chi \mid n \rangle \langle n \mid \psi \rangle$, and it is therefore natural to say that the operator $G(z)$ has a simple pole whose residue is the operator $|n\rangle\langle n|$ (which is just the projection operator onto the nth energy level).

This example shows (what is always true) that knowledge of the Green's operator $G(z)$, for all z, of a Hamiltonian H is equivalent to knowledge of a complete solution of the corresponding eigenvalue problem. The Green's operator $G(z)$ is analytic except on the spectrum of H (which in this case was assumed to be purely discrete, and in general is some set of real numbers). For each discrete eigenvalue, $G(z)$ has a pole whose position is precisely the eigenvalue and whose residue determines the corresponding eigenvector (or subspace of eigenvectors in the case of degeneracy).

When H has a continuous spectrum (as it always does in scattering theory) the Green's operator has a singularity there as well. To see this we consider the free Green's operator $G^0(z)$. In terms of the angular-momentum eigenvectors $|E, l, m\rangle$ we have an expansion analogous to (8.2),

$$G^0(z) = \int_0^\infty dE \sum_{l,m} \frac{|E, l, m\rangle\langle E, l, m|}{z - E}$$

or

$$\langle \chi | G^0(z) | \psi \rangle = \int_0^\infty \frac{dE}{z - E} \left\{ \sum_{l,m} \langle \chi \mid E, l, m \rangle \langle E, l, m \mid \psi \rangle \right\} \tag{8.3}$$

As long as z is not a positive real number the denominator in this integral does not vanish. In this case it is easily seen that the matrix elements, and hence the operator, are well defined and analytic. However, these analytic functions have a *branch cut* running along the real axis from 0 to ∞; that is, the value of $\langle \chi | G^0(z) | \psi \rangle$ at any $E_0 > 0$ approached from above ($z = E_0 + i0$) is different from its value at the same point approached from below ($z = E_0 - i0$) (see Fig. 8.1). Thus, from (8.3)[4]:

$$\langle \chi | G^0(E_0 + i0) | \psi \rangle - \langle \chi | G^0(E_0 - i0) | \psi \rangle$$

$$= \int_0^\infty dE \left(\frac{1}{E_0 - E + i0} - \frac{1}{E_0 - E - i0} \right) \{ \sum \cdots \}$$

$$= -2\pi i \sum_{l,m} \langle \chi \mid E_0, l, m \rangle \langle E_0, l, m \mid \psi \rangle$$

---

[3] The argument for this is as follows: (1) Each term of the series is analytic for $z \neq E_1$, $E_2, \ldots$. (2) For z in any closed region not including $E_1, E_2, \ldots$, the series is uniformly convergent. (3) Since a uniformly convergent series of analytic functions is analytic, the conclusion follows.

[4] Note that here and elsewhere any expression containing the argument $x \pm i0$ is understood to be evaluated at $x \pm i\epsilon$ with $\epsilon > 0$ and the limit $\epsilon \to 0$ then taken.

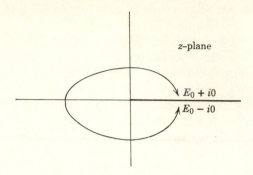

FIGURE 8.1.   The matrix elements $\langle \chi | \, G^0(z) \, | \psi \rangle$, and, hence, the operator $G^0(z)$, are analytic for z not on the positive real axis. Because their values as z approaches $E_0 \pm i0(E_0 > 0)$ are unequal, the positive real axis is a branch cut.

since the factor in parentheses $(\cdot \cdot \cdot)$ is just $(-2\pi i)$ times $\delta(E - E_0)$ (see Messiah, 1961, p. 469). Once again we can see that knowledge of the Green's operator for all z is equivalent to a complete solution of the eigenvalue problem of the corresponding Hamiltonian. The operator $G^0(z)$ is analytic except when z is in the spectrum of H^0. The singularity at the spectrum, which in this case is continuous, is a branch cut and the discontinuity of $G^0(z)$ across this cut at $E > 0$ is just $(-2\pi i)$ times the operator

$$\sum_{l,m} |E, l, m \rangle \langle E, l, m|$$

[which is the (improper) projector onto the states of energy E].

These two examples show what one should expect in general. In particular, we could apply the same arguments to the Hamiltonian H of a typical scattering problem, with bound states at $E_1, E_2, \ldots$, all less than zero, and a continuous spectrum running from 0 to ∞. It should be reasonably clear that $G(z)$ is then analytic except for poles at the bound states $z = E_1, E_2, \ldots$ and a branch cut from 0 to ∞ (Fig. 8.2).[5]

Knowledge of $G(z)$ for all z is equivalent to a complete solution of the eigenvalue problem of H. Needless to say, finding $G(z)$ is precisely as hard as solving the eigenvalue problem, and, generally, one cannot hope to find $G(z)$. For this reason it is useful to have an equation that relates the unknown $G(z)$ to some known operator. The usual choice for the latter is the free Green's operator $G^0(z)$. The equation relating G and G^0 is called the

---

[5] It should perhaps be emphasized that we do not claim to have *proved* this result, only to have made it plausible. To prove it one must first show that the spectrum is as claimed (see, for example, Faddeev, 1957, and Kato, 1959), and then appeal to standard theorems from the theory of spectral decompositions.

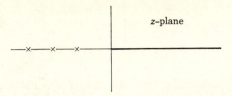

FIGURE 8.2.   The Green's operator $G(z)$ is analytic except for poles at the bound states $z = E_1, E_2, \ldots$ and a cut from 0 to ∞.

resolvent equation or *Lippmann–Schwinger equation for $G(z)$*. It is derived from the simple operator identity

$$A^{-1} = B^{-1} + B^{-1}(B - A)A^{-1}$$

If we set $A = z - H$ and $B = z - H^0$ this becomes:

$$G(z) = G^0(z) + G^0(z)VG(z) \qquad (8.4)$$

or, if we interchange A and B,

$$G(z) = G^0(z) + G(z)VG^0(z) \qquad (8.5)$$

It can be shown that, for z not in the spectrum of H, either of these equations characterizes $G(z)$; that is, if we find an operator that satisfies either equation it is indeed the Green's operator[6] $G(z) = (z - H)^{-1}$. As we shall see, these two equations constitute one of the essential foundation stones of the time-independent scattering formalism.

The free Green's operator $G^0(z)$ is, of course, explicitly known. In the momentum representation, for example, it is diagonal and is characterized by the relation:

$$G^0(z) |\mathbf{p}\rangle = (z - H^0)^{-1} |\mathbf{p}\rangle = \frac{1}{z - E_p} |\mathbf{p}\rangle \qquad (8.6)$$

Obviously, this allows us to calculate the matrix elements of $G^0(z)$ in any other representation. For example, in Chapter 10 we shall use (8.6) to calculate the spatial matrix elements (or Green's function) $\langle \mathbf{x} | G^0(z) | \mathbf{x}' \rangle$.

---

[6] A clear discussion of this and some other properties of $G(z)$ can be found in Faddeev (1965) Section 2.

To conclude this section on the definition and general properties of Green's operators we note that since $H = H^\dagger$ it is easily seen that[7]

$$G(z^*) = [G(z)]^\dagger \tag{8.7}$$

and similarly for $G^0(z)$. This important identity asserts that the value of the function G at the point z^* is the same as the adjoint of its value at the point z.

8-b.   The T Operator

In scattering theory it is convenient to introduce another operator $T(z)$, which is defined in terms of $G(z)$ as

$$T(z) = V + VG(z)V \tag{8.8}$$

It is clear that as an analytic function of z the operator $T(z)$ has the same properties as $G(z)$. That is, $T(z)$ is analytic for all z not in the spectrum of H. When z approaches the energy of a bound state $T(z)$ has a pole, and on the positive real axis $T(z)$ has a branch cut.

An important identity relating the operators T, G, and G^0 is obtained by multiplying (8.8) on the left by G^0

$$G^0 T = (G^0 + G^0 VG)V$$

or

$$G^0(z)T(z) = G(z)V \tag{8.9}$$

since, according to the Lippmann–Schwinger equation, $(G^0 + G^0 VG) = G$. Similarly, multiplying on the right, we find that

$$T(z)G^0(z) = VG(z) \tag{8.10}$$

As a first application of these identities we can find an expression for G in terms of T. According to the Lippmann–Schwinger equation $G = G^0 + G^0 VG$. Thus, replacing VG by TG^0 we find that

$$G(z) = G^0(z) + G^0(z)T(z)G^0(z)$$

This means that knowledge of T implies knowledge of G. Because the converse is obviously true (by the definition of T), we see that the information contained in T is precisely equivalent to that in G.

---

[7] We shall generally use the notation $G(z)^\dagger$ to denote the adjoint of the operator $G(z)$; and similarly, $f(z)^*$ for the complex conjugate of the number $f(z)$. However, we shall occasionally add a bracket, as in (8.7), to emphasize the meaning.

The defining equation $T = V + VGV$ gives T in terms of the unknown Green's operator G. If we use the identity (8.9) to replace GV by G^0T, we obtain the equation

$$T(z) = V + VG^0(z)T(z) \qquad (8.11)$$

for T in terms of the known operator G^0. This equation is known as the *Lippmann–Schwinger equation for* $T(z)$, and is the starting point of many methods for calculating T. In particular, when V is *sufficiently weak* one can hope to obtain a reliable solution by iteration, starting with the so-called Born approximation $T \approx V$. Inserting this into the right-hand side of (8.11) gives as the second approximation $T \approx V + VG^0V$; and continuing this procedure produces the infinite series,

$$T = V + VG^0V + VG^0VG^0V + \cdots \qquad (8.12)$$

This series (which may or may not converge) is known as the Born series and will be the subject of Chapter 9.

Finally we note that the identity (8.7) for G leads to a corresponding result for T,

$$T(z^*) = [T(z)]^\dagger \qquad (8.13)$$

8-c.   Relation to the Møller Operators

In this and the next section we establish the relevance of the operators $G(z)$ and $T(z)$ to scattering theory. Specifically, we establish expressions for the Møller operators $\Omega_\pm$ and the scattering operator S in terms of $G(z)$ and $T(z)$. While our primary interest is in the S operator, it is instructive and useful to discuss the Møller operators first.

Recall that if a particle entered a collision with the in asymptote labelled by the normalized vector $|\psi_{\text{in}}\rangle = |\phi\rangle$, then its actual state at $t = 0$ is given by

$$|\psi\rangle = \Omega_+ |\psi_{\text{in}}\rangle = \Omega_+ |\phi\rangle \equiv |\phi+\rangle \qquad [\text{if } |\psi_{\text{in}}\rangle = |\phi\rangle]$$

Similarly, if the out asymptote were going to be $|\psi_{\text{out}}\rangle = |\phi\rangle$, then the actual state at $t = 0$ would have to be

$$|\psi\rangle = \Omega_- |\psi_{\text{out}}\rangle = \Omega_- |\phi\rangle \equiv |\phi-\rangle \qquad [\text{if } |\psi_{\text{out}}\rangle = |\phi\rangle]$$

The Møller operators were defined as the limits as $t \to \mp\infty$ of $U(t)^\dagger U^0(t)$. In the course of proving that these limits exist, we rewrote $U^\dagger U^0$ as the integral of its derivative, which led to the result (which we write for Ω_-)

$$|\phi-\rangle \equiv \Omega_- |\phi\rangle = \lim_{t \to \infty} U(t)^\dagger U^0(t) |\phi\rangle$$

$$= |\phi\rangle + i \int_0^\infty d\tau\, U(\tau)^\dagger V U^0(\tau) |\psi\rangle \qquad (8.14)$$

We found that this integral was not only convergent, but was absolutely convergent; that is, if the integrand is replaced by its norm, the integral still converges. Now, it is a simple matter to verify that if an integral of the type $\int_0^\infty d\tau\,|\psi_\tau\rangle$ is absolutely convergent, then it satisfies the identity

$$\int_0^\infty d\tau\,|\psi_\tau\rangle = \lim_{\epsilon \downarrow 0}\int_0^\infty d\tau\,e^{-\epsilon\tau}\,|\psi_\tau\rangle \tag{8.15}$$

where the notation $\epsilon \downarrow 0$ is used to emphasize that ϵ tends to zero through *positive* values. This identity is easy to understand: The factor $e^{-\epsilon\tau}$ is approximately one when τ is small ($\tau \ll 1/\epsilon$), but damps out the integrand when τ is large ($\tau \gtrsim 1/\epsilon$). But since the integral converges, the contribution from τ sufficiently large is in any case insignificant. Thus, if we choose ϵ small enough, the damping factor does not change the value of the integral. (The reader can easily fill in the details of this argument—see Problem **8.1**.) Using this identity we can rewrite the result (8.14) as:

$$|\phi-\rangle \equiv \Omega_-\,|\phi\rangle = |\phi\rangle + i\lim_{\epsilon \downarrow 0}\int_0^\infty d\tau\,e^{-\epsilon\tau}\mathsf{U}(\tau)^\dagger V\mathsf{U}^0(\tau)\,|\phi\rangle \tag{8.16}$$

In the corresponding expression for $|\phi+\rangle \equiv \Omega_+\,|\phi\rangle$, the integral runs from 0 to $-\infty$ and the appropriate damping factor is therefore $e^{+\epsilon\tau}$ ($\epsilon > 0$).

Before applying this result let us make some comments. First, the validity of the result depends only on our demonstration that the original integral (8.14) was suitably convergent. Thus, (8.16) is rigorously true, and in particular, the limit exists for all proper[8] $|\phi\rangle$. Second, the usefulness of the result is that the original integral (8.14) is convergent, but only because of the comparatively subtle oscillations associated with the spreading of wave packets. On the other hand, the integral (8.16) is exponentially convergent. Of course this exponential convergence is achieved at the cost of having the limit $\epsilon \downarrow 0$ to handle. However, it turns out that this is a price worth paying. In particular, we shall find that with the help of (8.16) we can replace the proper vector $|\phi\rangle$ by the improper plane-wave state $|\mathbf{p}\rangle$—an essential step in the discussion of the stationary scattering states in Chapter 10.

Finally, we can interpret the expression (8.16) by noting that, roughly speaking, it differs from (8.14) by a change of potential from V to[9] $Ve^{-\epsilon|\tau|}$. (By writing $|\tau|$ we cover both Ω_+ and Ω_-.) Thus, we can regard the equivalence of (8.14) and (8.16) as expressing the fact that the motion of any given scattering orbit would be unchanged if we were to replace V by $Ve^{-\epsilon|\tau|}$, for sufficiently small ϵ. This is exactly what one would expect, since we know that for any given orbit the projectile eventually moves so far away that the

---

[8] Strictly speaking we should say that it exists for all $|\phi\rangle$ in the domain of H.

[9] This is not quite true since a change in V will also change $\mathsf{U}(t)$.

potential ceases to have any effect. Thus, for any given orbit we can choose ϵ so small that V and $Ve^{-\epsilon|\tau|}$ do not differ materially until such a time when the potential is in any case irrelevant. The result that one can do scattering theory with V replaced by $Ve^{-\epsilon|\tau|}$ is known as the *adiabatic theorem* and is sometimes used as the starting point for an approach based more heavily on the time-independent scattering formalism than ours.

Returning to our main calculation, we insert into (8.16) a complete set of states $|\mathbf{p}\rangle$ to give:

$$|\phi-\rangle \equiv \Omega_- |\phi\rangle = |\phi\rangle + i \lim_{\epsilon \downarrow 0} \int d^3p \int_0^\infty d\tau [e^{-\epsilon\tau} U(\tau)^\dagger V U^0(\tau)] \, |\mathbf{p}\rangle\langle \mathbf{p} \,|\, \phi\rangle \quad (8.17)$$

Now, the free evolution operator $U^0(\tau)$ acting on $|\mathbf{p}\rangle$ gives just $\exp(-iE_p\tau)$. Thus, the operator in brackets in the integrand can be replaced by

$$[\cdots] = \exp[-i(E_p - i\epsilon - H)\tau]V$$

and the integral over τ performed,

$$\int_0^\infty d\tau[\cdots] = -i(E_p - i\epsilon - H)^{-1}V = -iG(E_p - i\epsilon)V$$

(Because of the damping factor, there is no contribution from the upper limit of integration.) Substituting back into (8.17) we obtain:

$$|\phi-\rangle \equiv \Omega_- |\phi\rangle = |\phi\rangle + \lim_{\epsilon \downarrow 0} \int d^3p \, G(E_p - i\epsilon)V \, |\mathbf{p}\rangle\langle \mathbf{p} \,|\, \phi\rangle \qquad (8.18)$$

An exactly similar analysis gives a corresponding expression for $|\phi+\rangle$. In this case, since the appropriate damping factor is $e^{+\epsilon\tau}$, the argument of the Green's operator is $(E_p + i\epsilon)$;

$$|\phi+\rangle \equiv \Omega_+ |\phi\rangle = |\phi\rangle + \lim_{\epsilon \downarrow 0} \int d^3p \, G(E_p + i\epsilon)V \, |\mathbf{p}\rangle\langle \mathbf{p} \,|\, \phi\rangle \qquad (8.19)$$

Note that the signs of the term $\pm i\epsilon$ in these expressions for the Møller operators are the *same* as the subscripts $\pm$ on $\Omega_\pm$. This is in fact the original reason for this choice of the subscripts.

We shall return to these expressions for $\Omega_\pm$ in terms of $G(z)$ in Chapter 10, where they will be an essential tool for the discussion of the stationary scattering states. Here we remark that, since we could replace the factors GV in (8.18) and (8.19) by G^0T, we have successfully expressed the Møller operators $\Omega_\pm$ in terms of either $G(z)$ or $T(z)$. For the moment, our main objective is to express the *scattering operator* S in terms of $T(z)$. This we do in the next section, using the same techniques as used in this section.

8-d.   Relation to the Scattering Operator

Having expressed $\Omega_\pm$ in terms of $T(z)$ we can now do the same for $S = \Omega_-^\dagger \Omega_+$. We could use the results (8.18) and (8.19) for $\Omega_\pm$ but it is just as easy to proceed directly from first principles. The calculation is unfortunately rather messy, but the result is absolutely central. Our starting point is the equation

$$\langle\chi| S |\phi\rangle = \langle\chi| \Omega_-^\dagger \Omega_+ |\phi\rangle$$
$$= \lim_{\substack{t\to+\infty\\t'\to-\infty}} \langle\chi| (e^{iH^0 t}e^{-iHt})(e^{iHt'}e^{-iH^0 t'}) |\phi\rangle \qquad (8.20)$$

The order in which we take the limits $t \to +\infty$ and $t' \to -\infty$ is immaterial (as the interested reader can easily check). In particular, we can take the two limits simultaneously; that is, we can set $t' = -t$ and simply let $t \to +\infty$, to give

$$\langle\chi| S |\phi\rangle = \lim_{t\to\infty} \langle\chi| [e^{iH^0 t}e^{-2iHt}e^{iH^0 t}] |\phi\rangle$$

We now employ the familiar trick of writing this expression as the integral of its derivative, in this case

$$\frac{d}{dt}[\cdots] = -i\{e^{iH^0 t}Ve^{-2iHt}e^{iH^0 t} + e^{iH^0 t}e^{-2iHt}Ve^{iH^0 t}\}$$

This gives

$$\langle\chi| S |\phi\rangle = \langle\chi | \phi\rangle - i\int_0^\infty dt\langle\chi| \{\cdots\} |\phi\rangle$$
$$= \langle\chi | \phi\rangle - i\lim_{\epsilon\downarrow 0}\int_0^\infty dt e^{-\epsilon t}\langle\chi| \{\cdots\} |\phi\rangle$$

where the introduction of the damping factor is justified exactly as in the previous section. If we now replace the proper vectors $|\chi\rangle$ and $|\phi\rangle$ by momentum eigenstates $|\mathbf{p}'\rangle$ and $|\mathbf{p}\rangle$ the free evolution operators in the integrand simplify and we get

$$\langle\mathbf{p}'| S |\mathbf{p}\rangle = \delta_3(\mathbf{p}' - \mathbf{p}) - i\lim\int_0^\infty dt\langle\mathbf{p}'| \{Ve^{i(E_{p'}+E_p+i\epsilon-2H)t}$$
$$+ e^{i(E_{p'}+E_p+i\epsilon-2H)t}V\} |\mathbf{p}\rangle$$
$$= \delta_3(\mathbf{p}' - \mathbf{p}) + \tfrac{1}{2}\lim \langle\mathbf{p}'| \left\{VG\left(\frac{E_{p'}+E_p}{2} + i\epsilon\right)\right.$$
$$\left.+ G\left(\frac{E_{p'}+E_p}{2} + i\epsilon\right)V\right\} |\mathbf{p}\rangle$$

If we next replace VG by TG^0, and GV by G^0T, the free Green's operators acting on the momentum eigenstates can be replaced by their eigenvalues to give

$$= \delta_3(\mathbf{p}' - \mathbf{p}) + \lim \left\{ \frac{1}{E_{p'} - E_p + i\epsilon} + \frac{1}{E_p - E_{p'} + i\epsilon} \right\}$$

$$\times \langle \mathbf{p}'| \, T\left(\frac{E_{p'} + E_p}{2} + i\epsilon\right) |\mathbf{p}\rangle$$

Finally, we recognize the quantity in braces as one of the standard representations of the delta function, $\{\cdots\} = -2\pi i \, \delta(E_{p'} - E_p)$, and we obtain:

$$\boxed{\langle \mathbf{p}'| \, \mathsf{S} \, |\mathbf{p}\rangle = \delta_3(\mathbf{p}' - \mathbf{p}) - 2\pi i \, \delta(E_{p'} - E_p) \lim_{\epsilon \downarrow 0} \langle \mathbf{p}'| \, T(E_p + i\epsilon) \, |\mathbf{p}\rangle} \qquad (8.21)$$

This is one of the central results of time-independent scattering theory. Comparing it with the decomposition of $\langle \mathbf{p}'| \, \mathsf{S} \, |\mathbf{p}\rangle$ in terms of the on-shell T matrix,

$$\langle \mathbf{p}'| \, \mathsf{S} \, |\mathbf{p}\rangle = \delta_3(\mathbf{p}' - \mathbf{p}) - 2\pi i \, \delta(E_{p'} - E_p) \, t(\mathbf{p}' \leftarrow \mathbf{p})$$

we see that

$$\boxed{\begin{aligned} t(\mathbf{p}' \leftarrow \mathbf{p}) &= \lim_{\epsilon \downarrow 0} \langle \mathbf{p}'| \, T(E_p + i\epsilon) \, |\mathbf{p}\rangle \\ &\equiv \langle \mathbf{p}'| \, T(E_p + i0) \, |\mathbf{p}\rangle \end{aligned}} \quad \cdots [E_{p'} = E_p] \qquad (8.22)$$

where we have again used the notation $f(E + i0)$ to denote the limit of $f(E + i\epsilon)$ as $\epsilon \downarrow 0$. This shows that the *on-shell* T matrix $t(\mathbf{p}' \leftarrow \mathbf{p})$, which is defined only for $E_{p'} = E_p$, is in fact the $\mathbf{p}'$, $\mathbf{p}$ matrix element of the operator $T(z)$ for the particular values $z = E_p + i0$ and $E_{p'} = E_p$. For this reason $\langle \mathbf{p}'| \, T(z) \, |\mathbf{p}\rangle$ is known as the *off-shell* T matrix. It should be emphasized that this off-shell matrix is more general than $t(\mathbf{p}' \leftarrow \mathbf{p})$ in two ways. First, the variable z is an arbitrary complex number unrelated to either $\mathbf{p}'$ or $\mathbf{p}$, and second, $\mathbf{p}'$ and $\mathbf{p}$ are themselves quite independent. It just so happens that when $|\mathbf{p}'| = |\mathbf{p}|$ and $z = E_p + i0$ the off-shell matrix coincides precisely with $t(\mathbf{p}' \leftarrow \mathbf{p})$.

For the reader who has not met the off-shell T matrix before some comments are in order. First, the fact that $\langle \mathbf{p}'| \, T(z) \, |\mathbf{p}\rangle$ is defined and nonzero for $E_{p'} \neq E_p$ may at first seem surprising, since it perhaps appears to violate conservation of energy. However, it is the *scattering operator* S that conserves energy. The matrix elements $\langle \mathbf{p}'| \, \mathsf{S} \, |\mathbf{p}\rangle$ certainly are zero for $E_{p'} \neq E_p$, as is guaranteed by the delta function $\delta(E_{p'} - E_p)$ in (8.21). Since the delta function is zero wherever $E_{p'} \neq E_p$, there is no reason why the factor $\langle \mathbf{p}'| \, T(E_p + i0) \, |\mathbf{p}\rangle$ should vanish there, and in fact it does not.

Second, since it is the S matrix that is physically relevant, and since it is only the *on-shell* values of the T matrix that actually contribute to the S matrix in (8.21), one may reasonably ask why one bothers to define an off-shell T matrix at all. To put this differently, we could obviously add to the off-shell T matrix absolutely any function of $\mathbf{p}'$, $\mathbf{p}$ and z—provided only that it vanishes on-shell—without changing the result (8.21). In other words, the definition of the off-shell extension of the T matrix is almost entirely arbitrary. So why do we bother to define it?[10] The answer is, briefly, that it turns out to be *convenient* to define the off-shell T matrix and that the convenient definition is that given above. With this definition, the off-shell T matrix proves to be a powerful tool for calculating scattering amplitudes and for establishing their general analytic properties.

We can explain the situation informally as follows: the S matrix $\langle \mathbf{p}'| \, S \, |\mathbf{p}\rangle$ is a highly singular function of its arguments because of the two delta functions in (8.21). The off-shell T matrix $\langle \mathbf{p}'| \, T(z) \, |\mathbf{p}\rangle$ is a smooth function of its arguments and satisfies an equation (the Lippmann–Schwinger equation) that can be easily handled. This equation is an integral equation for $\langle \mathbf{p}'| \, T(z) \, |\mathbf{p}\rangle$,

$$\langle \mathbf{p}'| \, T(z) \, |\mathbf{p}\rangle = \langle \mathbf{p}'| \, V \, |\mathbf{p}\rangle + \int d^3p'' \, \frac{\langle \mathbf{p}'| \, V \, |\mathbf{p}''\rangle}{z - E_{p''}} \, \langle \mathbf{p}''| \, T(z) \, |\mathbf{p}\rangle \qquad (8.23)$$

It is, in fact, impossible to discuss this equation without reference to the off-shell values of $\langle \mathbf{p}'| \, T(z) \, |\mathbf{p}\rangle$. Thus, knowing that we are interested in the on-shell T matrix, we can determinedly take $|\mathbf{p}'| = |\mathbf{p}|$ and set $z = E_p + i0$. In this case the left hand side is precisely the required on-shell T matrix. However, the integral on the right still runs over *all* $\mathbf{p}''$. Thus if we wish to solve (8.23) as an integral equation for the T matrix, we must at the very least solve it for the function $\langle \mathbf{p}'| \, T(E_p + i0) \, |\mathbf{p}\rangle$ for *all* $\mathbf{p}'$ and only afterwards set $|\mathbf{p}'| = |\mathbf{p}|$. That is, any approach to the T matrix via the Lippmann–Schwinger equation involves discussion of this "half-off-shell" T matrix at the very least.[11]

Another use for the off-shell T matrix will emerge when we discuss multiparticle scattering. It turns out that the physically relevant *on*-shell T matrix for multiparticle processes can be related, in various approximations, to certain *off*-shell two-body T matrices. For example, we shall see in the

---

[10] In fact there is a school of thought (associated with the so-called analytic S-matrix theory) that argues that a reasonable physical theory should involve only on-shell quantities. Whether this approach will yield a complete and satisfactory scattering theory remains to be seen.

[11] This explains why we discuss $\langle \mathbf{p}'| \, T(z) \, |\mathbf{p}\rangle$ for $E_{p'} \neq E_p$, but not why we are interested in complex values of z; that is, one might hope always to hold z fixed at $E_p + i0$. However, the *meaning* of $T(E + i0)$ is the limit of $T(z)$ as z approaches the real axis at E from above. Thus, in a rigorous discussion of the Lippmann–Schwinger equation, one must first work with z complex and then take this limit.

scattering of one particle a off a bound state (bc) of b and c, that the amplitude can be written approximately as a weighted sum of the amplitudes for scattering of a off b and of a off c separately. However, as a scatters off b (bound to c), energy and momentum can be taken up by c and vice versa. Thus, the appropriate two-body amplitudes are off-shell.

Finally, it may be well to reiterate what was said at the beginning of this chapter: that our method of introducing the off-shell T matrix—to produce the definition $T = V + VGV$ out of a vacuum and then "discover" its relation to the on-shell T matrix—is certainly not the way in which it was originally discovered. Historically, scattering theory was developed (in analogy with bound state theory) in terms of the stationary "scattering eigenfunctions" of Chapter 10. It was in terms of these functions that the off-shell T matrix was introduced. The eigenfunctions satisfy the time-independent Schrödinger equation, which could be converted into an equivalent integral equation, which in turn led to the Lippmann–Schwinger for T. Thus, the Lippmann–Schwinger equation for T is historically viewed as the reflection of the Schrödinger equation for the scattering eigenfunctions.

PROBLEMS

8.1. Given that $\int_0^\infty d\tau \, \|\psi_\tau\| < \infty$, prove the identity

$$\int_0^\infty d\tau \, |\psi_\tau\rangle \equiv \lim_{\epsilon \downarrow 0} \int_0^\infty d\tau e^{-\epsilon\tau} \, |\psi_\tau\rangle$$

(Write $\int_0^\infty$ as $\int_0^T + \int_T^\infty$; choose T so large that $\int_T^\infty$ is small on both sides of the identity. Then choose ϵ so small that the two integrals $\int_0^T$ are close.)

8.2. Consider the separable potential

$$V = \lambda \, |\zeta\rangle\langle\zeta|$$

where $|\zeta\rangle$ is a normalized vector specified by its momentum wave function $\zeta(\mathbf{p})$.
(a) Show that $T(z)$ is given explicitly by

$$T(z) = \frac{\lambda \, |\zeta\rangle\langle\zeta|}{1 - \lambda \, \Delta(z)}$$

where

$$\Delta(z) = \langle\zeta| \, G^0(z) \, |\zeta\rangle = \int d^3p \, \frac{|\zeta(\mathbf{p})|^2}{z - E_p}$$

[Hint: First use the definition $T = V + VGV$ to show that $T = \alpha(z) \, |\zeta\rangle\langle\zeta|$ where $\alpha(z)$ is some number. Then use the Lippmann–Schwinger equation $T = V + VG^0T$ to find $\alpha(z)$.]

(b) Show that the Born series for T is convergent to the correct answer for λ small but divergent for λ large.

(c) By considering the poles of $T(z)$ show that V has either one or no bound states.

(d) If $\zeta(\mathbf{p})$ is spherically symmetric (i.e., $|\zeta\rangle$ is an eigenvector of $\mathbf{L}$ with eigenvalue zero), find the partial-wave amplitudes.

9 The Born Series

9-a   The Born Series

9-b   The Born Approximation

9-c   The Yukawa Potential

9-d   Scattering of Electrons off Atoms

9-e   Interpretation of the Born Series in Terms
         of Feynman Diagrams

In Chapter 8 it was shown that the on-shell T matrix $t(\mathbf{p}' \leftarrow \mathbf{p})$ is just the $\mathbf{p}', \mathbf{p}$ matrix element of the operator $T(z)$ in the limit $z \to E_p + i0$. In this chapter we discuss an important method for the actual computation of $T(z)$: the so-called Born series. As was mentioned in Chapter 8, this is obtained when one tries to solve the Lippman–Schwinger equation $T = V + VG^0T$ by iteration, using as first approximation the *Born approximation* $T \approx V$. One iteration gives the second Born approximation $T \approx V + VG^0V$, while the final answer is the Born series,

$$T(z) = V + VG^0(z)V + VG^0(z)VG^0(z)V + \cdots \tag{9.1}$$

It should be emphasized that the Born series does not always converge, and still less does it necessarily converge so rapidly that its first one or two terms provide a reliable approximation. Nonetheless, there are circumstances (notably high energies or weak potentials) where the series does converge and does so very rapidly. It is from this circumstance that the method derives its importance.

The applications of the Born series are many and varied. It can be used theoretically to establish certain general properties of the scattering amplitude, in which case it is usually sufficient to know only that the series *does* converge. It can be used as a practical method of computing amplitudes, in which case it must converge so rapidly that only a very few terms are important. In fact, in most cases even the second term is so cumbersome that the method is used only if the first term alone—the Born approximation—is dominant.[1] In some fields (e.g., atomic physics) there are situations where the series can be shown to converge so rapidly that the Born approximation alone is in excellent agreement with experiment. At the opposite extreme there are situations, notably in strong interaction physics, where almost nothing is known about convergence of the Born series, and the Born approximation is used simply for lack of any practicable alternative.

In this chapter we discuss some general conditions for convergence of the Born series. As specific examples we consider two simple models. In Section 9-c we discuss the scattering of two nucleons interacting via a Yukawa potential, and in Section 9-d the scattering of an electron off an atom (the latter being treated as a static charge distribution that defines the potential seen by the electron). Finally, in Section 9-e we discuss the interpretation of the Born series in terms of Feynman diagrams.

Further discussion of the Born series and some of its generalizations (the partial-wave version, the distorted-wave Born series, the multichannel version, etc.) can be found in several of the chapters following.

9-a.   The Born Series

To better understand the significance of the Born series it is useful to introduce a parameter λ multiplying the potential and to consider in place of the actual Hamiltonian $H = H^0 + V$ the family of Hamiltonians, $H = H^0 + \lambda V$. The number λ (which must be real for H to be Hermitian[2]) can be considered as a strength parameter or coupling constant, whose value in the actual problem of interest is $\lambda = 1$. The operators $G(z)$ and $T(z)$ are then functions of λ as well as z, but we shall not indicate this explicitly.

The Born series results from the *assumption* that $T(z)$ can be expanded as a

---

[1] An important exception is the Born series in quantum electrodynamics, where the techniques of Feynman graphs have made practicable the evaluation of several terms.

[2] We shall later have occasion to discuss *complex* values of λ for which the operator $H^0 + \lambda V$, although a well defined mathematical quantity, is no longer the Hamiltonian of any physical system. We shall then refer to real values of λ as "physical values" and complex values as "unphysical." For the present we consider only physical values.

power series in λ,

$$T(z) = \sum_{0}^{\infty} \lambda^n T^{(n)}(z) \qquad (9.2)$$

Substituting this expansion into the Lippmann–Schwinger equation,

$$T = \lambda V + \lambda V G^0 T \qquad (9.3)$$

and equating corresponding powers of λ we find that $T^{(0)} = 0$, $T^{(1)} = V$ and

$$T^{(n)} = V G^0 T^{(n-1)} = (V G^0)^{n-1} V$$

The power series (9.2) is therefore:

$$T(z) = \lambda V + \lambda^2 V G^0(z) V + \cdots \qquad (9.4)$$

which, for the actual value $\lambda = 1$, is just the series (9.1) discussed above.

An alternative route to the same answer is to rewrite the Lippmann–Schwinger equation (9.3) as:

$$(1 - \lambda V G^0) T = \lambda V$$

In this form it is clear that the problem of finding T is equivalent to that of inverting the operator $(1 - \lambda V G^0)$. Provided this inverse exists, the solution for T is just

$$T = (1 - \lambda V G^0)^{-1} \lambda V$$

In particular, if we assume that the inverse can be expanded as a binomial series

$$(1 - \lambda V G^0)^{-1} = 1 + \lambda V G^0 + (\lambda V G^0)^2 + \cdots$$

we recover the Born series (9.4).

From the series expansion for $T(z)$ one can immediately write down a corresponding Born series for the quantity of primary importance, the on-shell T matrix,

$$\boxed{\begin{aligned} t(\mathbf{p}' \leftarrow \mathbf{p}) &= \langle \mathbf{p}' | \, T(E_p + i0) \, | \mathbf{p} \rangle \\ &= \lambda \, \langle \mathbf{p}' | \, V \, | \mathbf{p} \rangle + \lambda^2 \, \langle \mathbf{p}' | \, V G^0(E_p + i0) V \, | \mathbf{p} \rangle + \cdots \end{aligned}} \qquad (9.5)$$

This is, of course, the result of direct physical interest, and the important questions to consider concerning the series are: Does the series converge? In particular does it converge for the value $\lambda = 1$ (corresponding to the actual Hamiltonian $H = H^0 + V$)? Does it converge sufficiently fast to give a useful method of computation?

Before giving some exact answers to these questions we make two comments about what one might expect to be true. The Born series is a series in powers of the operator $\lambda V G^0$ and we can anticipate that it should converge whenever this operator is "small" (in some appropriate sense). First, this suggests that convergence should be good *for λ sufficiently small*—a result that would make precise the idea that the series should converge for weak potentials. Second, the operator $G^0(z) = (z - H^0)^{-1}$ is obviously "small" (in some sense) when z *is large*, and we can anticipate that the series (9.5) should converge well *for the energy E_p sufficiently large*. These ideas are made precise in the following results, both of which apply to any potential V satisfying the conditions discussed in Section 2-b (i.e., $V = O(r^{-3/2 + \epsilon})$ for r small, $V = O(r^{-3 - \epsilon})$ for r large, and V is piecewise continuous in between). For the proofs, which are quite complicated, we refer the interested reader to the literature.[3]

Convergence for Weak Coupling. For any potential V there is a $\bar{\lambda} > 0$ such that the Born series (9.5) converges to the correct on-shell T matrix for all values of $\mathbf{p}$ and $\mathbf{p}'$, provided $|\lambda| < \bar{\lambda}$. The smaller λ the more rapid the convergence, and for λ sufficiently small, the first term alone (the Born approximation) dominates.

This result is of considerable theoretical interest. For example, it guarantees the existence of infinitely many potentials (namely any V multiplied by a sufficiently small λ) for which the amplitude is given by the Born series. However, it is of no practical use for a given potential V unless it happens to include the actual value $\lambda = 1$. In this connection it should be mentioned that if λ is small enough for convergence at all energies, then the potential λV supports no bound states. Turning this statement around, we see that if a given potential *does* support a bound state, then the corresponding Born series certainly does not converge for all momenta.[4]

Convergence at High Energies. For a given V there is an energy $\bar{E}$ above which the Born series converges for the actual value $\lambda = 1$. The higher the energy (above $\bar{E}$) the more rapid the convergence, and for E sufficiently large the Born approximation dominates.

---

[3] A rather complete account can be found in Newton (1966) Sections 9.1 and 10.3, or see the original paper of Klein and Zemach (1958). The important condition that the integral (10.64) (of Newton) be finite is certainly satisfied under the conditions mentioned above.
[4] See Newton (1966) p. 138. The reason for this can be roughly understood as follows: If there *is* a bound state of $H = H^0 + \lambda_0 V$ at energy E_0 less than zero, then $T(z)$ has a pole at $z = E_0$, $\lambda = \lambda_0$. This means that a power series in λ cannot possibly converge, at least for $z = E_0$, $\lambda = \lambda_0$. It can be shown that the situation cannot improve if we then move z up to a neighborhood of zero, and, hence, that the series diverges for some interval above $E = 0$ (and $\lambda = \lambda_0$). Thus, if the Born series *does* converge for all energies ($0 \leqslant E < \infty$) there cannot have been any bound states.

The chief practical importance of this second result is that at sufficiently high energies, for any given potential, the Born approximation is certainly reliable. The only reservation is that "sufficiently high energies" may be so high that our description of the system in terms of a potential has become unrealistic (e.g., because of relativistic effects).

9-b.   The Born Approximation

The two results just stated guarantee that there are many situations where the Born series is convergent and, better still, where it converges so rapidly that only its first term is important. When this is so, we have for the on-shell T matrix (returning λ to its actual value $\lambda = 1$)

$$t(\mathbf{p}' \leftarrow \mathbf{p}) \approx t^{(1)}(\mathbf{p}' \leftarrow \mathbf{p}) = \langle \mathbf{p}' | \, V \, | \mathbf{p} \rangle$$

and, hence, for the amplitude

$$f(\mathbf{p}' \leftarrow \mathbf{p}) \approx f^{(1)}(\mathbf{p}' \leftarrow \mathbf{p}) = -(2\pi)^2 m \langle \mathbf{p}' | \, V \, | \mathbf{p} \rangle$$

$$= -\frac{m}{2\pi} \int d^3x \, e^{-i\mathbf{p}' \cdot \mathbf{x}} V(\mathbf{x}) e^{i\mathbf{p} \cdot \mathbf{x}}$$

or

$$\boxed{f^{(1)}(\mathbf{p}' \leftarrow \mathbf{p}) = -\frac{m}{2\pi} \int d^3x \, e^{-i\mathbf{q} \cdot \mathbf{x}} V(\mathbf{x})} \tag{9.6}$$

In this last expression we have introduced the vector $\mathbf{q} \equiv \mathbf{p}' - \mathbf{p}$. This is the momentum transferred to the projectile by the target and is known as the *momentum transfer*. It has magnitude (Fig. 9.1)

$$q = 2p \sin \frac{\theta}{2}$$

The Born approximation (9.6) is probably the single most important approximate expression for the amplitude. Perhaps its most striking feature is that it depends on the momenta $\mathbf{p}$ and $\mathbf{p}'$ only through their difference $\mathbf{q} = \mathbf{p}' - \mathbf{p}$; that is, the Born approximation is a function of the momentum transfer only. In fact, apart from a numerical factor, $f^{(1)}(\mathbf{p}' \leftarrow \mathbf{p})$ is simply

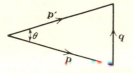

FIGURE 9.1.   The momentum transfer, $q = \mathbf{p}' - \mathbf{p}$.

the Fourier transform of the potential, evaluated at $\mathbf{q}$. For any given potential, calculation of the amplitude in Born approximation is a simple question of evaluating this Fourier transform.

If the potential is spherically symmetric, the angular integral in the Born approximation is easily carried out to give

$$f^{(1)}(\mathbf{p}' \leftarrow \mathbf{p}) = -2m \int_0^\infty r^2 \, dr \, \frac{\sin qr}{qr} \, V(r) \qquad (9.7)$$

In the next two sections we shall evaluate and discuss this integral for various specific potentials. In the meantime we mention some important general properties that hold for any spherical potential.

The Forward Amplitude.   If $\mathbf{p}' = \mathbf{p}$ then $\mathbf{q} = 0$ and it follows that:

$$f^{(1)}(\mathbf{p} \leftarrow \mathbf{p}) = -2m \int_0^\infty r^2 \, dr V(r) \qquad (9.8)$$

that is, the forward amplitude has a fixed value independent of energy.[5] This property is clearly visible in Fig. 9.2, which shows the cross section in Born approximation for the Yukawa potential [see (9.13)]

$$V(r) = \frac{\gamma e^{-\mu r}}{r}$$

Zero Energy.   At zero energy $q = 0$. Thus, $f^{(1)}(\mathbf{p}' \leftarrow \mathbf{p})$ is again given by (9.8) and is independent of θ; that is, the scattering (in Born approximation) becomes isotropic at low energy (see Fig. 9.2). This is also true of the exact amplitude $f(\mathbf{p}' \leftarrow \mathbf{p})$. However, since the Born approximation is not generally reliable at low energies, this property of $f^{(1)}$ is largely irrelevant.

The Forward Peak.   When q becomes large the Born amplitude (9.7) goes to zero, as is clear both from the factor q in the denominator and the oscillatory factor $\sin qr$ in the numerator. Since

$$q = 2p \sin \frac{\theta}{2}$$

this means that at high energy (when the Born approximation is certainly good) the amplitude falls off as θ increases from 0 to π. The higher the energy the faster the amplitude decreases until eventually all that remains is a narrow peak in the forward direction, as illustrated in Fig. 9.2.

---

[5] We assume as usual that V is $0(r^{-3-\epsilon})$ as $r \to \infty$, and the integral (9.8) is therefore finite. Notice that if V is greater than r^{-3} (as for example when V is a Coulomb potential) the integral diverges and the forward amplitude is infinite.

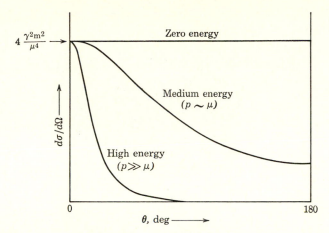

FIGURE 9.2. Typical behavior for the CM differential cross section in the Born approximation (see (9.13)).

Unitarity. One final property of the Born approximation requires comment. We have seen that $f^{(1)}(\mathbf{p}' \leftarrow \mathbf{p})$ is proportional to $\langle \mathbf{p}'| V |\mathbf{p} \rangle$ and since V is Hermitian, it follows that the *forward amplitude in Born approximation is real*.[6] According to the optical theorem the forward amplitude has an imaginary part proportional to the total cross section, and it can therefore *never* be real unless there is no scattering at all. Since the optical theorem holds because S is unitary, it is usual to describe this apparent conflict by saying that the *Born approximation violates unitarity*.

First we note that it is easy to see how this apparent contradiction arises. The Born series expresses f as a series in powers of λ starting with the linear term $\lambda f^{(1)}$. From the optical theorem,

$$\mathrm{Im}\, f(\mathbf{p} \leftarrow \mathbf{p}) = \frac{p}{4\pi} \int d\Omega \, |f|^2$$

it is clear that the imaginary part of the forward amplitude is of order λ^2. It is not at all surprising then, that if we calculate f to order λ only, we completely miss its imaginary part.

---

[6] The general result is obviously that $f^{(1)}(\mathbf{p}' \leftarrow \mathbf{p})$ is *Hermitian*, in the sense of (9.10) below. For simplicity we focus attention on the forward amplitude, for which Hermiticity is equivalent to reality. If the potential is spherically symmetric it is clear from (9.7) that $f^{(1)}(\mathbf{p}' \leftarrow \mathbf{p})$ is *real* for all $\mathbf{p}'$ and $\mathbf{p}$; nonetheless the general result is that in Born approximation the amplitude is Hermitian.

To understand the situation better we rewrite the Born series as a series for the S operator,

$$S = 1 + \lambda S^{(1)} + \lambda^2 S^{(2)} + \cdots$$

Substituting into the unitary equation $S^\dagger S = 1$ and collecting terms we find that

$$1 + \lambda[S^{(1)\dagger} + S^{(1)}] + \lambda^2[S^{(1)\dagger}S^{(1)} + S^{(2)\dagger} + S^{(2)}] + \cdots = 1 \qquad (9.9)$$

This equation can only hold if the coefficients of λ, λ^2, ... are all separately zero. In particular, considering the coefficient of λ, we see that

$$S^{(1)\dagger} = -S^{(1)}$$

or, taking matrix elements (remember the factor i in the decomposition of S in terms of f)

$$f^{(1)}(\mathbf{p} \leftarrow \mathbf{p}')^* = f^{(1)}(\mathbf{p}' \leftarrow \mathbf{p}) \qquad (9.10)$$

Setting $\mathbf{p}' = \mathbf{p}$ we see that, as a consequence of unitarity, the forward Born amplitude *has to* be real. Conversely, if (9.10) holds, then it is clear that the unitarity equation (9.9) *is* satisfied at least to order λ.

The trouble starts when we consider the term of order λ^2 in (9.9). Obviously if $S^{(1)}$ is nonzero, then $S^{(2)}$ must *also* be nonzero for the coefficient of λ^2 to vanish. Now the Born approximation consists in neglecting $S^{(2)}$ and higher terms in the series for S. But if $S^{(2)}$ is taken to be zero, then the coefficient of λ^2 in (9.9) cannot be zero and we find that to order λ^2 our approximation fails to be unitary. This situation is really not at all surprising. The actual S operator *is* unitary and satisfies the unitary equation (9.9) to *all* orders in λ. However, if we approximate S by neglecting terms of order λ^2, then it is natural that the resulting approximate expression satisfies the unitary equation to order λ only. Of course, if the Born series is rapidly convergent, the terms neglected are small and the violation of (9.9) is also small.

Our apparent conflict is seen to reduce to the not very surprising statement that if the Born series converges slowly, and we neglect all terms but the first, then our approximation will be poor and, in particular, will not satisfy unitarity. However, we can turn this result into a useful test by noting that if the Born approximation is found seriously to violate unitarity, then the approximation is certainly unreliable. This test is most useful in connection with the partial-wave Born series and we defer any further discussion to Chapter 11.

9-c.   The Yukawa Potential

As a first application of the Born series we consider the *Yukawa potential*

$$V(r) = \gamma \frac{e^{-\mu r}}{r} \qquad (9.11)$$

This potential was proposed by Yukawa as a model for the nucleon–nucleon interaction on the basis of his meson theory. It also appears in atomic physics, as a simple model for the screened Coulomb field produced by an atom. However, it is as a model for the nuclear interactions that the potential has been most extensively investigated, and it is in this context that we discuss it first.

The Fourier transform (9.7) for the Yukawa potential is easily evaluated to give for the Born approximation

$$f^{(1)}(\mathbf{p}' \leftarrow \mathbf{p}) = \frac{-2\gamma m}{\mu^2 + q^2} \tag{9.12}$$

where, of course, $q = 2p \sin \theta/2$. In fact, it is this result that suggests the relevance of the Yukawa potential (9.11) for the inter-nucleon interaction. The Born amplitude (9.12) arising from the Yukawa potential coincides precisely with the Born amplitude for nucleons interacting via a scalar meson field, if μ is identified as c times the meson mass, and γ is proportional to the square of the meson–nucleon coupling constant ($\mu \sim 140$ MeV/c and $\gamma \sim -2c/3$ if the meson is taken to be the pion[7]). Since the Born approximation is certainly unreliable for these values of μ and γ (as we shall see), this may seem a rather poor reason for considering the Yukawa potential. Nonetheless, there are good reasons to think the Yukawa potential *is* a reasonable model, at least for the long-range part of the nuclear interaction.

The CM cross section corresponding to the Born amplitude (9.12) is

$$\frac{d\sigma}{d\Omega} = \frac{4\gamma^2 m^2}{(\mu^2 + 4p^2 \sin^2 \theta/2)^2} \tag{9.13}$$

This is the cross section that was plotted in Fig. 9.2; it displays all of the general properties of the Born approximation discussed in Section 9-b. The forward cross section ($\theta = 0$) has the value $4\gamma^2 m^2/\mu^4$ independent of the energy, the zero-energy cross section is independent of θ. As the energy increases the cross section falls off more and more rapidly as a function of θ, leaving only a forward peak of angular width $\sim \mu/p$ once $p \gg \mu$.

To get a better approximation for the amplitude and to get some idea of the reliability of the first Born approximation $f^{(1)}$, it is natural to try computing the second Born term $f^{(2)}$, which is given by (9.5) as

$$f^{(2)}(\mathbf{p}' \leftarrow \mathbf{p}) = -(2\pi)^2 m \langle \mathbf{p}'| \, VG^0(E_p + i0)V \, |\mathbf{p}\rangle$$

---

[7] The Yukawa potential corresponds to a scalar meson, while the pion is in fact pseudo-scalar. The corresponding pseudoscalar equivalent of the Yukawa potential is given, for example, in Mott and Massey (1965) p. 317, Eq. (74).

or, inserting a complete set of states $|\mathbf{p}''\rangle$,

$$f^{(2)}(\mathbf{p}' \leftarrow \mathbf{p}) = -(2\pi)^2 m \int d^3 p'' \frac{\langle \mathbf{p}'| \, V \, |\mathbf{p}''\rangle \langle \mathbf{p}''| \, V \, |\mathbf{p}\rangle}{E_p - E_{p''} + i0} \qquad (9.14)$$

The matrix elements of V are simple Fourier transforms of $V(\mathbf{x})$ and have already been calculated for the first Born approximation. Thus, all that remains is to carry out the integration over $\mathbf{p}''$. Unfortunately, this integral is usually very cumbersome. For a Yukawa the answer can be given in terms of elementary functions (see Newton, 1966, p. 291), but since the resulting expression is not especially illuminating (for our present purposes) we consider only the special case of forward scattering, for which $\mathbf{p}' = \mathbf{p}$. For this case, a simple integration (Problem **9.1**) gives

$$f^{(2)}(\mathbf{p} \leftarrow \mathbf{p}) = \frac{2\gamma^2 m^2}{\mu^2(\mu - 2ip)} \qquad (9.15)$$

If the Born approximation is to be reliable, then we must certainly have $|f^{(2)}| \ll |f^{(1)}|$; conversely if $f^{(2)}$ is much smaller than $f^{(1)}$ it is at least plausible that the Born approximation is good. Thus, a reasonable, although not rigorous, test of the validity of the Born approximation is to compute the ratio $|f^{(2)}/f^{(1)}|$ and to compare it with unity. For the Yukawa potential we find from (9.15) and (9.12) (with $\theta = 0$) that this ratio is

$$\left| \frac{f^{(2)}}{f^{(1)}} \right| = \frac{\gamma m}{(\mu^2 + 4p^2)^{1/2}} \qquad [\theta = 0] \qquad (9.16)$$

For the nucleon-nucleon case[8] $\gamma \sim -2c/3$, $m \sim 500$ MeV/c^2 and $\mu \sim 140$ MeV/c. Thus, at zero energy, for example, the ratio $|f^{(2)}/f^{(1)}|$ in the forward direction is about 2.5. This means that at zero energy the Born approximation is almost certainly hopelessly wrong and the Born series divergent.

As the energy increases, the ratio $|f^{(2)}/f^{(1)}|$ in (9.16) steadily improves. However, it will not be comfortably less than unity until $p^2 \gg \mu^2$, in which case the nucleons would have relativistic energies, and the description in terms of a nonrelativistic potential is almost certainly unrealistic.

To make the corresponding test for $\theta \neq 0$ it is necessary to compute $f^{(2)}(\mathbf{p}' \leftarrow \mathbf{p})$ for arbitrary $\mathbf{p}$ and $\mathbf{p}'$. However, it should be emphasized that the simple comparison of $f^{(1)}$ and $f^{(2)}$ is in any case only a very crude test for convergence of the whole series. (For example, it could happen that although $f^{(2)}$ is small compared to $f^{(1)}$, the third term $f^{(3)}$ is nonetheless large.) For the purposes of the test it is certainly senseless to go to much trouble to evaluate

---

[8] Remember that m is the reduced mass, which for nucleon–nucleon scattering is half the nucleon mass.

$f^{(2)}$ exactly, and any rough estimate of the ratio $|f^{(2)}/f^{(1)}|$ will do just as well. There are several ways of estimating this ratio; one method is mentioned in Problem **10.2**. Here we simply state that whether one uses the exact form of $f^{(2)}$ or any reasonable estimate, one obtains essentially the same result for any θ as that for the forward direction given in (9.16). That is, at low energies $(p < \mu)$ a rough test for the validity of the Born approximation for the Yukawa potential is:

$$\left| \frac{\gamma m}{\mu} \right| \ll 1 \qquad \text{(low energy, } p < \mu) \qquad (9.17)$$

while at high energy it is[9]:

$$\left| \frac{\gamma m}{p} \right| \ll 1 \qquad \text{(high energy, } p > \mu) \qquad (9.18)$$

This result means, in particular, that direct application of the Born approximation (9.12) is unlikely to be of any value in the nucleon-nucleon scattering problem. However, as will become apparent, the Born approximation may be useful for computation of the partial-wave amplitudes of high l, even when it is not useful for computation of the full amplitude. In fact, used in this way, the Born approximation has played an important role in the analysis of nucleon-nucleon scattering (MacGregor, Moravcsik, and Stapp, 1960).

9-d.   Scattering of Electrons off Atoms

Having discussed one problem where the Born approximation is not directly useful, we now discuss another where it certainly *is*—the scattering of electrons off atoms. This is in reality a multiparticle problem of the type we shall be considering in Chapters 16–22. However, it turns out that a reasonable approximate picture of the elastic scattering (i.e., scattering where the target remains in its ground state) is obtained by treating the target atom as a fixed charge distribution $\rho(\mathbf{x})$. In this "static" approximation (also called the "one-state approximation"—see Section 19-c) the incident electron sees a fixed potential,

$$V(\mathbf{x}) = -e \int d^3x' \, \frac{\rho(\mathbf{x}')}{|\mathbf{x} - \mathbf{x}'|} \qquad (9.19)$$

and the elastic scattering can be treated by the methods already developed. Obviously this simple model takes no account of the possible distortion or polarization of the atom by the incident electron and it makes no allowance

---

[9] This condition is usually quoted with an extra factor of $\ln(2p/\mu)$ (see Problem **10.3**). Obviously this makes little difference at any reasonable energy.

for inelastic processes (where the target moves to an excited state or is ionized). The model also ignores "exchange" effects arising from the indistinguishability of the incident and atomic electrons.

If we consider the scattering of electrons off a hydrogen atom first, then, in the approximation proposed, the charge density of the target is just

$$\rho(\mathbf{x}) = e\{\delta_3(\mathbf{x}) - |\phi(\mathbf{x})|^2\}$$

Here the delta function represents the charge of the point nucleus and the second term is the charge distribution of the bound electron, $\phi(\mathbf{x})$ being the hydrogen ground-state wave function,

$$\phi(\mathbf{x}) = (\pi a^3)^{-\frac{1}{2}} e^{-r/a}$$

where a is the Bohr radius ($a = 1/me^2$). The corresponding potential (9.19), which is spherical (and, of course, spin-independent), is:

$$V(r) = -e^2\left(\frac{1}{r} + \frac{1}{a}\right) e^{-2r/a} \tag{9.20}$$

We can now proceed to calculate the amplitude for this potential. In particular, we can try using the Born approximation, for which an elementary integration gives:

$$f^{(1)}(\mathbf{p}' \leftarrow \mathbf{p}) = -2m \int_0^\infty r^2 \, dr \, \frac{\sin qr}{qr} \, V(r)$$

$$= 2a \, \frac{8 + q^2 a^2}{(4 + q^2 a^2)^2} \tag{9.21}$$

It should be emphasized that this expression for the e–H elastic amplitude is an approximation in two ways. First, starting from the potential (9.20) we have calculated not the exact amplitude, but only the Born approximation. Second, the potential (9.20) is itself only an approximate model for a target that is in reality a two-body system. Any disagreement between the answer (9.21) and experiment may reflect the inadequacy of either or both of these approximations.

As regards the validity of the Born approximation, we need to estimate the size of the second Born term $f^{(2)}$, as discussed in the previous section. In the present case we can obtain a rough estimate if we approximate the *potential* by the Yukawa potential

$$V(r) = \gamma \, \frac{e^{-\mu r}}{r}$$

If $\gamma = -e^2$ and $\mu \sim 1/a$ this Yukawa potential has the same singularity at $r = 0$ and the same qualitative behavior as the potential (9.20). For the Yukawa potential we already have rough criteria for validity of the Born approximation. The criterion for validity at low energies is (9.17),

$$\left| \frac{\gamma m}{\mu} \right| \sim e^2 ma \ll 1$$

which, since $a = 1/me^2$, is obviously *not* satisfied. On the other hand, the condition for convergence at high energies is (9.18),

$$\left| \frac{\gamma m}{p} \right| = \frac{e^2 m}{p} \ll 1$$

or $pa \gg 1$. This condition requires that the incident wavelength be much less than the Bohr radius or that the energy be much greater than 10 eV. Thus, we can reasonably hope that the amplitude (9.21) will be reliable above energies of 100 eV or so.

Unfortunately, hydrogen does not occur naturally in its atomic form, and the measurement of the electron cross section on atomic hydrogen requires the use of an atomic beam of dissociated hydrogen molecules. The low intensity of such beams makes all measurements very difficult, and so far the elastic e–H cross section at the energies of interest has not been measured.

To compare our theory with experiment we must consider as a target an element that occurs naturally in the atomic state, of which the simplest is helium. In the static approximation the helium target is considered as a fixed charge distribution

$$\rho(\mathbf{x}) = e \left\{ 2\delta_3(\mathbf{x}) - \int d^3 x_2 \, |\phi(\mathbf{x}, \mathbf{x}_2)|^2 - \int d^3 x_1 \, |\phi(\mathbf{x}_1, \mathbf{x})|^2 \right\}$$

where $\phi(\mathbf{x}_1, \mathbf{x}_2)$ is the ground-state wave function of the two electrons. Here, the first term represents the nuclear charge, while the second and third are the charge densities of the two electrons. The wave function $\phi(\mathbf{x}_1, \mathbf{x}_2)$ is not exactly known. However, there are several ways to get good approximations to it. For example, if one does a variational calculation of the helium ground state using a trial function

$$\phi(\mathbf{x}_1, \mathbf{x}_2) = \chi(\mathbf{x}_1)\chi(\mathbf{x}_2)$$

where

$$\chi(\mathbf{x}) = \alpha(\pi a^3)^{-\frac{1}{2}}(e^{-zr/a} + \beta e^{-2zr/a})$$

then the best estimate for the energy is obtained when $\alpha = 1.48$, $\beta = 0.61$, $z = 1.45$, and is only 1.5% off the experimental value. Using this trial function we can compute the atomic charge density $\rho(\mathbf{x})$, the corresponding potential $V(\mathbf{x})$, and finally the Born approximation

$$f^{(1)}(\mathbf{p}' \leftarrow \mathbf{p}) = \frac{\alpha^2}{z^3} a \left[4 \frac{8z^2 + q^2 a^2}{(4z^2 + q^2 a^2)^2} + \frac{64\beta}{27} \frac{18z^2 + q^2 a^2}{(9z^2 + q^2 a^2)^2} + \frac{\beta^2}{2} \frac{32z^2 + q^2 a^2}{(16z^2 + q^2 a^2)^2} \right]$$

$$(9.22)$$

The same estimates as given above for hydrogen suggest that this Born approximation should begin to be reliable at energies above 100 eV or so. This expectation is well borne out in Fig. 9.3 where the Born approximation is compared with the experimentally measured cross sections at 500 eV. It will be seen that the agreement is excellent except in the neighborhood of

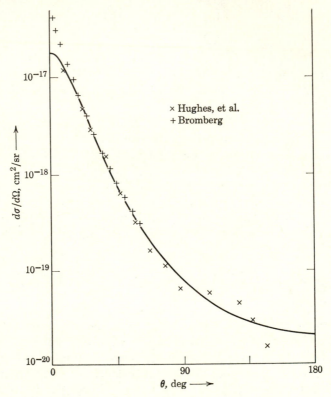

FIGURE 9.3.   **The electron–helium elastic cross section at 500 eV. The curve is the Born approximation (9.22), the experimental points are taken from Hughes, et al. (1932) and Bromberg (1969).**

the forward direction. (One should probably not take too seriously the smaller discrepancy near the backward direction where the cross section is extremely small and only the older measurements of Hughes et al., are available.) The quite appreciable discrepancy in the forward direction ($\theta \lesssim 15$ deg) is attributed to distortion of the atom by the electron and, to a smaller extent, to exchange effects. These are completely neglected in our static approximation.

9-e.   Interpretation of the Born Series in Terms of Feynman Diagrams

The Born series for the S matrix $\langle \mathbf{p}' | \, S \, | \mathbf{p} \rangle$ can be given a graphic interpretation, in which the nth term is viewed as the amplitude for making the "transition" $(\mathbf{p}' \leftarrow \mathbf{p})$ via a succession of $(n-1)$ "intermediate" or "virtual" states $(\mathbf{p}' \leftarrow \mathbf{k}_{n-1} \leftarrow \cdots \leftarrow \mathbf{k}_1 \leftarrow \mathbf{p})$—a process conveniently represented by the famous Feynman diagram. This interpretation arises naturally in an alternative derivation of the Born series using the interaction picture. Since both this alternative derivation and the interpretation that it suggests have played important roles in the recent history of scattering theory (nonrelativistic and relativistic) we use this last section to discuss them.

So far we have worked exclusively in the so-called Schrödinger picture of quantum mechanics. In this "picture" the state vectors representing an evolving system vary with time according to the Schrödinger equation with Hamiltonian H; on the other hand, the basic dynamical variables $\mathbf{X}$, $\mathbf{P}$, $\mathbf{S}$ are independent of time. By making a time-dependent unitary transformation we can change this situation and obtain a new "picture" of quantum mechanics. In particular, if $H = H^0 + V$, the *interaction*, or *Dirac picture* is defined by the unitary transformation of vectors and operators

$$|\psi_t\rangle_I = e^{iH^0 t} |\psi_t\rangle_S$$

and

$$A_I(t) = e^{iH^0 t} A_S(t) e^{-iH^0 t}$$

where the subscript I stands for "interaction," and we have temporarily added a subscript S to identify quantities of the Schrödinger picture. The physical state that was labelled by $|\psi\rangle_S$ in the Schrödinger picture is labelled by $|\psi\rangle_I$ in the interaction picture; the physical observable labelled by A_S in the former, is labelled by A_I in the latter. Since for any matrix element,

$$_I\langle \phi | \, A_I \, | \psi \rangle_I = {}_S\langle \phi | \, A_S \, | \psi \rangle_S$$

it is clear that the two pictures always lead to the same physical predictions.

In passing to the interaction picture we have factored out of the state vectors the time dependence of H^0. Thus, if for any reason the interaction V is negligible, the state vector in the interaction picture is a constant. In collision theory this means that as the scattered particle moves away and V ceases to have any effect the interaction picture state vectors tend to constants $|\psi_{\mp\infty}\rangle_I$. In fact, if we write

$$|\psi_t\rangle_I = e^{iH^0 t} |\psi_t\rangle_S = e^{iH^0 t} e^{-iHt} |\psi\rangle_S$$

it is clear that (if $|\psi\rangle_S$ is a scattering state) the limits $|\psi_{\mp\infty}\rangle_I$ in the interaction picture are precisely the in and out asymptotes $|\psi_{\text{in}}\rangle_S$ and $|\psi_{\text{out}}\rangle_S$ of the Schrödinger picture. Thus, in the interaction picture, a collision is described by a vector $|\psi_t\rangle_I$ with the following simple behavior. Well before the collision $|\psi_t\rangle_I$ evolves from some fixed limit $|\psi_{-\infty}\rangle_I = |\psi_{\text{in}}\rangle_S$, it begins to change only as the particles collide, and eventually it tends to some other limit $|\psi_\infty\rangle_I = |\psi_{\text{out}}\rangle_S$ when the collision is all over.[10]

The evolution operator from time t_0 to time t in the Schrödinger picture is

$$U_S(t, t_0) = e^{-iH(t-t_0)}$$

It follows that that of the interaction picture is

$$U_I(t, t_0) = e^{iH^0 t} e^{-iH(t-t_0)} e^{-iH^0 t_0} \tag{9.23}$$

From what has just been said it is clear that the S operator is just the limit[11]

$$S = U_I(\infty, -\infty) \tag{9.24}$$

Thus, any convenient method of calculating $U_I(t, t_0)$ should provide an expression for S. In particular, the procedure generally known as time-dependent perturbation theory yields the Born series for S, as we shall now see.

The evolution operator $U_I(t, t_0)$ of (9.23) satisfies the differential equation,

$$i \frac{\partial}{\partial t} U_I(t, t_0) = V_I(t) U_I(t, t_0)$$

where, of course,

$$V_I(t) = e^{iH^0 t} V e^{-iH^0 t} \tag{9.25}$$

(and V is the usual potential in the Schrödinger picture). Integrating with respect to t and using the fact that $U_I(t_0, t_0) = 1$ we obtain the integral equation,

$$U_I(t, t_0) = 1 - i \int_{t_0}^t d\tau V_I(\tau) U_I(\tau, t_0)$$

---

[10] The interaction picture is so clearly a natural and convenient way to describe single-channel scattering that the reader may well wonder why we have not used it sooner. The reasons are: (1) There is no convenient generalization of the interaction picture to multi-channel scattering, and (2) in relativistic quantum mechanics there is reason to doubt the existence of an interaction picture.

[11] Here we are being very cavalier about the various limiting processes. In fact $U_I(t, t_0)$ has the desired limit only if we first take its matrix element between proper states $|\chi\rangle$ and $|\phi\rangle$ [i.e., the limit (9.24) is a weak operator limit]. In the correct form $\langle\chi| S |\phi\rangle = \lim\langle\chi| U_I(t, t_0) |\phi\rangle$, the result (9.24) is just the limit (8.20) used to express $\langle\mathbf{p}'| S |\mathbf{p}\rangle$ in terms of the matrix elements of $T(z)$.

If the interaction is zero, $U_I = 1$. Thus, if the interaction is small, one can reasonably hope to solve this equation by iteration starting from $U_I = 1$ as the *zeroth* approximation. This procedure gives

$$U_I(t, t_0) = 1 - i \int_{t_0}^{t} d\tau V_I(\tau) + (-i)^2 \int_{t_0}^{t} d\tau' \int_{t_0}^{\tau'} d\tau V_I(\tau') V_I(\tau) + \cdots \quad (9.26)$$

If we now set $t = \infty$, $t_0 = -\infty$ to give $S = U_I(\infty, -\infty)$, take matrix elements, and use (9.25) to rewrite $V_I(t)$ in terms of V, we obtain

$$\langle \mathbf{p}' | S | \mathbf{p} \rangle = \delta_3(\mathbf{p}' - \mathbf{p}) - i \int_{-\infty}^{\infty} d\tau \langle \mathbf{p}' | e^{iE_{p'}\tau} V e^{-iE_p\tau} | \mathbf{p} \rangle$$

$$+ (-i)^2 \int_{-\infty}^{\infty} d\tau' \int_{-\infty}^{\tau'} d\tau \langle \mathbf{p}' | e^{iE_{p'}\tau'} V e^{-iH^0(\tau'-\tau)} V e^{-iE_p\tau} | \mathbf{p} \rangle$$

$$= \delta_3(\mathbf{p}' - \mathbf{p}) + \langle \mathbf{p}' | S^{(1)} | \mathbf{p} \rangle + \langle \mathbf{p}' | S^{(2)} | \mathbf{p} \rangle + \cdots \quad (9.27)$$

We shall refer to this result as the *perturbation expansion* of the S matrix.[12] Its first term is just the familiar no-scattering term; the amplitude that the particle propagates straight through unscattered. Before showing that the succeeding terms precisely reproduce the terms of the Born series, we pause to discuss their physical interpretation.

The first order term $\langle \mathbf{p}' | S^{(1)} | \mathbf{p} \rangle$ in the perturbation series (9.27) is the integral over all τ of the matrix element

$$-i \langle \mathbf{p}' | \exp(iE_{p'}\tau) V \exp(-iE_p\tau) | \mathbf{p} \rangle$$

It is natural to interpret this as the amplitude for the following process. The particle propagates freely in the state $| \mathbf{p} \rangle$ until time τ. [This gives the factor $\exp(-iE_p\tau) | \mathbf{p} \rangle$.] At time τ the particle is given a single jolt by V and is knocked instantaneously into the final state $| \mathbf{p}' \rangle$. (This gives the factor $-iV$.) It then moves away, once again propagating freely. [This gives the factor $\langle \mathbf{p}' | \exp(iE_{p'}\tau)$.] This sequence of events can be illustrated schematically by a space-time diagram as in Fig. 9.4. Finally, since the time τ is not measureable, the corresponding contribution to the S matrix is obtained by integrating this amplitude over all τ.

To interpret the second order term in the perturbation expansion (9.27) we insert a complete set of plane wave states $| \mathbf{k} \rangle$ to give

$$\langle \mathbf{p}' | S^{(2)} | \mathbf{p} \rangle = (-i)^2 \int_{-\infty}^{\infty} d\tau' \int_{-\infty}^{\tau'} d\tau \int d^3k \, e^{iE_{p'}\tau'} \langle \mathbf{p}' | V | \mathbf{k} \rangle$$

$$\times e^{-iE_k(\tau'-\tau)} \langle \mathbf{k} | V | \mathbf{p} \rangle e^{-iE_p\tau} \quad (9.28)$$

---

[12] We do not wish to imply any profound distinction between the result (9.27) and the usual Born series, which are in fact equivalent. Until we have established their equivalence, however, it is convenient to give them different names.

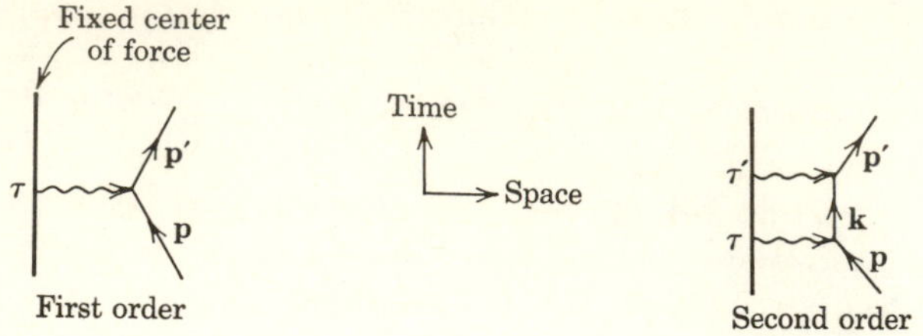

FIGURE 9.4.   **Schematic interpretation of the terms in the perturbation series for the S matrix.**

The integrand here can be interpreted as the amplitude for a process in which the particle propagates freely in the initial state $|\mathbf{p}\rangle$ until time τ; it is then knocked by V into an "intermediate state" $|\mathbf{k}\rangle$, in which it propagates freely between τ and τ'; at τ' it is again jolted by V, this time into the final state $|\mathbf{p}'\rangle$, in which it propagates away freely (see Fig. 9.4). Since neither the times τ and τ' nor the intermediate momentum $\mathbf{k}$ are observed, the corresponding contribution to the S matrix is obtained by integrating over all τ and τ', subject to $-\infty < \tau \leqslant \tau' < \infty$, and all $\mathbf{k}$.

The reader should have no difficulty in extending this discussion to the general nth order term. Before discussing this further, we turn to the question of the equivalence of the perturbation expansion (9.27) and the Born series. This is easily established: According to (9.27), the first order term in the perturbation series is

$$\langle \mathbf{p}'|\, S^{(1)}\,|\mathbf{p}\rangle = -i\int_{-\infty}^{\infty} d\tau\; e^{i(E_{p'}-E_p)\tau}\langle \mathbf{p}'|\, V\,|\mathbf{p}\rangle$$

$$= -2\pi i\, \delta(E_{p'} - E_p)\langle \mathbf{p}'|\, V\,|\mathbf{p}\rangle$$

This is precisely the contribution to $\langle \mathbf{p}'|\, S\,|\mathbf{p}\rangle$ of the first Born term

$$t^{(1)}(\mathbf{p}' \leftarrow \mathbf{p}) = \langle \mathbf{p}'|\, V\,|\mathbf{p}\rangle$$

in the decomposition

$$\langle \mathbf{p}'|\, S\,|\mathbf{p}\rangle = \delta_3(\mathbf{p}' - \mathbf{p}) - 2\pi i\, \delta(E_{p'} - E_p)\, t(\mathbf{p}' \leftarrow \mathbf{p})$$

In treating the second order term we must be a little careful. The point is that in writing (9.28) for $\langle \mathbf{p}'|\, S^{(2)}\,|\mathbf{p}\rangle$ we have shown a happy disregard for the important question of convergence of the integrals involved, and in fact, the integral over τ from $-\infty$ to τ' is meaningless. However, the reader should have no difficulty in convincing himself that if we had used proper initial and final states, the integral would have been convergent and would have been unchanged by the insertion of the now familiar damping factor $\exp(+\epsilon\tau)$.

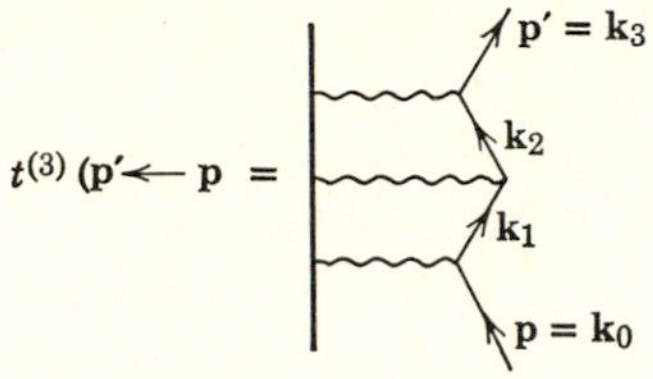

FIGURE 9.5.   The Feynman diagram for $t^{(3)}(\mathbf{p}' \leftarrow \mathbf{p})$.

In this form we can revert to improper states. The integral over τ then yields

$$\int_{-\infty}^{\tau'} d\tau \, e^{-i(E_p - E_k + i0)\tau} = \frac{i \, e^{i(E_k - E_p)\tau'}}{E_p - E_k + i0} \tag{9.29}$$

the integral over τ' gives $2\pi \, \delta(E_{p'} - E_p)$ and we arrive at the answer

$$\langle \mathbf{p}'| \, S^{(2)} \, |\mathbf{p}\rangle = -2\pi i \, \delta(E_{p'} - E_p) \int d^3 k \, \frac{\langle \mathbf{p}'| \, V \, |\mathbf{k}\rangle\langle \mathbf{k}| \, V \, |\mathbf{p}\rangle}{E_p - E_k + i0} \tag{9.30}$$

$$= -2\pi i \, \delta(E_{p'} - E_p)\langle \mathbf{p}'| \, V G^0(E_p + i0) V \, |\mathbf{p}\rangle$$

which is precisely the contribution of the second Born term $T^{(2)} = V G^0 V$, as expected. It is easy to see that the nth order term goes through in exactly the same way, and we conclude that the perturbation series (9.27) is indeed precisely the Born series.

It has become popular to view the diagrams in Fig. 9.4 as algebraic symbols that actually *are* the terms in the Born series for $t(\mathbf{p}' \leftarrow \mathbf{p})$. In this view one can write "equations" like that shown in Fig. 9.5 for the term

$$t^{(3)}(\mathbf{p}' \leftarrow \mathbf{p}) = \langle \mathbf{p}'| \, V G^0(E_p + i0) V G^0(E_p + i0) V \, |\mathbf{p}\rangle$$

$$= \int d^3 k_2 \int d^3 k_1 \, \frac{\langle \mathbf{p}'| \, V \, |\mathbf{k}_2\rangle\langle \mathbf{k}_2| \, V \, |\mathbf{k}_1\rangle\langle \mathbf{k}_1| \, V \, |\mathbf{p}\rangle}{(E_p - E_{k_2} + i0)(E_p - E_{k_1} + i0)} \tag{9.31}$$

The diagram of Fig. 9.5 is simply an alternative symbol for the expression (9.31)—a symbol that makes clear its interpretation as the amplitude for a process consisting of three jolts separated by free propagation in the intermediate states $\mathbf{k}_1$ and $\mathbf{k}_2$. As we have performed the integrations over times [in (9.29) for example] the factors of the form $\exp[-iE_k(\tau' - \tau)]$ that represented the free propagation of the intermediate state $\mathbf{k}$ have been replaced by the denominators $(E_p - E_k + i0)$. That is, in the expressions (9.30) and (9.31), the free propagation of the intermediate state $\mathbf{k}$ is represented by the denominator $(E_p - E_k + i0)$. For this reason the factor $1/(E_p - E_k + i0)$ [or more generally the operator $G^0(E_p + i0)$] is often referred to as the free (nonrelativistic) propagator.

The use of diagrams like that of Fig. 9.5 to represent terms in the Born series for the scattering amplitude was first introduced in quantum electro-dynamics by Feynman and they are therefore called *Feynman diagrams* or *Feynman graphs*. It is easy to see that one could devise a set of rules by means of which any term could simply be read off from the corresponding diagram. In fact, comparison of Fig. 9.5 and (9.31) shows that the following set of "Feynman rules" leads directly to the correct expression for $t^{(n)}(\mathbf{p}' \leftarrow \mathbf{p})$:

(1) Draw the appropriate Feynman diagram with n "vertices" (i.e., n interactions with the potential).
(2) Write down a factor $\langle \mathbf{k}_{r+1}| \, V \, |\mathbf{k}_r \rangle$ for each vertex (with $\mathbf{p} = \mathbf{k}_0$, $\mathbf{p}' = \mathbf{k}_n$).
(3) Write down a factor $1/(E_p - E_k + i0)$ for the free propagation of each intermediate state $\mathbf{k} = \mathbf{k}_1, \ldots, \mathbf{k}_{n-1}$.
(4) Integrate over all intermediate momenta $\mathbf{k}_1, \ldots, \mathbf{k}_{n-1}$.

In the present case (scattering of a single nonrelativistic particle) the struc-ture of the Born series is too simple for a prescription of this kind to be especially helpful. However, in multiparticle processes, and even more in quantum electrodynamics, the Born series are much more complicated, and the use of the corresponding Feynman rules leads to a tremendous simplifica-tion. In fact, it is no exaggeration to say that without the help of Feynman diagrams much of the dramatic progress of the fifties in quantum electro-dynamics would have been impossible.

Feynman diagrams have become so much a part of every physicist's thinking that one often hears the processes that they represent discussed as events which actually occur. It may therefore be worth concluding with a reminder that these diagrams are no more than an interpretation (albeit a persuasive interpretation) of the terms in a particular series expansion of the amplitude. According to the rules of quantum mechanics, the state vector actually propagates according to the full evolution operator $\mathsf{U}(t)$ at all times. The picture of free propagation [under $\mathsf{U}^0(t)$] interrupted by successive jolts from the potential V is a product of the particular method of solution, namely expansion of $\mathsf{U}(t)$ in powers of V. In particular, if the Born series diverges (as it certainly can do) there is no obvious reason for supposing that its individual terms and the corresponding diagrams will have any physical significance at all.

PROBLEMS

9.1.   Calculate the first and second Born approximations for the Yukawa potential $\gamma e^{-\mu r}/r$. In the case of the second Born term you can, if you wish, confine attention to the forward amplitude. (The general, nonforward, case can be found in Newton, 1966, p. 291.)

9.2. Calculate the first Born approximation for the square well

$$V(r) = \begin{cases} -V_0, & r < a \\ 0, & r \geqslant a \end{cases}$$

Show that it has all of the qualitative features discussed in the text.

9.3. We shall see in Chapter 11 that the scattering amplitude becomes iso-tropic at low energies—a property that we have seen is also true of the Born approximation. However, for many potentials the Born approximation differs from the exact amplitude by several orders of magnitude at low energies. Does this mean that for such potentials the isotropy of the Born approxima-tion at $E = 0$ is purely accidental? (The answer is no. Why?)

10

The Stationary Scattering States

10-a   **Definition and Properties of the Stationary Scattering States**

10-b   **Equations for the Stationary Scattering Vectors**

10-c   **The Stationary Wave Functions**

10-d   **A Spatial Description of the Scattering Process**

At the beginning of Chapter 8 it was stated that the time-independent scattering formalism is concerned with the two operators $G(z)$ and $T(z)$ and the stationary scattering states. The two operators were the subject of Chapters 8 and 9, the states are the subject of this.

The stationary scattering states are improper eigenvectors of the Hamiltonian $H = H^0 + V$ and are denoted by $|\mathbf{p}+\rangle$ and $|\mathbf{p}-\rangle$. In particular, the wave function $\langle \mathbf{x} \mid \mathbf{p}+\rangle$ is just the familiar scattering wave function, often denoted $\psi_{\mathbf{p}}^+(\mathbf{x})$, of elementary collision theory,

$$\langle \mathbf{x} \mid \mathbf{p}+\rangle \xrightarrow[r \to \infty]{} (2\pi)^{-3/2}\left(e^{i\mathbf{p}\cdot\mathbf{x}} + f\frac{e^{ipr}}{r} \right) \tag{10.1}$$

The scattering amplitude can be expressed in terms of $|\mathbf{p}+\rangle$ or $|\mathbf{p}-\rangle$ [by (10.1) for example] and any means of computing $|\mathbf{p}\pm\rangle$ therefore provides a method for calculating the amplitude. These methods fall into two (closely connected) categories: those using integral equations and those using differential equations. The integral method depends on the fact that the wave functions $\langle \mathbf{x} \mid \mathbf{p}\pm \rangle$ satisfy integral equations, closely related to the Lippmann–Schwinger equation for the operator $T(z)$. The differential method uses the fact that the vectors $|\mathbf{p}\pm \rangle$ are eigenvectors of H and, hence, that the wave functions $\langle \mathbf{x} \mid \mathbf{p}\pm \rangle$ satisfy the time-independent Schrödinger equation. As means of calculating the amplitude, the integral equations produce essentially the same results as the Lippmann–Schwinger equation discussed in Chapters 8 and 9. On the other hand the Schrödinger (differential) equation leads to a genuinely alternative approach. In practice, this approach is most useful when the potential is spherically symmetric, since in this case the wave function can be decomposed into angular-momentum eigenfunctions and the Schrödinger equation reduces to a set of ordinary differential equations. This approach will be the subject of Chapters 11 and 12.

In this chapter we define the stationary scattering states and discuss some of their general properties. Sections 10-a and 10-b give the definitions and some basic equations for the vectors $|\mathbf{p}\pm \rangle$. In Section 10-c we use the integral equations for the wave functions $\langle \mathbf{x} \mid \mathbf{p}\pm \rangle$ to prove the asymptotic form (10.1), and finally establish contact with the traditional approach based on this property. In Section 10-d we use the wave function $\langle \mathbf{x} \mid \mathbf{p}+ \rangle$ to give a coordinate-space description of the collision process. This supplements the momentum-space discussion of Chapter 3 and provides additional insight into the conditions under which cross sections can be measured.

10-a.   Definition and Properties of the Stationary Scattering States

We have from time to time used the notation $|\phi\pm \rangle$ to denote the image under $\Omega_\pm$ of any proper (normalizable) vector $|\phi \rangle$. The significance of $|\phi+ \rangle$, for example, is that if the in asymptote of some collision event is labelled by $|\psi_{\text{in}} \rangle = |\phi \rangle$, then the actual state at $t = 0$ is $|\psi \rangle = |\phi+ \rangle$. We now introduce two improper vectors $|\mathbf{p}\pm \rangle$ whose relationship to the plane wave $|\mathbf{p} \rangle$ is the same as that of $|\phi\pm \rangle$ to $|\phi \rangle$:

$$\boxed{|\mathbf{p}\pm \rangle \equiv \Omega_\pm |\mathbf{p} \rangle} \tag{10.2}$$

To understand the significance of this definition we consider an orbit with in asymptote labelled by $|\phi \rangle$, which we expand in terms of plane waves as

$$|\phi \rangle = \int d^3p\, \phi(\mathbf{p})\, |\mathbf{p} \rangle$$

In this case the actual state at $t = 0$ is

$$|\phi+ \rangle = \Omega_+ |\phi \rangle = \int d^3p\, \phi(\mathbf{p}) \Omega_+ |\mathbf{p} \rangle$$

$$= \int d^3p\, \phi(\mathbf{p})\, |\mathbf{p}+ \rangle \tag{10.3}$$

In other words, the *actual state* $|\phi+\rangle$ *has the same expansion in terms of* $|\mathbf{p}+\rangle$ *as does its in asymptote* $|\phi\rangle$ in terms of $|\mathbf{p}\rangle$. [In fact this is precisely what is *meant* by the definition (10.2) of the improper vector $|\mathbf{p}+\rangle$.] Similarly, if an orbit has out asymptote

$$|\chi\rangle = \int d^3p\,\chi(\mathbf{p})\,|\mathbf{p}\rangle$$

then the actual state at $t = 0$ is

$$|\chi-\rangle = \int d^3p\,\chi(\mathbf{p})\,|\mathbf{p}-\rangle \tag{10.4}$$

These two results give the primary significance of the improper vectors $|\mathbf{p}\pm\rangle$. They are the natural vectors for expanding the actual state at $t = 0$ to display explicitly its relation to the momentum expansions of the in and out asymptotes.

We have already mentioned that the vectors $|\mathbf{p}\pm\rangle$ are eigenvectors of the full Hamiltonian H. To prove this we have only to recall the important intertwining relations (3.1),

$$H\Omega_\pm = \Omega_\pm H^0$$

From these it follows that

$$H\,|\mathbf{p}\pm\rangle = H\Omega_\pm\,|\mathbf{p}\rangle = \Omega_\pm H^0\,|\mathbf{p}\rangle = E_p\Omega_\pm\,|\mathbf{p}\rangle$$

or

$$\boxed{H\,|\mathbf{p}\pm\rangle = E_p\,|\mathbf{p}\pm\rangle}$$

That is, both $|\mathbf{p}+\rangle$ and $|\mathbf{p}-\rangle$ are eigenvectors of H, the eigenvalue of H acting on $|\mathbf{p}\pm\rangle$ being the same as that of H^0 acting on $|\mathbf{p}\rangle$, namely, $E_p \equiv p^2/2m$. This means that the wave functions $\langle\mathbf{x}\,|\,\mathbf{p}\pm\rangle$ satisfy the time-independent Schrödinger equation (which provides one approach to the actual calculation of the stationary scattering states).

It is interesting to discuss the asymptotic condition in the light of this result. The physical connection between the proper vectors $|\phi+\rangle$ and $|\phi\rangle$ is that the free orbit $U^0(t)\,|\phi\rangle$ is the asymptote of the actual orbit $U(t)\,|\phi+\rangle$ as $t \to -\infty$,

$$U(t)\,|\phi+\rangle \xrightarrow[t\to-\infty]{} U^0(t)\,|\phi\rangle \qquad \text{[any proper } |\phi\rangle\text{]} \tag{10.5}$$

Concerning the improper vectors $|\mathbf{p}+\rangle$ and $|\mathbf{p}\rangle$, we have seen that they are eigenvectors of H and H^0 with the same eigenvalue E_p. From this it follows that:

$$U(t)\,|\mathbf{p}+\rangle = e^{-iE_p t}\,|\mathbf{p}+\rangle$$

and

$$U^0(t)\,|\mathbf{p}\rangle = e^{-iE_p t}\,|\mathbf{p}\rangle$$

That is, $|\mathbf{p}+\rangle$ and $|\mathbf{p}\rangle$ are *stationary states* with respect to the corresponding evolution operators, and clearly they do *not* themselves satisfy (10.5).

The explanation of this rather surprising result is this: The proper vectors of (10.5) can be expanded in terms of $|\mathbf{p}+\rangle$ and $|\mathbf{p}\rangle$ to give

$$U(t)\int d^3p\phi(\mathbf{p})\,|\mathbf{p}+\rangle \to U^0(t)\int d^3p\phi(\mathbf{p})\,|\mathbf{p}\rangle$$

or

$$\int d^3p\phi(\mathbf{p})[U(t)\,|\mathbf{p}+\rangle] \to \int d^3p\phi(\mathbf{p})[U^0(t)\,|\mathbf{p}\rangle] \tag{10.6}$$

We see that the vectors $|\mathbf{p}+\rangle$ and $|\mathbf{p}\rangle$ do satisfy the asymptotic condition when "smeared" by the integral $\int d^3p\phi(\mathbf{p})$ into proper states. In fact, since the improper vectors have physical meaning only when smeared into proper states, the relevant result is that they *do* satisfy the asymptotic condition in this sense. Nonetheless, if we try to remove the smearing integral from (10.6) and to discuss the improper wave functions themselves, they do not converge; and since they do not represent any real physical state there is really no reason why they should.[1]

As long as this situation is clearly understood, it is harmless and convenient to speak of $|\mathbf{p}+\rangle$ as the actual state at $t = 0$ that has evolved from the initial state $|\mathbf{p}\rangle$, and of $|\mathbf{p}-\rangle$ as the actual state that would evolve into the final state $|\mathbf{p}\rangle$.

The plane waves $|\mathbf{p}\rangle$ form an orthonormal basis of the Hilbert space $\mathscr{H}$, made up of eigenvectors of H^0. Since $\Omega_\pm$ map $\mathscr{H}$ onto the subspace $\mathscr{R}$ of scattering states, one would expect that the vectors $|\mathbf{p}+\rangle$ (or $|\mathbf{p}-\rangle$) should span $\mathscr{R}$. Indeed we have already seen in (10.3) that any scattering state can be expanded in terms of the $|\mathbf{p}+\rangle$ and, since

$$\langle\mathbf{p}'+\mid\mathbf{p}+\rangle = \langle\mathbf{p}'|\,\Omega_+^\dagger\Omega_+\,|\mathbf{p}\rangle = \langle\mathbf{p}'\mid\mathbf{p}\rangle$$
$$= \delta_3(\mathbf{p}' - \mathbf{p})$$

the vectors $|\mathbf{p}+\rangle$ are in fact an orthonormal basis of $\mathscr{R}$. Since

$$\mathscr{H} = \mathscr{R} \oplus \mathscr{B}$$

where the subspace $\mathscr{B}$ is spanned by the bound states ($|n\rangle$, say) it is clear that the stationary scattering states $|\mathbf{p}+\rangle$ *together with the bound states* $|n\rangle$ form an orthornormal basis of $\mathscr{H}$. (The same result holds for the $|\mathbf{p}-\rangle$ of course.)

---

[1] For the reader familiar with the Riemann–Lebesgue lemma it may help to draw an analogy with that result, which asserts that (for reasonable functions ϕ)

$$\int dp\phi(p)e^{ipt} \to 0 \qquad [t \to \infty]$$

In the language used above we can say that, when smeared by $\int dp\psi(p)$, e^{ipt} goes to zero as $t \to \infty$. Nonetheless, e^{ipt} by itself obviously does *not*.

This situation can be summarised by writing the corresponding resolutions of the identity:

$$1 = \int d^3p \, |\mathbf{p}\rangle\langle\mathbf{p}|$$

$$= \int d^3p \, |\mathbf{p}+\rangle\langle\mathbf{p}+| \; + \; \sum_n |n\rangle\langle n|$$

$$= \int d^3p \, |\mathbf{p}-\rangle\langle\mathbf{p}-| \; + \; \sum_n |n\rangle\langle n|$$

The existence of two distinct bases of eigenvectors of H (that with $|\mathbf{p}+\rangle$ and that with $|\mathbf{p}-\rangle$) reflects the well known fact that the positive energy spectrum of H is highly degenerate.

10-b.   Equations for the Stationary Scattering Vectors

It was shown in Section 8-c that for any proper $|\phi\rangle$ it was possible to express $|\phi\pm\rangle = \Omega_\pm \, |\phi\rangle$ in terms of $|\phi\rangle$ and the Green's operator as:

$$|\phi\pm\rangle = |\phi\rangle + \int d^3p \, G(E_p \pm i0)V \, |\mathbf{p}\rangle\langle\mathbf{p}\,|\,\phi\rangle \qquad (10.7)$$

If we expand $|\phi\pm\rangle$ and $|\phi\rangle$ in terms of the momentum wave function $\phi(\mathbf{p}) \equiv \langle\mathbf{p}\,|\,\phi\rangle$ this becomes

$$\int d^3p\, \phi(\mathbf{p}) \, |\mathbf{p}\pm\rangle = \int d^3p\, \phi(\mathbf{p})[|\mathbf{p}\rangle + G(E_p \pm i0)V \, |\mathbf{p}\rangle]$$

Since this holds for any $\phi(\mathbf{p})$ we conclude that:

$$\boxed{|\mathbf{p}\pm\rangle = |\mathbf{p}\rangle + G(E_p \pm i0)V \, |\mathbf{p}\rangle} \qquad (10.8)$$

The meaning of this important result needs comment. As usual in this book we have made no attempt to justify rigorously our handling of improper vectors. Given the result (10.7) from Chapter 8, the only sense in which (10.8) obviously follows from our argument is this: When both sides of (10.8) are smeared by the operation $\int d^3p\, \phi(\mathbf{p})$ the resulting proper vectors are equal [i.e., (10.7) holds]. However, the main usefulness of the result (and similar results below) is that they hold as *ordinary* equations for the corresponding wave functions [for example, the Lippmann–Schwinger equation (10.13) for the wave function $\langle\mathbf{x}\,|\,\mathbf{p}\pm\rangle$]. For the proofs that these equations really hold in the required sense, we refer the interested reader to the literature.[2] We shall continue to handle improper states without any attempt at rigor.

[2] See Faddeev (1965).

The relation (10.8) can be used to establish an expression for the on-shell T matrix in terms of $|\mathbf{p}+\rangle$ or $|\mathbf{p}-\rangle$. Since we already know that $t(\mathbf{p}' \leftarrow \mathbf{p})$ is the same thing as $\langle \mathbf{p}'| T(E_p + i0) |\mathbf{p}\rangle$ we begin by considering the vector

$$T(E_p \pm i0) |\mathbf{p}\rangle = [V + VG(\pm)V] |\mathbf{p}\rangle$$
$$= V[1 + G(\pm)V] |\mathbf{p}\rangle$$

Using (10.8) we then obtain the important identities[3]

$$\boxed{T(E_p \pm i0) |\mathbf{p}\rangle = V |\mathbf{p}\pm\rangle} \tag{10.9}$$

Multiplying the first of these by the bra $\langle \mathbf{p}'|$ gives[4]

$$\boxed{t(\mathbf{p}' \leftarrow \mathbf{p}) = \langle \mathbf{p}'| V |\mathbf{p}+\rangle} \tag{10.10}$$

To obtain an equivalent expression in terms of $|\mathbf{p}-\rangle$ we use the identity (8.13), $T(z)^{\dagger} = T(z^*)$ to rewrite the second identity (10.9) (with $\mathbf{p}$ replaced by $\mathbf{p}'$) as

$$\langle \mathbf{p}'| T(E_{p'} + i0) = \langle \mathbf{p}'-| V$$

This leads at once to the result:

$$\boxed{t(\mathbf{p}' \leftarrow \mathbf{p}) = \langle \mathbf{p}'-| V |\mathbf{p}\rangle} \tag{10.11}$$

These results allow one to calculate the scattering amplitude from either of the stationary states $|\mathbf{p}+\rangle$ or $|\mathbf{p}-\rangle$.

Returning to the equation (10.8) for $|\mathbf{p}\pm\rangle$ we note that, just as with the T operator in Section 8-b, it is convenient to replace the explicit expression in terms of $G(z)$ by an implicit expression in terms of $G^0(z)$. This can be done in two simple steps: we replace GV by G^0T, and then we use (10.9) to replace $T |\mathbf{p}\rangle$ by $V |\mathbf{p}\pm\rangle$. This gives

$$\boxed{|\mathbf{p}\pm\rangle = |\mathbf{p}\rangle + G^0(E_p \pm i0)V |\mathbf{p}\pm\rangle} \tag{10.12}$$

This equation, which is, of course, an integral equation for the wave function $\langle \mathbf{x} | \mathbf{p}\pm\rangle$, is called the *Lippmann–Schwinger equation for* $|\mathbf{p}\pm\rangle$.

---

[3] The importance of retaining the argument $(E \pm i0)$ in this equation should be emphasized. In fact one often finds (10.9) written as $T |\mathbf{p}\rangle = V |\mathbf{p}+\rangle$ with the claim that this *defines* the operator T. However, it is clear that (10.9) defines the action of $T(E \pm i0)$ only on those $|\mathbf{p}\rangle$ with $p = (2mE)^{\frac{1}{2}}$.

[4] From (10.9) it is clear that $\langle \mathbf{p}'| T(E_p + i0 |\mathbf{p}\rangle = \langle \mathbf{p}'| V |\mathbf{p}+\rangle$ for *any* $\mathbf{p}$; that is, $\langle \mathbf{p}'| V |\mathbf{p}+\rangle$ is an alternative expression for the "half-off-shell" T matrix.

The similarity of the Lippmann–Schwinger equations for G, T, and $|\mathbf{p}\pm\rangle$ can be emphasized if we display them simultaneously:

$$(1 - G^0 V)G = G^0$$
$$(1 - V G^0)T = V$$
$$(1 - G^0 V)\,|\mathbf{p}\pm\rangle = |\mathbf{p}\rangle$$

In particular, if we iterate the last of these, we obtain a Born series for $|\mathbf{p}+\rangle$,

$$|\mathbf{p}+\rangle = |\mathbf{p}\rangle + G^0 V\,|\mathbf{p}\rangle + \cdots$$

If we then use (10.10) to calculate the on-shell T matrix, we obtain

$$t(\mathbf{p}' \leftarrow \mathbf{p}) = \langle \mathbf{p}'|\,V\,|\mathbf{p}\rangle + \langle \mathbf{p}'|\,V G^0 V\,|\mathbf{p}\rangle + \cdots$$

which is precisely the Born series resulting from iterating the Lippmann–Schwinger equation for $T(z)$.

This last result makes clear that we should not expect to obtain any result from the Lippman–Schwinger equation for $|\mathbf{p}\pm\rangle$ that cannot be obtained from the corresponding equation for $T(z)$. However, we shall find that some results can be more compactly expressed in terms of $|\mathbf{p}\pm\rangle$, and that by using $|\mathbf{p}\pm\rangle$ we can gain additional insight into the significance of various results. For example, we can regard the Lippmann–Schwinger equation for $|\mathbf{p}+\rangle$ as expressing $|\mathbf{p}+\rangle$ as a sum of the "incident" plane wave $|\mathbf{p}\rangle$ plus the "distortion" $G^0 V\,|\mathbf{p}+\rangle$, caused by the potential. The Born approximation,

$$t(\mathbf{p}' \leftarrow \mathbf{p}) \approx \langle \mathbf{p}'|\,V\,|\mathbf{p}\rangle$$

simply amounts to ignoring the effects of this distortion (i.e., setting $|\mathbf{p}+\rangle \approx |\mathbf{p}\rangle$) in the exact expression $\langle \mathbf{p}'|\,V\,|\mathbf{p}+\rangle$.

10-c.   The Stationary Wave Functions

From the Lippmann–Schwinger equation for $|\mathbf{p}\pm\rangle$ it follows that the wave functions $\langle \mathbf{x} \mid \mathbf{p}\pm\rangle$ satisfy the integral equation

$$\langle \mathbf{x} \mid \mathbf{p}\pm\rangle = \langle \mathbf{x} \mid \mathbf{p}\rangle + \int d^3x' \langle \mathbf{x}|\,G^0(E_p \pm i0)\,|\mathbf{x}'\rangle V(\mathbf{x}')\langle \mathbf{x}' \mid \mathbf{p}\pm\rangle \quad (10.13)$$

In this section we use this equation to establish the behavior of the wave functions for large r; in particular, we shall establish the well-known result (10.1) that at large distances $\langle \mathbf{x} \mid \mathbf{p}+\rangle$ is just a plane wave plus a spherical outgoing wave.

To use the equation (10.13) we must calculate the Green's function $\langle \mathbf{x}|\,G^0(z)\,|\mathbf{x}'\rangle$. This function is probably already known to the reader, but

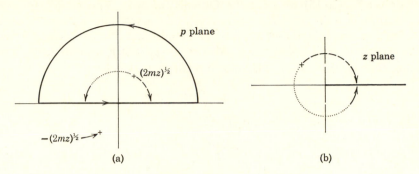

FIGURE 10.1.   (a) Plane of the integration variable p in (10.14);   (b) plane of z.

is in any case, easily calculated. We know that

$$G^0(z) \,|\mathbf{p}\rangle = \frac{1}{z - E_p} \,|\mathbf{p}\rangle$$

Thus, by inserting a complete set of states $|\mathbf{p}\rangle$ into the required spatial matrix element, we find that:

$$\langle \mathbf{x}| \, G^0(z) \, |\mathbf{x}'\rangle = \int d^3 p \langle \mathbf{x}| \, G^0(z) \, |\mathbf{p}\rangle \langle \mathbf{p} \,|\, \mathbf{x}'\rangle$$

$$= \frac{1}{(2\pi)^3} \int d^3 p \, \frac{e^{i\mathbf{p}\cdot(\mathbf{x}-\mathbf{x}')}}{z - E_p}$$

If we choose the direction of $(\mathbf{x} - \mathbf{x}')$ as polar axis, the exponent becomes $ip \,|\mathbf{x} - \mathbf{x}'| \cos \theta$ and the angular integral can be performed to give

$$= \frac{-i}{4\pi^2 \,|\mathbf{x} - \mathbf{x}'|} \int_0^\infty p \, dp \, \frac{e^{ip|\mathbf{x}-\mathbf{x}'|} - e^{-ip|\mathbf{x}-\mathbf{x}'|}}{z - E_p}$$

$$= \frac{im}{2\pi^2 \,|\mathbf{x} - \mathbf{x}'|} \int_{-\infty}^\infty dp \, \frac{p e^{ip|\mathbf{x}-\mathbf{x}'|}}{p^2 - 2mz} \tag{10.14}$$

This integral can be evaluated by contour integration. We first note that its value is unchanged if we close the contour with a large semicircle in the upper half plane of the complex variable p (see Fig. 10.1, part a). Its value is then given by Cauchy's theorem as $2\pi i$ times the sum of the residues at all poles inside the contour. Now, the integrand has two poles, at $p = \pm (2mz)^{1/2}$, of which only that at $+(2mz)^{1/2}$ (defined as the root with positive imaginary part) lies inside the contour. It follows that:

$$\langle \mathbf{x}| \, G^0(z) \, |\mathbf{x}'\rangle = - \frac{m}{2\pi} \cdot \frac{\exp\{i(2mz)^{1/2} \,|\mathbf{x} - \mathbf{x}'|\}}{|\mathbf{x} - \mathbf{x}'|} \tag{10.15}$$

For use in (10.13) we need the Green's function $\langle x| G^0(z) |x'\rangle$ for $z = E_p \pm i0$ (with p once again positive real). The corresponding square roots $(2mz)^{1/2}$ are indicated in Fig. 10.1. It can be seen that if z approaches the real axis at $E_p > 0$ from *above*, then $(2mz)^{1/2} \to +p$; but if it approaches from *below*, then $(2mz)^{1/2} \to -p$. That is, the relevant roots are:

$$[2m(E_p \pm i0)]^{1/2} = \pm p$$

Substitution of these results into (10.13) gives:

$$\langle x \mid p\pm \rangle = \langle x \mid p \rangle - \frac{m}{2\pi} \int d^3x' \, \frac{e^{\pm ip|x-x'|}}{|x - x'|} \, V(x')\langle x' \mid p\pm \rangle \qquad (10.16)$$

To see what happens when r is large, we suppose for simplicity that $V(x) \equiv 0$ for r greater than some a [although our result actually holds just as long as $V = O(r^{-3-\epsilon})$ as $r \to \infty$]. In this case the integral is confined to $r' < a$, and for large r we can expand $|x - x'|$ in powers of (r'/r),

$$|x - x'| = (x^2 - 2x \cdot x' + x'^2)^{1/2} = r\left[1 - \frac{x \cdot x'}{r^2} + O\left(\frac{r'}{r}\right)^2\right]$$

The expression (10.16) then becomes[5]:

$$\langle x \mid p\pm \rangle = \langle x \mid p \rangle - \frac{me^{\pm ipr}}{2\pi r} \int d^3x'$$

$$\times \exp(\mp ip\hat{x} \cdot x')V(x')\langle x' \mid p\pm \rangle\left[1 + O\left(\frac{a}{r} + \frac{pa^2}{r}\right)\right] \qquad (10.17)$$

$$\xrightarrow[r \to \infty]{} (2\pi)^{-3/2}\left[e^{ip \cdot x} - (2\pi)^2 m\langle \pm p\hat{x}| V |p\pm\rangle\frac{e^{\pm ipr}}{r}\right] \qquad (10.18)$$

Since $-(2\pi)^2 m \langle p'| V |p+\rangle = f(p' \leftarrow p)$, we can rewrite this (for the case of $|p+\rangle$) as:

$$\langle x \mid p+ \rangle \xrightarrow[r \to \infty]{} (2\pi)^{-3/2}\left[e^{ip \cdot x} + f(p\hat{x} \leftarrow p)\frac{e^{ipr}}{r}\right] \qquad (10.19)$$

This is the well known result quoted in (10.1). It establishes that our definition of the amplitude $f(p' \leftarrow p)$ in terms of $\langle p'| S |p\rangle$ is, in fact, the same as the traditional definition in terms of the asymptotic form of the stationary scattering wave function.

The interpretation of this asymptotic form as an "incident" plane wave plus a spherically spreading scattered wave is well known and is the basis of

---

[5] The correction term of order a/r comes from replacing the denominator $|x - x'|$ in (10.16) by r; that of order pa^2/r, from replacing $\exp(ip |x - x'|)$ by $\exp[i(pr - p\hat{x} \cdot x')]$.

the view that $\langle \mathbf{x} \mid \mathbf{p}+ \rangle$ represents an infinite steady beam of particles scatter-ing off V. From (10.18) it is clear that $\langle \mathbf{x} \mid \mathbf{p}- \rangle$ is a plane wave plus a spheric-ally *collapsing* wave, and is of no *direct* physical relevance.

At this point it may be helpful to review briefly the logic of the approach that is based on the wave function $\langle \mathbf{x} \mid \mathbf{p}+ \rangle$. This approach starts by seeking a stationary state of energy E_p defined by the boundary condition (10.19) (from the interpretation of which the result $d\sigma/d\Omega = |f|^2$ emerges immedi-ately, as discussed in the Introduction). In the standard way the Schrödinger differential equation plus the boundary condition is converted into an integral equation, which is just the Lippmann–Schwinger equation (10.13). The analysis leading from (10.13) to (10.18) can be repeated, and it establishes that $f = -(2\pi)^2 m \langle \mathbf{p}' \mid V \mid \mathbf{p}+ \rangle$. From this result and the Lippmann–Schwinger equation, one can then reconstruct all of the results of this and the previous two chapters.

In conclusion, it may be well to reiterate the view of the states $|\mathbf{p}\pm\rangle$ adopted in this book. This is that they are improper eigenvectors of H, in terms of which the proper scattering states can be expanded. The actual scattering state $|\psi\rangle$ at $t = 0$ has the same expansion in terms of $|\mathbf{p}+\rangle$ as does its in asymptote $|\psi_{\mathrm{in}}\rangle$ in terms of $|\mathbf{p}\rangle$; that is, if:

$$|\psi_{\mathrm{in}}\rangle = \int d^3 p\, \phi(\mathbf{p}) \mid \mathbf{p}\rangle \qquad (10.20)$$

then

$$|\psi\rangle = \int d^3 p\, \phi(\mathbf{p}) \mid \mathbf{p}+\rangle \qquad (10.21)$$

Similarly, $|\psi\rangle$ has the same expansion in terms of $|\mathbf{p}-\rangle$ as does its out asymptote $|\psi_{\mathrm{out}}\rangle$ in terms of $|\mathbf{p}\rangle$ [see (10.4)].

The particular importance of the states $|\mathbf{p}+\rangle$ (as against $|\mathbf{p}-\rangle$) lies in the fact that in actual experiments it is the in state $|\psi_{\mathrm{in}}\rangle$ that is well peaked in momentum; that is, $\phi(\mathbf{p})$ in (10.20) is sharply localized about some $\mathbf{p}_0$. To the extent that this is so, (10.21) shows that the improper state $|\mathbf{p}_0+\rangle$ can be regarded as a limiting case of the proper vector $|\psi\rangle$ and certain properties of $|\psi\rangle$ extracted directly from $|\mathbf{p}_0+\rangle$. (This is, of course, the justification of the historical approach.) The state $|\psi\rangle$ can equally be expanded in terms of the $|\mathbf{p}-\rangle$ but, since the corresponding expansion coefficient is not sharply peaked, the improper vectors $|\mathbf{p}-\rangle$ do not have the same intimate relationship to the actual state $|\psi\rangle$.

10-d.   A Spatial Description of the Scattering Process

Our description in Chapter 3 of the collision experiment was essentially a momentum-space description. In particular, the cross section was defined in terms of the number $N_{sc}(d\Omega)$ of particles whose final *momenta* lay in the cone defined by $d\Omega$. However, it is clear that for many experiments (e.g., one using counters) a more natural definition would involve the number of particles whose final *positions* lay in the cone $d\Omega$. While these two definitions are in fact equivalent (under suitable conditions), it would obviously be good to give an analysis directly in terms of the spatial wave functions, and this is what we now give. Our results will establish the equivalence of the two approaches.

We begin by discussing the motion of the particle before the collision occurs. The motion at *all* times is given by $\mathsf{U}(t)\,|\psi\rangle$, but since

$$\mathsf{U}(t)\,|\psi\rangle \xrightarrow[t \to -\infty]{} \mathsf{U}^0(t)\,|\psi_{\mathrm{in}}\rangle$$

there is a time t_0 (namely the time at which the actual collision begins) before which the motion is experimentally indistinguishable from that of $\mathsf{U}^0(t)\,|\psi_{\mathrm{in}}\rangle$. As before, we expand $|\psi_{\mathrm{in}}\rangle$ in plane waves as

$$|\psi_{\mathrm{in}}\rangle = \int d^3p\,\phi(\mathbf{p})\,|\mathbf{p}\rangle$$

where $\phi(\mathbf{p})$ is well peaked about the mean $\mathbf{p}_0$, which we take parallel to $\hat{\mathbf{3}}$. For all practical purposes the wave function of the particle at any time before the collision is then:

$$\psi_{\mathrm{in}}(\mathbf{x}, t) \equiv \langle\mathbf{x}|\,\mathsf{U}^0(t)\,|\psi_{\mathrm{in}}\rangle = (2\pi)^{-3/2}\int d^3p\,\phi(\mathbf{p})\,e^{i(\mathbf{p}\cdot\mathbf{x} - E_p t)} \qquad (10.22)$$

This is, of course, just a freely moving wave packet, whose motion is described in almost any textbook on quantum mechanics. In particular, over short time intervals $\psi_{\mathrm{in}}(\mathbf{x}, t)$ propagates rigidly with velocity $\mathbf{v}_0 = \mathbf{p}_0/m$—a result that is most easily seen by expanding E_p about the mean momentum $\mathbf{p}_0$. Since $\partial E_p/\partial\mathbf{p} = \mathbf{p}/m = \mathbf{v}$, this gives

$$E_p = E_0 + \mathbf{v}_0\cdot(\mathbf{p} - \mathbf{p}_0) + O\!\left(\frac{\Delta p^2}{m}\right)$$

$$= -E_0 + \mathbf{v}_0\cdot\mathbf{p} + O\!\left(\frac{\Delta p^2}{m}\right)$$

(where $E_0 \equiv E_{p_0}$). Substituted into (10.22), this gives

$$\psi_{\mathrm{in}}(\mathbf{x}, t) = (2\pi)^{-3/2}e^{iE_0 t}\int d^3p\;\phi(\mathbf{p})e^{i\mathbf{p}\cdot(\mathbf{x} - \mathbf{v}_0 t)}\left[1 + O\!\left(\frac{\Delta p^2 t}{m}\right)\right]$$

$$= e^{iE_0 t}\psi_{\mathrm{in}}(\mathbf{x} - \mathbf{v}_0 t, 0)\left[1 + O\!\left(\frac{t}{mb^2}\right)\right] \qquad (10.23)$$

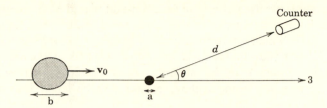

FIGURE 10.2.   Before the collision ($t \ll 0$) the wave packet approaches the target with velocity v_0. The target size a, the packet size b, and the distance d from target to counter must satisfy $a \ll b \ll d$. See (10.28).

where b denotes the original size of the incident wave packet (Fig. 10.2). (We have assumed that the packet is minimal; that is, $\Delta x \, \Delta p \sim 1$ or $\Delta p \sim 1/b$.[6]) This result shows clearly that, as long as $(t/mb^2) \ll 1$, the value of $\psi_{\text{in}}(\mathbf{x}, t)$ is exactly the same (apart from a phase) as its value at $t = 0$, *translated rigidly through* $\mathbf{v}_0 t$. Of course, for sufficiently large t the condition $(t/mb^2) \ll 1$ is always violated and, as is well known, the wave packet becomes distorted.

We now suppose, for definiteness that the incident packet is timed so that it would arrive in the plane of the target at $t = 0$; that is, the function $\psi_{\text{in}}(\mathbf{x}, 0)$ is centered in the plane $z = 0$.

Shortly before $t = 0$, the packet begins to feel the potential and the actual orbit $U(t) \, |\psi\rangle$ begins to diverge from the in asymptote $U^0(t) \, |\psi_{\text{in}}\rangle$. From this time onwards we must use the actual wave function, which is just

$$\psi(\mathbf{x}, t) \equiv \langle \mathbf{x}| \, U(t) \, |\psi\rangle = \int d^3p \phi(\mathbf{p}) e^{-iE_p t} \langle \mathbf{x} \mid \mathbf{p}+\rangle$$

since

$$|\psi\rangle = \Omega_+ \, |\psi_{\text{in}}\rangle = \int d^3p \phi(\mathbf{p}) \, |\mathbf{p}+\rangle$$

Since our object is to find the probability of the particle's being detected at a large distance from the target, we must evaluate $\psi(\mathbf{x}, t)$ for $t > 0$ and large r. Now, for r sufficiently large [$r \gg a$ and $r \gg pa^2$ according to (10.17)] we can replace $\langle \mathbf{x} \mid \mathbf{p}+\rangle$ by its asymptotic form to give:

$$\psi(\mathbf{x}, t) \xrightarrow[r \to \infty]{} (2\pi)^{-3/2} \int d^3p \phi(\mathbf{p}) e^{-iE_p t} \left[e^{i\mathbf{p} \cdot \mathbf{x}} + f(p\hat{\mathbf{x}} \leftarrow \mathbf{p}) \frac{e^{ipr}}{r} \right]$$

$$= \psi_{\text{in}}(\mathbf{x}, t) + \psi_{\text{sc}}(\mathbf{x}, t) \tag{10.24}$$

[6] For simplicity we suppose that the packet is more or less spherical. It is a simple matter, if one wishes, to distinguish between its length $b_{\parallel}$ (along $\mathbf{p}_0$) and breadth $b_{\perp}$ (normal to $\mathbf{p}_0$).

The first term is precisely the wave function that would have emerged if there had been no target; i.e., it is the unscattered wave. The second is the scattered wave, which we now manipulate as follows. First we assume, as in Chapter 3, that the amplitude is smooth compared to the wave function $\phi(\mathbf{p})$ and take f outside of the integral. (As discussed in Section 3-e, this is expected to require the incident wave packet to be larger than the target, $b \gg a$.) This gives:

$$\psi_{\mathrm{sc}}(\mathbf{x}, t) = (2\pi)^{-3/2}\frac{f(p_0\hat{\mathbf{x}} \leftarrow \mathbf{p}_0)}{r}\int d^3p\,\phi(\mathbf{p})e^{i(pr-E_pt)} \qquad (10.52)$$

Next, since $\phi(\mathbf{p})$ is well localized about $\mathbf{p}_0$, which is parallel to $\hat{\mathbf{3}}$, we can approximate pr by

$$pr = \mathbf{p}\cdot\hat{\mathbf{3}}r + O\left(\frac{\Delta p^2 r}{p}\right)$$

Substituting this into (10.25) gives:

$$\psi_{\mathrm{sc}}(\mathbf{x}, t) = \frac{f(p_0\hat{\mathbf{x}} \leftarrow \mathbf{p}_0)}{r}\,\psi_{\mathrm{in}}(r\hat{\mathbf{3}}, t)\left[1 + O\left(\frac{r}{p_0b^2}\right)\right] \qquad (10.26)$$

That is, the value of the scattered wave at any point $\mathbf{x}$ is just (f/r) times that of the *unscattered wave on the axis at the point* $r\hat{\mathbf{3}}$ (provided $r/p_0b^2 \ll 1$).[7]

We conclude that long after the collision the wave function is as shown in Fig. 10.3. There is an unscattered packet centered at a distance v_0t beyond the target, and there is a scattered wave that is nonzero in a spherical shell of radius v_0t.

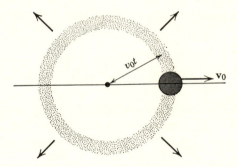

FIGURE 10.3.   After the collision ($t \gg 0$) the unscattered wave packet continues with velocity $\mathbf{v}_0$; the scattered wave moves outwards in a shell of radius v_0t.

[7] Two highly gratifying features of this result are apparent. First, it is clear that when $t \ll 0$, $\psi_{\mathrm{sc}}(\mathbf{x}, t)$ is identically zero (as it should be) since $\psi_{\mathrm{in}}(r\hat{\mathbf{3}}, t), (r > 0)$, is certainly zero until the incident packet reaches the plane of the target. Second, $\psi_{\mathrm{sc}}(\mathbf{x}, t)$ is zero for *all* t if $\psi_{\mathrm{in}}(\mathbf{x}, t)$ is zero on the axis. This expresses the very natural result that if the incident wave packet is aimed to miss the target, then there is no scattering.

Provided the counter is not in the path of the unscattered packet ($d \sin \theta \gg b$), the probability that it will register a count is

$$w(d\Omega \leftarrow \phi) = d\Omega \int_0^\infty r^2 \, dr \, |\psi_{sc}(\mathbf{x}, t)|^2$$

or, by (10.26)

$$w(d\Omega \leftarrow \phi) = d\Omega \, |f(\mathbf{p} \leftarrow \mathbf{p}_0)|^2 \int_{-\infty}^\infty dr \, |\psi_{in}(r\hat{\mathbf{3}}, t)|^2 \qquad (10.27)$$

where $\mathbf{p}$ points in the direction of $d\Omega$, and we can extend the integral to $-\infty$ since $\psi_{in}(\mathbf{x}, t)$ is zero on the negative z axis by now.

To obtain the cross section we must, as usual, repeat the experiment many times with the incident packet laterally displaced by random amounts $\boldsymbol{\rho}$. Under a displacement through $\boldsymbol{\rho}$ the spatial wave function $\psi_{in}(\mathbf{x}, t)$ becomes $\psi_{in}(\mathbf{x} - \boldsymbol{\rho}, t)$. Substituting into (10.27) we obtain the cross section:

$$\sigma(d\Omega \leftarrow \phi) = \int d^2\rho \, w(d\Omega \leftarrow \phi_\rho)$$

$$= d\Omega \, |f(\mathbf{p} \leftarrow \mathbf{p}_0)|^2 \int d^2\rho \int_{-\infty}^\infty dr \, |\psi_{in}(r\hat{\mathbf{3}} - \boldsymbol{\rho}, t)|^2$$

$$= d\Omega \, |f(\mathbf{p} \leftarrow \mathbf{p}_0)|^2 \int d^3x \, |\psi_{in}(\mathbf{x}, t)|^2$$

$$= d\Omega \, |f(\mathbf{p} \leftarrow \mathbf{p}_0)|^2$$

as expected.

It is of some interest to collect the various conditions for the validity of this description of the collision experiment. We first note that if the counter is at a distance d, the duration of the experiment is of the order $t \sim d/v_0$. The conditions are then as follows:

(1) No distortion of the packet (10.23) ... $t/mb^2 \ll 1$ or $d \ll p_0b^2$.
(2) Use of asymptotic form of $\langle \mathbf{x} \, | \, \mathbf{p}+ \rangle$ in (10.24) ... $a \ll d$ and $p_0a^2 \ll d$.
(3) Removal of f outside integral of (10.25) ... $a \ll b$.
(4) Replacement of p by $\mathbf{p} \cdot \hat{\mathbf{3}}$ in (10.26) ... $d \ll p_0b^2$.
(5) Unscattered wave not measured ... $b \ll d \sin \theta$ and, hence, $b \ll d$.

Collecting these together we find:

$$a \ll b \ll d \ll p_0b^2 \qquad (10.28)$$

and

$$p_0a^2 \ll d$$

These conditions are almost invariably met in practice. For example, for a 10 eV electron ($p \sim 10^{10}$ m^{-1}) collimated through 1-mm slits ($b \sim 10^{-3}$ m), which scatters off an atom ($a \sim 10^{-10}$ m) and is detected at one meter from the target $(d \sim 1$ m$)$ the inequalities are

$$10^{-10} \ll 10^{-3} \ll 1 \ll 10^{4}$$

and

$$10^{-10} \ll 1$$

PROBLEMS

10.1.   Prove that for the scattering of a single spinless particle off a potential $V(\mathbf{x})$, which is time-reversal invariant (i.e., real)

$$\mathsf{T}\,|\mathbf{p}\pm\rangle = |-\mathbf{p}\mp\rangle$$

or, in terms of wave functions, $\langle\mathbf{x}\,|\,\mathbf{p}\pm\rangle^{*} = \langle\mathbf{x}\,|-\mathbf{p}\mp\rangle$. [Review Section 6-e and particularly (6.24).]

10.2.   The Born approximation is obtained by writing

$$t(\mathbf{p}' \leftarrow \mathbf{p}) = \langle\mathbf{p}'|\,V\,|\mathbf{p}+\rangle \approx \langle\mathbf{p}'|\,V\,|\mathbf{p}\rangle$$

Thus, the criterion for its validity is that $|\mathbf{p}+\rangle$ not differ appreciably from $|\mathbf{p}\rangle$. If we write $t(\mathbf{p}' \leftarrow \mathbf{p})$ in terms of wave functions,

$$t(\mathbf{p}' \leftarrow \mathbf{p}) = (2\pi)^{-\frac{3}{2}}\int d^{3}x\, e^{-i\mathbf{p}'\cdot\mathbf{x}}V(\mathbf{x})\langle\mathbf{x}\,|\,\mathbf{p}+\rangle$$

we see that the criterion is really that the wave function $\langle\mathbf{x}\,|\,\mathbf{p}+\rangle$ differs little from the plane wave $\langle\mathbf{x}\,|\,\mathbf{p}\rangle$ in the region where $V(\mathbf{x})$ is appreciable. A rough version of this would be to require that the second term in the Born series for $\langle\mathbf{x}\,|\,\mathbf{p}+\rangle$ be much smaller than the first,

$$|\langle\mathbf{x}|\,G^{0}(E_{p} + i0)V\,|\mathbf{p}\rangle| \ll |\langle\mathbf{x}\,|\,\mathbf{p}\rangle|$$

wherever $V(\mathbf{x})$ is appreciable. Finally, since $V(\mathbf{x})$ is usually largest at the origin, we obtain a very rough criterion by setting $\mathbf{x} = 0$ in the above.
(a) Show that for a central potential this gives

$$\frac{m}{p}\left|\int_{0}^{\infty} dr(1 - e^{2ipr})V(r)\right| \ll 1$$

(b) Apply this to a square well of radius a and depth V_{0}. Find a criterion suitable for high energies and show that if $ma^{2}V_{0} \ll 1$, the Born approximation should be good at all energies.
(c) For the Yukawa potential, show that the condition approximately reproduces the rough criteria (9.17) and (9.18). [But note the extra factor of $\ln(\mu - 2ip)/\mu$. This would have been present in (9.18) if we had used the nonforward amplitude.]

10.3.   Give an alternative derivation of the result $t(\mathbf{p}' \leftarrow \mathbf{p}) = \langle \mathbf{p}' | \, V \, | \mathbf{p}+\rangle$ along the following lines:
(1) Write

$$\langle \mathbf{p}' | \, S \, | \mathbf{p} \rangle = \langle \mathbf{p}' | \, \Omega_-^\dagger \Omega_+ \, | \mathbf{p} \rangle = \langle \mathbf{p}'- | \, \mathbf{p}+\rangle$$

(2) Use equations (10.8) for $|\mathbf{p}\pm\rangle$ to rewrite $\langle \mathbf{p}'-|$ in terms of $\langle \mathbf{p}'+|$ and $\langle \mathbf{p}'|$.
(3) Show that

$$\langle \mathbf{p}' | \, S \, | \mathbf{p} \rangle = \delta_3(\mathbf{p}' - \mathbf{p}) - 2\pi i \; \delta(E_{p'} - E_p)\langle \mathbf{p}' | \, V \, | \mathbf{p}+\rangle$$

11 The Partial-Wave Stationary States

11-a   The Partial-Wave S Matrix

11-b   The Free Radial Wave Functions

11-c   The Partial-Wave Scattering States

11-d   The Partial-Wave Lippmann–Schwinger Equation

11-e   Properties of the Partial-Wave Amplitude

11-f   The Regular Solution

11-g   The Variable Phase Method

11-h   Iterative Solution for the Regular Wave Function

11-i   The Jost Function

11-j   The Partial-Wave Born Series

In this chapter and in Chapter 12, we continue our discussion of the stationary scattering states, but restrict attention to potentials that are spherically symmetric. In this case the vectors $|\mathbf{p}+\rangle$ can be expanded in terms of angular momentum eigenvectors, which we denote by $|E, l, m+\rangle$. (There is, of course, a corresponding expansion of $|\mathbf{p}-\rangle$ in terms of states $|E, l, m-\rangle$ but we shall not need to discuss it.) The wave function $\langle \mathbf{x} \mid E, l, m+\rangle$ is a product of the spherical harmonic $Y_l^m(\hat{\mathbf{x}})$ and a radial wave function $\psi_{l, p}(r)$ that satisfies the radial Schrödinger equation (a second order, ordinary, linear differential equation). We shall find that the asymptotic form of $\psi_{l,\,p}(r)$ is directly related to the scattering phase shift. This connection provides a simple and powerful approach to the calculation and analysis of scattering amplitudes.   It is this approach—the so-called partial-wave approach—that is the subject of the next two chapters.

The division between Chapters 11 and 12 is this: Chapter 11 contains a general introduction to the partial-wave method and at least a statement of most of the important results.   Chapter 12 is principally concerned with the important technique in which the variable p (the magnitude of the incident momentum) is allowed to become a complex variable, and the amplitude is considered as an analytic function of p.   This technique leads to simple proofs of several of the results stated in the present chapter and provides an introduction to several important new ideas: the connection between bound states, resonances, and poles of the amplitude; and the important concept of the dispersion relation.

11-a.   The Partial-Wave S Matrix

In Section 6-c we discussed the scattering of two spinless particles whose interactions are rotationally invariant and found the S matrix to be diagonal in the basis of eigenvectors $|E, l, m\rangle$,

$$\langle E', l', m'| \; \mathsf{S} \; |E, l, m\rangle = \delta(E' - E)\delta_{l'l}\delta_{m'm}\mathsf{s}_l(p)$$

Because S is unitary, the number $\mathsf{s}_l(p)$ [which it is now convenient to consider as a function of $p = (2mE)^{\frac{1}{2}}$ rather than E] has modulus one and can be written as

$$\mathsf{s}_l(p) = e^{2i\delta_l(p)}$$

This defines the *phase shift* $\delta_l(p)$ up to addition of an arbitrary multiple of π—an ambiguity that we shall call the *modulo π ambiguity* in $\delta_l(p)$.

Using the transformation matrix:

$$\langle \mathbf{p} \, | \, E, l \; m\rangle = (mp)^{-\frac{1}{2}} \, \delta(E_p - E)Y_l^m(\hat{\mathbf{p}}) \tag{11.1}$$

one can pass from the angular-momentum to the momentum basis.   In particular, we found for the amplitude $f(\mathbf{p}' \leftarrow \mathbf{p})$ that:

$$f(\mathbf{p}' \leftarrow \mathbf{p}) = \sum_l (2l + 1)f_l(p)P_l(\hat{\mathbf{p}}' \cdot \hat{\mathbf{p}}) \tag{11.2}$$

where $f_l(p)$ is the *partial-wave amplitude*,

$$f_l(p) = \frac{\mathsf{s}_l(p) - 1}{2ip} = \frac{e^{i\delta_l(p)} \sin \delta_l(p)}{p}$$

The partial wave series (11.2) for $f(\mathbf{p}' \leftarrow \mathbf{p})$ gives a double series for the differential cross section $d\sigma/d\Omega = |f|^2$.   When this is integrated over all angles, the cross terms cancel to give, for the total cross section,

$$\sigma = \sum_l \sigma_l(p) \tag{11.3}$$

where

$$\sigma_l(p) = 4\pi(2l + 1) \, |f_l(p)|^2 = 4\pi(2l + 1) \, \frac{\sin^2 \delta_l(p)}{p^2}$$

The maximum contribution from any partial wave to the total cross section is given by the "unitarity bound"

$$\sigma_l(p) \leqslant \frac{4\pi(2l + 1)}{p^2} \tag{11.4}$$

and $\sigma_l(p)$ attains this bound if and only if the phase shift δ_l is an odd multiple of $\pi/2$.

11-b.   The Free Radial Wave Functions

In this chapter we shall be concerned with the properties of the wave functions $\langle \mathbf{x} \mid E, l, m+ \rangle$, and as an essential preliminary we first review briefly the properties of the "free" wave functions $\langle \mathbf{x} \mid E, l, m \rangle$. These are, of course, products of the spherical harmonics $Y_l^m(\hat{\mathbf{x}})$ and functions of r. If we write these products in the form $(1/r) \, y(r) \, Y_l^m(\hat{\mathbf{x}})$, then the *radial wave function* $y(r)$ *satisfies the free radial Schrödinger equation:*

$$\left[\frac{d^2}{dr^2} - \frac{l(l + 1)}{r^2} + p^2 \right] y(r) = 0 \qquad [p = (2mE)^{\frac{1}{2}}] \tag{11.5}$$

The free radial Schrödinger equation is an ordinary, second-order, linear, differential equation and so has two linearly independent solutions, of which the physically relevant solution is that which vanishes at the origin. As $r \to 0$, the centrifugal term $l(l + 1)/r^2$ dominates the energy term p^2, and the solutions behave like solutions of the corresponding equation with $p = 0$; namely, like combinations of r^{l+1} and r^{-l}. Thus, the physically acceptable wave function is uniquely determined as that which behaves like r^{l+1}. This is, of course, the Riccati–Bessel function $\hat{j}_l(pr)$,

$$\hat{j}_l(z) \equiv z j_l(z) \equiv \left(\frac{\pi z}{2} \right)^{\frac{1}{2}} J_{l+\frac{1}{2}}(z)$$

$$= z^{l+1} \sum_{n=0}^{\infty} \frac{(-z^2/2)^n}{n! \, (2l + 2n + 1)!!} \tag{11.6}$$

where $j_l(z)$ is the spherical Bessel function and $J_\lambda(z)$ the ordinary Bessel function.[1] The precise behavior of $\hat{j}_l$ at the origin is obviously

$$\hat{j}_l(z) = \frac{z^{l+1}}{(2l+1)!!} [1 + O(z^2)] \qquad [z \to 0] \qquad (11.7)$$

It can be shown that

$$\int_0^\infty dr\, \hat{j}_l(p'r)\, \hat{j}_l(pr) = \frac{\pi}{2}\, \delta(p' - p)$$

and from this it is easy to check that the correctly normalized wave function $\langle \mathbf{x} \mid E, l, m \rangle$ is that given in (6.6), namely,

$$\langle \mathbf{x} \mid E, l, m \rangle = i^l \left(\frac{2m}{\pi p} \right)^{\frac{1}{2}} \frac{1}{r}\, \hat{j}_l(pr)\, Y_l^m(\hat{\mathbf{x}}) \qquad (11.8)$$

where the phase i^l is, of course, purely conventional.

For future reference we shall need some properties of the general solution to the free radial equation (11.5). Concerning the behavior at the origin we remark that any solution other than the Riccati–Bessel function $\hat{j}_l(pr)$ behaves like r^{-l} as $r \to 0$. A convenient choice for the second solution is the Riccati–Neumann function $\hat{n}_l(pr)$,[2]

$$\hat{n}_l(z) \equiv z n_l(z) \equiv (-)^l \left(\frac{\pi z}{2} \right)^{\frac{1}{2}} J_{-l-\frac{1}{2}}(z)$$

$$= z^{-l} \sum_{n=0}^\infty \frac{(-z^2/2)^n (2l - 2n - 1)!!}{n!} \qquad (11.9)$$

At the origin this has the behavior

$$\hat{n}_l(z) = z^{-l}(2l - 1)!![1 + O(z^2)] \qquad [z \to 0] \qquad (11.10)$$

When $r \to \infty$, the centrifugal term in the radial equation vanishes and the solutions behave like solutions of the corresponding equation with $l = 0$; namely, like combinations of $e^{\pm ipr}$. The solutions that actually behave like $e^{\pm ipr}$ are not $\hat{j}_l$ and $\hat{n}_l$ but are the Riccati–Hankel functions $\hat{h}_l^\pm(pr)$,

$$\hat{h}_l^\pm(z) = \hat{n}_l(z) \pm i\hat{j}_l(z)$$

These can be shown to have the asymptotic form

$$\hat{h}_l^\pm(z) = e^{\pm i(z - l\pi/2)}[1 + O(z^{-1})] \qquad [z \to \infty] \qquad (11.11)$$

---

[1] For a summary of properties of Bessel functions see Messiah (1961) p. 488. Proofs of all properties used here can be found in Watson (1958).

[2] In the series definition, $(-2n - 1)!! = (-1)^n/(2n - 1)!!$ and $(-1)!! = 1$.

from which one can immediately obtain the corresponding results for $\hat{j}_l$ and $\hat{n}_l$. For example,

$$\hat{j}_l(z) = \frac{[\hat{h}_l^+(z) - \hat{h}_l^-(z)]}{2i}$$

$$= \sin(z - \tfrac{1}{2}l\pi) + O(z^{-1}) \qquad [z \to \infty, \text{real}]$$

with a similar result (in which the sine is replaced by a cosine) for $\hat{n}_l(z)$.

It should be emphasized that there are several phase conventions and notations for the functions $\hat{j}$, $\hat{n}$ and $\hat{h}^{\pm}$. We follow the conventions of Messiah (1961), which have the convenient feature that the relationships among the four Riccati functions $\hat{j}$, $\hat{n}$ and $\hat{h}^{\pm}$ are precisely those of the trigonometric functions $\sin z$, $\cos z$ and $e^{\pm iz}$. In fact, for $l = 0$, the Riccati functions *are* the trigonometric functions, as can be seen in Table 11.1, where we give the Riccati functions for $l = 0, 1, 2$.

TABLE 11.1.   RICCATI FUNCTIONS

	$l = 0$	$l = 1$	$l = 2$
$\hat{j}_l(z)$	$\sin z$	$\dfrac{1}{z}\sin z - \cos z$	$\left(\dfrac{3}{z^2} - 1\right)\sin z - \dfrac{3}{z}\cos z$
$\hat{n}_l(z)$	$\cos z$	$\dfrac{1}{z}\cos z + \sin z$	$\left(\dfrac{3}{z^2} - 1\right)\cos z + \dfrac{3}{z}\sin z$
$\hat{h}_l^+(z)$	e^{iz}	$\left(1 + \dfrac{i}{z}\right)e^{i(z-\pi/2)}$	$\left(1 + \dfrac{3i}{z} - \dfrac{3}{z^2}\right)e^{i(z-\pi)}$
$\hat{h}_l^-(z)$	e^{-iz}	$\left(1 - \dfrac{i}{z}\right)e^{-i(z-\pi/2)}$	$\left(1 - \dfrac{3i}{z} - \dfrac{3}{z^2}\right)e^{-i(z-\pi)}$

Table 11.1 shows (as can be checked directly from the definitions) that the following useful identities hold:

$$\hat{j}_l(-z) = (-)^{l+1}\hat{j}_l(z)$$
$$\hat{n}_l(-z) = (-)^l\hat{n}_l(z)$$
$$\hat{h}_l^{\pm}(-z) = (-)^l\hat{h}_l^{\mp}(z)$$

Also, for real arguments $\hat{j}_l(x)$ and $\hat{n}_l(x)$ are real, whereas

$$[\hat{h}_l^{\pm}(x)]^* = \hat{h}_l^{\mp}(x) \qquad [x \text{ real}]$$

In conclusion we return to the states $|E, l, m\rangle$ and note the expansion of the momentum eigenstate $|\mathbf{p}\rangle$ in terms of the $|E, l, m\rangle$. Using the transformation matrix (11.1) we can write this down immediately as

$$|\mathbf{p}\rangle = \int dE \sum_{l,m} |E, l, m\rangle \langle E, l, m \mid \mathbf{p}\rangle$$

$$= (mp)^{-\frac{1}{2}} \sum_{l,m} |E_p, l, m\rangle Y_l^m(\hat{\mathbf{p}})^*$$

Multiplying by the bra $\langle \mathbf{x}|$ we obtain the corresponding expansion for the wave functions

$$\langle \mathbf{x} \mid \mathbf{p}\rangle = \left(\frac{2}{\pi}\right)^{\frac{1}{2}} \frac{1}{pr} \sum_{l,m} i^l \hat{j}_l(pr) Y_l^m(\hat{\mathbf{x}}) Y_l^m(\hat{\mathbf{p}})^*$$

$$= (2\pi)^{-\frac{3}{2}} \frac{1}{pr} \sum_l (2l + 1) i^l \hat{j}_l(pr) P_l(\hat{\mathbf{x}} \cdot \hat{\mathbf{p}}) \qquad (11.12)$$

which is just the well known expansion of a plane wave in terms of spherical harmonics [from which (11.1) is actually derived].

11-c.   The Partial-Wave Scattering States

The vectors $|E, l, m\rangle$ are eigenvectors of the angular momentum and the free Hamiltonian H^0. We now define corresponding eigenvectors $|E, l, m+\rangle$ of the angular momentum and the *full* Hamiltonian H. The relationship of $|E, l, m+\rangle$ to $|E, l, m\rangle$ is the same as that of $|\mathbf{p}+\rangle$ to $|\mathbf{p}\rangle$; namely,

$$|E, l, m+\rangle = \Omega_+ |E, l, m\rangle$$

In the same rough sense as applied to $|\mathbf{p}+\rangle$ and $|\mathbf{p}\rangle$, we can think of $|E, l, m+\rangle$ as the actual state at $t = 0$ arising from the in state $|E, l, m\rangle$. From the inter-twining relation $H\Omega_+ = \Omega_+ H^0$ it follows that $|E, l, m+\rangle$ is an eigenvector of H, with energy E; and since Ω_+ and $\mathbf{L}$ commute, it is an eigenvector of $\mathbf{L}^2$ and L_3.

The wave function $\langle \mathbf{x} \mid E, l, m+\rangle$ is of course a product of the spherical harmonic $Y_l^m(\hat{\mathbf{x}})$ and a function of r. By analogy with the expression (11.8) for the free wave function, we write this product as[3]

$$\langle \mathbf{x} \mid E, l, m+\rangle = i^l \left(\frac{2m}{\pi p}\right)^{\frac{1}{2}} \frac{1}{r} \psi_{l,p}(r) Y_l^m(\hat{\mathbf{x}}) \qquad (11.13)$$

which means that $\psi_{l,p}(r)$ reduces precisely to $\hat{j}_l(pr)$ when $V = 0$. The function $\psi_{l,p}(r)$ is determined (apart from its normalization) by the radial

[3] Strictly speaking the radial wave function should carry a superscript $+$; i.e., $\psi_{l,p}^+(r)$. However since we do not need to discuss the corresponding states $|E, l, m-\rangle$ we omit it.

Schrödinger equation

$$\left[\frac{d^2}{dr^2} - \frac{l(l+1)}{r^2} - U(r) + p^2\right]\psi_{l,p}(r) = 0$$

together with the boundary condition $\psi_{l,p}(0) = 0$. In writing this equation we have introduced the convenient notation

$$U(r) \equiv 2mV(r)$$

From the normalization

$$\langle E', l', m' + | E, l, m + \rangle = \delta(E' - E)\delta_{l'l}\delta_{m'm}$$

it follows that

$$\int_0^\infty dr\psi_{l,p'}(r)^*\psi_{l,p}(r) = \frac{\pi}{2}\delta(p' - p) \qquad (11.14)$$

and for this reason we shall refer to $\psi_{l,p}(r)$ as the *normalized radial function*, whenever we need to distinguish it from other wave functions.

Probably the most important property of the radial wave function $\psi_{l,p}(r)$ is that its asymptotic form as $r \to \infty$ is simply related to the partial-wave amplitude $f_l(p)$ or phase shift $\delta_l(p)$. To see this we first expand the wave function $\langle \mathbf{x} | \mathbf{p} + \rangle$ (whose behavior as $r \to \infty$ we already know) in terms of the partial-wave functions $\langle \mathbf{x} | E, l, m + \rangle$. From their definitions it is clear that the expansion of $|\mathbf{p}+\rangle$ in terms of $|E, l, m+\rangle$ is the same as that of $|\mathbf{p}\rangle$ in terms of $|E, l, m\rangle$ (to see this simply multiply the latter by Ω_+). The same is true then of the corresponding wave functions and it follows at once from (11.12) that

$$\langle \mathbf{x} | \mathbf{p}+\rangle = (2\pi)^{-\frac{3}{2}}\frac{1}{pr}\sum_l (2l+1)i^l\psi_{l,p}(r)P_l(\hat{\mathbf{x}}\cdot\hat{\mathbf{p}}) \qquad (11.15)$$

Now, for large r we know from (10.19) that

$$\langle \mathbf{x} | \mathbf{p}+\rangle \xrightarrow[r\to\infty]{} (2\pi)^{-\frac{3}{2}}\left[e^{i\mathbf{p}\cdot\mathbf{x}} + f(p\hat{\mathbf{x}} \leftarrow \mathbf{p})\frac{1}{r}e^{ipr}\right]$$

or, if we insert the expansion (11.12) for the plane wave and (11.2) for the amplitude,

$$\langle \mathbf{x} | \mathbf{p}+\rangle \xrightarrow[r\to\infty]{} (2\pi)^{-\frac{3}{2}}\frac{1}{pr}\sum_l (2l+1)[i^l j_l(pr) + pf_l(p)e^{ipr}]P_l(\hat{\mathbf{x}}\cdot\hat{\mathbf{p}})$$

Comparing this with (11.15) we conclude that

$$\psi_{l,p}(r) \xrightarrow[r \to \infty]{} \hat{j}_l(pr) + p f_l(p) e^{i(pr - l\pi/2)}$$

which is the desired result.

This last result can be rewritten in various useful forms. If we recall that $\hat{h}_l^+(z) \to \exp i(z - \frac{1}{2}l\pi)$, we can put it in the suggestive form

$$\boxed{\psi_{l,p}(r) \xrightarrow[r \to \infty]{} \hat{j}_l(pr) + p f_l(p) \hat{h}_l^+(pr)} \tag{11.16}$$

that is, because the potential is suitably short range, the radial function $\psi_{l,p}(r)$ behaves like a solution of the *free* radial equation as $r \to \infty$. The precise combination of free solutions, $\hat{j}_l + p f_l \hat{h}_l^+$, corresponds to the well known form

$$e^{i\mathbf{p} \cdot \mathbf{x}} + f(p\hat{\mathbf{x}} \leftarrow \mathbf{p}) \frac{e^{ipr}}{r}$$

of the full wave function $\langle \mathbf{x} \mid \mathbf{p}+ \rangle$. The term $\hat{j}_l$ is the "incident free wave function" of angular momentum l, while $p f_l \hat{h}_l^+$ is the scattered outgoing wave.

Alternatively, if we replace $\hat{j}_l(pr)$ by its asymptotic form $\sin(pr - \frac{1}{2}l\pi)$ and $p f_l$ by $e^{i\delta_l} \sin \delta_l$, some simple algebra gives

$$\boxed{\psi_{l,p}(r) \xrightarrow[r \to \infty]{} e^{i\delta_l(p)} \sin[pr - \frac{1}{2}l\pi + \delta_l(p)]} \tag{11.17}$$

In this form the result shows the significance of the name phase shift. At large r the actual radial function $\psi_{l,p}(r)$ is proportional to the free radial function

$$\hat{j}_l(pr) \to \sin(pr - \frac{1}{2}l\pi)$$

except that the phase of its oscillations is shifted by an amount $\delta_l(p)$. This important property (which is often used to *define* the phase shift[4]) is illustrated in Fig. 11.1, which shows the s-wave ($l = 0$) radial functions for a square well.

Since the kinetic energy inside the square well depicted in Fig. 11.1 is greater than that outside, the wave function emerges from the well with its phase in advance of the free wave function. Thus, an attractive potential "pulls the wave functions inward" and gives a positive phase shift. In the same way a repulsive potential "pushes the wave function outward" and gives a negative phase shift. These ideas are made precise in Section 11-g.

---

[4] The definition of δ_l in terms of the property (11.17) suffers the same modulo π ambiguity as does our definition $s_l = \exp(2i\delta_l)$.

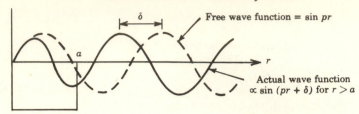

FIGURE 11.1.   **The *s*-wave radial function for an attractive square well.**

The form (11.17) also suggests a simple method for the actual computation of the phase shift. The radial equation is integrated outward from $r = 0$ (either analytically or numerically) with the boundary condition that the wave function vanish at $r = 0$ and with any convenient normalization. The solution generated in this way is proportional to $\psi_{l,p}(r)$ and, at large distances, behaves as

$$\text{constant} \times \sin(pr - \tfrac{1}{2}l\pi + \delta_l)$$

Thus, all that is necessary is to integrate out to a point well beyond the influence of the potential and then to compare the resulting phase with that of the free function $[\to \sin(pr - \tfrac{1}{2}l\pi)]$.

It is clear that the radial Schrödinger equation provides a self-contained framework for the calculation and analysis of the partial-wave amplitude and the phase shift. In fact, we can if we wish, completely divorce this framework from the realistic three-dimensional analysis on which it was based. The collision process is then thought of as proceeding within a single angular momentum state. If the asymptotic form (11.16) is rewritten [setting $\hat{j} = (\hat{h}^+ - \hat{h}^-)/2i$ and $pf_l = (s_l - 1)/2i$] as

$$\psi_{l,p}(r) \to \frac{i}{2}\,[\hat{h}_l^-(pr) - s_l(p)\hat{h}_l^+(pr)] \tag{11.18}$$

then the first term can be interpreted as a spherical wave moving inwards towards the origin, and the second as the resulting outgoing wave. The coefficient $s_l = \exp(2i\delta_l)$ measures the response of the target, and the fact that the potential cannot create or destroy particles restricts this response to a change in the phase (and not the modulus) of the outgoing wave. It is for this reason that $|s_l| = 1$ and δ_l is real.

11-d.   The Partial-Wave Lippmann–Schwinger Equation

Before we begin our study of the radial Schrödinger equation it is useful to note that we can convert the radial equation *plus* the boundary conditions into a single integral equation for $\psi_{l,p}(r)$. This can be done in two ways. In the first, we note that $\psi_{l,p}(r)$ is just the radial part of the wave function for the stationary state $|E, l, m+\rangle$, which satisfies the Lippmann–Schwinger equation

$$|E, l, m+\rangle = |E, l, m\rangle + G^0(E + i0)\, V\, |E, l, m+\rangle \qquad (11.19)$$

Since both $G^0(z)$ and V are rotationally invariant the corresponding equation for the wave function contains a common factor $Y_l^m(\hat{x})$ which can be factored out to give

$$\psi_{l,p}(r) = \hat{\jmath}_l(pr) + \int_0^\infty dr'\, G_{l,p}^0(r, r') U(r') \psi_{l,p}(r') \qquad (11.20)$$

Here, as the reader should check (see Problem **11.2**),

$$G_{l,p}^0(r, r') = -\frac{1}{p} \hat{\jmath}_l(pr_<) \hat{h}_l^+(pr_>) \qquad (11.21)$$

$r_<$ and $r_>$ being respectively the smaller and larger of r and r' (and, as above, $U \equiv 2mV$).

Alternatively, we can start with the radial, differential equation plus the boundary conditions $\psi_{l,p} = 0$ at $r = 0$ and $\psi_{l,p} \rightarrow \hat{\jmath}_l + pf_l\hat{h}_l^+$ as $r \rightarrow \infty$, and appeal to the standard theory of ordinary differential equations. This guarantees that the differential equation plus boundary conditions is equivalent to an integral equation of the form (11.20), where the Green's function (11.21) is simply the Green's function that yields the desired boundary conditions.[5]

We shall sometimes find it convenient to write the integral equation (11.20) in an operator notation as:

$$\psi_{l,p} = \hat{\jmath}_l + G_{l,p}^0 U \psi_{l,p} \qquad (11.22)$$

However we choose to arrive at this equation, it is clearly the partial-wave version of the Lippmann–Schwinger equation and many of the results discussed in connection with the full, three-dimensional Lippmann–Schwinger

---

[5] The transition procedure from differential equation plus boundary conditions to integral equation is:

(1) Write the radial equation as

$$\psi'' - \left[\frac{l(l + 1)}{r^2} - p^2 \right] \psi = U(r)\psi$$

(2) With the right hand side as the source term, "solve" this equation by standard Green's function techniques (see Courant and Hilbert, 1953, Chapter V; or Friedmann, 1956, Chapter 3). Since the source term is $U(r)\psi$ the result is an integral equation for ψ.

equation have corresponding partial–wave analogues. For example, if we decompose the result

$$f(\mathbf{p}' \leftarrow \mathbf{p}) = -(2\pi)^2 m \langle \mathbf{p}'| V |\mathbf{p}+\rangle$$

into partial waves we obtain at once (as the reader should check)

$$\boxed{f_l(p) = -\frac{1}{p^2} \int_0^\infty dr\, j_l(pr) U(r) \psi_{l,p}(r)} \qquad (11.23)$$

We obtain a Born series for $f_l(p)$ by iterating the partial-wave Lippmann–Schwinger equation. This gives, in the operator notation of (11.22),

$$\psi_{l,p} = \hat{j}_l + G^0_{l,p} U \hat{j}_l + \cdots$$

and, hence, from (11.23)

$$f_l(p) = -\frac{1}{p^2}\left(\int dr \hat{j}_l U \hat{j}_l + \int dr \hat{j}_l U G^0_{l,p} U \hat{j}_l + \cdots \right)$$

This is, of course, precisely the expansion one would obtain by resolving the full Born series

$$f(\mathbf{p}' \leftarrow \mathbf{p}) = -(2\pi)^2 m (\langle \mathbf{p}'| V |\mathbf{p}\rangle + \langle \mathbf{p}'| V G^0 V |\mathbf{p}\rangle + \cdots)$$

into partial waves, and many of our remarks about convergence of the full series apply to the partial-wave version. For example, the Born approximation

$$f_l(p) \approx -\frac{1}{p^2} \int dr\, \hat{j}_l U \hat{j}_l \qquad (11.24)$$

is expected to be good at high energies and for weak potentials. Just as with the full amplitude, there are certainly circumstances where the Born approximation is bad, and even where the Born series is divergent. An important new possibility is that even when the Born approximation is useless for the full amplitude it may be good for some values of l. We shall return to these points later.

To close this section we note that the Born approxmation (11.24) is always real. Since the exact amplitude

$$f_l(p) = \frac{1}{p} e^{i\delta_l} \sin \delta_l \qquad (11.25)$$

is only real if δ_l is small (modulo π) we see that a *necessary condition* for validity of the Born approximation is that the phase shift δ_l is small (modulo π). In fact, since f_l has the form (11.25) because S is unitary, this criterion is

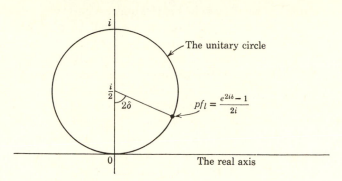

FIGURE 11.2.   The unitary circle. (Note that when inelastic processes become possible the elastic amplitude moves inside the unitary circle. See Chapter 17.)

just the partial-wave version of the condition of Chapter 9 that the Born approximation can be reliable only if it does not significantly violate unitarity.

This situation is illustrated in Fig. 11.2, which shows the "unitary circle," the locus of those values of the partial-wave amplitude consistent with unitarity. Since the exact value of pf_l lies on the unitary circle, while the Born result lies on the real axis, it is apparent that the Born approximation can be reliable only if pf_l (both exact and in Born approximation) is close to the origin.

11-e.   Properties of the Partial-Wave Amplitude

We begin our study of the partial-wave amplitude and phase shift with a brief survey of their general properties, omitting at this stage any formal proofs.

The amplitude and phase shift are determined by the radial wave function $\psi_{l,p}(r)$, which in turn is determined by the radial Schrödinger equation,

$$\left[\frac{d^2}{dr^2} - \frac{l(l+1)}{r^2} + p^2 - \lambda U(r)\right]\psi_{l,p}(r) = 0$$

Here, as above, $U(r) \equiv 2mV(r)$; we have introduced a strength parameter λ; and we assume as usual that V satisfies

I $V(r) = O(r^{-3-\epsilon})$ as $r \to \infty$
II $V(r) = O(r^{-\frac{3}{2}+\epsilon})$ as $r \to 0$
III $V(r)$ continuous for $0 < r < \infty$, except perhaps at a finite number of finite discontinuities

As usual, a number of results can be proved under slightly weaker conditions.[6] A new feature is that for certain results we shall need to make *stronger* assumptions on V. For example, the threshold behavior (11.28) below is true for all l only if $V(r)$ decreases faster than every power of $1/r$ as $r \to \infty$; this result is therefore *not* true for potentials such as the $1/r^4$ atomic polarization potential.

We first remark that if the potential term $\lambda U(r)$ in the radial equation is much smaller than the term $\{[l(l+1)/r^2] - p^2\}$ for all r, then the equation is essentially the *free* radial equation. In this case we would expect $\psi_{l,p}(r)$ to be close to the free function $\hat{j}_l(pr)$ and, hence, the amplitude $f_l(p)$ to be close to zero. This suggests several important results.

As $\lambda \to 0$ (i.e., the potential gets weaker) the amplitude $f_l(p)$ goes to zero and the phase shift $\delta_l(p)$ goes to an integral multiple of π. Similarly, at high energies (for a given potential and given l) $f_l(p)$ goes to zero[7] and $\delta_l(p)$ to an integral multiple of π. We shall see that the inherent modulo π ambiguity in the phase shift can be removed by requiring that $\delta_l(p)$ be a continuous function that goes to *zero* (not just $n\pi$) as $p \mapsto \infty$. Furthermore, if this is done, then it is also zero (not just $n\pi$) as $\lambda \to 0$.

As $l \to \infty$ (for a given potential and energy), we have the same result; $f_l \to 0$ and $\delta_l \to n\pi$. In this case the result is easily understood by regarding the term $l(l+1)/2mr^2$ as a repulsive "centrifugal potential." The larger l the more repulsive this centrifugal barrier becomes and the smaller the chance that the incident particle (of given energy) will penetrate into the region where the actual potential $V(r)$ is appreciable. This argument even provides us with a rough estimate of the maximum value of l for which δ_l is appreciable. If the target potential has approximate range a then the height of the centrifugal barrier at the edge of the potential is roughly $l^2/2ma^2$. [We ignore the difference between $l(l+1)$ and l^2.] If the energy is much less than this height then the particle is unlikely to penetrate the barrier and feel the potential.

---

[6] In fact, all results of this and the next chapter, for which we do not explicitly make *stronger* assumptions, are actually valid under the *weaker* conditions:

   I' $V = O(r^{-2-\epsilon})$ as $r \to \infty$.
   II' $V = O(r^{-2+\epsilon})$ as $r \to 0$.
   III As above.

The usefulness of this observation is not very great since the partial-wave formalism derives its physical justification from the time-dependent theory of Chapters 2–5, for which these weaker assumptions are not enough.

[7] In fact, because of the factor $1/p$ in our definition of $f_l = (1/p)e^{i\delta}\sin\delta$ it is *obvious* that $f_l \to 0$ as $p \to \infty$. The interesting result is that $pf_l = e^{i\delta}\sin\delta \to 0$ as well. From this point of view it would have been better to define f_l without the factor of $1/p$. However, for other reasons it is convenient to include this factor and we shall always understand the statement that the partial-wave amplitude goes to zero as $p \to \infty$ to mean $pf_l \to 0$.

Thus, for $l^2/2ma^2 \gg E = p^2/2m$, or

$$l \gg pa \qquad (11.26)$$

all phase shifts should be small.[8]

Under any of the above conditions (small λ, large p, or large l) the effect of the potential is small and we can expect that the Born approximation will be reliable. In particular, even at an energy where the Born approximation is useless when applied to the full amplitude $f(\mathbf{p'} \leftarrow \mathbf{p})$, it should nonetheless be good for l greater than some l_0. Under these circumstances the contribution to $f(\mathbf{p'} \leftarrow \mathbf{p})$ of all partial waves with $l > l_0$ can be calculated by computing the full Born approximation $f^{(1)}(\mathbf{p'} \leftarrow \mathbf{p})$ and then subtracting the Born contributions from $l = 0, 1, \ldots, l_0$. The lower partial waves are then calculated by some more exact procedure or are determined by fitting to the experimental data. This latter is the procedure used, for example, in nucleon–nucleon phase-shift analyses (see MacGregor, Moravcsik, and Stapp, 1960).

To analyze the low-energy behavior of the amplitude $f_l(p)$ it is convenient to consider the Lippmann–Schwinger equation for $\psi_{l,p}$,

$$\psi_{l,p}(r) = \hat{\jmath}_l(pr) + \int_0^\infty dr' G_{l,p}^0(r, r')U(r')\psi_{l,p}(r')$$

From the definition (11.21) and the small-argument expansions (11.7) and (11.10) for $\hat{\jmath}$ and $\hat{n}$, it is clear that for small p (and fixed r and r')

$$G_{l,p}^0(r, r') \xrightarrow[p \to 0]{} - \frac{(r_<)^{l+1}(r_>)^{-l}}{2l + 1}$$

that is, the Green's function becomes independent of p. If we return to the Lippmann–Schwinger equation, this suggests that as $p \to 0$ the radial wave function $\psi_{l,p}(r)$ will have the same p dependence as $\hat{\jmath}_l(pr)$, that is, p^{l+1}.

We now consider the amplitude

$$f_l(p) = -\frac{1}{p^2} \int_0^\infty dr\, \hat{\jmath}_l(pr)U(r)\psi_{l,p}(r) \qquad (11.27)$$

If the potential vanishes for r greater than some a (e.g., a square well) then as $p \to 0$ we can replace $\hat{\jmath}_l$ and $\psi_{l,p}$ by their small p expansions (both proportional to p^{l+1}) and we find that

$$\boxed{f_l(p) \xrightarrow[p \to 0]{} -a_l p^{2l}} \qquad (11.28)$$

---

[8] There is also the well known classical argument: In a classical collision the impact parameter is $\varrho = l/p$. If this is greater than the range a ($l > pa$), then the particle misses the target and is unscattered.

where a_l is some constant, known as the *scattering length*. (Only for the case $l = 0$ does a_l actually have the dimensions of length; nonetheless, for simplicity we shall use the name scattering length for all l.)

If the potential extends out to infinity this argument needs more care. Even when p is small the integral (11.27) has contributions for large r, for which we cannot use the small-argument expansions of $j_l(pr)$ and $\psi_{l,p}(r)$. We shall find that provided $V(r)$ falls off faster than every power of $1/r$ (e.g., a Yukawa), the result (11.28) still holds. If V falls off like $1/r^\nu$ then (11.28) needs to be modified for $l \geqslant (\nu - 3)/2$ (see Section 12-f). We shall see also that in certain exceptional cases, the scattering length a_l can be infinite, in which case the power behavior in (11.28) is reduced.

Accepting the result (11.28) for the moment, we see that as p approaches the threshold $p = 0$ all partial-wave amplitudes except the s wave vanish, and that the larger the value of l the more rapidly this vanishing occurs. This justifies our claim that at low energies only a few partial waves need to be considered. In particular, at very low energies only the s-wave amplitude is appreciable and it tends to a constant $-a_0$. It follows that the full amplitude also tends to the same constant,

$$f(\mathbf{p}' \leftarrow \mathbf{p}) = \sum (2l + 1)f_l(p)P_l(\cos \theta)$$
$$\xrightarrow[p \to 0]{} -a_0 \tag{11.29}$$

that is, at low energies the differential cross section is pure s-wave and isotropic, with the value $d\sigma/d\Omega = a_0^2$.

Since the amplitude $f_l = (1/p)e^{i\delta_l} \sin \delta_l$ behaves like $-a_l p^{2l}$ as $p \to 0$, it follows that $\sin \delta_l$ vanishes like p^{2l+1}. Therefore, the phase shift tends to a multiple of π as

$$\delta_l(p) \xrightarrow[p \to 0]{} n\pi - a_l p^{2l+1} \tag{11.30}$$

(from this it is clear that the scattering length a_l is always real). We have mentioned that the modulo π ambiguity in our definition of the phase shift will be removed by requiring that δ_l goes to zero (not just $n\pi$) at high energy. When this is done, it is *not* generally true that δ_l is zero at zero energy. On the contrary, we shall see in Chapter 12 that, according to *Levinson's theorem*, the difference $\delta_l(0) - \delta_l(\infty)$ is just π times the number n_l of bound states of angular momentum l. Thus, if $\delta_l(\infty)$ is taken to be zero, then

$$\delta_l(0) = n_l\pi \tag{11.31}$$

We shall see that this result holds for any local potential satisfying our usual assumptions, with the single exception that if the s-wave scattering length is infinite then for $l = 0$ the result (11.31) is replaced by $\delta_0(0) = (n_0 + \tfrac{1}{2})\pi$.

It is instructive to compute the amplitude and phase shift for some exactly soluble potentials and to consider the results in the light of the preceding discussion. In the problems at the end of this chapter two such potentials (the hard sphere and the square well) are suggested. In particular, the results given for the square well have been used in Fig. 11.3 to plot the phase shifts $\delta_l(p)$ and partial cross sections $\sigma_l(p)$ for a well of depth V_0 given by $(2ma^2V_0)^{1/2} = 4.8$ (where a is the well radius) for $l = 0, 1, 2,$ and 3. This particular depth was chosen so that there are two bound states with $l = 0$ and one each for $l = 1$ and $l = 2$. The well is not quite deep enough to bind an $l = 3$ bound state, and there are no bound states for $l \geqslant 3$. Several features of these plots deserve comment:

(1) In accordance with Levinson's theorem it will be seen that $\delta_0(0) = 2\pi$ and $\delta_1(0) = \delta_2(0) = \pi$, while $\delta_3(0) = 0$. As discussed above, all of the partial cross sections are zero at threshold except the s wave, which is well out of the picture at $\sigma_0(0) \approx 45\pi a^2$.

(2) In agreement with the threshold behavior (11.30) it can be seen that the larger l, the flatter the phase shifts at zero energy. Thus, for $0 \leqslant pa \leqslant 1$, σ_0 alone is more than 90% of the total cross section; in the interval $1 \leqslant pa \leqslant 2$, the same is true of $\sigma_0 + \sigma_1 + \sigma_2$. These intervals correspond well with the rough estimates (11.26) for the maximum l values that contribute significantly.

(3) At high energy the phase shifts go to zero. At $pa = 10$ all phase shifts are less than 60 deg and from this point onward they all drop steadily until at $pa = 50$ (for example) all are less than 15 deg.

(4) At about $pa = 2.7$ the s-wave phase shift passes through π and the corresponding cross section σ_0 vanishes. This happens because the radial wave function has exactly one more oscillation inside the well than does the corresponding free function. Outside the well the two functions are exactly in step and, hence, indistinguishable. As far as $l = 0$ scattering at this energy is concerned, the potential has no effect at all. For certain potentials this phenomenon can happen at an energy closer to threshold, where all other cross sections σ_l ($l > 0$) are still negligible. In this case there is no scattering whatever at this energy. This explains the famous *Ramsauer–Townsend effect*, in which certain gases are found to be completely "transparent" to electrons at one particular energy.

(5) Near $pa = 2.6$ the $l = 3$ phase shift increases rapidly through $\pi/2$ and the corresponding cross section σ_3 has a sharp peak. We shall discuss this resonance phenomenon in Chapter 13, where we shall see that it is closely related to the fact that the potential in question is nearly deep enough to bind an $l = 3$ bound state.

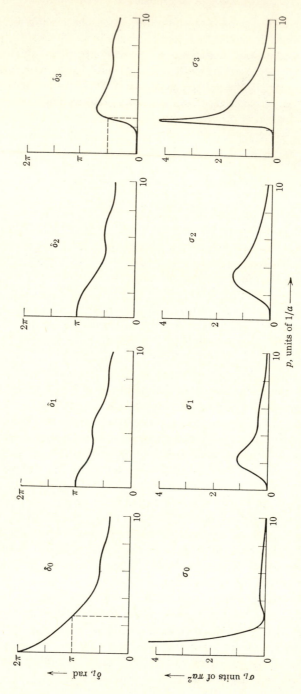

FIGURE 11.3. Phase shifts $\delta_l(p)$ and partial cross sections $\sigma_l(p)$ for a square well of depth V_0 given by $(2ma^2 V_0)^{1/2} = 4.8$.

11-f.   The Regular Solution

To facilitate further discussion of the radial equation and the partial-wave amplitude, it is convenient to introduce a new radial wave function $\phi_{l,p}(r)$, which differs from $\psi_{l,p}(r)$ in its normalization.

Recall that $\psi_{l,p}$ is (apart from the odd factor of π, etc.) the radial part of the wave function of the normalized state $|E, l, m+\rangle$. This is reflected in the normalization (11.14),

$$\int_0^\infty dr\, \psi_{l,p'}(r)^* \psi_{l,p}(r) = \frac{\pi}{2} \delta(p' - p)$$

and we can refer to $\psi_{l,p}(r)$ as the *normalized* radial function. The boundary conditions that define $\psi_{l,p}(r)$ as a solution of the radial equation are that $\psi_{l,p} = 0$ at $r = 0$ (or equivalently that $\psi_{l,p} \to$ constant $\times \hat{\jmath}_l$ as $r \to 0$), and that $\psi_{l,p} \to \hat{\jmath}_l + pf_l\hat{h}_l^+$ as $r \to \infty$. It develops that it is mathematically more convenient to discuss a solution that is defined by boundary conditions *at a single point*. It is for this reason that we define $\phi_{l,p}(r)$ as the solution that behaves exactly like $\hat{\jmath}_l(pr)$ as $r \to 0$,

$$\boxed{\phi_{l,p}(r) \xrightarrow[r \to 0]{} \hat{\jmath}_l(pr)}$$

There is no constant of proportionality in this definition; that is, the definition requires that ϕ vanish at $r = 0$ *and* fixes its normalization.[9]

With this definition $\phi_{l,p}(r)$, called the *regular solution*, is clearly real, since both the boundary conditions and the radial equation are real. Obviously $\phi_{l,p}(r)$ is proportional to $\psi_{l,p}(r)$, and it can be used to compute the amplitude just as well as can $\psi_{l,p}(r)$. The advantages of using $\phi_{l,p}(r)$ will emerge in the next two sections.

11-g.   The Variable Phase Method

As a first application of the regular solution $\phi_{l,p}(r)$ we discuss the so-called variable phase method. As a means of computing and analyzing phase shifts, this method is an important alternative to the straightforward integration of the radial Schrödinger equation. We give only a brief introduction to the method here and for simplicity confine our attention to s waves. (For more details see Calogero, 1967.)

---

[9] We use $\phi \to \hat{\jmath}$ to mean $\phi/\hat{\jmath} \to 1$. This is often written as $\phi \sim \hat{\jmath}$ in the mathematical literature; however, many physicists use $\phi \sim \hat{\jmath}$ to mean that $\phi/\hat{\jmath}$ tends to some (unspecified) constant.

For any given potential $V(r)$ we first introduce a new potential $V_\rho(r)$, which is defined as the potential obtained by truncating $V(r)$ at $r = \rho$ (Fig. 11.4); that is:

$$V_\rho(r) = \begin{cases} V(r) & r \leqslant \rho \\ 0 & r > \rho \end{cases}$$

We denote the s-wave regular solution for the actual potential V by $\phi(r)$ (omitting the labels $l = 0$ and p) and the solution for the truncated potential V_ρ by $\phi_\rho(r)$. The s-wave phase shift for V is δ, while that for V_ρ we denote $\delta(\rho)$.

The function $\delta(\rho)$ is the phase shift for the potential truncated at $r = \rho$. In particular, if $\rho = 0$, then the truncation leaves no potential at all and $\delta(0)$ is therefore an integral multiple of π. We now remove the inherent ambiguity in the phase shift by *defining* $\delta(0) = 0$. On the other hand, as $\rho \to \infty$ the truncated potential V_ρ gets closer and closer to the actual potential V and we shall see that $\delta(\rho)$ approaches the actual phase shift δ,

$$\delta(\rho) \xrightarrow[\rho \to \infty]{} \delta$$

We now set up a differential equation for $\delta(\rho)$ that will allow us to integrate $\delta(\rho)$ from its value at the origin (zero) out to its value at infinity (the actual phase shift of interest). For any ρ the two functions $\phi(r)$ and $\phi_\rho(r)$ are defined by the same boundary condition at $r = 0$ and satisfy the same differential equation for $0 \leqslant r \leqslant \rho$. It follows that

$$\phi(r) \equiv \phi_\rho(r) \qquad [0 \leqslant r \leqslant \rho]$$

(The same result is *not* true for the normalized wave functions ψ and ψ_ρ, since their normalization at any point r depends on the values of the potential

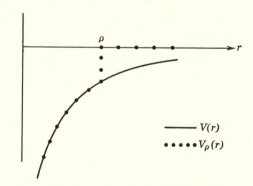

FIGURE 11.4.   The truncated potential $V_\rho(r)$.

for *all* points.) Beyond $r = \rho$, of course, the two functions differ. In particular, $\phi_\rho(r)$ satisfies the free ($l = 0$) radial equation and so can be written as:

$$\phi_\rho(r) = \alpha(\rho)\sin[pr + \delta(\rho)] \qquad [r \geqslant \rho] \qquad (11.32)$$

$\alpha(\rho)$ being some coefficient.

Both $\phi(r)$ and $\phi_\rho(r)$ are continuous; since they coincide for $r \leqslant \rho$, we can equate the value of $\phi(r)$ just inside $r = \rho$ with that of $\phi_\rho(r)$ just outside to give

$$\phi(\rho) = \alpha(\rho)\sin[p\rho + \delta(\rho)] \qquad (11.33)$$

In exactly the same way we can equate the derivative of $\phi(r)$ just inside $r = \rho$ with that of $\phi_\rho(r)$ just outside to give

$$\phi'(\rho) = p\alpha(\rho)\cos[p\rho + \delta(\rho)] \qquad (11.34)$$

The identities (11.33) and (11.34) hold for any choice of ρ and we can now replace ρ by r. It is a simple matter to substitute the two identities into the radial Schrödinger equation, to eliminate $\alpha(r)$, and to obtain the differential equation:

$$\delta'(r) = -\frac{1}{p}\,U(r)\sin^2[pr + \delta(r)] \qquad (11.35)$$

This is the *s*-wave *phase equation* and is the basis of the variable phase method.

Using the phase equation it is a simple matter to check that, provided only $V(r)$ is $0(r^{-1-\epsilon})$ as $r \to \infty$, the function $\delta(r)$ does have a limit as $r \to \infty$ and that this limit is the desired phase shift.

To recapitulate, we have defined the function $\delta(r)$ as the phase shift that results if the potential V is truncated at r. The radial equation for $\phi(r)$—a second-order, linear, differential equation—implies the first-order, nonlinear phase equation for $\delta(r)$. This equation, together with the boundary condition[10] $\delta(0) = 0$, determines $\delta(r)$. The limit of $\delta(r)$ as $r \to \infty$ is just the desired phase shift.

The fact that the phase equation is nonlinear may at first seem objectionable, but this is not really so. First, it is clear that in a numerical integration the nonlinearity is of no significance; that it is first order and that it directly yields the phase shift (as opposed to the wave function) are both marked advantages.[11] Second, it develops that even in theoretical problems the phase

---

[10] This boundary condition is actually quite subtle. Since $V(r)$ is in general infinite at $r = 0$, it is clear from the phase equation that $\delta'(r)$ remains finite at $r = 0$ only because $\delta(r)$, and hence the sine term, vanishes. To handle the phase equation near $r = 0$ in practice one must make a separate examination of $\delta(r)$ in that neighborhood. It turns out that $\delta(r)$ vanishes more quickly than linearly. Thus, close to $r = 0$ we can replace the phase equation by $\delta'(r) \approx -(1/p)U(r)\sin^2 pr$ as $r \to 0$.

[11] As emphasized by Calogero (1967) the numerical calculation of phase shifts using the phase equation is well within the power of a simple desk calculator.

equation is easily handled and provides some surprisingly transparent proofs of important results, as we now illustrate.

The phase equation shows clearly how the potential builds up the actual phase shift $\delta = \delta(\infty)$ starting from $\delta(0) = 0$. The derivative $\delta'(r)$ is simply the product of the potential with a certain negative quantity,

$$-(1/p)\sin^2[pr + \delta(r)]$$

Thus, a region where the potential is attractive makes a positive contribution to the phase shift, while a repulsive region contributes negatively. In particular, a *wholly attractive potential* $[V(r) \leqslant 0]$ gives a *positive phase shift*, while a *wholly repulsive potential* $[V(r) \geqslant 0]$ has a *negative phase shift*—a result anticipated in connection with Fig. 11.1.

A second result that follows almost instantly (see Problem **11.6**) from the phase equation is that if two potentials V_1 and V_2 satisfy

$$V_1(r) \geqslant V_2(r) \qquad [\text{all } r]$$

then $\delta_1(r) \leqslant \delta_2(r)$ for all r, and in particular,

$$\delta_1 \leqslant \delta_2 \tag{11.36}$$

Another important result concerns the limits of zero potential and high energy. Replacing V by λV and integrating the phase equation from zero to infinity, we find

$$\delta = -\frac{\lambda}{p} \int_0^\infty dr U(r)\sin^2[pr + \delta(r)] \tag{11.37}$$

If we suppose for a moment that $V(r)$ is less singular than $1/r$ at the origin, then this integral is especially easy to use since we can obtain a satisfactory bound by simply dropping the $\sin^2$ term:

$$|\delta| \leqslant \left|\frac{\lambda}{p}\right| \int_0^\infty dr \, |U(r)|$$

Clearly then,

$$\delta \to 0 \qquad [\text{as } \lambda \to 0 \text{ or } p \to \infty]$$

If $V(r)$ behaves like $1/r$, or worse as $r \to 0$, then we must be more careful since the factor $\sin^2[pr + \delta(r)]$ in the integral (11.37) is needed to ensure convergence at $r = 0$. However, it is a fairly straightforward matter to show that the same result still holds.

This last result illustrates an attractive feature of the variable phase method —that it works directly with the phase shift, which is defined without its usual *modulo* π ambiguity. Thus, from (11.37) it is clear not only that the amplitude is small for small λ or large p, but also that the same is true of the phase shift. In particular, we have justified our earlier claim that if the phase shift is chosen to be zero (not just $n\pi$) in the limit of high energy, then it is also zero (not just $n\pi$) for zero potential.

The variable phase method, which extends without difficulty to arbitrary angular momenta, can be used to provide alternative proofs of almost all the standard results relating to the partial-wave amplitude. In some cases (notably the three examples just cited) these proofs are both simpler and more transparent than more conventional arguments (see Calogero, 1967).

11-h.   Iterative Solution for the Regular Wave Function

The principal advantage of working with the regular wave function $\phi_{l,p}(r)$ emerges when we convert the differential radial equation plus boundary conditions into an integral equation and attempt to solve it by iteration. Standard methods give as the integral equation for $\phi_{l,p}(r)$[12]

$$\phi_{l,p}(r) = \hat{j}_l(pr) + \lambda \int_0^r dr' g_{l,p}(r, r')U(r')\phi_{l,p}(r') \qquad (11.38)$$

where

$$g_{l,p}(r, r') = \frac{1}{p} [\hat{j}_l(pr)\hat{n}_l(pr') - \hat{n}_l(pr)\hat{j}_l(pr')] \qquad (11.39)$$

is the Green's function appropriate to the boundary conditions for $\phi_{l,p}$, and we have replaced V by λV again.

The important point is that the new Green's function is zero for $r' > r$; that is, the integral in (11.38) stops at $r' = r$. (Integral equations with this property are called Volterra equations.) This reflects the fact that $\phi_{l,p}$ is defined by boundary conditions at the one point $r = 0$. It means that we can solve for $\phi_{l,p}$ by iteration, however strong the potential (something that is certainly not true of $\psi_{l,p}$). We state this important result as a theorem:

Theorem. The integral equation (11.38) for $\phi_{l,p}(r)$ can be solved by iteration for any value of λ; that is, for any λ the solution is given by the convergent power series (from which we temporarily omit the subscripts l and p)

$$\phi(r) = \sum_0^\infty \lambda^n \phi^{(n)}(r) \qquad (11.40)$$

where

$$\phi^{(0)}(r) = \hat{j}_l(pr)$$

[12] It is a simple matter to check by direct differentiation that any solution of (11.38) does satisfy the radial equation. [The identity $(d\hat{j}/dr)\hat{n} - \hat{j}(d\hat{n}/dr) = p$ is useful—see (12.16).] That the solution satisfies the right boundary conditions can be checked using the bounds established below. Strictly speaking, the Green's function $g_{l,p}$ should carry a superscript 0, since it is a Green's function for the free Hamiltonian

and

$$\phi^{(n)}(r) = \int_0^r dr' g(r, r') U(r') \phi^{(n-1)}(r')$$

$$= \int_0^r dr_n \int_0^{r_n} dr_{n-1} \cdots \int_0^{r_2} dr_1 g(r, r_n) U(r_n) g(r_n, r_{n-1}) \cdots U(r_1) \hat{j}(pr_1)$$

$$(11.41)$$

Proof: Our procedure is to use the integral (11.41) to *define* the function $\phi^{(n)}(r)$. Having established that this is a good definition (i.e., that the integral exists), we show that the series $\sum \lambda^n \phi^{(n)}(r)$ is convergent and can therefore be used to define a function $\phi(r)$. Finally, we verify that this function does satisfy the original integral equation. To save the trouble of handling the general Riccati–Bessel functions, we consider only the case $l = 0$, for which $\hat{j}_0(pr) = \sin pr$ and the Green's function $g_{0,p}(r, r')$ is just $p^{-1} \sin p(r - r')$. The general case is left as an exercise for the reader (Problem **11.7**).

For the case $l = 0$, the integral (11.41) that defines $\phi^{(n)}(r)$ is

$$\phi^{(n)}(r) = \frac{1}{p^n} \int_0^r dr_n \int_0^{r_n} dr_{n-1} \cdots \int_0^{r_2} dr_1$$

$$\times \sin p(r - r_n) U(r_n) \sin p(r_n - r_{n-1}) \cdots U(r_1) \sin pr_1 \quad (11.42)$$

These integrals are certainly well defined except possibly at the lower limits, $r_i = 0$, where $U(r_i)$ may be infinite. To check that they converge, we bound the sine functions as follows:

$$|\sin pr_1| \leqslant pr_1$$

$$|\sin p(r_2 - r_1)| \leqslant pr_2$$

$$\cdots \cdots \cdots$$

$$|\sin p(r - r_n)| \leqslant pr$$

[We shall return later to improve this bound for the last function, $\sin p(r - r_n)$.] This gives:

$$|\phi^{(n)}(r)| \leqslant pr \int_0^r dr_n \int_0^{r_n} dr_{n-1} \cdots \int_0^{r_2} dr_1 |U(r_n)r_n \cdots U(r_1)r_1| \quad (11.43)$$

Provided only that V is $0(r^{-2+\epsilon})$ at the origin (which is less than we have already assumed) these integrals are certainly convergent and $\phi^{(n)}(r)$ is therefore well defined.

To establish convergence of the series $\sum \lambda^n \phi^{(n)}$ we note that the integral in (11.43) runs over the "triangular" region

$$0 \leqslant r_1 \leqslant r_2 \leqslant \cdots \leqslant r_n \leqslant r$$

(This reflects the fact that the integral in the equation for $\phi(r)$ runs from zero to r, not from zero to infinity.) The integrand in (11.43) is a simple product $\prod_i |U(r_i)r_i|$ and is therefore unchanged by any permutation of the r_i. There are $n!$ such permutations, each of which would give a distinct region of integration, $0 \leqslant r_i \leqslant \cdots \leqslant r_k \leqslant r$. Adding all of these regions together produces the "square" region $\{0 \leqslant r_i \leqslant r\}$ where all of the r_i run independently from zero to r. It follows that the integral (11.43) is $1/n!$ times the same integral taken over the whole of this larger region. Thus

$$|\phi^{(n)}(r)| \leqslant \frac{1}{n!} pr \int_0^r dr_n \int_0^r dr_{n-1} \cdots \int_0^r dr_1 \prod_i |U(r_i)r_i|$$

$$= \frac{1}{n!} pr \left[\int_0^r dr_i |U(r_i)r_i| \right]^n$$

$$\leqslant \frac{1}{n!} pr\alpha^n \qquad \text{[all } r \text{ and } p\text{]} \qquad (11.44)$$

where $\alpha = \int_0^\infty dr\, |U(r)r|$. Inserting this bound into the series $\sum \lambda^n \phi^{(n)}$ we find that

$$|\phi(r)| \equiv \left| \sum_0^\infty \lambda^n \phi^{(n)}(r) \right| \leqslant pr \sum_0^\infty \frac{|\lambda\alpha|^n}{n!} = pre^{|\lambda\alpha|} \qquad (11.45)$$

that is, we have bounded the series for $\phi(r)$ by the series for $e^{|\lambda\alpha|}$. Since the latter is convergent for all λ, so is the former. Also, since the bound is independent of r and p (apart from the trivial factor pr) we see that the series $\phi = \sum \lambda^n \phi^{(n)}$ is uniformly convergent for r and p in any finite intervals.

It remains to be shown that the function $\phi(r)$ satisfies the original integral equation (for $l = 0$),

$$\phi(r) = \sin pr + \frac{\lambda}{p} \int_0^r dr' \sin p(r - r')U(r')\phi(r')$$

If we substitute the series $\phi = \sum \lambda^n \phi^{(n)}$ into the integral on the right we can interchange the sum and integral (since the series is uniformly convergent[13]) to give, on the right hand side:

$$\sin pr + \lambda \sum_0^\infty \lambda^n \left[\frac{1}{p} \int_0^r dr' \sin p(r - r')U(r')\phi^{(n)}(r') \right]$$

From the definition (11.42) of $\phi^{(n)}$ it is clear that the term in brackets $[\ldots]$ is just $\phi^{(n+1)}$. Thus, the right hand side is just:

$$\sin pr + \lambda \sum_0^\infty \lambda^n \phi^{(n+1)}(r) = \phi(r) \qquad \text{Q.E.D.}$$

---

[13] A little care is needed at the origin where $U(r)$ may be infinite. However the interested reader can check by splitting the integral as $\int_0^\epsilon + \int_\epsilon^r$ that the interchange is legitimate.

The expansion of the wave function $\phi(r)$ in powers of the potential (whatever its strength) will be our principal tool of investigation in the remaining sections of this and the next chapter. To conclude this section, we note that we shall need a more incisive bound on $\phi(r)$ than that of (11.45), and fill this need.

The reader will recall that in establishing the bounds on $\phi^{(n)}$ and ϕ we used the inequality

$$|\sin x| \leqslant x \qquad [x \geqslant 0]$$

Although this bound is quite adequate for small x, it is very wasteful for large x, where a better bound is $|\sin x| \leqslant 1$. A bound that is satisfactory for all x is obtained by combining these two to give

$$|\sin x| \leqslant \frac{\beta x}{1 + x} \qquad [x \geqslant 0] \tag{11.46}$$

where β is some constant whose precise value is unimportant. The right hand side of this inequality is a monotonic function of x and the inequality still holds if we replace the x on the right by any number greater than x.

Returning to the expression (11.42) for $\phi^{(n)}$ we now use this improved bound for the first sine factor to give:

$$|\sin p(r - r_n)| \leqslant \frac{\beta p r}{1 + p r}$$

Using the same bounds as before for all other factors we obtain the same results of (11.43) to (11.45), except that wherever the factor pr formerly appeared, we now have $\beta pr/(1 + pr)$. In particular, the bounds (11.44) and (11.45) for the s-wave functions $\phi^{(n)}$ and ϕ become:

$$|\phi^{(n)}(r)| \leqslant \beta \frac{pr}{1 + pr} \frac{\alpha^n}{n!}$$

$$|\phi(r)| \leqslant \beta \frac{pr}{1 + pr} e^{|\lambda \alpha|} \tag{11.47}$$

The reader can establish the corresponding bounds for arbitrary l with the guidance of Problem **11.7**.

11-i.   The Jost Function

Having established the power series expansion for the regular function $\phi_{l,p}(r)$ we next show how to extract a corresponding expression for the amplitude. To this end we must obviously examine the behavior of $\phi_{l,p}$ for large r.

As $r \to \infty$, $\phi_{l,p}$ must approach some combination of solutions of the free radial equation, $a\hat{h}^- + b\hat{h}^+$, say. Remembering that $\phi_{l,p}$ is real and inserting a factor $i/2$ for convenience, we can write this as

$$\phi_{l,p}(r) \xrightarrow[r \to \infty]{} \frac{i}{2} [f_l(p)\hat{h}_l^-(pr) - f_l(p)^*\hat{h}_l^+(pr)] \qquad (11.48)$$

where the coefficient $f_l(p)$ is called the *Jost function*.[14] Now, the regular solution $\phi_{l,p}$ is proportional to the normalized function $\psi_{l,p}$, whose asymptotic form is (11.18)

$$\psi_{l,p}(r) \xrightarrow[r \to \infty]{} \frac{i}{2} [\hat{h}_l^-(pr) - s_l(p)\hat{h}_l^+(pr)]$$

Comparing these two forms we see that

$$\boxed{s_l(p) = \frac{f_l(p)^*}{f_l(p)}} \qquad (11.49)$$

and

$$\boxed{\phi_{l,p}(r) = f_l(p)\psi_{l,p}(r)} \qquad (11.50)$$

The result (11.50) shows that the Jost function is just the ratio of the two solutions ϕ and ψ. The expression (11.49) for the S matrix shows clearly that $s_l(p)$ has modulus unity (i.e., S is unitary). Further, since $s_l = \exp(2i\delta_l)$, it is clear that the phase shift is just minus the phase of the Jost function, that is,

$$f_l(p) = |f_l(p)| \, e^{-i\delta_l(p)} \qquad (11.51)$$

Having expressed the S matrix (or phase shift) in terms of the Jost function, we now need an explicit expression for $f_l(p)$ in terms of $\phi_{l,p}(r)$. This we obtain by returning to the integral equation

$$\phi_{l,p}(r) = \hat{j}_l(pr) + \lambda \int_0^r dr' g_{l,p}(r, r')U(r')\phi_{l,p}(r')$$

If we write $\hat{j} = (i/2)(\hat{h}^- - \hat{h}^+)$, and from (11.39),

$$g_{l,p}(r, r') = \frac{i}{2p} [\hat{h}_l^-(pr)\hat{h}_l^+(pr') - \hat{h}_l^+(pr)\hat{h}_l^-(pr')]$$

then it is clear that as $r \to \infty$,

$$\phi_{l,p}(r) \to \frac{i}{2}\left\{ \left[1 + \frac{\lambda}{p} \int_0^\infty dr' \, \hat{h}_l^+(pr')U(r')\phi_{l,p}(r')\right]\hat{h}_l^-(pr) - \left[\cdots\right]^*\hat{h}_l^+(pr)\right\}$$

---

[14] Definitions differing by miscellaneous factors of p, π, and i are common. Ours is chosen so that $f_l(p) \equiv 1$ if $V = 0$. More important, the function that we call $f_l(p)$ has traditionally been called $f_l(-p)$. Our notation, which follows Newton (1966), is considerably more convenient.

Comparing this with (11.48) we see that the Jost function is given by

$$\boxed{\;f_l(p) = 1 + \frac{\lambda}{p} \int_0^\infty dr\, \hat{h}_l^+(pr) U(r) \phi_{l,p}(r)\;}$$

(11.52)

which is the desired expression.

It is a simple matter to check that this integral for $f_l(p)$ is convergent. For example, for the case $l = 0$,

$$f_0(p) = 1 + \frac{\lambda}{p} \int_0^\infty dr\, e^{ipr} U(r) \phi_{0,p}(r)$$

and so, with the bound (11.47),

$$|f_0(p) - 1| \leqslant \beta e^{|\lambda \alpha|} \left| \frac{\lambda}{p} \right| \int_0^\infty dr\, |U(r)| \frac{pr}{1 + pr}$$

(11.53)

(Remember that α and β are just constants whose actual value need not concern us.) This integral is certainly convergent at both $r = 0$ and ∞.

The bound (11.53) allows us to prove the important result that the Jost function tends to its free value, $f_l(p) = 1$, in the limits $\lambda \to 0$ or $p \to \infty$,

$$f_l(p) \to 1 \qquad [\text{as } \lambda \to 0 \text{ or } p \to \infty]$$

(For $l > 0$ see Problem **11.8**.) This result implies that, as we already know, $s_l \to 1$ (or the amplitude $f_l \to 0$) as $\lambda \to 0$ or $p \to \infty$. The proof for the case $\lambda \to 0$ is immediate from (11.53). For the case $p \to \infty$ there are two possibilities: If the potential is less singular than $1/r$ at the origin, we can drop the factor $pr/(1 + pr)$ in (11.53) to give

$$|f_0(p) - 1| \leqslant \beta e^{|\lambda \alpha|} \left| \frac{\lambda}{p} \right| \int_0^\infty dr\, |U(r)|$$

Clearly then $f_0(p)$ tends to 1 as $p \to \infty$, the difference going to zero like $1/p$.

If the potential goes like $1/r$ or worse as $r \to 0$, we cannot afford to drop the factor $pr/(1 + pr)$ from (11.53) since the remaining integral would diverge. In this case, one can show by splitting the integral $\int_0^\infty$ as $\int_0^1 + \int_1^\infty$ that $f_0(p) - 1$ vanishes like $1/p^\eta$, for some $\eta > 0$, as $p \to \infty$ (see Problem **11.8**).

The existence of the power series expansion for $\phi_{l,p}$ implies a corresponding expansion of the Jost function. This follows at once if we substitute $\phi = \sum \lambda^n \phi^{(n)}$ into the integral $f = 1 + (\lambda/p) \int_0^\infty h^+ U\phi$. Interchanging the summation and integral,[15] we find that

$$f_l(p) = 1 + \sum_1^\infty \lambda^n f_l^{(n)}(p)$$

(11.54)

---

[15] Again, if we take care of the neighborhoods of $r = 0$ and ∞ separately, the interchange is justified by the uniform convergence of $\sum \lambda^n \phi^{(n)}$.

where $f^{(n)} = (1/p) \int_0^\infty \hat{h}^+ U \phi^{(n-1)}$, and it is easily shown that this series is convergent for all λ. In the final section of this chapter we shall use this power series for the Jost function $f_l(p)$ and the series (11.40) for $\phi_{l,p}(r)$ to analyze the partial-wave Born series.

11-j. The Partial-Wave Born Series

The expansions of $\phi_{l,p}(r)$ and $f_l(p)$ in powers of the potential are convergent for all values of the real coupling parameter λ. In fact, nothing in the analysis of the last two sections depends on the reality of λ, and the power series for ϕ and f are actually convergent for all λ—*real or complex*. Of course complex values of λ do not correspond to any physical problem, since when λ is complex $H^0 + \lambda V$ is not Hermitian; but they do correspond to a well defined mathematical problem (The Schrödinger equation is a well-defined differential equation even if λ is complex.) and our result is that this mathematical problem has power series solutions for all λ. Now, a power series $g(\lambda) = \sum \lambda^n g^{(n)}$ is convergent for all λ if and only if $g(\lambda)$ is *analytic* for all λ. Therefore our result means that ϕ and f are *analytic functions of the complex variable λ for all λ*; that is, *they are entire functions* of λ. We can even reverse this statement and say that the *reason* the series for ϕ and f converge for all λ is that both functions were cleverly chosen to be entire in λ.

Let us now consider the corresponding series expansions of the normalized wave function $\psi_{l,p}(r)$ and the partial-wave S matrix $s_l(p)$. We have seen that

$$\psi_{l,p}(r) = \frac{\phi_{l,p}(r)}{f_l(p)}$$

and

$$s_l(p) = \frac{f_l(p)^*}{f_l(p)}$$

which shows that each of the functions ψ and s is the quotient of two entire functions of λ.[16] Thus, both are analytic functions of λ, *provided the denominator $f_l(p)$ is nonzero*. Now, since $f_l(p) = 1$ when $\lambda = 0$, there is certainly some region around $\lambda = 0$ where $f_l(p)$ is nonzero and where ψ and s are analytic. This in turn means that ψ and s can be expanded as power

---

[16] One must be careful in the case of s. The result $s(\lambda) = [f(\lambda)]^*/f(\lambda)$—where we display the dependence on λ explicitly, but omit the variables l and p—is true for *real* λ. Now, if $f(\lambda)$ is analytic in λ (which it is), then $[f(\lambda)]^*$ is *not*, while $[f(\lambda^*)]^*$ is (as we shall discuss in Chapter 12). When λ is real these two functions are the same and we can equally well write $s(\lambda) = [f(\lambda^*)]^*/f(\lambda)$. This latter is the appropriate definition of $s(\lambda)$ for complex λ, and it is in this form that $s(\lambda)$ is the quotient of two entire functions of λ.

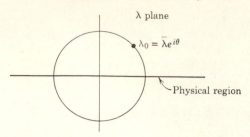

FIGURE 11.5.   **The complex plane of the coupling parameter** λ. **The point** λ_0 **is the zero of** f **closest to the origin.**

series in λ within some neighborhood of $\lambda = 0$; that is, the partial-wave Born series for ψ and s are convergent for λ sufficiently small (this proves one of the results anticipated in Section 11-e).

On the other hand, when $\lambda \neq 0$ the Jost function can have zeros, and generally does. At these points both ψ and s have poles. If we denote by $\lambda_0 = \bar{\lambda} e^{i\theta}$ the zero of f closest to the origin (for given l and p), then, by the well-known *circle of convergence theorem*, the following situation holds (see Fig. 11.5): Inside the circle $|\lambda| < \bar{\lambda}$, the Born series for ψ and s both converge, but for $|\lambda| > \bar{\lambda}$ they both diverge. The Born series for the actual potential ($\lambda = 1$) converge provided the circle $|\lambda| < \bar{\lambda}$ includes the point $\lambda = 1$.

This discussion makes clear the advantage of using the regular solution ϕ and Jost function f—the power series for ϕ and f, unlike those for ψ and s, converge for all λ. The discussion also illustrates the surprising usefulness of allowing the parameter λ, which is *a priori* real, to become complex. For example, the reason that the Born series for ψ converges for λ smaller than $\bar{\lambda}$ is that ψ is analytic for *all* λ, real or complex, with $|\lambda| < \bar{\lambda}$; the reason that it diverges for λ greater than $\bar{\lambda}$ is that ψ has a singularity at λ_0, somewhere on the circle $|\lambda| = \bar{\lambda}$. In fact, the point λ_0 is always complex. (We shall see that ψ and s cannot have singularities on the real axis.) Thus, it is only by moving away from the real, "physical," values of λ that we uncover the "reasons" for the convergence of the Born series for $|\lambda| < \bar{\lambda}$ and its divergence for $|\lambda| > \bar{\lambda}$.

To conclude, we note that we can now prove another result anticipated in Section 11-e: that the partial-wave Born series (for the actual value $\lambda = 1$) converges for sufficiently high energy. To prove this one has only to show that for p greater than some $\bar{p}$ there are no zeros of $f_l(p)$ inside or on the unit circle $|\lambda| = 1$. Now from (11.53) it is clear, not only that $f_l(p) \rightarrow 1$ as $p \rightarrow \infty$, but that it does so uniformly for λ inside any finite region. Thus, there is a $\bar{p}$ for which $|f_l(p)| > \frac{1}{2}$ (for instance) for all λ inside the unit circle, provided $p > \bar{p}$, and the desired result follows.

PROBLEMS

11.1. The radial wave function $\psi_{l,p}(r)$ should really carry a superscript $+$ because it is the radial part of $\langle \mathbf{x} \mid E, l, m+ \rangle$. Temporarily write $\psi_{l,p}^{\pm}(r)$ for the radial wave functions of $|E, l, m\pm\rangle$ [both defined as in (11.13] and prove that

$$\psi_{l,p}^{-}(r) = \psi_{l,p}^{+}(r)*$$

(Use time reversal invariance.) Notice that since ψ^{+} has the form (constant) $\times$ (real function of r) this means that ψ^{+} and ψ^{-} are proportional.

11.2. Show that the Lippmann–Schwinger equation (11.19)

$$|E, l, m+\rangle = |E, l, m\rangle + G^{0}(E + i0) \, V \, |E, l, m+\rangle$$

for $|E, l, m+\rangle$ is equivalent to the integral equation (11.20)

$$\psi = \hat{\jmath} + \int G_{l,p}^{0} U\psi$$

for the radial wave function, with $G_{l,p}^{0}(r, r') = -(1/p)\hat{\jmath}_{l}(pr_{<})\hat{h}_{l}^{+}(pr_{>})$.

There are several ways to do this. Obviously one needs to express the Green's function $\langle \mathbf{x}| G(z) |\mathbf{x}'\rangle$ as a sum of spherical harmonics of the general form $\sum_{l,m} Y_{l}^{m}(\hat{\mathbf{x}})F_{l}^{0}(r, r')Y_{l}^{m}(\hat{\mathbf{x}}')*$ (where F_{l}^{0} depends on z and differs from the desired $G_{l,p}^{0}$ by some constant). One way to do this is to insert a complete set of states $|E, l, m\rangle$ into the desired Green's function and then to proceed much as in Section 10-c. (After a little manipulation you will arrive at an integral of the form

$$\int_{-\infty}^{\infty} dp \, \frac{\hat{\jmath}_{l}(pr) \, \hat{\jmath}_{l}(pr')}{p^{2} - 2mz}$$

It is impossible to complete the contour for this integral in either the upper or lower half planes, since $\hat{\jmath}$ blows up in either direction. However, if you split the $\hat{\jmath}$ with larger argument as $\hat{\jmath} = (\hat{h}^{+} - \hat{h}^{-})/2i$, then you can do the two terms separately.)

11.3. (a) Decompose the vectors $|\mathbf{p}'\rangle$ and $|\mathbf{p}+\rangle$ into partial waves and insert into the expression $-(2\pi)^{2}m \, \langle \mathbf{p}'| \, V \, |\mathbf{p}+\rangle$, for the full amplitude. By comparing the result with the partial-wave series confirm that $f_{l}(p) = -(1/p^{2}) \times \int dr \hat{\jmath}_{l} U\psi_{l,p}$ as claimed in (11.23).
(b) Similarly, show that, when resolved into partial waves, the Born series for the full amplitude implies the Born series for the partial-wave amplitude.

11.4. A simple soluble potential is the so called hard sphere—$V(r)$ infinite for $r < a$ and zero for $r \geqslant a$. (This is not really a "reasonable potential" in our usual sense. It is equivalent to the boundary condition that the wave function vanish at $r = a$.)

(a) Find the amplitude $f_l(p)$ for this potential.

(b) Show that $f_l(p)$ has the advertised form $-a_l p^{2l}$ at low energy and, in particular, that the s-wave scattering length a_0, is equal to the sphere's radius a.

The trick in solving this (and the next) problem is to note that when $V = 0$, ψ must be some linear combination of free solutions (e.g., $a\hat{j}_l + b\hat{h}_l^+$) and that the precise form of this combination can be read off from the known asymptotic forms (11.16), etc.

The result (b) lets one interpret the s-wave scattering length of any potential as the radius of the "low-energy equivalent hard sphere;" that is, the hard sphere which reproduces the low-energy behavior of the potential.

11.5.   (a) Show that the amplitude for a square well of depth V_0 and radius a is

$$f_l(p) = \frac{1}{p} \frac{pj(ka)j'(pa) - kj'(ka)j(pa)}{kj'(ka)h(pa) - pj(ka)h'(pa)}$$

where we have abbreviated $\hat{j}_l(z)$ and $\hat{h}_l^+(z)$ to j and h; where $j'(z) = dj/dz$, and $k = (p^2 + 2mV_0)^{\frac{1}{2}}$ is the momentum inside the well.

(b) Show that in general $f_l(p) \to -a_l p^{2l}$ as $p \to 0$, and that $f_l(p) \to 0$ as $p \to \infty$.

(c) Consider in detail the case $l = 0$. Show that there are exceptional values of V_0 for which $f_0(0)$ is infinite and that these values correspond to depths at which a new bound state appears.

(d) At low energy s waves dominate, so that $f(\mathbf{p}' \leftarrow \mathbf{p}) \approx f_0(p)$. Compare the correct low-energy amplitude with the Born result obtained in Problem **9.2** and show that the Born approximation is good at low energies only for a sufficiently shallow well. Compare your results with the criterion of Problem **10.2b**.

11.6.   Use the variable phase method to show that if two potentials satisfy $V_1(r) \geqslant V_2(r)$ for all r then their s-wave phase shifts satisfy $\delta_1 \leqslant \delta_2$.

The first step is to show that $\delta_1(r) \leqslant \delta_2(r)$ when r is small. Here you can assume the potentials have the form $V \to \alpha r^\nu$ for r small, in which case $\delta(r) \to -\alpha p r^{\nu+3}/(\nu + 3)$. Then show that $\delta_1(r)$ and $\delta_2(r)$ cannot cross over. This is most easily done by supposing that they *do* cross at some r.

11.7.   To establish the existence and properties of the regular solution $\phi_{l,p}(r)$ and Jost function $f_l(p)$ for arbitrary l one needs bounds on the functions $\hat{j}_l(x)$, etc. analogous to the bound (11.46) for $\hat{j}_0 = \sin x$. The behavior of these functions for small and large x is determined by (11.7), (11.10), and (11.11).

(a) Use these results plus the fact that these functions are certainly continuous for all x to show that

$$|\hat{j}_l(x)| \leqslant \text{constant}\left(\frac{x}{1+x}\right)^{l+1}$$

and

$$|\hat{n}_l(x)| \leqslant \text{constant}\left(\frac{x}{1+x}\right)^{-l}$$

and hence that

$$|g_{l,p}(r, r')| \leqslant \text{constant}\left(\frac{pr}{1+pr}\right)^{l+1}\left(\frac{pr'}{1+pr'}\right)^{-l} \qquad [r \geqslant r']$$

(b) Establish the theorem of Section 11-h—that the regular solution ϕ can be found by iteration for any potential—for arbitrary l.

11.8. (a) Use the results of Problem **11.7** to show that, for any l, the Jost function satisfies the same bound as given for the s wave function, namely (11.53). (Note that the constant β is different for different l.)
(b) It follows (as in Section 11-i) that if $V(r)$ is less singular than $1/r$ at $r = 0$, then $f_l(p) \to 1$ at high energies, the difference going to zero as $1/p$. Suppose instead that $V(r) = O(r^{-2+\eta})$ $(0 < \eta < 1)$,[17] and show that $f_l(p) - 1$ goes to zero like $p^{-\eta}$ when $p \to \infty$. [Write the integral of (11.53) as $\int_0^\infty = \int_0^1 + \int_1^\infty$. The term $\int_1^\infty$ goes to zero like $1/p$ as before, while $\int_0^1$ can be shown to vanish like $1/p^\eta$.]

---

[17] This is, of course, included in our usual assumptions that $V(r) = O(r^{-3/2+\epsilon})$ as $r \to 0$.

12 Analytic Properties of the Partial-Wave Amplitude

12-a	Analytic Functions of a Complex Variable
12-b	Analytic Properties of the Regular Solution
12-c	Analytic Properties of the Jost Function and S Matrix
12-d	Bound States and Poles of the S Matrix
12-e	Levinson's Theorem
12-f	Threshold Behavior and Effective Range Formulas
12-g	Zeros of the Jost Function at Threshold

At the end of the last chapter we found it useful to let the *a priori* real parameter λ range into the complex plane. In a similar way we now allow the momentum p to become a complex variable. The physical significance of p as the magnitude of the incident momentum requires that it be a positive, real quantity. Nonetheless, p enters the radial Schrödinger equation—or, equivalently, the integral equation for $\phi_{l,p}(r)$—as a mere parameter, which can just as well be complex. We shall find that by studying the properties of the radial equation and its solutions for complex p we uncover a wealth of physically interesting results. In fact, the study of scattering amplitudes as analytic functions of complex momenta has proved to be one of the most powerful techniques of modern scattering theory.

12-a.   Analytic Functions of a Complex Variable

To begin, we briefly review some important properties of analytic functions. A function $f(z)$ is said to be an *analytic* (or holomorphic, or regular) in the complex variable z, on some region R, if $f(z)$ is differentiable at every point of R. (For example, $\sin z$ is analytic everywhere, $1/z$ is analytic except at $z = 0$ where it has a pole.) From the simple requirement of differentiability follows a remarkable wealth of results, of which the most important is Cauchy's theorem: that the integral $\oint dz\, f(z)$ around any closed path (in the region of analyticity) is zero. Cauchy's theorem leads to further results:

(1) An analytic function is actually infinitely differentiable.
(2) It can be expanded as a power series about any point in the region of analyticity.
(3) If $f(z)$ is analytic except at a number of poles, then the integral $\oint dz\, f(z)$ around any closed path is $2\pi i$ times the sum of residues at those poles inside the contour.

An important property of analytic functions is that two analytic functions that coincide on any line segment must coincide everywhere. This means, for example, that if $f_1(z)$ is defined on a line segment Γ, while $f_2(z)$ is analytic on a region R including Γ (Fig. 12.1, part a) and if $f_1(z) = f_2(z)$ on Γ, then we can regard $f_2(z)$ as the *unique analytic continuation* of $f_1(z)$ onto R. Similarly, if $f_1(z)$ and $f_2(z)$ are analytic on distinct but overlapping regions R_1 and R_2 (Fig. 12.1, part b) and if $f_1(z) = f_2(z)$ on the intersection $R_1 \cap R_2$, then each function provides a unique analytic continuation of the other, and we have, in fact, a single function analytic on the whole union $R_1 \cup R_2$.

If $f(z)$ is analytic on some region R, then the function $[f(z)]^*$ is, in general, *not* analytic.[1] However, the function

$$g(z) = [f(z^*)]^*$$

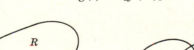

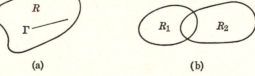

(a)   (b)

FIGURE 12.1.   (a) A region R containing the line segment Γ. (b) Two distinct, but overlapping, regions R_1 and R_2.

[1] Consider, for example, the analytic function z and its complex conjugate z^*. If z^* is to be analytic, it must have a derivative; that is, the quotient $(z^* - z_0^*)/(z - z_0)$ must have a unique limit as $z \to z_0$. But if $z - z_0$ is real, this quotient is always 1, while if $z - z_0$ is pure imaginary, the quotient is -1. The function z^* is therefore not differentiable.

is analytic (as is easily checked by directly computing its derivative) on the region R^*, made up of the complex conjugates of the points of R (Fig. 12.2).

This last observation leads us to the *Schwartz reflection principle:* if $f(z)$ is analytic in a region R that includes a segment of the real axis (as in Fig. 12.2), and if $f(z)$ is real on this segment, then $f(z)$ can be continued onto the region R^* and satisfies:

$$f(z) = [f(z^*)]^* \tag{12.1}$$

for all z in R and R^*. (That is, the value of f at any point z is the complex conjugate of its value at the point z^*.) The proof of this result is easy: The reality of $f(z)$ for real z implies that (12.1) holds on the real axis. Since the two sides of the identity are analytic functions, the identity holds for all z in their common region of analyticity, and is then used to continue $f(z)$ throughout R^*.

The Schwartz reflection principle plays an important role in scattering theory, sometimes appearing in slightly disguised form. For example, we shall find that the Jost function $f(p)$ is always analytic for p in the upper half plane $\{\operatorname{Im} p > 0\}$ and is real on the *imaginary* axis (Fig. 12.3). Making the substitution $p = ip'$ we can apply the Schwartz principle and conclude that:

$$f(p) = [f(-p^*)]^*$$

The functions of scattering theory usually appear as sums or integrals of other functions and we shall need tests to establish the analyticity of these sums and integrals. To conclude this section we give two such tests, which will be sufficient for our purposes:

Series of Analytic Functions.   If (1) the functions $f^{(1)}(z), f^{(2)}(z), \dots$ are all analytic in a region R, and (2) the series

$$f(z) = \sum_{1}^{\infty} f^{(n)}(z)$$

is uniformly convergent on R, then $f(z)$ is analytic in R. (This well known result is proved in almost any text on analysis—e.g., Titchmarsh, 1939, p. 95.)

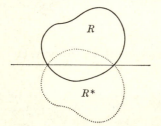

FIGURE 12.2.   When $f(z)$ **is analytic on** R, **then** $g(z) = [f(z^*)]^*$ **is analytic on** R^*.

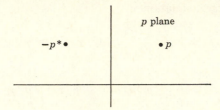

FIGURE 12.3.   **The Jost function is analytic for Im $p > 0$ and real on the imaginary axis. By the Schwartz principle $f(p) = [f(-p^*)]^*$.**

Integrals of Analytic Functions.   If (1) for each fixed r in some interval $[a, b]$ the function $g(z, r)$ is analytic in z on some region R, and (2) the function $g(z, r)$ is continuous in (z, r) on $R \times [a, b]$, then

$$f(z) = \int_a^b dr\, g(z, r)$$

is analytic on R.

This second result is proved, for example, in Titchmarsh (1939) p. 99. For our purposes it has to be extended in three simple ways. First, the integral of interest will usually have $r = 0$ as its lower limit and the integrand, which will contain a factor of $V(r)$, can be infinite at this point. Second, the upper limit of the integral may be infinite. In either case, the conclusion still holds provided one can establish *uniform convergence* at its end points. Third, the conclusion is obviously unaffected if the integrand has a finite number of discontinuities in r, since the integral can then be split up as a finite sum of terms each satisfying the conditions stated.

12-b.   Analytic Properties of the Regular Solution

We now return to the radial Schrödinger equation

$$\left[\frac{d^2}{dr^2} - \frac{l(l+1)}{r^2} - U(r) + p^2\right] y(r) = 0$$

where we have restored the coupling parameter λ to its actual value $\lambda = 1$. We now suppose p to be an arbitrary complex number. In the complex plane of p only those points on the positive real axis ($p \geqslant 0$) are relevant to the actual physical problem; for this reason we refer to the positive real axis as the *physical region*.

We seek a solution $\phi_{l,p}(r)$ that behaves like $j_l(pr)$ at the origin and, as before, the differential equation, together with the boundary condition, is equivalent to the integral equation

$$\phi_{l,p}(r) = \hat{j}_l(pr) + \int_0^r dr'\, g_{l,p}(r, r')U(r')\phi_{l,p}(r') \qquad (12.2)$$

where

$$g_{l,p}(r, r') = \frac{1}{p}\,[\hat{j}_l(pr)\hat{n}_l(pr') - \hat{n}_l(pr)\hat{j}_l(pr')]$$

It is easily seen from the power series definitions (11.6) and (11.9) of $\hat{j}$ and $\hat{n}$ that the free solution $\hat{j}_l(pr)$ and Green's function $g_{l,p}(r, r')$ are well defined and, in fact, analytic for all p. We now show that the same is true of the regular solution $\phi_{l,p}(r)$.

Theorem. For any p (real or complex) the equation (12.2) can be solved by iteration, and the solution $\phi_{l,p}(r)$ is an entire function of p.

Proof: The proof that ϕ can be obtained by iteration is almost identical to the corresponding proof of Section 11-h for the physical case $p \geqslant 0$. The only difference comes in the precise form of the bounds, and to see how this difference arises it is again sufficient to consider the case $l = 0$. Exactly as in (11.42), the s-wave solution has the form $\sum \phi^{(n)}(r)$ where

$$\phi^{(n)}(r) = \frac{1}{p^n} \int_0^r dr_n \int_0^{r_n} dr_{n-1} \cdots \int_0^{r_2} dr_1$$

$$\times \sin p(r - r_n)\, U(r_n)\sin p(r_n - r_{n-1}) \cdots U(r_1)\sin pr_1 \qquad (12.3)$$

We recall that when p was positive real, the sine functions were bounded using one of two inequalities, the more incisive of which was $|\sin x| < \beta x/(1 + x)$. When the arguments become complex, the essential new feature is that $\sin z$ grows like $e^{|\mathrm{Im}\, z|}$ when $\mathrm{Im}\, z$ becomes large, either positive or negative. Thus, the corresponding bound for arbitrary z is:

$$|\sin z| \leqslant \beta \frac{|z|}{1 + |z|}\, e^{|\mathrm{Im}\, z|} \qquad (12.4)$$

and from (12.3) it follows exactly as before that

$$|\phi^{(n)}(r)| \leqslant \beta \frac{|pr|}{1 + |pr|}\, e^{|\mathrm{Im}\, pr|} \frac{\alpha^n}{n!} \qquad (12.5)$$

where $\alpha = \int_0^\infty dr\, |U(r)r|$.

It follows, exactly as in the case $p \geqslant 0$, that the series $\phi = \sum \phi^{(n)}$ is convergent and satisfies the original integral equation. To see that it is analytic we first note that the bound (12.5) guarantees uniform convergence for p in any finite region. Thus, ϕ is analytic for all p provided the individual terms $\phi^{(n)}$ are. We already know that the zeroth term $\phi^{(0)} = \hat{j}_l(pr)$ is analytic in p, and the analyticity of the nth term

$$\phi^{(n)}(r) = \frac{1}{p} \int_0^r dr' \sin p(r - r') U(r') \phi^{(n-1)}(r')$$

follows by induction using the test given at the end of the last section.

Q.E.D.

We can view this result in two complementary ways. First, we can say that the regular solution $\phi_{l,p}(r)$ of (12.2) exists and is analytic for all complex p, and that when p returns to its physical region ($p \geqslant 0$) the function $\phi_{l,p}(r)$ reduces to the physically relevant solution discussed in the last chapter. Alternatively we can say that the physically relevant solution $\phi_{l,p}(r)$ with $p \geqslant 0$ has been shown to have a (unique) analytic continuation into the complex plane of p, and that the resulting continuation continues to satisfy the (now complex) radial Schrödinger equation.

For future reference we note that the bound (12.5) for $\phi^{(n)}$ leads to a corresponding bound for ϕ. For arbitrary l (for $l > 0$ see Problem **12.1**) this is

$$|\phi_{l,p}(r)| \leqslant \gamma_l \left(\frac{|pr|}{1 + |pr|}\right)^{l+1} e^{|\operatorname{Im} pr|} \tag{12.6}$$

where γ_l is some constant. Here the factor $|pr|^{l+1}$ reflects the familiar z^{l+1} behavior of $\hat{\jmath}_l(z)$ for small z, while the factor $e^{|\operatorname{Im} pr|}$ reflects the exponential growth of $\hat{\jmath}_l(z) \to \sin(z - \frac{1}{2}l\pi)$ as $|\operatorname{Im} z|$ becomes large.

Also, for future reference we note that if we move from any point p to $-p$ then the radial Schrödinger equation (which depends only on p^2) is unchanged. We can therefore expect some simple relation between the solutions $\phi_{l,p}$ and $\phi_{l,-p}$. In fact, as was noted in Section 11-b the free solution satisfies

$$\hat{\jmath}_l(-pr) = (-)^{l+1}\hat{\jmath}_l(pr)$$

Since we also know that $\hat{n}_l(-pr) = (-)^l\hat{n}_l(pr)$, it follows that

$$g_{l,-p}(r, r') = g_{l,p}(r, r')$$

and hence that $\phi_{l,p}(r)$ behaves just like the free solution

$$\phi_{l,-p}(r) = (-)^{l+1}\phi_{l,p}(r)$$

12-c.   Analytic Properties of the Jost Function and S Matrix

Having established the analytic properties of the wave function $\phi_{l,p}(r)$ we can now go on to discuss the Jost function $f_l(p)$ and finally the S matrix $s_l(p)$.

The Jost function f is given in terms of ϕ by (11.52) as

$$f_l(p) = 1 + \frac{1}{p} \int_0^\infty dr \, \hat{h}_l^+(pr) U(r) \phi_{l,p}(r) \tag{12.7}$$

In Section 11-i we saw that this integral is convergent for the physical values $p \geqslant 0$. We now show that this same integral defines the (unique) analytic continuation of $f_l(p)$ into at least a part of the complex plane.

When p becomes complex we have seen that in general $\phi_{l,p}(r)$ grows exponentially like $e^{|\mathrm{Im}\, pr|}$ for large r. When r is large the Riccati–Hankel function $\hat{h}_l^+(pr)$ behaves like $\exp i(pr - \frac{1}{2}l\pi)$ and so has magnitude $e^{-\mathrm{Im}\, pr}$; that is, it grows exponentially for $\mathrm{Im}\, p < 0$, but *decreases* exponentially for $\mathrm{Im}\, p > 0$. Thus, in discussing the integral (12.7) for f we must expect a sharp distinction between the lower and upper half planes. In the lower, the integral will in general diverge exponentially, while in the upper the growth of ϕ is cancelled by the decrease of $\hat{h}^+$ and the integral converges.

To see this precisely we insert into (12.7) the bounds (12.6) for ϕ and

$$|\hat{h}_l^+(pr)| \leqslant \beta_l \left(\frac{|pr|}{1 + |pr|}\right)^{-l} e^{-\mathrm{Im}\, pr}$$

(see Problem **12.1**) to give

$$|f_l(p) - 1| \leqslant \frac{\text{constant}}{|p|} \int_0^\infty dr \, |U(r)| \frac{|pr|}{1 + |pr|} e^{(|\mathrm{Im}\, p| - \mathrm{Im}\, p)r} \tag{12.8}$$

As anticipated, when $\mathrm{Im}\, p > 0$ the exponential factors cancel and the bound is the same as for p physical. Thus, the integral converges for all p in the upper half plane. In fact it is easily checked that the integral satisfies the conditions for the theorem of Section 12-a and $f_l(p)$ is therefore analytic in the upper half plane $\{\mathrm{Im}\, p > 0\}$.[2] When $\mathrm{Im}\, p < 0$ the integral (12.8) diverges at its upper limit unless we impose stronger conditions on the rate of decrease of the potential as $r \to \infty$.

Unfortunately, the region in which we have shown $f_l(p)$ to be analytic, $\{\mathrm{Im}\, p > 0\}$, does not include the physical region—the positive real axis. However, it is a simple matter to check that the integral (12.7) is *continuous* in $\{\mathrm{Im}\, p \geqslant 0\}$.[3] Thus the physically relevant Jost function $f_l(p)$ (p positive real) is continuously connected to the analytic function $f_l(p)$ in $\{\mathrm{Im}\, p > 0\}$.

---

[2] Notice that we cannot say that f is analytic in the *closed* half plane $\{\mathrm{Im}\, p \geqslant 0\}$ since it makes no sense to speak of analyticity on a closed region.

[3] To prove this we have only to note that the conditions of the theorems at the end of Section 12-a certainly guarantee continuity, and that the integral for $f(p)$ does satisfy these conditions in $\{\mathrm{Im}\, p \geqslant 0\}$.

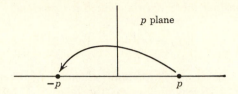

FIGURE 12.4.   Continuous movement from p to $-p$.

To summarize, if we make only our usual assumptions on the potential (p. 191) then we can establish the following situation for the Jost function $f_l(p)$:

$$\text{physical region} = \{p \text{ real} \geqslant 0\}$$
$$f_l(p) \text{ continuous in } \{\text{Im } p \geqslant 0\}$$
$$f_l(p) \text{ analytic in } \{\text{Im } p > 0\}$$

Before discussing the corresponding results for $s_l(p)$ we must prove one further property of $f_l(p)$; namely, that $f_l(p)$ is real when p is on the imaginary axis. Our proof depends on the discussion of the Schwartz reflection principle in Section 12-a. We consider the physical Jost function $f = 1 + (1/p) \int \hat{h}^+ U\phi$, with $p \geqslant 0$, and examine what happens when we move continuously from the point p to $-p$ (Fig. 12.4). We have already noted that

$$\phi_{l,-p}(r) = (-)^{l+1}\phi_{l,p}(r)$$

and it is easily checked that (for p real)

$$\hat{h}_l^+(-pr) = (-)^l \hat{h}_l^+(pr)^*$$

Inserting these relations into the integral for $f_l(p)$ and remembering that $\phi_{l,p}(r)$ is real when p is, we find that

$$f_l(-p) = [f_l(p)]^* \tag{12.9}$$

that is, continuing from p on the positive real axis to $-p$ changes $f_l(p)$ to its complex conjugate.[4] We can rewrite this result as

$$\boxed{f_l(p) = [f_l(-p^*)]^*} \tag{12.10}$$

in which form, as discussed in Section 12-a, the result must then be true for all p (in the region of analyticity). In particular, for p pure imaginary, $f_l(p)$ is real.

---

[4] This result is easily understood if we recall that f and f^* are the coefficients of ingoing and outgoing waves in the asymptotic expansion (11.48) of ϕ. Changing p to $-p$ interchanges the roles of ingoing and outgoing waves and therefore interchanges f and f^*.

We can now consider the S matrix, which for physical p was given by

$$s_l(p) = \frac{f_l(p)^*}{f_l(p)} \qquad [p \geqslant 0] \tag{12.11}$$

In this form $s_l(p)$ is certainly not analytic since $f_l(p)^*$ is not. However, remembering that p is real and using the identity (12.9) we can rewrite (12.11) in either of the forms

$$s_l(p) = \frac{f_l(p^*)^*}{f_l(p)} = \frac{f_l(-p)}{f_l(p)}$$

In either of these equivalent forms $s_l(p)$ is the quotient of two analytic functions. Unfortunately, the denominator is analytic for $\text{Im } p > 0$, while the numerator is analytic for $\text{Im } p < 0$. Thus, so far we have failed to establish any region of analyticity for $s_l(p)$ at all. In spite of this rather disappointing result we shall find that we already have enough information to prove several useful results, using just analyticity of f in $\{\text{Im } p > 0\}$. However, if we wish to establish analyticity of $s_l(p)$, then we must extend the region of analyticity of f into the lower half plane and to do this we must make more restrictive assumptions on the behavior of the potential as $r \to \infty$.

First, we note that if the potential is identically zero beyond some radius $r = a$ (e.g., a square well) then the integral (12.7) defining $f_l(p)$ actually runs only from zero to a. Thus, the divergences for large r do not occur and it is easily seen that in this case $f_l(p)$, as defined by (12.7), is analytic for *all* p. This means that $s_l(p) = f_l(-p)/f_l(p)$ is *meromorphic* [that is, analytic except at possible poles where $f_l(p) = 0$] for all p. Unfortunately, potentials that are identically zero for large r do not play an important role in physics,[5] and we must consider some other cases.

If the potential falls off exponentially when r is large, $V = O(e^{-\mu r})$, (e.g., a Yukawa), then inspection of the bound (12.8) shows that the integral defining $f_l(p)$ is also convergent and analytic in that part of the lower half plane with $\text{Im } p > -\mu/2$ [where the exponential decrease of $U(r)$ can offset the exponential growth of the factor $\exp(2 |\text{Im } p| r)$]. Thus, for an exponentially bounded potential $f_l(p)$ is analytic in $\{\text{Im } p > -\mu/2\}$ and $s_l(p)$ is meromorphic in the strip $\{\mu/2 > \text{Im } p > -\mu/2\}$.

Many potentials $V(r)$ are *analytic* functions of r in $\{\text{Re } r > 0\}$ and satisfy our usual assumptions on any ray $r = \rho e^{i\theta}$ in this half plane.[6] We shall refer

---

[5] The potentials of atomic physics usually behave like $1/r^\nu$ as $r \to \infty$, while the strong interactions between elementary particles lead to exponential tails like the Yukawa.

[6] An obvious example is the Yukawa, $\gamma e^{-\mu r}/r$. Notice that it does *not* satisfy our usual assumptions in the left half plane $\{\text{Re } r < 0\}$ since it blows up as $r \to \infty$ with $\text{Re } r < 0$.

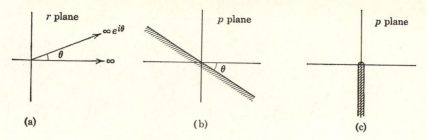

FIGURE 12.5.   (a) The original integral for $f_l(p)$ runs from 0 to ∞; that of (12.12) runs from 0 to $\infty e^{i\theta}$. (b) (12.12) is analytic in $\{\operatorname{Im}(pe^{i\theta}) > 0\}$, the half plane above the sloping line shown. (c) The union of all such regions with $-\pi/2 < \theta < \pi/2$ is the whole plane cut along the negative imaginary axis.

to these simply as analytic potentials. The free wave function $\hat{j}_l(pr)$ is analytic for all r; and it is easy to show that for such potentials $\phi_{l,p}(r)$ (as defined by the usual series of integrals) is analytic for Re $r > 0$ and satisfies the usual bound (12.6). We can therefore continue $f_l(p)$ into the lower half plane as follows: first take p anywhere on the positive imaginary axis; the integral defining $f_l(p)$ can now have its contour bent onto the line $r = \rho e^{i\theta}$,

$$f_l(p) = 1 + \frac{1}{p} \int_0^{\infty e^{i\theta}} dr \hat{h}^+ U \phi \qquad (12.12)$$

(Fig. 12.5, part a). This new integral has the same form as the original (12.7) except that the real variable r has become $r = \rho e^{i\theta}$. The same arguments as given before guarantee that it is analytic for Im $(pe^{i\theta}) > 0$, which defines the region shown in Fig. 12.5, part (b). We now have two integrals, (12.7) and (12.12), which define analytic functions in the overlapping regions $\{\operatorname{Im} p > 0\}$ and $\{\operatorname{Im} pe^{i\theta} > 0\}$. Since the two functions coincide in the region of overlap, each is a continuation of the other. Thus, $f_l(p)$ can be continued into the region of Fig. 12.5, part (b). Since this is true for any θ with $-\pi/2 < \theta < \pi/2$, we conclude that for an analytic potential $f_l(p)$ can be continued into the whole complex plane except perhaps for the negative imaginary axis (see Fig. 12.5, part c). Finally, we note that it is actually not essential that the potential be analytic in the whole of $\{\operatorname{Re} r > 0\}$. Thus if $V(r)$ were analytic in $\{\operatorname{Re} r > a\}$ (e.g., some nonanalytic short range potential plus a $1/r^\nu$ tail), then we could arrive at exactly the same conclusion using a contour as shown in Fig. 12.6.

If a potential is both analytic and exponentially bounded (e.g., a Yukawa or sum of Yukawas) we can combine the conclusions of the last two paragraphs and conclude that $f_l(p)$ is analytic in the region shown in Fig. 12.7, part (a), and is real on the imaginary axis from $-i\mu/2$ up to $+i\infty$. The

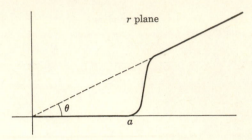

FIGURE 12.6.   Alternative contour for the integral (12.12) if $V(r)$ **is analytic in** $\{\text{Re } r > a\}$.

nature of the singularities on the negative imaginary axis depends on the details of the potential. For example, if V is the Yukawa $\gamma e^{-\mu r}/r$ then $f_i(p)$ has branch points at $p = -i\mu/2, -i\mu, \ldots$. If V is the pure exponential $\gamma e^{-\mu r}$ then $f_i(p)$ has poles at $p = -i\mu/2, -i\mu, \ldots$ (see Problem **12.2**). For any such potential the S matrix $s_i(p) = f_i(-p)/f_i(p)$ is meromorphic in the region shown in Fig. 12.7, part (b), and is real in the interval between $\pm i\mu/2$ on the imaginary axis.[7]

Two points in conclusion: First, we have seen that, while $f_i(p)$ is analytic in the upper half plane for any V satisfying our usual assumptions, some additional assumptions are needed to continue $f_i(p)$ into the lower half plane. However, it should be emphasized that almost any potential encountered in practice *does* allow continuation into the lower half plane. Loosely speaking, we can say that any "reasonable" potential has this property, and we shall often confine our attention to potentials which are "reasonable" in

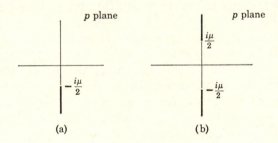

(a)   (b)

FIGURE 12.7.   (a) Region of analyticity of the Jost function for a potential like the Yukawa which is analytic and exponentially bounded. (b) Corresponding region in which $s_i(p)$ **is meromorphic (i.e., analytic except for poles).**

[7] It should be mentioned that having established this region of analyticity, we are almost ready to write down the so-called *partial-wave dispersion relation*. This is little more than a simple application of Cauchy's theorem to the function $f_i(p) = [s_i(p) - 1]/2ip$. However we defer discussion of dispersion relations to Chapter 15.

this sense. Second, it should be pointed out that there is a sense in which the analytic properties of $f_l(p)$ in the lower half plane (when it can be continued there) are less physically relevant. The point is that the properties in $\{\text{Im } p < 0\}$ are extremely sensitive to small changes in the potential at large distances. For example, we have seen that the Jost function for a large class of potentials (e.g., a Yukawa) has singularities on the negative imaginary axis. However, if we were to truncate such a potential at some point $r = a$, then all of these singularities would disappear [since $f_l(p)$ is entire for any truncated potential]. Now if a is very large (10 miles, for instance) it is clear that the difference between the two potentials is completely unobservable. In this sense it is clear that the singularities of $f_l(p)$ for the first potential are physically unimportant (for further discussion of this point see Problem **12.2**).

12-d.   Bound States and Poles of the S Matrix

After three sections of barren mathematics we can now begin to reap some physically interesting rewards. The first of these is the remarkable connection between the bound states of a potential and the poles of its S matrix in the half plane $\{\text{Im } p > 0\}$.

To understand this connection we recall the definition of the Jost function in terms of the asymptotic behavior of the wave function

$$\phi_{l,p}(r) \to \frac{i}{2} [f_l(p)\hat{h}_l^-(pr) - f_l(-p)\hat{h}_l^+(pr)] \tag{12.13}$$

This relation does not always hold once p moves up from the real axis, since $f_l(-p)$ does not necessarily continue there. However, let us suppose for a moment that the potential is such that it does, and suppose that at some point $\bar{p}$ ($\text{Im } \bar{p} > 0$) the Jost function vanishes, $f_l(\bar{p}) = 0$. In this case,

$$\phi_{l,\bar{p}}(r) \to \frac{-i}{2} f_l(-\bar{p})\hat{h}_l^+(\bar{p}r) \tag{12.14}$$

Now as $r \to \infty$

$$\hat{h}_l^{\pm}(pr) \to e^{\pm i(pr - l\pi/2)}$$

Thus, when $\text{Im } p > 0$, $\hat{h}^+$ decreases exponentially as $r \to \infty$ while $\hat{h}^-$ increases exponentially. It follows from (12.13) that in general ϕ contains both increasing and decreasing components. But at the point $\bar{p}$, where f vanishes and (12.14) holds, ϕ is purely decreasing. Since $\phi_{l,\bar{p}}(r)$ vanishes at $r = 0$ and decreases exponentially as $r \to \infty$, it is a normalizable solution of the radial Schrödinger equation (with $p = \bar{p}$ and angular momentum l). That is, the Hamiltonian has a proper eigenstate of energy $\bar{p}^2/2m$ and angular momentum l. Since all eigenvalues of H are real, $\bar{p}$ must in fact be pure imaginary, $\bar{p} = i\alpha$ ($\alpha > 0$) and the energy of the bound state is $-\alpha^2/2m$.

Conversely if H has a bound state with energy $-\alpha^2/2m$ and angular momentum l then at the point $p = i\alpha$ the solution $\phi_{l,p}$ must be exponentially decreasing, and according to (12.13), $f_l(i\alpha)$ must be zero.

Provided $f_l(-p)$ is also analytic at $p = \bar{p} = i\alpha$ we can equivalently say that the bound states correspond to poles of $s_l(p) = f_l(-p)/f_l(p)$. Indeed we can understand this result directly if we recall that $s_l(p)$ is simply the ratio of outgoing to ingoing wave in ϕ. When $\text{Im}\, p > 0$ outgoing waves decrease as $r \to \infty$, while incoming waves increase. Since at a bound state ϕ is purely decreasing, the ratio $s_l(p)$ has to be infinite.

To put these results on a more precise footing we must unfortunately introduce some more notation. We first define two new solutions of the radial equation denoted $\chi^{\pm}_{l,p}(r)$ and defined by the condition

$$\chi^{\pm}_{l,p}(r) \xrightarrow[r \to \infty]{} \hat{h}^{\pm}_l(pr)$$

Thus, $\chi^{\pm}$ are solutions of the radial equation which, as $r \to \infty$, are pure outgoing and incoming waves respectively. Generally they are *not* zero at $r = 0$, and are not proportional to the physical scattering functions ψ or ϕ.[8]

The integral equation for $\chi^{\pm}$ is

$$\chi^{\pm}_{l,p}(r) = \hat{h}^{\pm}_l(pr) - \int_r^{\infty} dr'\, g_{l,p}(r, r') U(r') \chi^{\pm}_{l,p}(r')$$

with g given by (11.39) as before. This equation can be solved by iteration just as was that for ϕ. The only difference is that the integrals now run from r to ∞ and so the exponential increase of $\hat{h}^{\pm}$ in the lower and upper half planes can cause divergences. It is a simple exercise (which we leave to the reader) to check that, under our usual assumptions on V, $\chi^{+}_{l,p}(r)$ exists and is continuous for all p in $\{\text{Im}\, p \geqslant 0\}$ and is analytic in $\{\text{Im}\, p > 0\}$. The result for χ^- is identical except that the upper half plane is replaced by the lower. With suitable additional restrictions on the potential for large r both solutions can be continued into larger regions exactly as was $f_l(p)$.

At least for real p (where both $\chi^{\pm}$ exist) the regular solution ϕ can of course be expanded in terms of the two solutions $\chi^{\pm}$ as $\phi = a\chi^- - b\chi^+$, say. By considering the expansion of ϕ for large r, we can identify the coefficients a and b with the Jost functions (times $i/2$), to give

$$\phi_{l,p}(r) \equiv \frac{i}{2}[f_l(p)\chi^-_{l,p}(r) - f_l(-p)\chi^+_{l,p}(r)] \tag{12.15}$$

[8] These solutions χ are often denoted by the much overworked letter f and are called the Jost solutions.

We must now introduce a second piece of notation. This is the *Wronskian*, which for any two functions $\alpha(r)$ and $\beta(r)$ is defined as

$$W(\alpha, \beta) = \alpha(r)\beta'(r) - \alpha'(r)\beta(r) = \det\begin{pmatrix} \alpha & \beta \\ \alpha' & \beta' \end{pmatrix}$$

where α' denotes $d\alpha/dr$. If the two functions are linearly dependent then $W(\alpha, \beta) = 0$ and conversely. If the two functions satisfy the same radial equation (the free equation or full equation, each for given l and p), then it is easily checked that $dW/dr = \alpha\beta'' - \alpha''\beta = 0$; that is, the Wronskian is independent of r. Thus, for example, the Wronskian of $\hat{j}_l(pr)$ and $\hat{n}_l(pr)$ is a constant and, since it is constant, it can be evaluated for large r where the known asymptotic forms give

$$W(\hat{j}_l, \hat{n}_l) = -p \tag{12.16}$$

In the same way we find that

$$W(\chi^+_{l,p}, \chi^-_{l,p}) = -2ip$$

If we now use the identity (12.15) to evaluate the Wronskian of ϕ and χ^+ we find that, since $W(\chi^+, \chi^+) = 0$,

$$W(\chi^+, \phi) = p f_l(p)$$

or

$$\boxed{f_l(p) = \frac{1}{p}\, W(\chi^+_{l,p}, \phi_{l,p})} \tag{12.17}$$

Since ϕ is entire in p while χ^+ is analytic in $\{\text{Im } p > 0\}$ this relation continues into the upper half plane and we find at once the following result. The Jost function $f_l(p)$ vanishes at some point $\bar{p}$ ($\text{Im } \bar{p} > 0$) if and only if $W(\chi^+, \phi) = 0$; that is, $f_l(\bar{p}) = 0$ if and only if the regular solution $\phi_{l,\bar{p}}$ is proportional to the exponentially decreasing solution $\chi^+_{l,\bar{p}}(r)$,

$$\phi_{l,\bar{p}}(r) = \lambda \chi^+_{l,\bar{p}}(r)$$

and, hence, yields a proper eigenfunction of the Hamiltonian. Thus, exactly as anticipated, $f_l(p)$ can vanish in $\{\text{Im } p > 0\}$ only on the imaginary axis, and it does so there if and only if $p = i\alpha$ where $-\alpha^2/2m$ is the energy of a bound state of angular momentum l.

This result is clearly true quite independently of the possibility that $f_l(-p)$ can also be continued into $\{\text{Im } p > 0\}$. If, however, this is possible and $p = i\alpha$ is in the region of analyticity of $f_l(-p)$, then $f_l(-i\alpha)$ cannot be zero [since if it were, then from (12.15) it would follow that $\phi \equiv 0$, which is impossible]. It therefore follows that in this case $s_l(p) = f_l(-p)/f_l(p)$ has a pole at the position of the bound state, $p = i\alpha$.

It should be emphasized that while the correspondence between bound states and zeros of $f_l(p)$ is always valid, that between bound states and poles of $s_l(p)$ depends on the possibility of continuing $f_l(-p)$ to the point $p = i\alpha$. Nonetheless, one almost always hears the result of this section referred to as the correspondence between bound states and poles of $s_l(p)$; the reasons for this are twofold. First, in the case of the full amplitude (as opposed to the partial-wave amplitude) the complication of possible singularities of $f_l(-p)$ does not arise, and the bound states do always correspond, one-to-one, to poles of the amplitude (as we shall see in Chapter 15). Second, in relativistic scattering the existence of $s_l(p)$ is beyond doubt, whereas it is not at all clear that a Jost function can always be defined.

We conclude this section with a result that we shall need later: that the zero of $f_l(p)$ at a bound state is always a *simple* zero. To see this we recall that if $f_l(\bar{p}) = 0$ then $\phi = \lambda \chi^+$ at $\bar{p}$ (with λ obviously nonzero). It is easily shown (see Problem **12.3**) that if $f_l(\bar{p}) = 0$ then

$$\frac{df_l}{dp}(\bar{p}) = \int_0^\infty dr\, \chi_{l,\bar{p}}^+(r)\phi_{l,\bar{p}}(r) = \lambda \int_0^\infty dr\, \chi_{l,\bar{p}}^+(r)^2$$

Since this is certainly nonzero, the zero of $f_l(p)$ has to be simple.

12-e.   Levinson's Theorem

Having established that, for a given potential, there is a one-to-one correspondence between the zeros of the Jost function f_l in $\{\text{Im } p > 0\}$ (all of which are simple) and the l-wave bound states, we are almost ready to prove Levinson's theorem. Before we do so however, we need two more results.

First, we prove that the Jost function *cannot* have zeros on the real axis, except perhaps at the origin. To see this we have only to note that for p real the expansion (12.15) of ϕ in terms of $\chi^\pm$ can be written

$$\phi = \frac{i}{2}\left[f_l(p)\chi^- - f_l(p)^*\chi^+ \right]$$

Thus, if $f_l(p) = 0$ for p real, then $f_l(p)^* = 0$ and the solution is identically zero. However, from the boundary condition

$$\phi \xrightarrow[r \to 0]{} \hat{j}_l(pr) \to \frac{(pr)^{l+1}}{(2l + 1)!!}$$

it is clear that, except when $p = 0$, the solution ϕ cannot be identically zero; and the desired result follows. We shall discuss the exceptional case that $f_l(0) = 0$ in Section 12-g.

Second, as $p \to \infty$ anywhere in $\{\text{Im } p \geqslant 0\}$ the Jost function goes to one. We saw in Chapter 11 (Section 11-i and Problem **11.8**) that this is certainly true for p physical, and indeed that $f_l(p) - 1$ vanishes like $1/p^\eta$ for some $\eta > 0$. The inequality (12.8) shows that for any p in $\{\text{Im } p \geqslant 0\}$ the Jost function satisfies the same bound as that used in Section 11-i to discuss the physical case. Accordingly, the same result holds anywhere in $\{\text{Im } p \geqslant 0\}$,

$$f_l(p) - 1 \underset{|p| \to \infty}{=} O\left(\frac{1}{p^\eta}\right) \qquad \text{[uniformly in } \{\text{Im } p \geqslant 0\}, \text{ some } \eta > 0] \quad (12.18)$$

An immediate corollary is that there is some radius $|p| = \rho$ beyond which $f_l(p)$ has no zeros in $\{\text{Im } p \geqslant 0\}$. Since an analytic function can have only a finite number of zeros inside any finite radius, we conclude that $f_l(p)$ has at most a finite number of zeros in $\{\text{Im } p \geqslant 0\}$;[9] that is, that there is only a finite number of bound states of angular momentum l.

We are now ready to establish Levinson's theorem.

Levinson's Theorem. For any spherical potential (subject to our usual assumptions) the phase shift $\delta_l(p)$ satisfies

$$\boxed{\delta_l(0) - \delta_l(\infty) = n_l \pi}$$

where n_l denotes the number of bound states of angular momentum l. This holds except in the exceptional case when the s-wave Jost function vanishes at threshold, $f_0(0) = 0$, in which case the same result holds for $l > 0$, but for $l = 0$ we have

$$\delta_0(0) - \delta_0(\infty) = (n_0 + \tfrac{1}{2})\pi \qquad (12.19)$$

Proof: We suppose here that $f_l(0) \neq 0$, and shall return to the special case $f_l(0) = 0$ in Section 12-g. We consider the integral

$$I = \oint dp \, \frac{\dot{f_l}(p)}{f_l(p)}$$

---

[9] This argument needs to be given rather carefully close to the real axis, where we have not established analyticity for an arbitrary potential. The point is that an analytic function (which is not identically zero) can vanish only a finite number of times in any closed bounded region *inside* its region of analyticity. Thus, it could be that $f_l(p)$ has an infinite sequence of zeros in $\{\text{Im } p > 0\}$ (on the imaginary axis, of course) accumulating at the point $p = 0$. However, $f_l(p)$ is continuous onto the real axis. Thus, provided $f_l(0) \neq 0$ such a sequence is obviously impossible. For the special case that $f_l(0) = 0$ we need the result of Section 12-g that when $f_l(p)$ vanishes at $p = 0$ it does so either linearly or quadratically; this means that it is nonzero in a neighborhood of $p = 0$ and again a sequence of zeros accumulating at $p = 0$ is impossible.

where the contour is the infinite semicircle shown in Fig. 12.8 and $\dot{f}$ denotes the derivative df/dp. The integrand $\dot{f}_l(p)/f_l(p)$ is analytic on and inside the contour except that at each simple zero of $f_l(p)$ it has a pole with residue one. Thus, by Cauchy's theorem,

$$I = 2\pi i n_l \tag{12.20}$$

Alternatively, we can rewrite I as

$$I = \oint d \ln f_l(p) = \int_{-\infty}^{\infty} d \ln f_l(p) \tag{12.21}$$

since from (12.18) it is clear that the contribution of the large semicircle is zero. Now, on the positive real axis

$$f_l(p) = |f_l(p)| \, e^{-i\delta_l(p)} \qquad [p \geqslant 0]$$

and, hence,

$$\ln f_l(p) = \ln |f_l(p)| - i \, \delta_l(p)$$

On the negative real axis we use the relation $f_l(-p) = f_l(p)^*$ to give

$$\ln f_l(-p) = \ln |f_l(p)| + i \, \delta_l(p) \qquad [p > 0]$$

Inserting these results into (12.21) we find that the contributions of the real parts from the positive and negative axes cancel and that

$$I = -2i \int_0^{\infty} d \, \delta_l(p)$$

$$= 2i[\delta_l(0) - \delta_l(\infty)] \tag{12.22}$$

Comparing (12.20) and (12.22) we arrive at the desired result and, apart from the exceptional case $f_l(0) = 0$, our proof is complete.

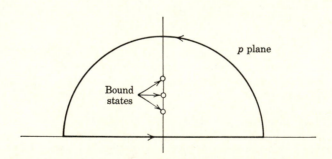

FIGURE 12.8.   Contour used to prove Levinson's theorem.

12-f. Threshold Behavior and Effective-Range Formulas

In this section we discuss the behavior of the partial-wave amplitude close to the threshold $p = 0$ and supply proofs of some results anticipated in the previous chapter.

The partial-wave amplitude is given by

$$f_l(p) = \frac{s_l(p) - 1}{2ip} = \frac{f_l(-p) - f_l(p)}{2ipf_l(p)} \tag{12.23}$$

For an arbitrary potential this expression can only be used on the real axis since it is only there that both $f_l(p)$ and $f_l(-p)$ are defined. Nonetheless, this will prove sufficient to prove the most physically interesting result: namely, the dominant behavior of $f_l(p)$ as $p \to 0$ through physical values. However, if the potential is exponentially bounded, then both $f_l(p)$ and $f_l(-p)$ are analytic in some strip on either side of the real axis and we can use (12.23) in a whole neighborhood of $p = 0$. This will simplify matters considerably; in particular, in this case $f_l(p)$ will have a power series expansion about $p = 0$.

We begin by inserting the integral (12.7) for $f_l(p)$ into the numerator of (12.23). With some simple manipulation this gives

$$f_l(p) = \frac{-1}{p^2 f_l(p)} \int_0^\infty dr\, \hat{j}_l(pr)U(r)\phi_{l,p}(r)$$

which, since $\phi/f = \psi$, the reader will recognize as our old friend

$$f_l(p) = -\frac{1}{p^2} \int_0^\infty dr\, \hat{j}_l(pr)U(r)\psi_{l,p}(r)$$

For real p we know that both $\hat{j}$ and ϕ are bounded by

$$|\hat{j}_l(pr)| \quad \text{and} \quad |\phi_{l,p}(r)| \leqslant \text{constant}\left(\frac{pr}{1 + pr}\right)^{l+1}$$

Therefore, *provided $f_l(0)$ is nonzero*,

$$|f_l(p)| \leqslant \frac{\text{constant}}{f_l(0)} \frac{1}{p^2} \int_0^\infty dr\, |U(r)| \left(\frac{pr}{1 + pr}\right)^{2l+2} \tag{12.24}$$

in some neighborhood of $p = 0$.

Let us first suppose that the potential is exponentially bounded. In this case we can simply drop the denominator $(1 + pr)^{2l+2}$ from this bound to give

$$|f_l(p)| \leqslant \frac{\text{constant}}{f_l(0)} p^{2l} \int_0^\infty dr\, |U(r)|\, r^{2l+2} \tag{12.25}$$

which establishes that

$$f_l(p) = O(p^{2l}) \qquad [p \to 0] \qquad (12.26)$$

Since $f_l(p)$ is analytic near $p = 0$, it has a power series expansion whose lowest term must then be $O(p^{2l})$; that is,

$$f_l(p) = -a_l p^{2l} + b_l p^{2l+1} + \cdots$$

where a_l is some finite constant (possibly zero), which will be recognized as the scattering length introduced earlier. This expansion is just the result anticipated in (11.28), which, we see, holds for any exponentially bounded potential provided $f_l(0) \neq 0$. When $f_l(0) = 0$ it is clear that the scattering length a_l will be infinite—a situation to which we shall return in Section 12-g.

If our potential is *not* exponentially bounded, but we suppose instead that it satisfies

$$V(r) = O\left(\frac{1}{r^\nu}\right) \qquad [r \to \infty]$$

then the integral in (12.25) is convergent only for $2l + 2 < \nu - 1$. Thus, for $l < (\nu - 3)/2$ the result (12.26) still holds. But for $l > (\nu - 3)/2$ we must go back to the bound (12.24) and retain the denominator $(1 + pr)^{2l+2}$ to ensure convergence. By splitting the integral as $\int_0^1 + \int_1^\infty$ we find in place of (12.26) that

$$f_l(p) = O(p^{\nu-3}) \qquad [p \to 0, \, l > (\nu - 3)/2]$$

Having established the dominant behavior of $f_l(p)$ at threshold, we can go on to discuss its power series expansion at that point. Of course the power series expansion exists only if $f_l(p)$ is analytic near $p = 0$, and for this reason we now restrict our attention to exponentially bounded potentials.

It turns out that the expansion of the amplitude itself in powers of p is less useful than that of a related quantity—the K matrix (also called the reaction, or reactance matrix). We shall discuss the general theory of the K matrix in Chapter 14, but the partial-wave version can be briefly described as follows. In place of the partial-wave S matrix $s_l(p)$ we work with a quantity $k_l(p)$ defined so that

$$s_l(p) = \frac{1 + ik_l(p)}{1 - ik_l(p)} \qquad (12.27)$$

In the physical region the unitarity of s_l implies that k_l is *real*. (One of the advantages of the K matrix is that in any approximation scheme, the unitarity of S is guaranteed by insisting that k_l be real, or, more generally, that the K matrix be Hermitian.) We can solve (12.27) to give

$$k_l(p) = i\frac{1 - s_l(p)}{1 + s_l(p)} = \tan \delta_l(p) \qquad (12.28)$$

From the first equality it is clear that $k_l(p)$ is analytic wherever $s_l(p)$ is analytic [except for poles wherever $s_l(p) = -1$] and, in particular, it is analytic around $p = 0$.

An important property of the K matrix emerges if we continue from p to $-p$. Since $s_l(p) = f_l(-p)/f_l(p)$, it follows when p goes to $-p$ that $s_l(p)$ becomes $1/s_l(p)$ and, hence, from (12.27) that $k_l(p)$ changes sign

$$k_l(-p) = -k_l(p)$$

that is, $k_l(p)$ is an *odd* function of p and therefore its power-series expansion contains only odd powers of p.

As $p \to 0$, we know that $\delta_l \to -a_l p^{2l+1}$; this implies the same result for $k_l = \tan \delta_l$,

$$k_l(p) \xrightarrow[p \to 0]{} -a_l p^{2l+1}$$

Thus, it is convenient to consider either of the functions k_l/p^{2l+1} or p^{2l+1}/k_l, both of which are analytic near $p = 0$ (unless $a_l = \infty$ or 0), are *even* functions of p, and therefore have power-series expansions in p^2.[10] In practice (in n–p scattering, for example), it is the expansion of p^{2l+1}/k_l that proves most useful,

$$\frac{p^{2l+1}}{k_l(p)} = p^{2l+1} \cot \delta_l(p) = \frac{-1}{a_l} + \frac{r_l}{2} p^2 + O(p^4) \tag{12.29}$$

This is known as the *effective range expansion*, since for $l = 0$ the coefficient r_0 is usually approximately related to the range of the potential. The effective range *approximation* (or shape independent approximation) is obtained if we retain only the first two terms on the right of (12.29). The s-wave effective range approximation has been one of the most important methods of parametrizing low-energy scattering data (e.g., n–p scattering up to about 10 MeV). Some further discussion of its use can be found in Problems **12.4** and **13.1**.

When the potential has an $1/r^{\nu}$ tail, the Jost function and related functions s_l, f_l, k_l usually have a branch-point singularity at $p = 0$, and for this reason an effective range expansion in powers of p^2 is *not* possible. For example, the corresponding s-wave expansion for a potential with an $1/r^4$ tail is[11]

$$p \cot \delta_0(p) = \frac{-1}{a_0} + bp + cp^2 \ln p + O(p^2)$$

---

[10] Another way to say this is that these functions are actually analytic functions of the energy $E = p^2/2m$ at $E = 0$; this is the form in which this important property of the K matrix is usually used

[11] See O'Malley, Spruch, and Rosenberg (1961).

12-g.   Zeros of the Jost Function at Threshold

In the last two sections we have had to exclude from our discussion the special case that the Jost function vanishes at threshold, $f_l(0) = 0$. In this section we discuss this possibility and establish how the previous results are modified when it occurs.

We know that zeros of $f_l(p)$ in $\{\text{Im } p > 0\}$ correspond to bound states, and we must first consider whether the same is true of zeros at $p = 0$. To this end we consider the radial Schrödinger equation, which for $p = 0$ is

$$\left[\frac{d^2}{dr^2} - \frac{l(l+1)}{r^2} - U(r)\right]y(r) = 0$$

It is clear that with p zero, the centrifugal term dominates as $r \to 0$ *or* $r \to \infty$ and it follows that the solutions behave like combinations of r^{l+1} and $1/r^l$ for *small* or *large* r. For a general potential the solution that is finite (r^{l+1}) at $r = 0$ will contain both terms at infinity and there will be no proper eigenfunction of zero energy. However, for certain potentials the solution that is finite (r^{l+1}) at $r = 0$ may be pure $1/r^l$ at infinity. For the case $l = 0$ this possibility does not lead to a bound state because a solution that behaves like $1/r^l$ (i.e., a constant) as $r \to \infty$ is *not* normalizable. On the other hand, for $l > 0$ such a solution *is* normalizable and we conclude that bound states of zero energy *are* possible for $l > 0$.

We must next determine whether a zero-energy bound state (when it occurs) corresponds to a zero of the Jost function at the origin. Unfortunately our notation is not well suited to handle the point $p = 0$ because all of our standard solutions, $\hat{j}_l(pr)$, $\phi_{l,p}(r)$, etc., behave badly there. For example, as $p \to 0$, $\hat{j}_l(pr)$ vanishes and therefore ceases to be an interesting solution of the free equation. The reason is, of course, that $\hat{j}_l(pr)$ has a factor p^{l+1} for small arguments, and if we factor out this p^{l+1}, then

$$\frac{\hat{j}_l(pr)}{p^{l+1}} \xrightarrow[p \to 0]{} \frac{r^{l+1}}{(2l+1)!!}$$

That is, $\hat{j}_l/p^{l+1}$ reduces to the r^{l+1} solution of the free zero-energy equation as $p \to 0$. In the same way $\hat{n}_l(pr)$ blows up as $p \to 0$, but if we multiply it by p^l then $p^l \hat{n}_l$ (and likewise $p^l \hat{h}_l^{\pm}$) reduces to the $1/r^l$ solution as $p \to 0$.

It is easily shown that by removing the appropriate powers of p from the solutions ϕ and χ^+ of the full radial equation we can handle them in the same way. Thus, when $p \to 0$, $\phi/p^{l+1} = \tilde{\phi}$ (say) approaches the solution that behaves like r^{l+1} at the origin; while $p^l \chi^+ = \tilde{\chi}^+$ approaches the solution that behaves like $1/r^l$ at infinity. Now in terms of $\tilde{\phi}$ and $\tilde{\chi}^+$ the Jost function is:

$$f_l(p) = \frac{1}{p} W(\chi^+, \phi) = W(\tilde{\chi}^+, \tilde{\phi})$$

and from this it is clear that $f_l(0) = 0$ if and only if the solution $\tilde{\phi}$ ($\to r^{l+1}$ as $r \to 0$) is proportional to the solution $\tilde{\chi}^+$ ($\to 1/r^l$ as $r \to \infty$). Thus, for $l > 0$ the zero-energy Jost function vanishes if and only if there is a zero energy bound state.

To proceed further we need to know *how* $f_l(p)$ vanishes at $p = 0$ when it does. To simplify matters we suppose that the potential is exponentially bounded so that $f_l(p)$ is actually analytic near $p = 0$. In this case we can insert into the integral

$$f_l(p) = 1 + \frac{1}{p}\int \hat{h}^+ U\phi$$

$$= 1 + \left(\frac{1}{p}\int \hat{n}U\phi\right) + i\left(\frac{1}{p}\int \hat{j}U\phi\right) \tag{12.30}$$

the expansions of $\hat{n}, \hat{j}$ and ϕ in powers of p. Now, $\hat{n}$ has the form $1/p^l$ times a power series in p^2, while both $\hat{j}$ and ϕ have the form p^{l+1} times a power series in p^2. Thus, from (12.30) we see that $f_l(p)$ has the form:

$$f_l(p) = 1 + [\alpha_l + \beta_l p^2 + O(p^4)] + i[\gamma_l p^{2l+1} + O(p^{2l+3})] \tag{12.31}$$

(with all coefficients $\alpha_l, \dots$, real).

The Jost function $f_l(p)$ vanishes at $p = 0$ if it happens that the coefficient α_l in (12.31) has the value -1. When this happens

$$f_l(p) = [\beta_l p^2 + O(p^4)] + i[\gamma_l p^{2l+1} + O(p^{2l+3})] \tag{12.32}$$

For the s-wave case we see that the leading term comes from the second bracket and

$$f_0(p) = i\gamma_0 p + O(p^2) \tag{12.33}$$

while for any $l > 0$ it comes from the first bracket and

$$f_l(p) = \beta_l p^2 + O(p^3 \text{ or } p^4) \qquad [l > 0] \tag{12.34}$$

Thus, if the Jost function has a zero at $p = 0$ then for $l = 0$ this zero is simple, while for $l > 0$ it is double.[12]

We are now ready to see how the results of the previous two sections are modified when $f_l(0) = 0$. We consider first Levinson's theorem and observe that if $f_l(0) = 0$ the contour used in proving the theorem must be modified

---

[12] What this simple argument fails to show is that the coefficients γ_0 in (12.33) and β_l in (12.34) are always nonzero. This can be shown by evaluating the first and second derivatives of $f_l(p)$ as described at the end of Section 12-d. This alternative method can also be applied even when $f_l(p)$ cannot be continued into the lower half plane, to show that in this case as $p \to 0$ from the upper half plane $f_l(p)$ vanishes linearly or quadratically (according as $l = 0$ or $l > 0$). See Newton (1960), Eqs. (4.25) and (4.25').

as in Fig. 12.9 to avoid the pole of the integrand in

$$I = \oint dp \, \frac{\dot{f_l}(p)}{f_l(p)}$$

Exactly as before,

$$I = 2\pi i \times (\text{number of zeros in } \{\text{Im } p > 0\}) \qquad (12.35)$$

However, because of the small semicircle Γ_ϵ of radius ϵ, we now find that

$$I = 2i[\delta_l(0) - \delta_l(\infty)] + \lim_{\epsilon \to 0} \int_{\Gamma_\epsilon} dp \, \frac{\dot{f_l}(p)}{f_l(p)} \qquad (12.36)$$

The contribution of Γ_ϵ depends on l and is easily seen to be

$$\lim_{\epsilon \to 0} \int_{\Gamma_\epsilon} dp \, \frac{\dot{f_l}(p)}{f_l(p)} = \begin{cases} -\pi i & [l = 0] \\ -2\pi i & [l > 0] \end{cases}$$

since the zero of $f_l(p)$ is simple or double according as $l = 0$ or $l > 0$.

For the s-wave case we know that there are no bound states of zero energy and, hence, that the number of bound states n_0 is the same as the number of zeros in $\{\text{Im } p > 0\}$. Comparing (12.35) and (12.36) we find that

$$\delta_0(0) - \delta_0(\infty) = (n_0 + \tfrac{1}{2})\pi$$

For $l > 0$ we know that when $f_l(0) = 0$ there *is* a bound state of zero energy. This means that the number of bound states n_l, is the number of zeros in $\{\text{Im } p > 0\}$ *plus one*. Comparing (12.35) and (12.36) we find:

$$\delta_l(0) - \delta_l(\infty) = n_l \pi$$

which is the same result as before.

The behavior of the amplitude $f_l(p)$ near threshold when $f_l(0) = 0$ is easily found. Remember that

$$f_l(p) = \frac{-1}{p^2 f_l(p)} \int_0^\infty j_l U \phi_l$$

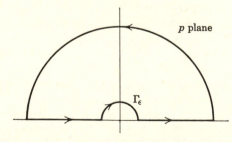

FIGURE 12.9.   Contour used to prove Levinson's theorem if $f_l(0) = 0$.

When $f_l(0) \neq 0$ we found from this that

$$f_l(p) = -a_l p^{2l} + O(p^{2l+1}) \tag{12.37}$$

For the case $l = 0$, we know that when $f_0(p)$ vanishes at $p = 0$ it does so like $i\gamma_0 p$. In this case the result (12.37) becomes

$$f_0(p) = i\frac{a_0}{p} + O(1)$$

where a_0 is some real constant.[13] Thus, when $f_0(0) = 0$, the s-wave amplitude, and the corresponding cross section $\sigma_0 = 4\pi|f_0|^2$, are infinite at threshold—a phenomenon sometimes called a *zero energy resonance*. When this occurs, it is clear that the s-wave scattering length, defined as $a_0 = -f_0(0)$, is also infinite. In practice, $f_0(0)$ is unlikely to be *exactly* zero and one does not expect to find a scattering length that is actually infinite. However, it is perfectly possible for $f_0(p)$ to have a zero *close* to threshold and, hence, for $f_0(0)$ to be small. In this case the zero-energy cross section is anomalously large. An example of this phenomenon is illustrated by the square well cross sections of Fig. 11.3 where $\sigma(0)$ is more than 40 times the geometrical cross section πa^2; a well known real example is given by the n–p singlet system, which has a zero of f so close to threshold that the zero-energy cross section (70 barns) is at least 200 times any reasonable estimate of the geometrical size.

For $l > 0$, $f_l(0)$ vanishes like p^2 (if at all) and the behavior (12.37) becomes

$$f_l(p) = a_l p^{2l-2} + O(p^{2l-1})$$

that is, the power behavior of $f_l(p)$ ($l > 0$) is reduced by two when $f_l(0) = 0$. In particular, this means the p-wave amplitude remains nonzero as $p \to 0$.

PROBLEMS

12.1. (a) Using the methods of Problem **11.7**, prove that

$$|\hat{j}_l(z)| \leqslant \text{constant}\left(\frac{|z|}{1+|z|}\right)^{l+1} e^{|\text{Im } z|}$$

for all z, *real or complex*. Establish corresponding bounds for $\hat{n}_l(z)$, $\hat{h}_l^+(z)$ and the Green's function. The bound for $\hat{h}$ is given in Section 12-c; that for the Green's function is the same as in Problem **11.7** except for a factor $\exp\{|\text{Im } p| (r - r')\}$.

---

[13] This also follows from Levinson's theorem, which asserts that $\delta_0(0) = \pi/2$ (modulo π) if $f_0(0) = 0$. This shows directly that $f_0(p) - (1/p)\exp(i\delta_0)\sin\delta_0$ goes to infinity and becomes imaginary.

(b) Establish the theorem of Section 12-b (that the regular solution ϕ can be found by iteration for any potential and any complex p) for arbitrary l. Establish that the solution is bounded as in (12.6).

12.2.   To illustrate the relative unimportance of the lower half plane $\{\text{Im } p < 0\}$ do the following:

(a) Calculate the s-wave Jost function $f(p)$ for the *weak* exponential $\gamma e^{\mu r}$ (the point of the weakness is that you can approximate f by the first two terms in its "Born series," $f = 1 + (1/p) \int \hat{h}^{+} U \hat{f}$). In the same approximation calculate the s-wave Jost function $f^{a}(p)$ for the same potential *cut off* at $r = a$.

(b) Show that f has a pole at $p = -i\mu/2$ but that the cut off function f^{a} does not, however large we make the cut off.

(c) Show that in $\{\text{Im } p \geqslant 0\}$, $\Delta f = f^{a} - f$ can be made uniformly small by making a large. In the lower half plane it is clear that Δf is infinitely large at $p = -i\mu/2$ however large we make the cut off.

12.3.   (a) Show that if the Jost function has a zero in the upper half plane, then this zero is simple. Hints:

(1) Suppose $f(\bar{p}) = 0$ and hence $\phi = \lambda \chi^{+}$ at $\bar{p}$.

(2) Differentiate $f = (1/p)W(\chi^{+}, \phi)$, to give $\dot{f}(\bar{p}) = (1/p)\{W(\chi^{+}, \dot{\phi}) + W(\dot{\chi}^{+}, \phi)\}$ (where $\dot{f} = df/dp$, etc.)

(3) The Wronskian $W(\chi^{+}, \dot{\phi})$ can be evaluated using the relation $dW(\chi_{\bar{p}}^{+}, \phi_{p})/dr = (\bar{p}^{2} - p^{2})\chi_{\bar{p}}^{+}\phi_{p}$, which follows from the radial equation. Integrate from zero to r to give $W(\chi_{\bar{p}}^{+}, \phi_{p})$, then differentiate with respect to p and finally set $p = \bar{p}$. The second Wronskian can be treated similarly but integrating from r to infinity.

(4) This gives $\dot{f}(\bar{p}) = -2 \int_{0}^{\infty} dr \chi^{+} \phi = -2\lambda \int_{0}^{\infty} dr(\chi^{+})^{2}$.

(5) Since $\bar{p}$ is pure imaginary, χ^{+} is i^{l} times a real function and this integral is nonzero.

(b) Prove that the residue (as a function of E) of the partial-wave amplitude f_{l} at a bound-state pole is $(-)^{l+1}\gamma^{2}/2m$, where the real number γ is the "asymptotic normalization" of the bound-state wave function, defined as follows: Let $\eta(r)$ be the normalized, real, bound-state radial function, and let the bound-state energy be $-\alpha^{2}/2m$; then γ is defined by the relation $\eta(r) \to \gamma e^{-\alpha r}$ as $r \to \infty$. Note that $\eta(r)$ is obviously proportional to χ^{+} and, in fact,

$$\eta(r) = i^{l}\chi^{+}(r)\bigg/ \int_{0}^{\infty} dr(\chi^{+})^{2}$$

12.4.   The effective range formula (12.29) is often arrived at by a method quite different from that of Section 12-f, as follows (consider the case $l = 0$): Let u_{p} denote a radial wave function proportional to $\psi_{0, p}$ but normalized such that $u_p \to \cos pr + \cot \delta \sin pr$ as $r \to \infty$. Let v_p denote the solution of the *free* equation that coincides with u_p for large r; that is, $v_p \equiv \cos pr + \cot \delta \sin pr$. Use the fact that $p \cot \delta(p) \to -1/a$ as $p \to 0$ to show that $v_0 = 1 - r/a$. Hence show that

$$p \cot \delta(p) = \frac{-1}{a} + p^2 \int_0^\infty dr(v_p v_0 - u_p u_0)$$

Hint: Use the relation $dW(u_p, u_0)/dr = p^2 u_p u_0$ (which follows from the radial equation as in the last problem) to show that

$$[W(u_p, u_0)]_a^b = p^2 \int_a^b dr \, u_p u_0$$

with a similar relation for v_p. Hence evaluate $[W(v_p, v_0) - W(u_p, u_0)]_0^\infty$.

The effective range expansion follows if we expand v_p and u_p in powers of p. In particular the effective range is just

$$r_0 = 2 \int_0^\infty dr(v_0^2 - u_0^2)$$

Since $v_0 \approx u_0$ outside the potential, while $v_0 = 1$, $u_0 = 0$ at the origin, this suggests that r_0 is roughly related to the range of the potential.

If the potential is deep and short range then both u_p and v_p would be expected to be slowly varying functions of p inside the well. This suggests that the effective range *approximation* $p \cot \delta \approx -(1/a) + (r_0/2)p^2$, which corresponds to setting $v_p \approx v_0$ and $u_p \approx u_0$, would be good.

13 Resonances

13-a   **Resonances and Poles of the S Matrix**

13-b   **Bound States and Resonances**

13-c   **Time Delay**

13-d   **Decay of a Resonant State**

Probably the most striking phenomenon in the whole range of scattering experiments is the resonance. Resonances are observed in atomic, nuclear, and particle physics. In their simplest form they lead to sharp peaks in the total cross section as a function of energy, like those shown in the two examples of Fig. 13.1. Obviously, effects as dramatic as those shown demand explanation and, once understood, can be expected to provide useful information about the underlying interactions.

There are many different theoretical approaches to the resonance phenomenon, all of them having in common that the sharp variation of the cross section at a resonant energy E_R is in some way related to the existence of a *nearly bound state* of the projectile–target system with energy E_R. When the projectile is sent in with energy E_R it can be temporarily captured into this "metastable" state; and it is this possibility that is considered the cause of the violent variations in the cross section.

We shall approach the problem of resonances via the analytic properties of the amplitude established in Chapter 12. In Chapter 12 we saw that poles of the S matrix $s_l(p)$ [or more precisely, zeros of the Jost function $f_l(p)$] in the upper half plane $\{\text{Im } p > 0\}$ correspond to the bound states of angular momentum l. What we shall show in this chapter is that if $s_l(p)$ has a pole in the *lower* half plane {Im $p < 0$}, then, under certain circumstances, this can correspond to a *resonance* of angular momentum l—a concept we shall define more precisely in a moment.

The fact that poles of $s_l(p)$ can correspond to resonances when Im $p < 0$ and to bound states when Im $p > 0$ obviously suggests a close connection between resonances and bound states; we shall explore this connection in Section 13-b. First, in Section 13-a, we shall show how it is that poles of

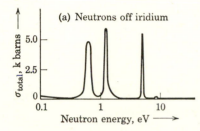
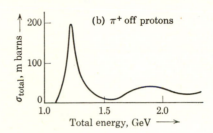

FIGURE 13.1.   Resonances in the scattering of (a) neutrons off iridium, and (b) pions off protons.

$s_l(p)$ [or zeros of $f_l(p)$] cause resonances. Finally, in Sections 13-c and 13-d we shall examine in detail the evolution of the scattering of a wave packet whose energy is in the neighborhood of a resonance.

It should be emphasized at the outset that the correspondence between resonances and the poles of $s_l(p)$, or zeros of $f_l(p)$, in {Im $p < 0$} is by no means as exact as that between bound states and the zeros of $f_l(p)$ in {Im $p > 0$}. First, to discuss *anything* in the lower half plane it is obviously necessary that the potential be "reasonable," in the sense that it allows continuation into the lower half plane.

Second, we shall see that there can be zeros of f_l in {Im $p < 0$} (specifically, those far from the real axis) that lead to no observable resonance effect; and, conversely, one can construct potentials that display resonance effects without having any zeros of f_l (or poles of s_l) in {Im $p < 0$}. Nonetheless, the explanation of resonances as poles of s_l in {Im $p < 0$} is so satisfactory that it is generally accepted that all (or at least almost all) resonances are correctly described in this way.

Finally, we should note that most real resonances occur in multichannel systems of the type to be discussed in Chapters 16 and those following. However, many of the essential features of the resonance phenomenon show up in the simpler, one-channel situation, and it is for this reason that we begin our discussion of resonances here. We shall take up the general multichannel case in Chapter 20.

13-a.   Resonances and Poles of the S Matrix

In this section we shall suppose that the Jost function has a zero (and hence s_l a pole) in the lower half plane close to the real axis, and then explore its physically observable consequences. It is essential that the potential be such as to allow continuation into $\{\mathrm{Im}\, p < 0\}$, and throughout this chapter *we shall assume that this is the case.* Having made this assumption, we remark that, since $f_l(-p)$ is always analytic in $\{\mathrm{Im}\, p < 0\}$, and since $f_l(\pm p)$ cannot both vanish at the same point, the Jost function $f_l(p)$ vanishes at some point $\bar{p}$ in $\{\mathrm{Im}\, p < 0\}$ *if and only if* $s_l(p) = f_l(-p)/f_l(p)$ has a pole. Thus, there is a complete equivalence between zeros of f_l and poles of s_l in $\{\mathrm{Im}\, p < 0\}$, and our discussion can be couched in terms of either. Since the discussion in terms of zeros of f_l is a little simpler we begin with it.

We assume that the Jost function $f(p)$ (from which we now omit the subscript l) has a zero at some point

$$\bar{p} = p_R - ip_I \qquad [p_I > 0]$$

where R stands for either "real" or "resonance" and I for "imaginary." Exactly as in the case of bound states, a zero of f implies that the regular solution ϕ is proportional to the outgoing solution χ^+; but since χ^+ *increases* exponentially with r when $\mathrm{Im}\, p < 0$, this does *not* give a proper eigenfunction of the Hamiltonian. We shall suppose for the moment that the zero of f at $\bar{p}$ is a simple zero. However, it should be emphasized that, while the bound-state zeros of f *have* to be simple in $\{\mathrm{Im}\, p > 0\}$, there is no general principle that guarantees the same in $\{\mathrm{Im}\, p < 0\}$; and we shall, in fact, return briefly to the exotic possibility of multiple zeros in Problem **13.5**.

Since $f(p)$ has a simple zero at $\bar{p}$, there is some neighborhood of $\bar{p}$ in which we can approximate $f(p)$ as

$$f(p) \approx \left(\frac{df}{dp}\right)_{\bar{p}}(p - \bar{p}) \tag{13.1}$$

Now, if $\bar{p}$ is sufficiently close to the axis, then there is a real interval centered on p_R in which this approximation is good; and we now examine the phase shift in this interval. To do this we need only recall that, for physical p, the phase shift is minus the phase of $f(p)$. Thus from (13.1) we see that

$$\delta(p) \approx -\arg\left(\frac{df}{dp}\right)_{\bar{p}} - \arg(p - \bar{p}) \qquad [p \text{ real, close to } p_R] \tag{13.2}$$

$$\equiv \delta_{bg} + \delta_{res}(p)$$

where

$$\delta_{res}(p) = -\arg(p - \bar{p}) \tag{13.3}$$

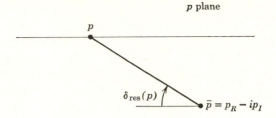

FIGURE 13.2   Resonant part $\delta_{res}(p)$ of phase shift.

The quantity δ_{bg} is called the *background phase shift* and is just minus the phase of $(d\not f/dp)_{\bar{p}}$. The quantity $\delta_{res}(p)$ is called the *resonant part of the phase shift* and is (apart from an irrelevant π) the angle shown in Fig. 13.2.

It is clear from Fig. 13.2 that as p increases past the location of the zero, the resonant part of the phase shift $\delta_{res}(p)$ increases from 0 to π. Also, the closer the zero is to the real axis, the more suddenly this increase occurs. (For example, δ_{res} increases from $\pi/4$ to $3\pi/4$ in the interval of width $2p_I$ between $p = p_R \pm p_I$.) Thus, near a zero of $\not f(p)$ that is close to the real axis, the complete phase shift $\delta = \delta_{bg} + \delta_{res}$ increases suddenly from the value δ_{bg} to $\delta_{bg} + \pi$.

We shall take this rapid increase of $\delta_l(p)$ by π, as given by (13.2), to be our definition of a resonance of angular momentum l. Before we examine the physical consequences of this phenomenon it must be mentioned that there is considerable confusion in the literature concerning this definition. The term resonance is sometimes taken to be exactly synonymous with a zero of $\not f$ (or pole of s) in $\{\text{Im } p < 0\}$. However, it is clear that a zero sufficiently far from the real axis will produce no physically observable effects; and, as we have already mentioned, all of the observable effects usually associated with a resonance can occur without there being any zero of $\not f(p)$ in $\{\text{Im } p < 0\}$.[1] Thus, the definition of a resonance as a zero of $\not f(p)$ in $\{\text{Im } p < 0\}$ does not correspond precisely to any observable phenomenon.

The behavior of the partial cross section $\sigma_l(p)$,

$$\sigma_l(p) = \frac{4\pi(2l + 1)}{p^2} \sin^2 \delta_l(p)$$

near a resonance depends on the value of the background δ_{bg}. Four different possibilities are shown in Fig. 13.3. The first and simplest, shown as the curves below (a), is when the background δ_{bg} is zero, and $\delta(p)$ therefore

[1] An elegant example of a potential that produces all the symptoms of a resonance and for which s(p) is analytic in $\{\text{Im } p < 0\}$ but has no poles there can be found in Calucci and Ghirardi (1968).

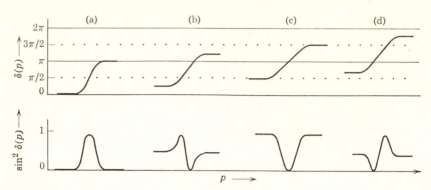

FIGURE 13.3. **Four possible resonances. The $\delta(p)$ plots show the resonant phase shifts for $\delta_{bg} = 0$, $\pi/4$, $\pi/2$, and $3\pi/4$. The $\sin^2 \delta(p)$ plots show the corresponding behavior of the partial cross section [apart from a factor $4\pi(2l+1)/p^2$].**

increases suddenly from 0 to π near $p = p_R$. In this case the partial cross section is small on either side of the resonance and has a sharp peak as δ passes through $\pi/2$ and σ_l attains its unitarity limit. This type of resonance is often called a pure Breit–Wigner resonance. The opposite extreme occurs if $\delta_{bg} = \pi/2$, and is shown in (c); here $\delta(p)$ increases from $\pi/2$ to $3\pi/2$. Thus, on either side of the resonance the cross section has its maximum value, and at resonance it has a sharp *minimum* as δ passes through π. Two intermediate cases ($\delta_{bg} \approx \pi/4$ and $3\pi/4$) are shown in (b) and (d).[2]

Needless to say the curves shown in Fig. 13.3 are idealizations, which are never precisely realized in practice. This is because the behavior (13.2) for $\delta(p)$ comes from the approximation $f'(p) \approx$ constant $\times (p - \bar{p})$. A convenient way to express the exact situation is to write

$$\delta(p) = \delta_{bg}(p) + \delta_{res}(p)$$

where $\delta_{res}(p)$ is as before, but $\delta_{bg}(p)$, instead of being a constant, is a slowly varying function of p. This means, for example, that at an actual resonance the phase shift increases suddenly by *approximately* (but not exactly) π.

If there is a resonance of angular momentum l and if all other phase shifts are slowly varying in the neighborhood of the resonance, then the rapid variations of Fig. 13.3 will also be visible in the total cross section $\sigma = \sum \sigma_l$. (This was the case with the resonances shown in Fig. 13.1.) If all other

---

[2] Perhaps it should be emphasized that any value of δ_{bg} is possible and many different values are observed in practice. In the literature of high energy physics one sometimes finds the claim that for elastic scattering δ_{bg} must be small. This is false, and any value of δ_{bg} is consistent with pure elastic scattering. On the other hand we shall see in the next section that for a resonance *sufficiently close to threshold* δ_{bg} is always small.

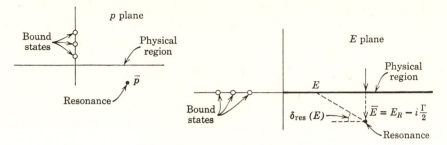

FIGURE 13.4.   Planes of the complex variables p and $E = p^2/2m$.

phase shifts are also *small*, then the scattering will be completely dominated by the resonant angular momentum. In this case the angular momentum of the resonance can be easily identified by examining the angular distribution of the scattering in the neighborhood of resonance. Unfortunately, it often happens (especially at high energies) that several phase shifts are varying appreciably at the same time. In this case a resonance can only be detected after an analysis of angular distributions has sorted out the various partial waves.

One can, of course, give an analytic expression for the partial cross section $\sigma_l(p)$ in the neighborhood of a resonance. Because this is usually done in terms of the energy $E = p^2/2m$ rather than p, we first say a few words about the connection between these two variables. Since the mapping from p to E is two-to-one, a function like the Jost function, when considered as a function of E, is a function on a *two-sheeted Riemann surface*. The correspondence between the p plane and the two sheets of E is traditionally chosen as shown in Fig. 13.4. The *first sheet* or *"physical sheet"* of E corresponds to the upper half plane {Im $p > 0$}, the *second sheet* to the lower half plane {Im $p < 0$}. For any p, the points $\pm p$ correspond to two different points on the Riemann surface of E (one on the first sheet, the other beneath it on the second sheet) both with the same numerical value $E = p^2/2m$. The bound states, which are zeros of f on the positive imaginary axis of p, lie on the negative real axis of the first sheet of E. This first sheet has a cut from 0 to ∞, the upper rim of which corresponds to the physical region $p \geqslant 0$. If one continues through this cut, one comes to the second sheet. Thus, the resonance zeros of f are found by continuing from the first sheet downward through the cut.

If we regard the Jost function as a function of E [which we write for simplicity as $f(E)$], then the resonance zero at $\bar{p}$ becomes a zero on the second sheet, close to the physical region, at

$$\bar{E} = \frac{\bar{p}^2}{2m} = E_R - i\frac{\Gamma}{2}$$

(where the notation $-\Gamma/2$ for the imaginary part of $\bar{E}$ is conventional). We can now recast our original argument in terms of E. Close to the resonance we write $f(E) \approx (df/dE)_{\bar{E}}(E - \bar{E})$ and, as before, we find that for real E close to E_R,

$$\delta(E) \approx \delta_{bg} + \delta_{res}(E) \qquad [E \text{ real, close to } E_R]$$

The resonant part of the phase shift $\delta_{res}(E)$ is the angle shown in Fig. 13.4 and some elementary trigonometry gives

$$\sin \delta_{res}(E) = \frac{\Gamma/2}{[(E - E_R)^2 + (\Gamma/2)^2]^{1/2}}$$

In particular, in the simple case where the background phase δ_{bg} is zero and $\delta = \delta_{res}$, this gives for the partial cross section

$$\sigma_l(E) \propto \sin^2 \delta_l(E) = \frac{(\Gamma/2)^2}{(E - E_R)^2 + (\Gamma/2)^2} \qquad (13.4)$$

This is the famous *Breit–Wigner formula* that gives the peak shown in Fig. 13.3, part (a). As a function of energy the peak is centered at[3] E_R and has width Γ. When the background δ_{bg} is nonzero the Breit–Wigner is replaced by a more general formula, which we leave the reader to derive.

Finally, we remark that in relativistic scattering theory the existence of a Jost function is not firmly established; and it is therefore important that we could just as well derive our results in terms of poles of $s(p)$ rather than zeros of $f(p)$. Thus, if we suppose that $s(p)$ has a simple pole in $\{\text{Im } p < 0\}$ at a point $\bar{p}$ close to the real axis, then the most general form for $s(p)$ compatible with unitarity is:

$$s(p) = e^{2i\delta(p)} \approx e^{2i\delta_{bg}} \frac{p - \bar{p}^*}{p - \bar{p}} \qquad [p \text{ close to } p_R]$$

where δ_{bg} is any real number. From this it follows that $\delta \approx \delta_{bg} + \delta_{res}$, exactly as before. We also see that, because $s(p)$ is unitary, any pole of $s(p)$ at $\bar{p}$ must always be accompanied by a zero of $s(p)$ at $\bar{p}^*$. [This is also clear if we write $s(p) = f(p^*)^*/f(p)$.] This result lets us conclude this section with a simple moral: It is misleading to think of a resonance as arising from *large values* of the amplitude caused by the proximity of a pole. Any tendency for the pole to make $s(p)$ large is precisely offset by the tendency for the zero to make it small. In reality, the essential feature of the resonance is the rapid change in the *phase* of $s(p)$ (caused by both the pole and zero) and, as is clear from Fig. 13.3, this can give the amplitude at resonance any value from its maximum (i/p) to its minimum (0).

---

[3] The partial cross section $\sigma_l(E)$ is actually proportional to $(1/E)\sin^2\delta_l(E)$. The factor $1/E$ pushes the peak to the left.

13-b.   Bound States and Resonances

The fact that the zeros of f (or poles of s) in $\{\operatorname{Im} p > 0\}$ correspond to bound states, while those in $\{\operatorname{Im} p < 0\}$ may represent resonances suggests that there should be some connection between bound states and resonances. We can now show that, at least in certain cases, this is indeed so. We begin by re-introducing the variable coupling parameter λ and considering the scattering of the Hamiltonian $H = H^0 + \lambda V$. We saw in Chapter 11 that the Jost function is entire in λ when p is physical, $p \geqslant 0$; in Chapter 12 we saw that it is analytic in p (except for possible singularities in $\{\operatorname{Im} p < 0\}$) when $\lambda = 1$. In fact, it is easy to show that it is actually an analytic function— which we write as $f(\lambda, p)$—of both λ *and* p for all λ, and all p except for the usual singularities in $\{\operatorname{Im} p < 0\}$.

We now suppose that for some real value λ_0 there is a zero of f on the boundary of the upper and lower half planes: that is, a zero at threshold. We shall see that if we vary λ this zero moves. If we vary λ to make the potential more attractive, the zero moves up into $\{\operatorname{Im} p > 0\}$ and becomes a bound state. If we make the potential less attractive, the zero moves down into $\{\operatorname{Im} p < 0\}$ and becomes a resonance if $l > 0$, or a so-called "virtual state" for $l = 0$.[4]

We first remark that when we vary λ the zero of f certainly must move, because if it remains fixed for λ in some interval, then (since f is analytic) it must remain fixed for *all* λ, which contradicts the known fact that $f \neq 0$ for $\lambda = 0$. To see exactly *how* the zero moves as λ varies we make a double power series expansion of $f(\lambda, p)$ in λ *and* p about the point $\lambda = \lambda_0$, $p = 0$ (we assume that the potential is exponentially bounded, so that f is analytic at $p = 0$),

$$f(\lambda, p) = \sum_{m,n} \alpha_{mn} p^m (\lambda - \lambda_0)^n$$

For the case $l = 0$ we saw in (12.33) that the zero at threshold is *simple*, and comparison with (12.33) shows that the lowest terms of the double power series have the form:

$$f(\lambda, p) = i\xi p + \eta(\lambda - \lambda_0) + \cdots \qquad [l = 0]$$

(with ξ and η real). Obviously when $\lambda = \lambda_0$ this has the simple zero at $p = 0$, while for λ close to λ_0 the same zero is to be found at

$$\bar{p} \approx -i\frac{\eta}{\xi}(\lambda - \lambda_0) \qquad [l = 0]$$

[4] Some difference between the cases $l = 0$ and $l > 0$ was certainly to be expected, since in the former case the zero at threshold is simple while in the latter it is double.

That is, as λ moves past λ_0 (with real values) the zero moves linearly down the imaginary axis as shown in Fig. 13.5, part (a). The corresponding motion of the zero on the Riemann surface of E is shown in Fig. 13.5, part (b).

In Figure 13.5, part (a), the zero of the s-wave Jost function is shown starting in $\{\text{Im } p > 0\}$ as a bound state. As the potential weakens the bound state moves down the imaginary axis and becomes progressively less well bound until, when $\lambda = \lambda_0$, it becomes a zero-energy resonance. As we continue to weaken the potential the zero moves on down the imaginary axis. It clearly does not remain close to the real axis and, hence, does not become a resonance of the type discussed in Section 13-a. (Of course, it may eventually move away from the negative imaginary axis into the right half plane, and then become a resonance. This is the origin of the s-wave resonances discussed in Problem **13.2**.)

The zero of the s-wave Jost function on the negative imaginary axis is sometimes referred to as a *virtual state*. Obviously a virtual state is not a proper bound state. Its physical significance is that a virtual state close to the origin implies that $f(0)$ is *small* and, hence, that the scattering length is large. Thus, a virtual state close to threshold (just like an s-wave bound state near threshold) causes a large cross section at low energy. (An example is the famous virtual state in the n–p singlet system; for further discussion of this phenomenon see Problem **13.1**.)

We consider now the case $l > 0$. In this case we know from (12.34) that the zero at threshold is a double zero, and inspection of (12.34) shows that the double power series for $f(\lambda, p)$ has the form

$$f(\lambda, p) = \xi p^2 + \eta(\lambda - \lambda_0) + \cdots \qquad [l > 0]$$

When λ is close to λ_0 this has *two* zeros at

$$\bar{p} \approx \pm i \left(\frac{\eta}{\xi}\right)^{\frac{1}{2}} (\lambda - \lambda_0)^{\frac{1}{2}} \qquad [l > 0] \qquad\qquad (13.5)$$

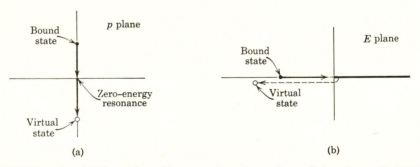

FIGURE 13.5. (a) Transition of an s-wave bound state into a "virtual state." (b) On the Riemann surface of E the bound state is on the physical sheet, the virtual state on the second sheet.

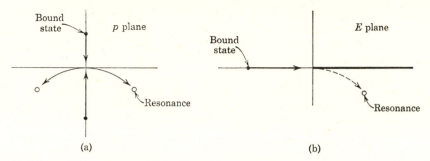

FIGURE 13.6.   (a) Transition of a bound state with $l > 0$ into a resonance. (b) Corresponding motion in the E plane (showing only the physically relevant zeros).

As λ moves through λ_0, these two zeros coalesce at the origin and part company again as shown in Fig. 13.6, part (a).

When the potential is slightly more attractive than at λ_0 the two zeros lie on the imaginary axis, one above the origin representing a bound state, and the other below.   As $\lambda \to \lambda_0$, the zeros move inwards until, when $\lambda = \lambda_0$, both are at the origin and there is a zero-energy bound state. As λ moves past λ_0 the zeros leave the origin tangentially to the real axis. [The change of direction by 90 deg comes from the square root in (13.5).] They cannot move into the upper half plane or onto the real axis, and they therefore move into the lower half plane as shown.[5] In particular, the zero on the right becomes a resonance and continues as such until it eventually moves too far away from the axis to be observable.

A closer examination of the power series (12.31) shows that the imaginary part of the resonance position comes from the term $i\gamma_l p^{2l+1}$ and, hence, that it grows as

$$\text{Im } \bar{p} \text{ (resonance)} \propto (\lambda - \lambda_0)^l \qquad [l > 0]$$

Thus, the higher the angular momentum, the more closely the path of the zero osculates the real axis and the sharper the corresponding resonance.

An important property of these low-energy resonances is that their background phase δ_{bg} is always small and, hence, that the corresponding cross section shows a pure Breit–Wigner peak of the type shown in Fig. 13.3, part (a). To understand this, recall that the background phase is given by

$$\delta_{bg} = -\arg\left(\frac{df}{dE}\right)_E$$

---

[b] From the relation $f'(p) = f'(-p^*)^*$ it follows that if f is zero at p, it is also zero at $-p^*$. For this reason the two zeros move out symmetrically from the imaginary axis, as shown.

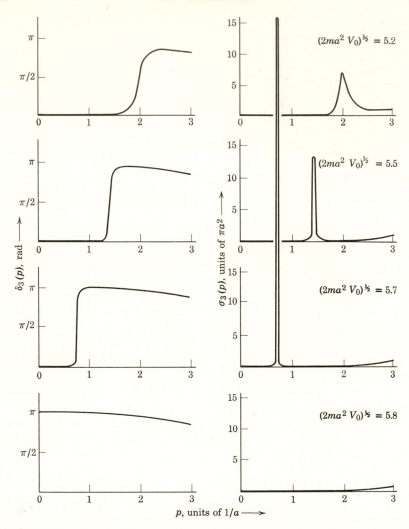

FIGURE 13.7. The $l = 3$ phase shifts and cross sections for four successively deeper square wells. The top three wells are too shallow to support an $l = 3$ bound state—the fourth just can.

(It is simpler to work in terms of E rather than p.) Now, while the zero is a bound state on the negative real axis $(df/dE)_E$ has to be real (since f is). Since $(df/dE)_E$ varies continuously (and is nonzero when the zero is at threshold) it follows that it must be predominantly real just after the zero passes through the threshold and becomes a resonance. That is, $\delta_{bg} = -\arg(df/dE)_E$ is small as long as the resonance remains close to threshold.

We can now understand completely the $l = 3$ resonance of the square well of Fig. 11.3. This well was chosen just too shallow to bind an $l = 3$ bound state, which means there has to be a resonance zero of f close to threshold. Since it is close to threshold, its background phase is small and the resonance is manifested as a simple Breit–Wigner peak in σ_3. If we were to deepen the well, the resonance would move still closer to threshold. As it moves in, it would get closer to the real axis and hence become sharper; at the same time, since the height at resonance is proportional to $1/p_R^2$, it would get progressively higher. Finally, when the well becomes deep enough the resonance should move into $\{\operatorname{Im} p > 0\}$ and become a bound state. This behavior is well borne out in Fig. 13.7, which shows the phase shifts and cross sections for four successively deeper wells. The first three (all a little deeper than that of Fig. 11.3) are still insufficient to support a bound state, the fourth is just sufficient. The complete disappearance of the resonance in the fourth pair of diagrams is indeed a striking phenomenon. Notice how, in accordance with Levinson's theorem, the zero-energy phase shift jumps by π at the moment the resonance disappears.

In conclusion, we remark that for certain systems it is possible that all resonances are "would-be bound states" of the type just discussed. In this case bound states and resonances could be regarded as essentially the same thing; both would be zeros of f (or poles of s), the only distinction being that the bound states happen to lie in $\{\operatorname{Im} p > 0\}$, the resonances in $\{\operatorname{Im} p < 0\}$. A "democratic" arrangement of this type has been much discussed as a natural scheme for the elementary particles. However, appealing as this scheme undoubtedly is, we should perhaps close by repeating that there certainly are systems for which the situation is more complicated, for which some resonances do not correspond to poles of s, and for which some poles of s do not correspond to resonances.

13-c.   Time Delay

The reader will recall that in our discussions of scattering cross sections in Chapters 3 and 10, an essential assumption was always that the amplitude $f(\mathbf{p}' \leftarrow \mathbf{p})$ is a slowly varying function of its arguments, as compared to the incident momentum-space wave function $\phi(\mathbf{p})$. In Sections 13-a and 13-b we have seen that the amplitude can be a very rapidly varying function of energy in the neighborhood of a resonance, and it is therefore appropriate that we reexamine our discussion of the collision process to see what can happen near a resonance.

We saw in Section 10-d that the wave function $\psi(\mathbf{x}, t)$ describing a collision can be written as

$$\psi(\mathbf{x}, t) = \psi_{\text{in}}(\mathbf{x}, t) + \psi_{\text{sc}}(\mathbf{x}, t)$$

where $\psi_{\text{in}}(\mathbf{x}, t)$ is the freely evolving unscattered wave packet with momentum-space wave function $\phi(\mathbf{p})$, and

$$\psi_{\text{sc}}(\mathbf{x}, t) \xrightarrow[r \to \infty]{} (2\pi)^{-3/2} \frac{1}{r} \int d^3 p \, \phi(\mathbf{p}) f(p\hat{\mathbf{x}} \leftarrow \mathbf{p}) e^{i(pr - Et)} \qquad (13.6)$$

In Section 10-d we assumed that $f(p\hat{\mathbf{x}} \leftarrow \mathbf{p})$ was sufficiently smooth that it could be taken outside of the integral to give

$$\psi_{\text{sc}}(\mathbf{x}, t) \xrightarrow[r \to \infty]{} \frac{1}{r} f(p_0 \hat{\mathbf{x}} \leftarrow \mathbf{p}_0) \psi_{\text{in}}(r\hat{\mathbf{3}}, t) \qquad (13.7)$$

that is, the scattered wave at any point is proportional to the unscattered wave on the axis at the same distance from 0, times the appropriate amplitude—*provided f* is smoothly varying. Our task now is to see how this result is modified if $f(\mathbf{p}' \leftarrow \mathbf{p})$ is resonating.

It turns out that the behavior of the scattered wave depends critically on the width Γ of the resonance; more precisely, it depends on the relative sizes of the resonance width Γ and the energy spread ΔE of the incident packet. To some extent this dependence is to be expected. For example, if the resonance width is much smaller than the uncertainty in the incident energy ($\Gamma \ll \Delta E$) it is clear that the experimental resolution of energy will be too poor to see the rapid variations of cross section discussed in Section 13-a; at the opposite extreme, if $\Delta E \ll \Gamma$ then the resolution certainly can be good enough to observe the bumps and dips precisely.

In principle, the width ΔE of the incident packet can be assigned at will by the experimenter; in practice, however, one has quite little control over ΔE. This means that if a resonance is very narrow, the wave packets that are actually available will be very broad compared to the resonance ($\Delta E \gg \Gamma$); while if the resonance is wide the available packets may satisfy $\Delta E \ll \Gamma$. In this section we consider a collision for which $\Delta E \ll \Gamma$, and in 13-d we examine the opposite extreme where $\Delta E \gg \Gamma$.

We suppose, accordingly, that there is a resonance of angular momentum l, energy E_R, and width Γ; and that we have available wave packets whose spread in energy is much less than Γ,

$$\Delta E \ll \Gamma$$

For simplicity we suppose for the moment that all other phase shifts are negligible and, hence, that near E_R the full amplitude has the form

$$f(\mathbf{p}' \leftarrow \mathbf{p}) = (2l + 1)f_l(p)P_l(\hat{\mathbf{p}}' \cdot \hat{\mathbf{p}}) = (2l + 1)\frac{e^{i\delta_l} \sin \delta_l}{p} P_l(\hat{\mathbf{p}}' \cdot \hat{\mathbf{p}}) \quad (13.8)$$

where $\delta_l(p)$ has the resonant behavior discussed in the last two Sections. Since the wave packet $\phi(\mathbf{p})$ is sharply peaked compared to Γ we can treat the

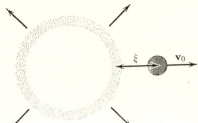

FIGURE 13.8.   The scattered wave lags a distance ξ behind the unscattered wave. (Note that ξ is actually less than the original packet size and the two packets should overlap. See (13.12).)

scattered wave (13.6) almost exactly as before. However, before we take the amplitude outside of the integral we approximate its rapidly changing phase by writing

$$\exp[i\delta_l(p)] \approx \exp\{i[\delta_l(p_0) + \delta_l'(p_0)(p - p_0)]\}$$
$$= \text{constant} \times \exp\{i[\delta_l'(p_0)p]\}$$

If we now retain this explicit dependence of the phase on p but otherwise proceed as before, we find from (13.6) that

$$\psi_{sc}(\mathbf{x}, t) \xrightarrow[r\to\infty]{} \text{constant} \frac{P_l(\cos\theta)}{r} \int d^3p\, \phi(\mathbf{p}) e^{i[p(r+\delta_l')-Et]}$$

$$\approx \text{constant} \frac{P_l(\cos\theta)}{r} \psi_{in}[(r + \delta_l')\hat{\mathbf{3}}, t]$$

$$= \frac{1}{r} f(p_0\hat{\mathbf{x}} \leftarrow \mathbf{p_0})\psi_{in}[(r + \delta_l')\hat{\mathbf{3}}, t] \tag{13.9}$$

(apart from an unimportant, constant phase factor). The important difference between this result and that of (13.7) is that the spherically spreading wave ψ_{sc} is proportional to the unscattered wave, not at the point $r\hat{\mathbf{3}}$, but at $(r + \delta_l')\hat{\mathbf{3}}$. That is, the scattered wave moves outwards always lagging a distance

$$\xi = \delta_l'(p_0) \equiv \frac{d\delta_l}{dp} = v_0 \frac{d\delta_l}{dE}$$

behind the unscattered packet, as shown in Fig. 13.8.

This result is usually expressed by saying that the scattered wave is *delayed* by a time $\tau = \xi/v_0$ or

$$\boxed{\tau = \frac{1}{v_0} \frac{d\delta_l}{dp} = \frac{d\delta_l}{dE}} \tag{13.10}$$

In this form it makes clear that the derivative of the phase shift cannot take on arbitrarily negative values: If $\delta_l' < 0$ then the scattered wave appears *ahead* of the unscattered wave. Now, it is clear that under no circumstances should the scattered wave appear before the incident wave has hit the target. Thus, the maximum possible time advance is of the order of a/v_0 where a denotes the target size. This means that τ must be greater than $-a/v_0$ or,

$$\boxed{\frac{d\delta_l}{dp} \gtrsim -a} \tag{13.11}$$

That is, the phase shift can never *decrease* with a slope steeper than $-a$. There is, of course, no such restriction on the *increase* of $\delta_l(p)$, since the delay can have any positive value. Indeed, it is generally agreed that a real resonance corresponds to the temporary capture of the projectile in a metastable state, which then decays and re-emits the projectile after some (positive) time τ. Thus at a real resonance $d\delta_l/dp$ is always positive and can be arbitrarily large.

It is important to note that, *apart from the time delay τ*, the scattered wave (13.9) of the present discussion (for which $\Delta E \ll \Gamma$) behaves just like that of the nonresonant case (13.7). In particular, to evaluate the cross section we integrate $|\psi_{sc}|^2$ from $r = 0$ to ∞ and clearly obtain exactly the usual result $d\sigma/d\Omega = |f|^2$. Thus, in this case the measured cross section will indeed display the bumps and dips discussed in Section 13-a.

It should also be noted that the picture of Fig. 13.8 is actually somewhat misleading, since in the kind of experiment discussed, the distance ξ between the scattered and unscattered waves is always much less than the size of the original packet, and the two waves therefore overlap almost completely. The reason is easy to see: The distance by which the scattered wave lags is

$$\xi = v_0 \frac{d\delta_l}{dE} \sim \frac{v_0}{\Gamma}$$

since $d\delta_l/dE \sim 1/\Gamma$, while the original packet size satisfies

$$\text{packet size} \gtrsim \frac{1}{\Delta p} = \frac{v_0}{\Delta E}$$

But in the experiment discussed $\Delta E \ll \Gamma$ and, hence,

$$\text{packet size} \gg \xi \tag{13.12}$$

It follows that this type of experiment (with $\Delta E \ll \Gamma$), while suitable for measuring the cross section as a function of energy across the resonance bump, is *not* a practical method for measurement of the time delay. In any case, if the delay associated with a given resonance is appreciable, then the corresponding width Γ is small and it will be impossible, in practice, to produce an incident wave packet for which ΔE is narrow compared to Γ. For this reason we now go on to consider a resonance whose width is smaller than that of the available incident wave packets.

13-d. Decay of a Resonant State

We are led by the preceding discussion to consider next a resonance which is so narrow that the available packets are much wider than Γ,

$$\Delta E \gg \Gamma$$

For the moment we suppose that the resonance is a pure Breit–Wigner resonance ($\delta_{bg} = 0$) in which case:

$$s_l = \frac{E - E_R - i\Gamma/2}{E - E_R + i\Gamma/2}$$

or

$$f_l = -\frac{1}{2p} \cdot \frac{\Gamma}{E - E_R + i\Gamma/2}$$

Thus, if we assume (as before) that all other phase shifts are negligible and substitute this amplitude into (13.8) and (13.6) we find

$$\psi_{\text{sc}}(\mathbf{x}, t) \xrightarrow[r \to \infty]{} \text{constant} \, \frac{\Gamma}{r} \int d^3p \, \frac{\phi(\mathbf{p}) P_l(\hat{\mathbf{x}} \cdot \hat{\mathbf{p}})}{p(E - E_R + i\Gamma/2)} \, e^{i(pr - Et)}$$

We can now perform the angular integral, which involves only $\phi(\mathbf{p})$ and $P_l(\hat{\mathbf{x}} \cdot \hat{\mathbf{p}})$ and just projects out the angular momentum l component $\phi_l(E)$ of $\phi(\mathbf{p})$.[6] Thus,

$$\psi_{\text{sc}}(\mathbf{x}, t) \to \text{constant} \, \frac{\Gamma Y_l^0(\hat{\mathbf{x}})}{r} \int_0^\infty p^{1/2} \, dp \, \frac{\phi_l(E) e^{i(pr - Et)}}{E - E_R + i\Gamma/2}$$

We next approximate the exponent by writing

$$p \approx p_R + \frac{dp}{dE}(E - E_R) = p_R + \frac{E - E_R}{v_R}$$

and hence

$$\exp[i(pr - Et)] \approx \exp[i(p_R r - E_R t)] \exp\left[-i(E - E_R)\left(t - \frac{r}{v_R}\right)\right]$$

---

[6] Since $\phi(\mathbf{p})$ is well peaked about $\mathbf{p}_0$, which points along the z axis, the expansion of $\phi(\mathbf{p})$ in spherical harmonics contains only $m = 0$ terms, $\phi(\mathbf{p}) = (mp)^{-1/2} \sum \phi_l(E) Y_l^0(\hat{\mathbf{p}})$. Thus, when $\phi(\mathbf{p}) P_l(\hat{\mathbf{x}} \cdot \hat{\mathbf{p}})$ is integrated over all angles the result is proportional to $p^{-1/2} \phi_l(E) Y_l^0(\hat{\mathbf{x}})$.

At the same time we use the assumption that $\Delta E \gg \Gamma$ to take the wave function $\phi_l(E)$ outside of the integral. [This is not permissible when $t - r/v_R$ is small, since the remaining integral is then too slowly convergent; thus, our approximation is unreliable in the immediate neighborhood of $t - r/v_R = 0$.] This gives

$$\psi_{sc}(\mathbf{x}, t) = \text{constant } \Gamma Y_l^0(\hat{\mathbf{x}}) \phi_l(E_R) \frac{e^{i(p_R r - E_R t)}}{p_R^{\frac{1}{2}} r}$$

$$\times \int_0^\infty dE \, \frac{e^{-i(E-E_R)(t-r/v_R)}}{E - E_R + i\Gamma/2} \qquad (13.13)$$

for r large. Finally, the integral can be extended to $-\infty$ without significantly affecting its value, and can then be evaluated by contour integration (Problem **13.3**) to give

$$|\psi_{sc}(\mathbf{x}, t)|^2 = 2\pi m \Gamma^2 \, |Y_l^0(\hat{\mathbf{x}})|^2 \, |\phi_l(E_R)|^2 \frac{e^{-\Gamma(t-r/v_R)}}{p_R r^2} \, \theta\left(t - \frac{r}{v_R}\right) \qquad (13.14)$$

where

$$\theta(x) = \begin{cases} 0 & \text{for } x < 0 \\ 1 & \text{for } x > 0 \end{cases}$$

The result (13.14) lets us see what is observed by a counter at any fixed position (see Fig. 13.9). For times before r/v_R, $(t < r/v_R)$ the scattered wave is zero, reflecting the fact that the particle, whose velocity is about v_R, cannot yet have reached the point of observation. At the time $t = r/v_R$ the wave front arrives. [Note that since our approximation is bad when $t \approx r/v_R$ we cannot predict the exact shape of the wave front. In fact, this depends on the details of $\phi(\mathbf{p})$.] At times after r/v_R, $(t > r/v_R)$ one sees an outgoing wave whose angular dependence is $|Y_l^0(\hat{\mathbf{x}})|^2$ and whose intensity drops exponentially with time like $e^{-\Gamma t}$. This behavior is, of course, precisely the characteristic behavior of a metastable state of energy E_R and angular momentum l, which is formed at $t = 0$ and then decays exponentially with mean life $1/\Gamma$. Our result, therefore, justifies the claim that an l-wave

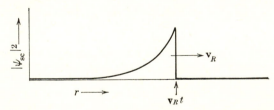

FIGURE 13.9. Profile of the scattered intensity (13.14) as a function of r in any fixed direction.

resonance of width Γ *is* a metastable state of angular momentum l and mean life $1/\Gamma$.

It is interesting to note that if we use (13.14) to compute the differential cross section then, because of the factor $|\phi_l(E_R)|^2$, the result obviously depends explicitly on the shape of the incident wave packet $\phi(\mathbf{p})$ (see Problem **13.4**). In particular, we see that, as one should expect, in the neighborhood of a very sharp resonance the differential cross section is *not* given by the famous formula $d\sigma/d\Omega = |f|^2$.[7]

We can summarize the findings of the last two sections as follows: If we wish to measure the resonant bumps and dips in the cross section as a function of energy, then it is clear that we must use packets satisfying

$$\Delta E \ll \Gamma \qquad (13.15)$$

to give the required energy resolution. In Section 13-c we saw that when this condition is satisfied, the cross section is indeed given by the usual formula and can be measured in the usual way to show the bumps and dips. If, on the other hand, we wish to measure the exponential decay of the associated metastable state, then it is obviously essential that the time at which the state is formed be well-defined compared to the time of decay; since the time of formation is uncertain by about $\Delta x/v$ (Δx being the original packet size) this requires that

$$\frac{\Delta x}{v} \ll \frac{1}{\Gamma}$$

or

$$\Gamma \ll \frac{v}{\Delta x} \lesssim v\Delta p = \Delta E$$

and, hence,

$$\Gamma \ll \Delta E \qquad (13.16)$$

As we have just seen, we can indeed measure the exponential decay when this condition is satisfied. The conditions (13.15) and (13.16) show clearly that

---

[7] It should also be remarked that because of the factor $\Gamma\,|\phi_l(E)|^2$ the cross section is extremely small. This is what one would expect—a narrow resonance decays slowly and so would be expected to be formed with low probability. Thus, very narrow metastable states are not usually formed directly as described above. For example, the unstable nucleus ^{210}Po is a narrow resonance ($\Gamma \sim 10^{-18}$ eV!) at 5.4 MeV in the α-^{206}Pb system. However it would never be formed in practice by 5.4 MeV collisions of alpha particles on ^{206}Pb. Instead it could be formed, for example, by the inelastic process

$$d + {}^{209}\mathrm{Bi} \rightarrow n + {}^{210}\mathrm{Po}$$

This does not affect our conclusions concerning its exponential decay.

the two types of measurement envisaged are complimentary (in the sense of Bohr) and mutually exclusive in any single experiment.

To conclude, we remark that we have so far restricted our attention to the simplest situations, and we should therefore examine how our discussion can be generalized. In both of the last two sections we have assumed that all of the nonresonant phase shifts are negligible. If this is not the case, then we have only to write the amplitude, and, hence, the scattered wave ψ_{sc} as the sum of two terms, one corresponding to the resonant partial wave, the second to all other partial waves. The analysis of the resonant part then goes through as before.

In this section an additional important assumption was that the resonance is pure Breit–Wigner ($\delta_{bg} = 0$). If this is not the case then we can write the resonant amplitude as

$$f_l = \frac{s_l - 1}{2ip} = \frac{\exp[2i(\delta_{bg} + \delta_{\mathrm{res}})] - 1}{2ip}$$

$$= \exp(2i\delta_{bg}) \frac{\exp(2i\delta_{\mathrm{res}}) - 1}{2ip} + \frac{\exp(2i\delta_{bg}) - 1}{2ip}$$

$$\equiv f_l^{\mathrm{res}} + f_l^{bg}$$

The term f_l^{res} is, apart from a fixed phase factor, exactly the same as what was discussed above; while the term f_l^{bg} is a constant and can therefore be grouped together with the nonresonant partial waves. Thus, in general, the scattered wave in the neighborhood of a narrow resonance consists of two pieces: The first, the "prompt" part, comes from the nonresonant partial waves plus f_l^{bg}, and emerges undelayed and undistorted in profile. The other comes from f_l^{res} and has the long exponential tail corresponding to the decay of the resonant metastable state.

Of course, there are several more complicated possibilities. As mentioned earlier there are no theoretical arguments against a multiple pole of $s_l(p)$ in the lower half plane. Such a multiple pole could be analyzed by exactly the same methods (see Problem **13.5**) and would lead to a different energy dependence for the cross section (e.g., a double pole can cause a double peak) and a slightly different decay law.

Finally, it is perfectly possible that $s_l(p)$ has several poles close to the real axis and close to one another. In this case no single pole dominates at any one energy and they must all be taken into account (perhaps statistically if there are many of them). Just as a single pole corresponds to a single metastable energy level, so we can regard the several poles as corresponding to several levels, all close together; and the "one-level" Breit–Wigner formula (13.4) is then replaced by the appropriate "many level" formula.

PROBLEMS

13.1. (a) Suppose that the s-wave Jost function has a zero at $p = i\alpha$ (α real) close to the origin. Approximating $\mathcal{f}(p)$ as $\beta(p - i\alpha)$ find the behavior of the s-wave amplitude and cross section close to threshold. Show that the cross section is the same whether the zero is on the positive or negative imaginary axis. (That is, as far as scattering is concerned there is no way to distinguish a bound state from a virtual state.) Show that the scattering length a is just $1/\alpha$.

(b) The approximation made above is equivalent to retaining only the first term in the effective range expansion (12.29). Show that if one keeps the first *two* terms then knowledge of the zero of $\mathcal{f}$ at $p = i\alpha$ gives a relation between the scattering length and the effective range $(1/a) + (r_0/2)\alpha^2 + \alpha = 0$. *Hint:* if $\mathcal{f}(i\alpha)$ is zero, then $s(i\alpha)$ has a pole and $\cot \delta(i\alpha) = i$.

This problem is particularly relevant to the n–p system. The triplet n–p has a bound state—the deuteron—close to threshold and knowledge of the deuteron binding energy gives a relation between the triplet scattering length and effective range. The singlet n–p has a virtual state close to threshold, whose position can therefore be found by measuring the singlet scattering length and effective range. It turns out that the singlet virtual state is much closer to threshold than the triplet bound state and so singlet scattering dominates at low energy.

13.2. Show that the s-wave phase shift for a square well is

$$\delta_0(p) = -pa + \arctan\left(\frac{p}{k} \tan ka\right)$$

where $k = (p^2 + 2mV_0)^{\frac{1}{2}}$ and V_0 and a are the well depth and radius. Hence, show that for a very deep well there are regularly spaced, low-energy ($E \ll V_0$) s-wave resonances, with energies E_R given by $k_R = (2n + 1)\pi/2a$. Show that the resonances have width $\Gamma = p_R/ma$ and check that for a sufficiently deep well this width is much less than the spacing between the resonances. Notice that the resonances have a background phase shift $\delta_{bg} = -p_R a$.

13.3. In order to arrive at the exponential decay law (13.14), it was necessary to evaluate an integral of the form

$$\int_{-\infty}^{\infty} dz \, \frac{e^{-iz\tau}}{z + i\Gamma/2}$$

Use contour integration to show that this integral is 0 for $\tau < 0$ but $-2\pi i e^{-\Gamma \tau/2}$ for $\tau > 0$.

13.4. Starting from (13.14) for $|\psi_{sc}|^2$ and using our standard methods to get a cross section show that at a narrow resonance

$$\frac{d\sigma}{d\Omega} = \frac{2\pi^2(2l+1)}{p_R^2}\,\Gamma w(E_R)\,|Y_l^0(\hat{\mathbf{p}})|^2$$

where $w(E) = (p/m)\int d\Omega\,|\phi(\mathbf{p})|^2$ is the probability per unit energy that the incident packet have energy E. Notice that $\Gamma w(E_R)$ can be interpreted as the probability that the incident particle has energy within the interval Γ about E_R.

13.5. (a) Along the lines of the discussion in Section 13-a, describe the behavior of the phase shift and cross section in the neighborhood of a *double zero* of the Jost function in $\{\mathrm{Im}\,p < 0\}$ close to the real axis. Do the same for two single zeros close to the axis and to one another. Treat the case $\delta_{bg} = 0$ first.

(b) Consider a sharp resonance, which is given by a pole of order n close to the real axis. Show that its decay is given by the exponential $e^{-\Gamma t}$ *times a polynomial* of degree $2(n-1)$ in t. [All that is needed is to evaluate the integral (13.13) for the case of a pole of order n.]

14 Additional Topics in Single-Channel Scattering

14-a   **Coulomb Scattering**

14-b   **Coulomb Plus Short-Range Potentials**

14-c   **The Distorted-Wave Born Approximation**

14-d   **Variational Methods**

14-e   **The K Matrix**

In this chapter we introduce an assortment of topics not previously discussed. In Sections 14-a and 14-b we discuss the irritating special case of Coulomb scattering, which was excluded by the conditions assumed thus far. In Section 14-c we describe the distorted-wave Born approximation, which is often useful as a substitute for the Born approximation when the potential is too strong for the usual Born approximation to be reliable. In Section 14-d we discuss a completely different approach to computation of the amplitude—the so-called variational approach. Finally, in Section 14-e, we give a brief introduction to the K matrix, giving its definition and various applications.

14-a. Coulomb Scattering

This section discusses the scattering of two charged particles interacting via the Coulomb potential

$$V(r) - \frac{e_1 e_2}{r}$$

As $r \to \infty$, this potential falls off too slowly to satisfy the conditions used to prove any of our basic results—the asymptotic condition, the asymptotic form of the stationary wave functions, etc. In fact, the scattering by the Coulomb potential does *not* satisfy the results. The reason is that the potential falls off so slowly that it continues to influence the particles even as they move far apart. Thus (for instance), a scattering orbit never behaves freely even as $t \to \pm \infty$, and the asymptotic condition

$$U(t)\, |\psi\rangle \xrightarrow[t \to \pm \infty]{} U^0(t)\, |\psi_{\text{out/in}}\rangle \qquad (14.1)$$

does not hold.

There are two ways out of this difficulty. In the first, one recognizes that the asymptotic condition (14.1) is a much stronger condition than is used experimentally. In practice, one never measures the complete state vector $U(t)\,|\psi\rangle$ to check that it is evolving freely as $t \to \pm \infty$. In reality the most that is done is to measure a small set of observables such as the momenta, the kinetic energy, and the spin (all of which commute with H^0) and to check that these approach constants. It is certainly true that if the asymptotic condition (14.1) holds, then these observables will become constant as $t \to \pm \infty$; but it is by no means the case that if these observables are found to be asymptotically constant, then the asymptotic condition (14.1) necessarily holds. This suggests that one could construct a scattering theory based on a weaker form of the asymptotic condition than that of (14.1), and that this more general scattering theory would encompass the conventional short-range theory as developed in this book, *and* be able to handle long-range potentials like the Coulomb potential. This point of view was successfully pioneered by Dollard (1964) and has been further developed by Amrein, et al. (1970).

The second possibility, and the one we shall follow, is to argue that pure Coulomb potentials never really occur in nature. For example, in the famous Rutherford experiment the Coulomb field of the gold nuclei was completely shielded at a range of a few Angstroms by the atomic electrons. Even in interplanetary space the field of a single charge would be screened beyond about a meter by the free charges present. Since the Coulomb potentials in nature are invariably *screened Coulomb potentials*, and since such potentials are "well-behaved," a realistic and reasonable way to avoid the above-mentioned difficulties should be always to impose some appropriate cutoff.

To better understand some of the difficulties of the Coulomb potential, and to see what happens when it is cut off, we first discuss the radial equation for a single partial wave. The reader will recall that for short range potentials (in fact for any potential better than $1/r$ at infinity) the radial wave functions have the free asymptotic form:

$$\psi_{l,p}(r) \to \text{constant} \times \sin(pr - \tfrac{1}{2}l\pi + \delta_l) \qquad (14.2)$$

However, when one considers the Coulomb potential, any solution of the radial equation,

$$\left[\frac{d^2}{dr^2} - \frac{l(l+1)}{r^2} - \frac{2me_1e_2}{r} + p^2\right]y(r) = 0$$

has the asymptotic form,[1]

$$y(r) \rightarrow \text{constant} \times \sin\left[pr - \frac{me_1e_2}{p}\ln r + \text{constant}\right]$$

That is, however large r becomes, the solutions continue to pick up phase logarithmically. In particular, the solution that is zero at $r = 0$ is traditionally denoted by $F_l(pr)$ and has the asymptotic form:

$$F_l(pr) \xrightarrow[r \to \infty]{} \sin(pr - \gamma \ln 2pr - \tfrac{1}{2}l\pi + \sigma_l) \qquad (14.3)$$

where we have introduced the dimensionless quantity

$$\gamma \equiv \frac{me_1e_2}{p}$$

which serves as a strength parameter in the Coulomb problem. The term σ_l is the so-called *Coulomb phase shift*, which can be shown to have the value[2]:

$$\sigma_l = \arg \Gamma(l + 1 + i\gamma) \qquad (14.4)$$

However, it should be emphasized that, as is clear from (14.3), the Coulomb wave function does not have a phase shift in the ordinary sense. If the Coulomb potential is switched off ($\gamma = 0$), then $\sigma_l = 0$ and $F_l(pr)$ reduces to the Ricatti–Bessel function $\hat{\jmath}_l(pr)$.

Suppose that we now cut off the Coulomb potential sharply at $r = \rho$; that is, we replace it by the potential

$$V_\rho(r) = \begin{cases} \dfrac{e_1e_2}{r} & r \leqslant \rho \\ 0 & r > \rho \end{cases} \qquad [\rho \text{ large}] \qquad (14.5)$$

[1] The reason for the ln r term is easily understood by substituting into the radial equation. It will be seen that the ln r term exactly cancels the $1/r$ Coulomb term in the equation. Alternatively, the result emerges clearly from the variable-phase method of Section 11-g (see Problem **14.1**.)

[2] For details of Coulomb wave functions, amplitudes, and phase shifts we refer the reader to Messiah (1961) Chapter XI and Appendix B.1.

The potential $V_\rho(r)$ is short-range and therefore has a well defined phase shift, which we can easily calculate for any given l. For $r \leqslant \rho$ the radial wave function is proportional to the Coulomb function F_l, and for r close to ρ (provided ρ is large) F_l can be replaced by its asymptotic form (14.3). Matching the wave function and its derivative across $r = \rho$ we find that (to order $1/\rho$) the phase shift for the cutoff potential is:

$$\delta_l(\rho) = \sigma_l - \gamma \ln 2p\rho \tag{14.6}$$

We see that the phase shift $\delta_l(\rho)$ for the cutoff Coulomb potential depends critically on the value of the cutoff radius ρ and that it has no limit as $\rho \to \infty$. In fact, the phase shift can be given any value whatever (modulo 2π) by the choice of ρ. This result seems inauspicious for our proposed program of using cutoff Coulomb potentials. However, it could still be that, even though the amplitude is critically dependent on ρ, the quantity that is actually measured (the cross section) is not. We shall argue that this is so provided ρ is taken suitably large, and provided the measurements avoid the immediate vicinity of the forward direction (both of which conditions are normally realized in practice).

To compute the cross section we must compute the full amplitude $f(\mathbf{p}' \leftarrow \mathbf{p})$, and for the purposes of orientation we consider first the Born approximation. Recall that for any short-range spherical potential this approximation gives

$$f(\mathbf{p}' \leftarrow \mathbf{p}) = -(2\pi)^2 m \langle \mathbf{p}' | \, V \, | \mathbf{p} \rangle = -\frac{2m}{q} \int_0^\infty dr \; r \sin qr \; V(r)$$

where $\mathbf{q} = \mathbf{p}' - \mathbf{p}$ is the momentum transfer. If this result is applied to the Coulomb potential $e_1 e_2 / r$, we obtain the divergent integral

$$\int_0^\infty dr \sin qr$$

This is hardly surprising; from our discussion of the radial wave function it should be fairly clear that for the Coulomb potential an amplitude (in the ordinary sense) does not exist. Hence the failure of the Born approximation to produce an answer.

We next apply the Born approximation to a shielded Coulomb potential and we consider two different methods of shielding: an *exponential shielding*,

$$V_\rho(r) = \frac{e_1 e_2}{r} e^{-r/\rho} \qquad [\rho \text{ large}] \tag{14.7}$$

and the sharp cutoff discussed previously. The Born approximation for either potential is easily evaluated and gives (of course) a well defined answer.

For the exponential shielding we find

$$f_\rho(\mathbf{p'} \leftarrow \mathbf{p}) = \frac{-2me_1e_2}{q^2 + \rho^{-2}} \underset{\rho\,\text{large}}{\approx} -\frac{2me_1e_2}{q^2} \qquad \text{[exponential cutoff]} \quad (14.8)$$

It will be seen that this has a well defined limit as $\rho \to \infty$ except when $q = 0$. This means that, except in the neighborhood of the forward direction, the amplitude for the exponentially shielded Coulomb (with ρ large) is indistinguishable from the amplitude $-2me_1e_2/q^2$, which is independent of ρ.

However, if we evaluate the amplitude for the sharply cutoff case we find:

$$f_\rho(\mathbf{p'} \leftarrow \mathbf{p}) = -\frac{2me_1e_2}{q^2}(1 - \cos q\rho) \qquad \text{[sharp cutoff]} \quad (14.9)$$

In this case we get a well defined amplitude, but one which has no limit as $\rho \to \infty$ and, in fact, oscillates indefinitely.

At first sight the contrast between (14.8) and (14.9) seems distressing. Not only does the amplitude for a shielded Coulomb potential depend on *where* the shielding is made, it depends on *how* the shielding is achieved. However, a little reflection will show that this is to be expected, since the pure Coulomb potential has no amplitude at all (in the conventional sense). Further, an examination of the difference between (14.8) and (14.9) shows us the way out of our difficulties. The difference between the two amplitudes is the term in (14.9) $(2me_1e_2/q^2)\cos q\rho$, which is a *rapidly oscillating function of* $\mathbf{p}$ *and* $\mathbf{p'}$ when ρ is large.[3] To see the significance of this oscillatory term we must return to the calculation in Section 3-e of the differential cross section in terms of the amplitude.

The reader will recall that the cross section was defined in terms of a large number of collisions with incident wave packets suitably randomized as to shape and impact parameter. The first step in the calculation was to evaluate the out wave function $\psi_{\text{out}}(\mathbf{p})$ for any given in wave packet $\psi_{\text{in}}(\mathbf{p})$ as in (3.16),

$$\psi_{\text{out}}(\mathbf{p}) = \int d^3p' \langle \mathbf{p}|\, \mathsf{S}\, |\mathbf{p'}\rangle \psi_{\text{in}}(\mathbf{p'})$$

$$= \psi_{\text{in}}(\mathbf{p}) + \frac{i}{2\pi m}\int d^3p'\,\delta(E_p - E_{p'})f(\mathbf{p} \leftarrow \mathbf{p'})\psi_{\text{in}}(\mathbf{p'})$$

$$= \psi_{\text{in}}(\mathbf{p}) + \frac{ip}{2\pi}\int d\Omega_{p'} f(\mathbf{p} \leftarrow \mathbf{p'})\psi_{\text{in}}(\mathbf{p'})$$

[3] The important point in the contrast between (14.8) and (14.9) is *not* that one method of shielding is smooth and the other sharp; but that the potential that is shielded is the long-range Coulomb potential. If we were to apply the two different cutoffs to a short-range potential like the Yukawa, then both of the resulting amplitudes would have limits as $\rho \to \infty$, the two limits would be the same, and this common limit would be the amplitude of the original potential.

Here, the first term is the unscattered incident wave, and to avoid seeing this term we had to restrict measurements to nonforward directions. The integral represents the scattered wave (which we wish to measure) and, in the case of the cutoff Coulomb amplitude (14.9), consists of two terms, the second of which contains the oscillatory factor

$$\cos q\rho = \cos\left(2p\rho \sin \frac{\theta}{2}\right)$$

In typical situations where Coulomb scattering is measured, the term $p\rho$ is very large. (For p–p scattering at a few MeV with screening provided at $\rho \sim 1$ Å by the atomic electrons of the target, $p\rho \sim 10^5$).[4] Since the angular width of a typical incident packet is of the order of a degree or so, this oscillatory term integrates out to zero; that is, the oscillatory term makes no contribution to $\psi_{out}(\mathbf{p})$. Therefore, we can proceed to compute the cross section using just the first term $-2me_1e_2/q^2$ in the amplitude, and arrive at the same answer as for the exponentially cutoff potential. That is, in evaluating cross sections for the ordinary Coulomb scattering experiment, the difference between our two methods of screening has no observable effect.

This example well illustrates how, even though the amplitude for a screened Coulomb potential critically depends on the nature of the screening, the observed cross section does not, being given in either case by the effective amplitude $-2me_1e_2/q^2$. It should be emphasized that we are *not* claiming that there is *no* difference between the potentials arising from different methods of cutoff. Obviously there *is* a difference, which can, in principle, be measured. We are only claiming that in a typical real experiment there is no observable difference, provided the cutoffs are made sufficiently far out.

It should also be emphasized that the amplitudes (14.8) and (14.9) for the screened potentials are both finite in the forward direction ($q = 0$), while the "pure Coulomb limit" $-2me_1e_2/q^2$ is infinite. Thus, in the very close neighborhood of $\theta = 0$ there is always a large difference between the pure Coulomb and its screened relations. This is to be expected since, roughly speaking, those particles scattered near $\theta = 0$ are the particles with large impact parameters, and these are (of course) sensitive to the method of screening. In particular, with the pure Coulomb potential, particles are slightly scattered however large their impact parameter, and one should expect an infinite cross section at $\theta = 0$.

---

[4] There are situations where $p\rho$ is *not* large (e.g., in electron collisions with a neutral atom at a few eV where $1/p \sim \rho \sim 1$ Å), and in such cases our conclusion that the cross section is independent of cutoff would not follow. However, these are situations where one would not *expect* to measure Coulomb scattering and one *would* expect effects of the screening (the atomic electrons in this example) to be observable.

So far we have examined only the Born approximation. However, one can show that the same results hold for the exact amplitude. That is, the exact amplitude depends critically on how and where the Coulomb potential is shielded, but the observed cross sections do not, provided the cutoff radius is large. Here we shall discuss just one particular method of shielding—the sharp cutoff at $r = \rho$.

We first examine the full three-dimensional stationary wave functions. In the case of a short-range potential these are determined as solutions of the time-independent Schrödinger equation with the asymptotic form

$$\langle \mathbf{x} \mid \mathbf{p}+ \rangle \rightarrow (2\pi)^{-\frac{3}{2}} \left[e^{ipz} + f(p\hat{\mathbf{x}} \leftarrow \mathbf{p}) \frac{e^{ipr}}{r} \right]$$

(For convenience we have put $\mathbf{p}$ along the z axis.) For the pure Coulomb potential no such solutions exist. However, the Schrödinger equation can be exactly solved (in terms of hypergeometric functions) and there is, in particular, a solution

$$\langle \mathbf{x} \mid \mathbf{p}+_{\mathrm{c}} \rangle = \alpha e^{ipz} F[-i\gamma \mid 1 \mid ip(r-z)]$$

Here $F(a \mid b \mid z)$ is the confluent hypergeometric function,[5] α is a normalization factor, and $\gamma = me_1e_2/p$. This solution has the asymptotic form as[6] $(r - z) \rightarrow \infty$

$$\langle \mathbf{x} \mid \mathbf{p}+_{\mathrm{c}} \rangle \rightarrow (2\pi)^{-\frac{3}{2}} \Big(\exp\{i[pz + \gamma \ln p(r-z)]\}$$

$$+ f_{\mathrm{C}}(p\hat{\mathbf{x}} \leftarrow \mathbf{p}) \frac{\exp[i(pr - \gamma \ln 2pr)]}{r} \Big) \quad (14.10)$$

where

$$f_{\mathrm{C}}(\mathbf{p}' \leftarrow \mathbf{p}) = - \frac{2me_1e_2}{q^2} \exp \left[2i \left(\sigma_0 - \gamma \ln \sin \frac{\theta}{2} \right) \right] \quad (14.11)$$

is called the *Coulomb amplitude* and σ_0 is the s-wave Coulomb phase shift.

We see that the Coulomb stationary wave functions differ from their short-range counterparts by the inevitable logarithmic phase factors in both the incident and scattered waves. It is clear from this that there is no amplitude (in the conventional sense) for the Coulomb potential. However, the coefficient $f_{\mathrm{C}}(\mathbf{p}' \leftarrow \mathbf{p})$ is known as the Coulomb amplitude and, as we shall see

---

[5] See Messiah (1961) p. 421, ff.
[6] This asymptotic form holds as $(r - z)$, not r, goes to infinity. However, since $(r - z) = r(1 - \cos \theta)$, these two are the same thing except in the forward direction (which we must avoid anyway). Also, we have indicated just the dominant behavior of each term; the first correction to the "incident" wave is $O(1/r)$ and should, strictly speaking, be included in (14.10).

directly, plays a role analogous to that of the conventional short-range amplitude.

If we turn now to the cutoff Coulomb potential then the wave function inside $r = \rho$ can be related to the Coulomb functions, while that outside has the usual form for any short range potential. It can then be shown that the corresponding amplitude has the form

$$f_\rho(\mathbf{p}' \leftarrow \mathbf{p}) = e^{-2i\gamma \ln 2p\rho} f_C(\mathbf{p}' \leftarrow \mathbf{p}) + f_{\text{osc}}(\mathbf{p}' \leftarrow \mathbf{p}) \qquad (14.12)$$

where f_{osc} oscillates rapidly like $\cos(p\rho\theta)$. Precisely the analysis applied to the Born approximation shows that this oscillatory term makes no contribution to the cross section as measured in typical experiments. Since only the first term contributes, the phase factor in front disappears and we arrive at the simple answer

$$\frac{d\sigma}{d\Omega} = |f_C(\mathbf{p}' \leftarrow \mathbf{p})|^2 = \left(\frac{2me_1e_2}{q^2}\right)^2 = \frac{m^2 e_1^2 e_2^2}{4p^4 \sin^4\theta/2} \qquad (14.13)$$

This result is the celebrated Rutherford formula. It is a remarkable fact that the exact quantum mechanical treatment yields the same answer as both the classical calculation and the quantum-mechanical Born approximation.[7] We have derived (14.13) here as the exact cross section for a cutoff Coulomb potential, provided the cutoff is far out and provided measurements avoid the immediate neighborhood of the forward direction.

A second remarkable feature of the result (14.13) is that it is the answer one would get by using the amplitude f_C in the asymptotic form (14.10) and simply ignoring the unwanted logarithmic phases. This is a common feature in scattering involving Coulomb potentials—that one can simply ignore the logarithmic phases and use the Coulomb wave functions as if they were the wave functions of an ordinary short-range potential. We shall see further examples of this later.

14-b.   Coulomb Plus Short-Range Potentials

An enormous number of important processes—scattering of charged pions off protons, of protons off nuclei, of electrons off ions, etc.—involve the scattering by Coulomb forces that are modified at small distances by some short-range force. Since the resulting potential has the same long-range tail

---

[7] The exact Coulomb amplitude (14.11) differs from the Born approximation $-2me_1e_2/q^2$ only by a phase factor. Since this factor is 1 when the strength parameter γ is 0, the Born approximation is, as one would expect, the first term in an expansion of the exact amplitude in powers of the potential.

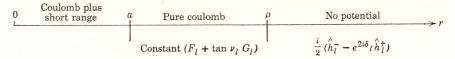

FIGURE 14.1.   Three regions defined by the Coulomb plus short-range potential cut-off at ρ. The wave functions in the outer regions are indicated.

as the pure Coulomb, it presents all of the problems discussed above. Fortunately the problems are susceptible to the same solutions.

We consider a potential

$$V(r) = \frac{e_1 e_2}{r} + V_{\rm sr}(r)$$

where, for simplicity, we suppose that the short-range potential $V_{\rm sr}(r)$ is identically zero for r greater than some a. As before, our procedure will be to cut off the potential at some large ρ. We shall then compare the resulting wave function with that for the case $V_{\rm sr} = 0$. To this end we shall use a partial-wave analysis, beginning with the case $V_{\rm sr} = 0$.

The solution of the pure Coulomb radial equation that vanishes at $r = 0$ is the function $F_l(pr)$ with the asymptotic form,

$$F_l(pr) \xrightarrow[r \to \infty]{} \sin(pr - \gamma \ln 2pr - \tfrac{1}{2}l\pi + \sigma_l)$$

where σ_l is the Coulomb phase shift. As a second solution of the pure Coulomb radial equation we choose the function traditionally denoted $G_l(pr)$ with the asymptotic form,

$$G_l(pr) \xrightarrow[r \to \infty]{} \cos(pr - \gamma \ln 2pr - \tfrac{1}{2}l\pi + \sigma_l)$$

This second solution is divergent like $1/r^l$ as $r \to 0$ and plays no role in the pure Coulomb problem.

If we now add to the Coulomb potential the term $V_{\rm sr}(r)$, then the resulting potential (with the cutoff at ρ) defines three distinct regions as shown in Fig. 14.1. In the interval $0 < r < a$ the potential is Coulomb plus short range, in $a < r < \rho$ it is pure Coulomb, and beyond the cutoff at ρ there is no potential at all. In the pure Coulomb region $a < r < \rho$ the radial wave function can be written as

$$\psi_{l,p}(r) = \text{constant } [F_l(pr) + \tan \nu_l(p)G_l(pr)] \qquad (14.14)$$

The angle ν_l measures the effect of the additional short-range potential (if $V_{\rm sr} = 0$ then $\nu_l = 0$). Just as in the ordinary short-range case, where the phase shifts δ_l are negligible for l greater than some $l_0 \approx pa$, so in this case $\nu_l \approx 0$, for $l > l_0 \approx pa$. Thus, the modification of the wave function due to the short-range force is confined to some finite number of partial waves, $0 \leqslant l \leqslant l_0$. We now look at these waves more closely.

Because ρ is large, it follows that close to ρ we can replace F_l and G_l (for any of the l values of interest) by their asymptotic forms. Thus, (14.14) becomes:

$$\psi_{l,p}(r) = \text{constant} \times \sin(pr - \gamma \ln 2pr - \tfrac{1}{2}l\pi + \sigma_l + \nu_l) \qquad [r \lessgtr \rho]$$

Matching the wave function across the cutoff at ρ we see immediately that the new phase shift is (to order $1/\rho$)

$$\delta_l(\rho) = \sigma_l + \nu_l - \gamma \ln 2p\rho$$

That is, ν_l is just the *additional phase shift introduced by the short-range forces* (which is generally *not* the same thing as the phase shift due to the short-range forces alone). Thus, the radial wave function for $r > \rho$ has the form

$$\psi_{l,p}(r) = \frac{i}{2}[\hat{h}_l^- - e^{2i\delta_l}\hat{h}_l^+]$$

$$= \frac{i}{2}\left\{\hat{h}_l^- - e^{2i(\sigma_l + \nu_l - \gamma \ln 2p\rho)}\hat{h}_l^+\right\} \qquad [r > \rho]$$

and the *change in the radial function* due to the addition of the short-range force is:

$$= \frac{1}{2i}[e^{2i(\sigma_l + \nu_l - \gamma \ln 2p\rho)} - e^{2i(\sigma_l - \gamma \ln 2p\rho)}]\hat{h}_l^+$$

$$\xrightarrow[r \to \infty]{} \frac{1}{2i} e^{2i(\sigma_l - \gamma \ln 2p\rho)}[e^{2i\nu_l} - 1](-i)^l e^{ipr}$$

Summing over all l we find for the *change in the full wave function* $\langle \mathbf{x} \mid \mathbf{p}+\rangle$:

$$\xrightarrow[r \to \infty]{} (2\pi)^{-3/2} \frac{1}{2ipr} e^{-2i\gamma \ln 2p\rho} \sum_l (2l+1)e^{2i\sigma_l}[e^{2i\nu_l} - 1]P_l(\cos\theta)e^{ipr}$$

Thus, the *change in the amplitude* due to the addition of the short-range forces is

$$= \frac{1}{2ip} e^{-2i\gamma \ln 2p\rho} \sum_l (2l+1)e^{2i\sigma_l}[e^{2i\nu_l} - 1]P_l(\cos\theta)$$

$$= e^{-2i\gamma \ln 2p\rho}f_{\text{add}}(\mathbf{p}' \leftarrow \mathbf{p}) \qquad (14.15)$$

The amplitude for the pure Coulomb potential (cutoff at ρ) is given by (14.12). Adding to this the additional amplitude (14.15), we find for the total amplitude (Coulomb plus short range)

$$f_\rho(\mathbf{p}' \leftarrow \mathbf{p}) = e^{-2i\gamma \ln 2p\rho}[f_C(\mathbf{p}' \leftarrow \mathbf{p}) + f_{\text{add}}(\mathbf{p}' \leftarrow \mathbf{p})] + f_{\text{osc}}(\mathbf{p}' \leftarrow \mathbf{p})$$

By the same reasoning as before the corresponding cross section is just

$$\frac{d\sigma}{d\Omega} = |f_C + f_{add}|^2 \qquad (14.16)$$

where f_C is the Coulomb amplitude and f_{add}, which represents the additional effects of the short-range forces, is given by (14.15) as:

$$f_{add}(\mathbf{p}' \leftarrow \mathbf{p}) = \frac{1}{2ip} \sum_l (2l + 1)e^{2i\sigma_l}(e^{2i\nu_l} - 1)P_l(\cos\theta) \qquad (14.17)$$

It is important to remember that f_{add} is generally *not* just the amplitude due to short-range forces alone.[8]

The calculation of the full effective amplitude $f_C + f_{add}$ is now a straightforward matter—at least in principle. The pure Coulomb part is explicitly known, while the additional short-range part is given by the partial-wave series (14.17). In this series the Coulomb phase shifts σ_l are explicitly known, while the additional short-range phase shifts can be determined by integrating the radial equation out to a point beyond the short-range potential and comparing with the form $F_l + \tan\nu_l G_l$.

The fact that the cross section (14.16) contains interference between f_C and f_{add} often allows measurements that are impossible with just short-range forces. Under the most favorable conditions it can happen that the term f_{add} is just the amplitude for the short-range forces alone and in this case the situation is especially simple. For example, in proton–proton scattering at a few MeV, only the *s*-wave phase shift ν_0 is important and at the same time the Coulomb shift σ_0 is very small. This means that ν_0 is precisely δ_0, the pure nuclear phase shift, and that

$$f_{add} = f_{nuc} = \frac{1}{p}e^{i\delta_0}\sin\delta_0$$

This gives (ignoring the identity of the protons and their spins)

$$\frac{d\sigma}{d\Omega} = |f_C + f_{nuc}|^2 = |f_C|^2 + |f_{nuc}|^2 + 2\operatorname{Re}f_C^* f_{nuc}$$

$$= \left(\frac{d\sigma}{d\Omega}\right)_C + \frac{\sin^2\delta_0}{p^2} - \frac{me_1e_2}{p^3\sin^2\theta/2}\sin\delta_0\cos\left(\delta_0 - 2\gamma\ln\sin\frac{\theta}{2}\right)$$

which, in contrast to the pure nuclear cross section, depends on the *sign* of the nuclear phase shift δ_0 and allows this sign to be determined.

---

[8] The answer (14.17) is just what one would get for the change in the amplitude when short-range forces are added if one simply ignored the logarithmic phases and treated $(\sigma_l + \nu_l)$ and σ_l as ordinary phase shifts.

14-c.   The Distorted-Wave Born Approximation

The Born series is an expansion of the amplitude in powers of the potential, and is of use only when the potential is weak enough to give very rapid convergence. When the potential is too strong—as it almost always is in nuclear physics, for example—an alternative is needed; one such alternative is the *distorted-wave Born approximation* or DWBA (or just the distorted-wave approximation).

The DWBA can be applied whenever the potential can be split into two parts,

$$V = V_{\mathrm{I}} + V_{\mathrm{II}}$$

in such a way that the amplitude for the potential V_{I} is exactly known (or at least some good approximation is available) and that the effects of the term V_{II} are small. Under these circumstances it should be no surprise that an expansion of the amplitude in powers of V_{II} is useful. In particular, the first two terms of such an expansion are called the DWBA and provide one of the most useful approximations in nuclear physics.

Most realistic examples of the use of the DWBA involve multichannel scattering, as we shall discuss in Chapter 21. However, it is easy to think of some single-channel situations to which one could apply the method. In the scattering of electrons off nuclei, one could take V_{I} to be the exactly soluble Coulomb potential and V_{II} to represent the deviation from the pure Coulomb inside the nucleus. In electron-proton scattering, V_{I} could again be the Coulomb potential while V_{II} would be the spin–orbit interaction. In proton–proton scattering at a few MeV, the s-wave amplitude could be computed by taking V_{I} to be the strong nuclear force (represented by a square well for example) and the small V_{II} to be the Coulomb potential (all other partial waves would be pure Coulomb). With any of these examples in mind we can proceed to describe the method.

As usual, we denote by $t(\mathbf{p}' \leftarrow \mathbf{p})$ and $|\mathbf{p}\pm\rangle$ the T matrix and stationary scattering states appropriate to the full potential $V = V_{\mathrm{I}} + V_{\mathrm{II}}$. We denote by $t_{\mathrm{I}}(\mathbf{p}' \leftarrow \mathbf{p})$ and $|\mathbf{p}\pm_{\mathrm{I}}\rangle$ the corresponding quantities for scattering by the potential V_{I} alone. For the purposes of the present discussion these latter quantities are supposed to be known, either exactly or in some good approximation.

We first derive an exact expression for the full T matrix, starting with the standard result

$$t(\mathbf{p}' \leftarrow \mathbf{p}) = \langle \mathbf{p}'| \, V \, |\mathbf{p}+\rangle \qquad (14.18)$$

The Lippmann–Schwinger equation for V_{I} alone lets us rewrite the bra $\langle \mathbf{p}'|$. Thus, since

$$|\mathbf{p}'-_{\mathrm{I}}\rangle = |\mathbf{p}'\rangle + G^0(E' - i0)V_{\mathrm{I}} |\mathbf{p}'-_{\mathrm{I}}\rangle$$

it follows that

$$\langle \mathbf{p}'| = \langle \mathbf{p}'-_{\mathrm{I}}| - \langle \mathbf{p}'-_{\mathrm{I}}| \, V_{\mathrm{I}} G^0(E' + i0)$$

Substitution into (14.18) gives

$$t(\mathbf{p}' \leftarrow \mathbf{p}) = \langle \mathbf{p}'-_{\mathrm{I}}| \, V \, |\mathbf{p}+\rangle - \langle \mathbf{p}'-_{\mathrm{I}}| \, V_{\mathrm{I}} G^0(E + i0) V \, |\mathbf{p}+\rangle$$

Using the Lippmann–Schwinger equation for the complete potential V, the last factor $G^0 V \, |\mathbf{p}+\rangle$ can be rewritten as $|\mathbf{p}+\rangle - |\mathbf{p}\rangle$ to give

$$t(\mathbf{p}' \leftarrow \mathbf{p}) = \langle \mathbf{p}'-_{\mathrm{I}}| \, V_{\mathrm{II}} \, |\mathbf{p}+\rangle + \langle \mathbf{p}'-_{\mathrm{I}}| \, V_{\mathrm{I}} \, |\mathbf{p}\rangle$$

or, since the last term is just $t_{\mathrm{I}}(\mathbf{p}' \leftarrow \mathbf{p})$,

$$\boxed{t(\mathbf{p}' \leftarrow \mathbf{p}) = t_{\mathrm{I}}(\mathbf{p}' \leftarrow \mathbf{p}) + \langle \mathbf{p}'-_{\mathrm{I}}| \, V_{\mathrm{II}} \, |\mathbf{p}+\rangle} \qquad (14.19)$$

This exact result is known variously as the *two-potential formula*, *Watson's theorem*, or the *Gell-Mann–Goldberger result*. It gives the T matrix for the complete potential $V = V_{\mathrm{I}} + V_{\mathrm{II}}$ as the sum of the exact T matrix for the potential V_{I} alone, plus a certain matrix element of V_{II}. This matrix element involves the "minus" scattering state $\langle \mathbf{p}'-_{\mathrm{I}}|$ for the potential V_{I} and the "plus" state $|\mathbf{p}+\rangle$ for the complete potential.

The DWBA is obtained from the exact two-potential formula by dropping terms that are higher than linear in the small potential V_{II}. Since the first term $t_{\mathrm{I}}(\mathbf{p}' \leftarrow \mathbf{p})$ is independent of V_{II} and the second term contains one factor V_{II} explicitly, this means one simply ignores all effects of V_{II} in the state $|\mathbf{p}+\rangle$. That is, we set $|\mathbf{p}+\rangle \approx |\mathbf{p}+_{\mathrm{I}}\rangle$ to give

$$\boxed{t(\mathbf{p}' \leftarrow \mathbf{p}) \approx t_{\mathrm{I}}(\mathbf{p}' \leftarrow \mathbf{p}) + \langle \mathbf{p}'-_{\mathrm{I}}| \, V_{\mathrm{II}} \, |\mathbf{p}+_{\mathrm{I}}\rangle} \quad [DWBA] \quad (14.20)$$

The DWBA can be compared with the ordinary Born approximation

$$t(\mathbf{p}' \leftarrow \mathbf{p}) \approx \langle \mathbf{p}'| \, V \, |\mathbf{p}\rangle \qquad \text{[Born approximation]}$$

for a single potential V. The differences are: First, that in the DWBA, part of the potential (V_{I}) is accounted for to all orders in the term $t_{\mathrm{I}}(\mathbf{p}' \leftarrow \mathbf{p})$; and second, that the second term is the matrix element of V_{II}, not between plane waves, but between "distorted" waves—that is, the scattering states appropriate to the potential V_{I}. In this terminology the potential V_{I} has two distinct roles: to scatter the projectile, and to distort the waves seen by the second potential V_{II}. The second term is then interpreted as the Born approximation for scattering by V_{II} in the presence of the distorting potential V_{I}.

Mathematically, the Born approximation arises from an expansion of the amplitude, as a function of the potential, about the value $V = 0$; the

DWBA arises from a corresponding expansion about $V = V_I$. More precisely, the Born approximation comes from replacing V by λV, expanding in powers of λ, and dropping terms of order λ^2 and higher; the DWBA comes from replacing V by $V_I + \lambda V_{II}$, and again expanding in powers of λ and dropping terms of order λ^2 and higher.

One can of course decompose the DWBA into partial waves (if the potentials are spherical). In place of the Born approximation (11.24)

$$f_l(p) \approx - \frac{2m}{p^2} \int_0^\infty dr\, \hat{j}_l(pr) V(r) \hat{j}_l(pr)$$

we find the almost obvious analogue,

$$f_l(p) \approx f_l^I(p) - \frac{2m}{p^2} \int_0^\infty dr \psi_{l,p}^I(r) V_{II}(r) \psi_{l,p}^I(r) \tag{14.21}$$

Here the only possible surprise is the absence of a star on the first factor $\psi_{l,p}^I(r)$. The reason for this is that it is the "minus" state $\langle \mathbf{p}' -_I|$ which appears in the full DWBA (14.20), and that the radial wave function for the "minus" state is just the complex conjugate of that for the "plus" state (see Problem **11.1**).

Rather than applying the DWBA to any realistic situation, we consider an artificial problem in which both V_I and V_{II} are square wells with the same radius. In this case, $V = V_I + V_{II}$ is itself a square well and can, of course, be exactly solved. If we let V_I be deep and V_{II} be shallow, then we can apply the DWBA and compare the results with the exact answer as well as with the ordinary Born approximation and with the scattering of V_I alone. The simplest case, that of the s-wave amplitude, is easily computed analytically (see Problem **14.2**), and some results are shown in Table 14.1.

This table shows the s-wave cross sections at various momenta for the case that V_I and V_{II} are wells of radius a and depths $9\pi^2/ma^2$ and $\pi^2/4ma^2$, respectively. The first line shows the exact results. The second line shows the DWBA, in which V_I is treated exactly but V_{II} to first order. The third line shows the results when V_I is treated exactly but V_{II} completely neglected. The last line shows the Born approximation, in which both V_I and V_{II} are

TABLE 14.1.   THE S-WAVE CROSS SECTION FOR A SQUARE WELL COMPUTED BY FOUR DIFFERENT METHODS

	p							
	0	1	3	7	13	25	60	100
Exact	3.89	2.76	0.0493	0.0707	0.000164	0.00615	0.000517	0.000077
DWBA	3.89	2.76	0.0489	0.0719	0.000168	0.00619	0.000517	0.000077
V_I	4.00	2.81	0.0371	0.0754	0.000509	0.00626	0.000495	0.000073
Born	3704.—	2479.—	113.—	3.—	0.3—	0.02—	0.000637	0.000084

treated to first order. The momenta are in units of $1/a$ and the cross sections in units of πa^2. It will be seen that the complete neglect of V_{II} gives results in reasonable agreement (better than 10% at most energies) with the exact results for $V = V_I + V_{II}$. However, the inclusion of V_{II} by means of the DWBA significantly improves matters and the agreement between the exact result and DWBA is excellent (better than 1% at most energies). Finally, it will be seen that at all but high energies the ordinary Born approximation is ludicrously unreliable.

14-d. Variational Methods

The Born approximation and DWBA depend for their usefulness on the effects of the potential (or the relevant part of the potential) being small. The partial wave method depends on the number of appreciable phase shifts being small. We now describe a third approach, the variational approach, which depends on one's ability to guess a trial wave function as a reasonable approximation to the actual wave function of the problem at hand. We begin with a brief review of the corresponding method for computing the energy of a ground state.

In the variational calculation of the ground state for a particle with Hamiltonian H, the first step is to guess a trial wave function $\zeta(\mathbf{x})$, which reproduces as closely as possible the actual ground state wave function $\psi_0(\mathbf{x})$. One then computes the functional[9]

$$\epsilon[\zeta] = \frac{\langle \zeta | H | \zeta \rangle}{\langle \zeta | \zeta \rangle}$$

The variational method hinges on the following three properties of this functional: First, $\epsilon[\zeta]$ is independent of the normalization of $\zeta(\mathbf{x})$ and if $\zeta_{ex}(\mathbf{x})$ is proportional to the exact ground state wave function, then $\epsilon[\zeta_{ex}]$ is precisely the ground state energy E_0,

$$\epsilon[\zeta_{ex}] = E_0 \qquad [\zeta_{ex} = c\psi_0]$$

This suggests that for any reasonable guess $\zeta(\mathbf{x})$ we can expect $E_0 \approx \epsilon[\zeta]$ to be a reasonable approximation. Second, $\epsilon[\zeta]$ is *stationary* with respect to variations of $\zeta(\mathbf{x})$ about the exact wave function. That is, if

$$\zeta(\mathbf{x}) = \zeta_{ex}(\mathbf{x}) + \delta\zeta(\mathbf{x})$$

then

$$\epsilon[\zeta] = \epsilon[\zeta_{ex}] + O(\delta\zeta)^2$$

[9] A *functional* $\alpha[f]$ is a *number* depending on a *function* $f(x)$; e.g.,

$$\alpha[f] = \int_0^1 dx\, f(x)$$

This means, in an obvious sense, that the error in the estimate $E_0 \approx \epsilon[\zeta]$ is minimal. Finally, the functional $\epsilon[\zeta]$ is not only stationary at the exact wave function, it is a minimum; that is, $\epsilon[\zeta] \geqslant E_0$, for all $\zeta(\mathbf{x})$. This means that any trial function provides one with an upper bound on E_0, and allows one to determine uniquely the best function from any family of trial functions as that one for which $\epsilon[\zeta]$ is a minimum.[10]

The variational calculations of scattering amplitudes are similar to those of binding energies, although they are usually more complicated, and do not always have the desirable feature of providing a minimum (or maximum) principle. Of the many possible methods we describe just two: the methods of Schwinger and Kohn. Both methods can be used to compute the full amplitude $f(\mathbf{p}' \leftarrow \mathbf{p})$, in which case the trial functions $\zeta(\mathbf{x})$ are chosen so as to approximate the three dimensional wave functions $\langle \mathbf{x} \mid \mathbf{p}+ \rangle$ or $\langle \mathbf{x} \mid \mathbf{p}- \rangle$. However, a simpler and more practical procedure is to apply the methods to the individual partial-wave amplitudes, in which case the trial functions $\zeta(r)$ are chosen to approximate the radial wave functions $\psi_{l,p}(r)$ (possibly multiplied by some convenient constant). It is the partial-wave version of the variational approach that we describe below.

The Schwinger Method.   The partial-wave amplitude is given in terms of the radial wave function $\psi_{l,p}(r)$ by the integral

$$f_l(p) = -\frac{1}{p^2} \int_0^\infty dr \, j_l(pr) U(r) \psi_{l,p}(r)$$

[As usual, $U(r) \equiv 2mV(r)$.] If one could make a guess $\zeta(r)$ for $\psi_{l,p}(r)$, then the simplest corresponding guess for the amplitude is obviously $f_l \approx -(1/p^2) \int dr j_l U \zeta$. This guess gives the correct answer when ζ is the exact radial wave function. However, it is certainly not stationary, and we therefore seek a more complicated expression that is stationary.

The Schwinger method starts from the functional

$$\alpha[\zeta] = -\frac{1}{p^2} \frac{\left(\int dr \, j_l U \zeta \right)^2}{\int dr \, \zeta (U - U G^0 U) \zeta} \tag{14.22}$$

---

[10] It should be emphasized that to uniquely determine a best estimate from a family of trial functions it is essential that we have a minimum principle, not just a stationary principle. Thus, suppose a functional $\epsilon[\zeta]$ is stationary, but not minimum or maximum, for arbitrary variations of ζ about some desired ζ_{ex}. Then in any chosen family of trial functions the functional $\epsilon[\zeta]$ may have several stationary points, *none* of which provide the best estimate. Unfortunately, most variational methods in collision theory do not provide minimum principles, and hence cannot be used for a systematic search for a best estimate. (See Geltman, 1969, p. 58.)

where $G^0 \equiv G^0_{l,p}$ is the free Green's operator (11.21) for angular momentum l and $\zeta(r)$ is any trial wave function. We first remark that $\alpha[\zeta]$ is independent of the normalization of $\zeta(r)$ and there is no loss of generality in choosing $\zeta(r)$ to be real. (Recall that the exact radial function has the form of some constant times a real function of r.) We next show that if $\zeta_{ex}(r)$ is proportional to the exact radial function,

$$\zeta_{ex}(r) = c\psi_{l,p}(r)$$

then $\alpha[\zeta_{ex}]$ is just the desired amplitude. This follows from the fact that $\zeta_{ex}(r)$ satisfies the Lippmann–Schwinger equation

$$\zeta_{ex} = c\hat{\jmath}_l + G^0 U \zeta_{ex}$$

and that the denominator in (14.22) is therefore

$$\int dr \zeta_{ex} U(1 - G^0 U)\zeta_{ex} = \int dr \zeta_{ex} U c\hat{\jmath}_l = c^2 \int dr \, \hat{\jmath}_l U \psi_{l,p}$$

This exactly cancels one factor from the numerator to give

$$\alpha[\zeta_{ex}] = -\frac{1}{p^2}\int dr \, \hat{\jmath}_l U \psi_{l,p} = f_l(p)$$

Finally, it can easily be shown that $\alpha[\zeta]$ is stationary for arbitrary real variations of $\zeta(r)$ about the exact function; that is, if

$$\zeta(r) = \zeta_{ex}(r) + \delta\zeta(r)$$

then

$$\alpha[\zeta] = \alpha[\zeta_{ex}] + O(\delta\zeta)^2$$

(see Problem **14.3**). From this it follows that

$$f_l(p) = \alpha[\zeta_{ex}] = \alpha[\zeta] + O(\delta\zeta)^2$$

or

$$f_l(p) = -\frac{1}{p^2}\frac{\left(\int dr \, \hat{\jmath}_l U \zeta\right)^2}{\int dr \, \zeta(U - UG^0 U)\zeta} + O(\delta\zeta)^2 \tag{14.23}$$

This is Schwinger's variational expression for the partial-wave amplitude. If our guess $\zeta(r)$ for the radial function is reasonable, the approximation obtained by neglecting the term $O(\delta\zeta)^2$ should be good.

One choice for the trial function $\zeta(r)$, which should be good when the potential is weak, is the free function $\hat{j}_l(pr)$. With this choice the Schwinger expression gives:

$$f_l(p) = -\frac{1}{p^2} \frac{\int dr\, \hat{j}_l U \hat{j}_l}{1 - \dfrac{\int dr\, \hat{j}_l U G^0 U \hat{j}_l}{\int dr\, \hat{j}_l U \hat{j}_l}} \tag{14.24}$$

The second term in the denominator of (14.24) will be recognized as the ratio of the first two terms in the Born series for $f_l(p)$. If it happens that the Born series is rapidly convergent, then this number is small and we can expand (14.24) to give

$$f_l(p) \approx -\frac{1}{p^2}\left(\int dr\, \hat{j}_l U \hat{j}_l + \int dr\, \hat{j}_l U G^0 U \hat{j}_l + \cdots \right)$$

This shows that under those conditions where the Born approximation is good, the Schwinger expression (with a free trial function) is equivalent to the second Born approximation. In general, it would be expected to be better.

These results can be well illustrated with the case of the square well potential. For simplicity we consider just the s-wave amplitude at zero energy, which is minus the s-wave scattering length. For this, the exact result is (see Problem **11.5**):

$$a_0 \equiv -f_0(0) = 1 - \frac{\tan U_0^{\frac{1}{2}}}{U_0^{\frac{1}{2}}} \quad \text{[exact]}$$

(Here we have taken a well of unit radius and of depth V_0 with $U_0 \equiv 2mV_0$.) Straightforward calculations of the first and second Born approximations and the Schwinger expression (14.24) give (see Problem **14.4**):

$$\left. \begin{array}{ll} a_0 \approx -\dfrac{U_0}{3} & \text{[first Born]} \\[3ex] a_0 \approx -\dfrac{U_0}{3} - \dfrac{2U_0^2}{15} & \text{[second Born]} \\[3ex] a_0 \approx \dfrac{-U_0/3}{(1 - 2U_0/5)} & \text{[Schwinger]} \end{array} \right\} \tag{14.25}$$

All of these results are displayed as functions of well depth in Fig. 14.2. It will be seen that, as one would expect, all three approximations are good

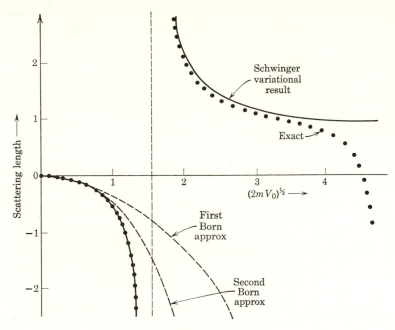

FIGURE 14.2.   The s-wave scattering length as a function of well depth for a square well of unit radius and depth V_0.

when the well is very shallow. As the well deepens, the first Born approximation and then the second Born approximation deviate from the exact result, the former being within 10% out to $U_0^{1/2} \sim 0.5$ and the latter until $U_0^{1/2} \sim 0.8$. The variational expression is within $\frac{1}{4}$% of the exact answer in this same range ($U_0^{1/2}$ less than 0.8) and remains satisfactory out to depths with $U_0^{1/2} \sim 3$. Both the exact result and the variational expression become infinite with vertical asymptotes, the former at $U_0^{1/2} = 1.57$ (i.e., $\pi/2$), the latter at $U_0^{1/2} = 1.58$. (Note that to the left of the vertical asymptote the exact and variational results are indistinguishable on the scale of Fig. 14.2.) Finally by about $U_0^{1/2} \sim 3$ the variational expression ceases to be reliable (discrepancy greater than 10%), reflecting the fact that by the time the well is this deep the free trial functions have become completely unrealistic.

The Kohn Method.   The Schwinger variational method depends on the fact that the exact wave function satisfies the Lippmann–Schwinger integral equation. There are several methods based on the Schrödinger differential equation, one of the most important of which is that due to Kohn. This method provides a stationary expression not for the amplitude itself but for the tangent of the phase shift.

We begin as before with a real trial radial function $\zeta(r)$. However, *we restrict the class of functions considered to those $\zeta(r)$ satisfying:*

$$\zeta(0) = 0 \qquad (14.26)$$

and having the asymptotic form:

$$\zeta(r) \xrightarrow[r \to \infty]{} \frac{1}{p} \sin(pr - \tfrac{1}{2}l\pi) + \tau \cos(pr - \tfrac{1}{2}l\pi) \qquad (14.27)$$

Here τ is any constant (which can be different for different trial functions) and the factor $1/p$ is inserted for convenience in discussing the zero energy limit. The corresponding exact radial function $\zeta_{\mathrm{ex}}(r)$ satisfies the radial equation which we write as

$$D\zeta_{\mathrm{ex}}(r) = 0$$

where

$$D = -\frac{d^2}{dr^2} + \frac{l(l+1)}{r^2} + U(r) - p^2$$

Since

$$\zeta_{\mathrm{ex}}(r) \xrightarrow[r \to \infty]{} \mathrm{constant} \times \sin(pr - \tfrac{1}{2}l\pi + \delta_l)$$

we see that the coefficient τ_{ex} of (14.27) for the exact function is just

$$\tau_{\mathrm{ex}} = \frac{1}{p} \tan \delta_l$$

where δ_l is the (exact) phase shift.

We now consider the functional

$$\beta[\zeta] = \tau - \int_0^\infty dr\, \zeta(r) D\zeta(r)$$

Clearly, since $D\zeta_{\mathrm{ex}} = 0$,

$$\beta[\zeta_{\mathrm{ex}}] = \tau_{\mathrm{ex}} = \frac{1}{p} \tan \delta_l$$

and we can now show that $\beta[\zeta]$ is stationary for arbitrary variations of $\zeta(r)$ about $\zeta_{\mathrm{ex}}(r)$, subject to the conditions (14.26) and (14.27). We write

$$\zeta(r) = \zeta_{\mathrm{ex}}(r) + \delta\zeta(r)$$

and then, to first order in $\delta\zeta$, the difference between $\beta[\zeta]$ and $\beta[\zeta_{\mathrm{ex}}]$ is

$$\delta\beta = \delta\tau - \int_0^\infty dr[\zeta_{\mathrm{ex}} D(\delta\zeta) + (\delta\zeta)D\zeta_{\mathrm{ex}}] \qquad (14.28)$$

Integrating the first term of the integral by parts gives

$$\delta\beta = \delta\tau + [\zeta_{\mathrm{ex}}\, \delta\zeta' - \zeta_{\mathrm{ex}}'\, \delta\zeta]_0^\infty - 2\int_0^\infty dr(\delta\zeta)D\zeta_{\mathrm{ex}}$$

Since $D\zeta_{ex} = 0$ the integral vanishes, and since both ζ_{ex} and $\delta\zeta$ vanish at $r = 0$ the lower limit in the bracket is zero. The upper limit is easily seen to give $-\delta\tau$, which exactly cancels the remaining term. Thus   to order $(\delta\zeta)$ we have $\delta\beta = 0$ or

$$\beta[\zeta] = \beta[\zeta_{ex}] + O(\delta\zeta)^2$$

Finally, since $\beta[\zeta_{ex}] = (1/p)\tan\delta_l$, this gives

$$\frac{1}{p}\tan\delta_l = \tau - \int_0^\infty dr\, \zeta(r)D\zeta(r) + O(\delta\zeta)^2 \qquad (14.29)$$

which is the Kohn expression for the phase shift.

If we take as trial function the *free* radial function

$$\zeta(r) = \frac{1}{p}\,\hat{j}_l(pr) \xrightarrow[r\to\infty]{} \frac{1}{p}\sin(pr - \tfrac{1}{2}l\pi)$$

then for this function $\tau = 0$ and

$$D\zeta = \frac{1}{p}\left(-\frac{d^2}{dr^2} + \frac{l(l+1)}{r^2} + U - p^2\right)\hat{j}_l = \frac{1}{p}\,U\hat{j}_l$$

Thus, the Kohn expression reduces to

$$\frac{1}{p}\tan\delta_l \approx -\frac{1}{p^2}\int_0^\infty dr\,\hat{j}_l U \hat{j}_l$$

The expression on the right will be recognized as the Born approximation for the amplitude $f_l(p)$. Now, if it happens that the Born approximation is good, then δ_l is certainly small and hence $(1/p)\tan\delta_l \approx f_l(p)$; that is, under those conditions where the Born approximation is good, the Kohn expression with the free trial function $\hat{j}_l$ is equivalent to the first Born approximation. Since the corresponding Schwinger expression is equivalent, under the same conditions, to the *second* Born approximation, this suggests that for a given trial function the Schwinger principle will generally give better results than that of Kohn. However, it should be emphasized that the Schwinger expression (14.23) contains a double integral involving the Green's function $G^0_{l,p}(r, r')$, which for anything but the simplest trial functions is very inconvenient to compute. Thus, the advantages of Schwinger's expression for a *given* trial function may be offset by the fact that Kohn's expression can be used with considerably more sophisticated functions.

A Minimum Principle.   So far neither of the stationary principles that we have described is a minimum (or maximum) principle. This means that they do not allow a systematic search for a best estimate from any family of trial functions. To conclude this section we show that under certain conditions the Kohn expression can yield a minimum principle. Specifically, consider a potential with *no bound states;* and consider the case $l = 0$ at zero energy, in which case $(1/p) \tan \delta_l$ is just minus the scattering length a_0.

We first note that the correction term $O(\delta\zeta)^2$ in the Kohn result (14.29) is just $\int dr(\delta\zeta)D(\delta\zeta)$. [This is just the term omitted from the first order expression (14.28) for $\delta\beta$.] Thus, the zero energy limit of the Kohn result is exactly

$$a_0 = -\tau + \int_0^\infty dr\zeta D\zeta - \int_0^\infty dr(\delta\zeta)D(\delta\zeta)$$

For $l = 0$ at zero energy, the operator D is:

$$D = -\frac{d^2}{dr^2} + U(r)$$

which is just the *s*-wave Hamiltonian (times $2m$). In particular, if H has no bound states, this operator is positive definite and[11]

$$\int dr(\delta\zeta)D(\delta\zeta) \geqslant 0$$

It follows that

$$a_0 \leqslant -\tau + \int_0^\infty dr\zeta(r)D\zeta(r) \qquad \text{[no bound states]} \qquad (14.30)$$

That is, for the case that H supports no bound states, the Kohn expression for the scattering length is not just stationary at the correct value, it is actually a minimum. This means that among any given family of trial functions, that which provides the best estimate for a_0 is uniquely determined as that for which the right hand side of (14.30) is a minimum.

For the generalization of this last result to the case that there are some bound states, and more generally for more details on variational methods, we refer the reader to the literature.[12]

14-e. The K Matrix

An important technique in the mathematical study of unitary operators is the so-called Caley transform, which expresses any unitary operator in

---

[11] Care is needed here since $\delta\zeta(r) \to$ constant as $r \to \infty$ and so is not square-integrable. However, the result is easily proved by approximating $\delta\zeta(r)$ by some cutoff function such as $\delta\zeta(r)e^{-\epsilon r}$ with ϵ small.

[12] See Newton (1966) Section 11.3, and (for a comprehensive survey of variational methods) Demkov (1963).

terms of a certain related Hermitian operator.[13] In scattering theory the
unitary operator of interest is the S operator and its Caley transform, which
we denote by M, is defined as

$$M = i(1 - S)(1 + S)^{-1} \qquad (14.31)$$

This definition can be inverted to give

$$S = (1 + iM)(1 - iM)^{-1} \qquad (14.32)$$

It is immediately clear from these two equations that if S is unitary then M is
Hermitian and vice versa. This result is, in fact, one of the most important
properties of the operator M. It is usually much easier to check that a given
operator is Hermitian than it is to check for unitarity. Thus, a convenient
way to calculate an approximate S that is guaranteed to be unitary is to write
S in terms of M and then devise a scheme for calculating an approximate
Hermitian M.

Since S conserves energy the same is true of M. If, for example, we consider
scattering of two spinless particles, this means we can write

$$\langle \mathbf{p}' | M | \mathbf{p} \rangle = \delta(E_{p'} - E_p) k(\mathbf{p}' \leftarrow \mathbf{p})$$

Here the quantity $k(\mathbf{p}' \leftarrow \mathbf{p})$ is variously called the K *matrix*, the *reaction* or
reactance matrix, and *Heitler's matrix;* it was first introduced by Heitler
and plays an important role in the Wigner theory of reactions.[14]

It should be emphasized that, like the on-shell T matrix $t(\mathbf{p}' \leftarrow \mathbf{p})$, the
K matrix $k(\mathbf{p}' \leftarrow \mathbf{p})$ is defined only on the energy shell $E_{p'} = E_p$. In fact, it is
useful to compare its definition with that of the T matrix,

$$S = 1 + R$$

and

$$\langle \mathbf{p}' | R | \mathbf{p} \rangle = -2\pi i\, \delta(E_{p'} - E_p) t(\mathbf{p}' \leftarrow \mathbf{p})$$

We see that (apart from a conventional $-2\pi i$) the K matrix $k(\mathbf{p}' \leftarrow \mathbf{p})$ stands
to the operator M as does the T matrix $t(\mathbf{p}' \leftarrow \mathbf{p})$ to R. The operators R and
M conserve energy (both being functions of S) and so their matrix elements

[13] Actually the Caley transform as defined here applies only to unitary operators that do
not have -1 as a proper eigenvalue. Fortunately the S operator is in this class.
[14] The terminology and notation for the K matrix is very confused. The name reaction
matrix is sometimes used for what we call the T matrix. Our notation, in which $k(\mathbf{p}' \leftarrow \mathbf{p})$
stands to the operator M as does $t(\mathbf{p}' \leftarrow \mathbf{p})$ to the operator $R = S - 1$, is something less
than perfect. However, without recourse to an intolerable number of alphabets, no more
rational scheme seems possible.

contain the factor $\delta(E_{p'} - E_p)$. When this delta function is factored out one is left with the on-shell matrices $t(\mathbf{p}' \leftarrow \mathbf{p})$ and $k(\mathbf{p}' \leftarrow \mathbf{p})$, both defined only when $E_{p'} = E_p$.

The S matrix, in addition to being unitary, is often also symmetric. For example, in the scattering of two spinless particles it is symmetric in the momentum representation, $\langle \mathbf{p}' | S | \mathbf{p} \rangle = \langle \mathbf{p} | S | \mathbf{p}' \rangle$, provided the interactions are invariant under either rotations or PT. When this is the case the K matrix is both Hermitian and symmetric, and hence also real. In this case unitarity of S can be guaranteed simply by making sure that the K matrix is real and symmetric.

The K matrix can naturally be related directly to the T matrix. From the definitions of the operators R and M, it is easily seen that

$$R = 2iM + iMR \tag{14.33}$$

Taking matrix elements of this equation and factoring out the common energy delta function we find (for the case of spinless particles),

$$-\pi t(\mathbf{p}' \leftarrow \mathbf{p}) = k(\mathbf{p}' \leftarrow \mathbf{p}) - \pi i \int d^3 p'' k(\mathbf{p}' \leftarrow \mathbf{p}'') \, \delta(E_{p''} - E_p) t(\mathbf{p}'' \leftarrow \mathbf{p})$$

a relation known as Heitler's damping equation. If $k(\mathbf{p}' \leftarrow \mathbf{p})$ has been calculated by any convenient approximation method (chosen to ensure that it is Hermitian), then the corresponding $t(\mathbf{p}' \leftarrow \mathbf{p})$ is found by solving Heitler's equation and is then automatically consistent with the unitarity of S.

All of these ideas can be well (though almost trivially) illustrated in the angular momentum representation for the case of two spinless particles with a rotationally invariant interaction. In this representation S, and hence M and R, are completely diagonal,

$$\langle E', l', m' | \ S \ | E, l, m \rangle = \delta(E' - E) \, \delta_{l'l} \, \delta_{m'm} \, s_l(E)$$
$$\langle \cdot \quad \cdot \quad \cdot | M | \cdot \quad \cdot \quad \cdot \rangle = \qquad \cdot \qquad \cdot \quad \cdot \quad k_l(E)$$
$$\langle \cdot \quad \cdot \quad \cdot | R | \cdot \quad \cdot \quad \cdot \rangle = \qquad \cdot \qquad \cdot \quad \cdot \quad 2ip f_l(E)$$

Since the operator M is Hermitian and diagonal, its matrix elements have to be real; that is, the numbers $k_l(E)$ are real. The expression (14.32) for S in terms of M reduces to

$$s_l = \frac{1 + ik_l}{1 - ik_l}$$

from which it is quite clear that s_l is unitary (that is, of modulus one) if and only if k_l is real. Remembering that $s_l = \exp(2i\delta_l)$ we can write the inverse of this equation as

$$k_l = i \frac{1 - s_l}{1 + s_l} = \tan \delta_l$$

that is, in the partial-wave representation the K-matrix elements k_l are just the tangents of the phase shifts. The Heitler equation becomes

$$pf_l = k_l + ik_l pf_l$$

which can be solved to give the familiar result:

$$f_l = \frac{1}{p}\frac{k_l}{1 - ik_l} = \frac{1}{p}\frac{\tan\delta_l}{1 - i\tan\delta_l} = \frac{1}{p}e^{i\delta_l}\sin\delta_l \qquad (14.34)$$

Finally, we note that it is possible to relate the K matrix to certain "standing-wave" stationary states in the same way that the T matrix can be related to the states $|\mathbf{p}\pm\rangle$. One first introduces a free Green's operator

$$G^0_{(s)}(E) = \tfrac{1}{2}[G^0(E + i0) + G^0(E - i0)]$$

From the known form (10.15) of the spatial matrix elements

$$\langle \mathbf{x}| G^0(E \pm i0) |\mathbf{x}'\rangle$$

it follows that

$$\langle \mathbf{x}| G^0_{(s)}(E) |\mathbf{x}'\rangle = -\frac{m}{2\pi}\frac{\cos p\,|\mathbf{x} - \mathbf{x}'|}{|\mathbf{x} - \mathbf{x}'|}$$

that is, for large $\mathbf{x}$ this Green's function is an equal mixture of incoming and outgoing waves and is therefore called the *standing-wave Green's function*. In terms of $G^0_{(s)}(E)$ one can define *standing-wave* stationary states by means of a Lippmann–Schwinger equation:

$$|\mathbf{p}(s)\rangle = |\mathbf{p}\rangle + G^0_{(s)}(E)V\,|\mathbf{p}(s)\rangle$$

and is then easily shown that

$$k(\mathbf{p}' \leftarrow \mathbf{p}) = -\pi\langle \mathbf{p}'| V\,|\mathbf{p}(s)\rangle$$

which is the K-matrix analogue of the familiar result $t(\mathbf{p}' \leftarrow \mathbf{p}) = \langle \mathbf{p}'| V |\mathbf{p}+\rangle$.

PROBLEMS

14.1. Try applying the variable phase method of Section 11-g to the Coulomb potential. Show that the phase function $\delta(r)$ has no limit (that is, that the radial function continues to pick up phase however large we make r) and, hence, that a phase shift cannot be defined (at least for s waves).

14.2. Consider the s-wave scattering by a square well of unit radius and depth V_0. Write down the exact amplitude $f_0(p, V_0)$ for arbitrary V_0 (see Problem **11.5**) and the corresponding radial function $\psi_{0,p}(r)$. Now consider two square wells (at the same center) both of unit radius, well V_I of depth V_0 and well V_{II} of small depth ϵ. Their sum $V = V_I + V_{II}$ is, of course, a square well of depth $V_0 + \epsilon$. Compute the amplitude $f_0(p, V_0 + \epsilon)$ using DWBA with V_I treated exactly and V_{II} to first order. Verify that your answer is just the first two terms in a Taylor expansion of $f_0(p, V_0 + \epsilon)$ about V_0. [To normalize the radial function $\psi_{0,p}(r)$ it is sufficient to write $\psi_{0,p}(r) = \exp(i\delta_0)\sin(pr + \delta_0)$ for r outside the well.]

14.3. Show that the Schwinger functional

$$\alpha[\zeta] = -\frac{1}{p^2} \frac{\left(\int dr \, \hat{\jmath} U \zeta\right)^2}{\int dr \, \zeta(U - UG^0U)\zeta}$$

is stationary for arbitrary real variations of $\zeta(r)$ about some $\zeta_0(r)$ if and only if $\zeta_0(r)$ is proportional to the exact radial wave function. Hint: Write $\zeta(r) = \zeta_0(r) + \delta\zeta(r)$ and then show that to first order $\delta\alpha$ has the form $\delta\alpha = \int dr \, \delta\zeta\eta$ where $\eta(r)$ has the form

$$\eta = aU\hat{\jmath} + bU(1 - G^0U)\zeta_0$$

(for certain numbers a and b). Thus, $\delta\alpha$ can be zero for arbitrary $\delta\zeta$ if and only if $\eta(r) = 0$, which means that $(1 - G^0U)\zeta_0$ is proportional to $\hat{\jmath}$. That is, $\delta\alpha = 0$ if and only if ζ_0 satisfies the Lippmann–Schwinger equation.

14.4. Calculate the s-wave scattering length for a square well of unit radius and depth V_0 in first and second Born approximations and by the Schwinger method with a free trial function. [See (14.25).] Check that all three answers agree for a sufficiently shallow well.

15 Dispersion Relations and Complex Angular Momenta

15-a   Partial-Wave Dispersion Relations

15-b   Forward Dispersion Relations

15-c   Nonforward Dispersion Relations

15-d   The Mandelstam Representation

15-e   Complex Angular Momenta

15-f   Regge Poles

15-g   The Watson Transform

In Chapters 11, 12, and 13 we have seen the surprising usefulness of allowing certain physically real parameters (coupling strengths, energies, and momenta) to move away from the real axis and to become complex variables. In fact, the analysis of scattering amplitudes as analytic functions of complex variables has become one of the most powerful tools of modern collision theory. The technique has had applications in atomic and nuclear physics; and, above all, in relativistic particle physics, where it can reasonably be said to have accounted for 30–50% of all theoretical research during the past 10 years or so.[1]

This chapter is an introduction to those properties of the analytically continued scattering amplitude that have been most important in relativistic particle physics.[2] The discussion is primarily intended to provide the reader

---

[1] Jackson (1970) has made a survey of papers in particle theory during 1968–1969 and found that 35% of all papers were primarily concerned with dispersion relations, complex angular momenta, and related topics.

[2] The whole chapter can be omitted or postponed without affecting the subsequent chapters.

with a bridge to the several more detailed texts available on the subject.[3] In Sections 15-a to 15-d we give an introduction to dispersion relations, which are integral identities obtained by applying Cauchy's theorem to the analytically continued scattering amplitude. We discuss dispersion relations for the partial-wave amplitude as a function of energy, and for the full amplitude as a function of energy and momentum transfer, and conclude with a brief discussion of the famous double dispersion relation known as the Mandelstam representation. In Sections 15-e to 15-g we discuss the partial-wave amplitude as an analytic function of energy and *angular momentum*, leading up to the important idea of Regge poles.

Before beginning our main discussion we mention two general points. First, in the literature of high-energy physics one finds two different attitudes toward dispersion relations and complex angular momenta. One view is that one should be able to construct a complete relativistic dynamics based on these techniques (a point of view often known as analytic S-matrix theory); at the other extreme is the view that the techniques are merely useful tools that can be used in the analysis and interpretation of data (and should presumably be derivable from some more basic principles—probably relativistic quantum field theory). As will be discussed later, it is still not clear whether the first, and more ambitious, view is really justified. Therefore, it should be emphasized that whether or not the first view is correct, the usefulness of dispersion relations and complex angular momenta on the basis of the second view alone amply justifies their study.

Second, it should be said that the status of the results of this chapter is somewhat anomalous, since we shall be using nonrelativistic quantum mechanics to discuss results whose principal application is in relativistic problems. In particular, an important part of some proofs will be to establish the behavior of the amplitude as $E \to \infty$; but when $E \to \infty$ all real scattering problems necessarily become relativistic. Thus, from a physical point of view our analysis is logically inconsistent. Nonetheless, and in spite of this apparent inconsistency, there is reason to believe that our results do give a good approximation in many nonrelativistic problems.[4] Further, in the study of relativistic problems it is very useful to know that there do exist internally consistent model theories—even theories which violate some known requirements—where the results of interest can be rigorously proved.

---

[3] The most up to date text concerned with these ideas is the monograph of Collins and Squires (1968). Older and more elementary texts include Chew (1962), Frautschi (1965), Jacob and Chew (1964), Newton (1964), Omnes and Froissart (1963), and Squires (1963).
[4] This is really quite a subtle point. Dispersion relations, even for low-energy amplitudes, involve integrals of the amplitude over *all* energies, including relativistic energies. However, it may happen that the contribution to these integrals from large E is negligible (both for the correct relativistic amplitude and the incorrect, nonrelativistic amplitude). In this case, even though a nonrelativistic proof is incorrect, the result will be valid.

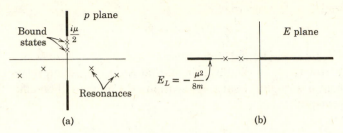

FIGURE 15.1.   Regions of analyticity of f_l **as a function of (a) the momentum, and (b) the energy.**

15-a.   Partial-Wave Dispersion Relations

The partial-wave dispersion relation follows from the analytic properties of the partial-wave amplitude $f_l(E)$ established in Chapter 12. Since dispersion relations have been most extensively used in relativistic scattering, in which context the Yukawa potential is generally considered to give the most realistic nonrelativistic analogue, we shall suppose that our potential is a Yukawa or sum of Yukawas. For such potentials the results of Chapter 12 can be summarized as follows.

As a function of p the partial-wave amplitude

$$f_l = \frac{s_l - 1}{2ip} = \frac{1}{2ip}\frac{f_l(-p) - f_l(p)}{f_l(p)} \tag{15.1}$$

is meromorphic (analytic except for poles) in the region shown in Fig. 15.1, part (a); it has poles on the positive imaginary axis and is real on the imaginary axis between the two branch cuts. Therefore, as a function of E on the physical sheet (corresponding to $\{\text{Im } p > 0\}$), it is meromorphic in the region of Fig. 15.1, part (b) with two cuts: a right-hand cut running from 0 to ∞, and a left-hand cut from $E_L \equiv -\mu^2/8m$ to $-\infty$. It has poles at the bound state energies and is real on the real axis for $E_L < E < 0$.

As $p \to \infty$ in the upper half plane the Jost function $f_l(p)$ tends to 1. In the case of the Yukawa potential we can extend this result into the lower half plane by using the contour distortion described in Section 12-c. Since $f_l(p) \to 1$ as $p \to \infty$ in all directions, it follows from (15.1) that the amplitude f_l goes to zero as p (or E) goes to infinity in any direction.

Finally, we recall that (see Problem **12.3**) at each bound-state pole the partial-wave amplitude $f_l(E)$ has residue (as a function of E) $\Gamma = (-)^{l+1}\gamma^2$, where γ is related to the asymptotic form of the normalized wave function for the bound state concerned,

$$\eta(r) \xrightarrow[r \to \infty]{} \gamma e^{-\alpha r}$$

where $-\alpha^2/2m$ is the bound state energy.[5]

We are now ready to establish the partial-wave dispersion relation, which results from a straightforward application of Cauchy's theorem. We consider the integral

$$I = \oint_C dE' \frac{f_l(E')}{E' - E}$$

where C is the closed contour consisting of the four sections $C_1, \ldots, C_4$ shown in Fig. 15.2 and E has any fixed complex value. The integrand is analytic inside the contour except for simple poles at $E' = E$ and at the bound state energies $E' = E_n$. Thus, by Cauchy's theorem

$$I = 2\pi i \left[f_l(E) + \sum_n \frac{\Gamma_n}{E_n - E} \right] \tag{15.2}$$

If we let the two semicircular parts of the contour C_2 and C_4 go to infinity, then, since $f_l(E)$ goes to zero (like $E^{-\eta}$ for some $\eta > 0$), their contribution disappears. The contribution from the section C_1 can then be written as:

$$\int_{C_1} dE' \frac{f_l(E')}{E' - E} = \int_0^\infty \frac{dE'}{E' - E} [f_l(E' + i0) - f_l(E' - i0)] \tag{15.3}$$

Since $f_l(E)$ is real on the real axis when $E_L < E < 0$ it satisfies the Schwartz reflection principle and, in particular,

$$f_l(E' - i0) = [f_l(E' + i0)]^*$$

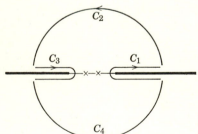

FIGURE 15.2.   Contour used to prove the partial-wave dispersion relation.

---

[5] In relativistic scattering the numbers γ are the fundamental coupling constants. For example, in π–N scattering the nucleon appears as a bound-state pole in the partial-wave amplitude for $l = 1$, $j = \frac{1}{2}$, and isospin $\frac{1}{2}$. The residue at this pole is the square of the π–N coupling constant.

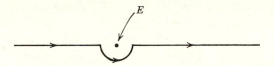

FIGURE 15.3. Distortion of contour to show what happens to integral (15.5) when E moves onto the axis.

Thus the quantity in the bracket of (15.3) is just $2i$ times the imaginary part of $f_l(E' + i0)$. Since exactly the same considerations apply to the contour C_3, we can write the whole integral as

$$ I = 2i \left\{ \int_{-\infty}^{E_L} + \int_0^{\infty} \right\} dE' \, \frac{\text{Im} \, f_l(E' + i0)}{E' - E} $$

Comparing this with (15.2) we conclude that

$$ f_l(E) = \sum \frac{\Gamma_n}{E - E_n} + \frac{1}{\pi} \left\{ \int_{-\infty}^{E_L} + \int_0^{\infty} \right\} dE' \, \frac{\text{Im} \, f_l(E')}{E' - E} \qquad (15.4) $$

where we follow tradition and make the convention that, in the integrand, the real E' is understood to be on the *upper* edge of the cuts (i.e., E' stands for $E' + i0$). This relation is the so-called partial-wave dispersion relation.[6] It expresses the partial-wave amplitude $f_l(E)$ for arbitrary E in terms of its imaginary part on the two cuts and the residues Γ_n.

To obtain the physical amplitude, we must bring E into the positive real axis from above. As we do this it is easily seen by distorting the contour as shown in Fig. 15.3, that the integral along the positive axis becomes:

$$ \int_0^{\infty} dE' \, \frac{\text{Im} \, f_l(E')}{E' - E} \xrightarrow[\substack{E \to \text{positive} \\ \text{real}}]{} P \int_0^{\infty} dE' \, \frac{\text{Im} \, f_l(E')}{E' - E} + i\pi \, \text{Im} \, f_l(E) \qquad (15.5) $$

where $P \int$ denotes the Cauchy principal value integral,

$$ P \int_0^{\infty} dE' \, \frac{g(E')}{E' - E} = \lim_{\epsilon \to 0} \left\{ \int_0^{E-\epsilon} + \int_{E+\epsilon}^{\infty} \right\} dE' \, \frac{g(E')}{E' - E} $$

[6] The original dispersion relation, derived by Kramers and Kronig, was concerned with scattering of light in a refractive medium and was an integral relation similar to (15.4) for the refractive index $n(\omega)$ as a function of frequency. Since the frequency dependence of $n(\omega)$ determines the dispersion, the relation was called a dispersion relation.

In this case the imaginary part of the dispersion relation (15.4) becomes an identity while the real part reads:

$$\text{Re}\, f_l(E) = \sum \frac{\Gamma_n}{E - E_n} + \frac{1}{\pi}\left\{\int_{-\infty}^{E_L} + P\int_0^{\infty}\right\} dE' \frac{\text{Im}\, f_l(E')}{E' - E}$$

[E physical]   (15.6)

This expresses the real part of the physical partial-wave amplitude in terms of an integral over its imaginary part, together with $\text{Im}\, f$ on the left-hand cut (which is, of course, unphysical) and the residues Γ_n at the bound states.

The importance of the partial-wave dispersion relation in relativistic scattering theory can be briefly described as follows: In relativistic scattering it is generally believed that the partial-wave amplitude should have essentially the same analytic structure as that derived for the Yukawa potential, and should therefore satisfy a dispersion relation of the general form (15.6). Now, the imaginary part of f_l on the left-hand cut can usually be approximated in terms of some small number of parameters. (This is facilitated by the fact that in relativistic scattering the value of the amplitude for a process $a + b \rightarrow c + d$ on its left-hand cut is closely related to the amplitudes for the so-called "crossed" processes $a + \bar{c} \rightarrow \bar{b} + d$ and $a + \bar{d} \rightarrow c + \bar{b}$ where $\bar{b}$, $\bar{c}$, and $\bar{d}$ denote the antiparticles of b, c, and d.) The residues Γ_n at the bound states can be expressed in terms of the basic coupling constants. Finally, on the right-hand cut we have[7] $\text{Im}\, f_l = p\,|f_l|^2$—which lets one eliminate either $\text{Re}\, f_l$ or $\text{Im}\, f_l$ (or some suitable combination) from the dispersion relation (15.6). Thus, the partial-wave dispersion relation becomes an integral equation for a single real function (either $\text{Im}\, f_l$ or $\text{Re}\, f_l$ or some combination) in terms of functions that are known or approximated. In principle, this equation could be solved to give the partial-wave amplitude, and in practice it can at least be used to fit scattering data and determine some of the parameters used.[8]

---

[7] This relation is often described as a consequence of unitarity because it follows from the equation $s_l = \exp(2i\delta_l)$. In relativistic problems the amplitude is usually defined a little differently so that the relation has an extra factor of energy. Obviously this doesn't affect the general principle.

[8] Three further comments:

(1) One common method works neither with $\text{Im}\, f_l$ nor $\text{Re}\, f_l$. Instead, f_l is written as n/d where $d \equiv f_l(p)$, and the unitarity condition is used to eliminate either d or n, leaving an integral equation for n or d. This is the so-called "n over d" method.

(2) Working with the n over d method it should not be necessary to include the bound-state poles as input; the calculation should *predict* the bound states as zeros of d.

(3) In the relativistic case the problem is complicated by the inevitable possibility of inelastic processes.

15-b.   Forward Dispersion Relations

The amplitude of primary interest is, of course, the full amplitude $f(E, \theta)$, and it is natural to ask whether it too can be regarded as an analytic function of its variables. It develops that the full amplitude is indeed an analytic function of the appropriate variables, and the study of its analytic properties has become an extensive and fruitful part of collision theory. We begin by discussing the special case of the forward amplitude $f(E, 0)$, which we temporarily abbreviate as:

$$f(E, 0) \equiv f(E)$$

The analytic properties of $f(E)$ can be investigated using the results already established for the partial-wave amplitude; however, it turns out that the problem is just as easily attacked from first principles.

The forward amplitude can be written in terms of the on-shell T matrix as

$$f(E) = -(2\pi)^2 m \langle \mathbf{p} | \, T(E + i0) \, | \mathbf{p} \rangle \qquad (15.7)$$

where $E = p^2/2m$ (in order to be on shell) and we assume that V is spherically symmetric (so that the direction of $\mathbf{p}$ is arbitrary). The reader will recall that the operator $T(z)$ is analytic in the whole plane of z except on a cut along the positive real axis and at poles at the bound-state energies. This immediately suggests that we should be able to continue the amplitude (15.7) up from the real axis of E and that it will be analytic in E except on a right-hand cut $\{0 < E < \infty\}$ and at the bound-state poles. Unfortunately, this conclusion—which is in fact correct—is not quite as obvious as it seems. The point is that in (15.7) the magnitude p of the momentum $\mathbf{p}$ is tied to the energy as $p = (2mE)^{\frac{1}{2}}$. If we make E complex, then p also becomes complex; and if we wish to consider $f(E)$ as an analytic function of E, then we must recognize that it depends on E both because of the E in the argument of T *and* because of the initial and final momentum eigenstates.

Since there is no meaning to a momentum eigenstate $|\mathbf{p}\rangle$ with complex momentum, our first task is to rewrite the on-shell T matrix in (15.7) to display its p dependence explicitly. This is easily done by writing the matrix element in terms of wave functions (remember that $T = V + VGV$):

$$f(E) = -\frac{m}{2\pi} \int d^3x V(r)$$

$$-\frac{m}{2\pi} \int d^3x' \int d^3x \, e^{-i\mathbf{p}\cdot\mathbf{x}'} V(r') \langle \mathbf{x}' | \, G(E + i0) \, | \mathbf{x} \rangle V(r) e^{i\mathbf{p}\cdot\mathbf{x}} \quad (15.8)$$

The first term here is the Born approximation f_{Born}, which is independent of energy (in the forward direction) and, hence, certainly analytic. Accordingly we write,

$$f(E) = f_{\text{Born}} + \tilde{f}(E)$$

and can concentrate on the second term:

$$\tilde{f}(E) = -\frac{m}{2\pi} \int d^3x' \int d^3x \, e^{ip \, |\mathbf{x}-\mathbf{x}'| \cos \theta} V(r')V(r)\langle \mathbf{x}'| \, G(E + i0) \, |\mathbf{x}\rangle \quad (15.9)$$

where θ is the angle between $(\mathbf{x} - \mathbf{x}')$ and the fixed direction of $\mathbf{p}$.

The analytic properties of $\tilde{f}(E)$ are fairly easy to establish. Having devoted considerable space to detailed proofs for the partial-wave amplitude, we shall content ourselves with just a brief outline of the argument. For full details we refer the reader to the literature (Faddeev, 1958; Klein and Zemach, 1958). The integrand in (15.9) depends on E through two terms: (1) the exponential $\exp(ip \, |\mathbf{x} - \mathbf{x}'| \cos \theta)$, which is analytic in E except for a branch point at $E = 0$, and (2) the Green's function $\langle \mathbf{x}'| \, G(E + i0) \, |\mathbf{x}\rangle$, which is analytic except at the branch cut for $0 \leqslant E < \infty$ and at the bound state poles. Provided the integral (15.9) is suitably convergent, the analytic properties of the integral are the same as those of the integrand; and for any potential satisfying our usual assumptions (p. 191) this is, in fact, the case. Thus, it follows that $\tilde{f}(E)$—and also $f(E) = f_{\text{Born}} + \tilde{f}(E)$—is analytic except on a right-hand cut and at the bound-state poles, as shown in Fig. 15.4.[9]

To establish the forward dispersion relation we need three further results:

(1) As $E \to \infty$ in any direction $\tilde{f}(E) \to 0$; hence,

$$f(E) \to f_{\text{Born}} \quad (15.10)$$

This is the same result quoted in Section 9-a; for its proof see Klein and Zemach (1958).

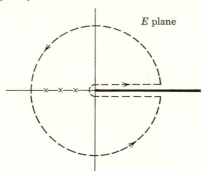

FIGURE 15.4.   Region of analyticity of the forward amplitude showing contour used to prove the forward dispersion relation.

[9] We shall return later to the remarkable fact that $f(E)$ has no left-hand cut, even though the individual partial-wave amplitudes do (for a Yukawa potential at least).

(2) For E on the negative real axis, $f(E)$ is real. This follows immediately from (15.8) since for $E < 0$, the Green's operator is Hermitian. It means that for any E,

$$f(E) = [f(E^*)]^* \qquad (15.11)$$

(3) At the bound state poles $f(E)$ has residue

$$\Gamma = (-)^{l+1}(2l + 1)\gamma^2$$

where γ is the asymptotic normalization of the corresponding normalized radial function, as discussed in the previous section. This can be proved directly from (15.9) but is easily understood by looking at the partial-wave series:

$$f(E, \theta) = \sum (2l + 1)f_l(E)P_l(\cos \theta)$$

Since we already know that each $f_l(E)$ has poles at the energies of all l-wave bound states, and that the residues have the form $(-)^{l+1}\gamma^2$, the result immediately follows.[10]

We are now ready to establish the forward dispersion relation. According to (15.10) it is $\tilde{f} = f - f_{\text{Born}}$ that goes to zero as $E \to \infty$. We therefore apply Cauchy's theorem to the function $\tilde{f}(E')/(E' - E)$ using the contour shown in Fig. 15.4. As we make the circular contour go to infinity its contribution vanishes, and by the result (15.11) the contribution from the upper and lower lips of the cut can be written as an integral over $\text{Im} f(E')$. (This works exactly as in the partial-wave case. Remember that f_{Born} is real.) Thus, we arrive at the forward dispersion relation

$$f(E) = f_{\text{Born}} + \sum_n \frac{\Gamma_n}{E - E_n} + \frac{1}{\pi} \int_0^\infty dE' \frac{\text{Im} f(E')}{E' - E} \qquad (15.12)$$

The remarkable features of the forward dispersion relation are that: (1) It is true quite independent of the details of the underlying interactions, and (2) it often happens that all quantities involved can be measured or calculated. If we let E move down to the physical region, the amplitude $f(E)$ becomes the physically measurable forward amplitude. The Born amplitude can often be simply calculated, and in most practical problems the number of bound states is small and the numbers Γ_n can either be calculated or put in as parameters. Finally, and most important, the integrand of the dispersion relation is the imaginary part of the forward amplitude (divided by

---

[10] To *prove* the result this way we must first prove that the partial-wave series is convergent in the neighborhood of the bound-state poles.

$E' - E$) which, according to the optical theorem, is

$$\text{Im} f(E) = \frac{p}{4\pi} \sigma(E)$$

where $\sigma(E)$ is the measureable total cross section.

If all of the quantities involved are either measured or calculated, then the forward dispersion relation provides one with a powerful consistency check. If not all of the quantities are known, then the relation can be used to find some or all of the unknowns. An interesting example of this type of application is in the scattering of electrons off hydrogen. Here there is just one bound state (the H^- ion) and so, for the particular case of zero energy, where the amplitude is minus the scattering length, (15.12) reads[11]

$$a = a_{\text{Born}} + \frac{\Gamma}{E_0} - \frac{1}{2\pi^2} \int_0^\infty dp \, \sigma(p) \tag{15.13}$$

In 1960 it happened that there were two conflicting measurements of the e–H cross section $\sigma(p)$. Since approximate values of the scattering lengths and residue Γ could be calculated, while the energy E_0 was known, it was possible to insert the various data into (15.13) and to use the forward dispersion relation to pick out the correct set of measurements. In high-energy physics, forward dispersion relations for pion–nucleon scattering have been used to determine the π–N coupling constant, whose square is the residue at the one pole in this problem.

15-c.   Nonforward Dispersion Relations

For several years it has been hoped that one could construct a theory of relativistic scattering based on dispersion relations; and to this end it is essential to have relations more general than the forward dispersion relation (15.12). In particular, if one is to be able to calculate the amplitude $f(E, \theta)$ for arbitrary E and θ, one must obviously have a relation for the *nonforward* amplitude. This immediately poses a problem concerning the choice of independent variables.

Until now we have written the amplitude as $f(E, \theta)$, regarding E and θ as the two independent variables. However, we could equally well have written it as $f(E, q)$, for example, and thought of it as a function of E and the momentum transfer q. If we now wish to discuss the analytic properties of f,

---

[11] The real e–H problem is complicated by the fact that the particles have spin and the target is itself a two-body system. Nonetheless, the amplitudes are still believed to satisfy forward dispersion relations and the essential features of their use are illustrated by our simplified account (see Krall and Gerjuoy, 1960).

then it makes a real difference which variables we take to be independent. To understand this we can consider the Born approximation:

$$f_{\text{Born}} = -\frac{2m}{q} \int_0^\infty r\, dr\, \sin qr V(r) = \text{function of } q \qquad (15.14)$$

If we consider f as a function of the two variables E and q, then the Born approximation is *independent* of E; and f_{Born} is *obviously* analytic in E. If, however, we treat f as a function of E and θ, then,

$$f_{\text{Born}} = \text{function of } \left[2(2mE)^{\frac{1}{2}} \sin\frac{\theta}{2} \right]$$

that is, as a function of E and θ, f_{Born} is *not* independent of E, and to prove that it is analytic in E would require a detailed examination of the integral (15.14).

This simple example would seem to indicate that the best choice of independent variables would be the energy E and momentum transfer q; this conclusion is essentially correct. In fact, the most convenient variables for most purposes are E and q^2, the principal reason being that in relativistic processes E and q^2 are intimately related by the process of "crossing" mentioned earlier. Accordingly we consider the amplitude as a function $f(E, q^2)$ of E and q^2 and examine first its analytic properties as a function of E with q^2 fixed. For the special case that $q^2 = 0$ we have already shown that $f(E, q^2)$ is analytic in E and satisfies the forward dispersion relation (15.12). To establish corresponding results for $q^2 \neq 0$ it turns out that one must make more restrictive assumptions on the potential. To be definite we consider a Yukawa

$$V(r) = \gamma\, \frac{e^{-\mu r}}{r}$$

(although the results we shall quote also hold for sums or even integrals of Yukawas). In this case, an analysis quite similar to that of the preceding section shows that, for any fixed q^2 with $0 \leqslant q^2 < 4\mu^2$, the amplitude $f(E, q^2)$ has the same analytic properties as for $q^2 = 0$, is real for $E < 0$, and tends to f_{Born} as $E \to \infty$. Therefore, it satisfies the *nonforward dispersion relation*:

$$f(E, q^2) = -\frac{2\gamma m}{\mu^2 + q^2} + \sum \frac{\Gamma_n P_{ln}(\cos\theta)}{E - E_n}$$

$$+ \frac{1}{\pi} \int_0^\infty dE'\, \frac{\text{Im}\, f(E', q^2)}{E' - E} \qquad (15.15)$$

where the first term is just the Born approximation for the Yukawa potential and the angle θ is defined by $\cos\theta \equiv 1 - q^2/4mE$.

The methods of the last section establish the relation (15.15) for[12] $0 \leqslant q^2 < 4\mu^2$. However, using the techniques described in the next section the relation can be continued to all q^2; thus, the amplitude for a Yukawa potential satisfies the nonforward dispersion relation (15.15) for any fixed q^2 with $0 \leqslant q^2 < \infty$.

Disappointingly, the nonforward dispersion relation with $q^2 \neq 0$ proves much less useful than its forward equivalent. In the first place, we cannot use the optical theorem to replace $\mathrm{Im}\, f(E', q^2)$ by the experimentally measurable total cross section, as we could in the case $q^2 = 0$. Second, the amplitude that appears in the integrand is not even the physical amplitude, even though it has E' real. The point is that the integral runs from $E' = 0$ to ∞ with q^2 fixed. For any given physical energy E the possible values of the momentum transfer are limited to

$$0 \leqslant q^2 \leqslant 4p^2 = 8mE$$

thus, for any given momentum transfer, the physical energy E must satisfy

$$\frac{q^2}{8m} \leqslant E < \infty$$

Thus, that part of the integral (15.15) that runs from 0 to $q^2/8m$ involves not the physical amplitude, but rather its analytic continuation to unphysical energies.

It is because of these difficulties that we must examine the properties of $f(E, q^2)$ as an analytic function of E *and* q^2 and establish the "double dispersion relation" known as the Mandelstam representation. Before we do this, however, there is an apparent contradiction that deserves comment.

We saw that the partial-wave amplitudes $f_l(E)$ for a Yukawa potential are analytic except at: (1) a right-hand branch cut for $0 \leqslant E < \infty$, (2) bound-state poles, and (3) a left-hand cut for $-\infty < E \leqslant -\mu^2/8m$. We have now seen that the full amplitude $f(E, q^2)$ is, for each fixed q^2, analytic in E with a right-hand cut and with bound-state poles, *but with no left-hand cut*. Since

$$f(E, q^2) = \sum (2l + 1)f_l(E)P_l(\cos\theta) \qquad (15.16)$$

the question naturally arises how the left-hand branch points that are present in all of the $f_l(E)$ can fail to appear in $f(E, q^2)$.

Mathematically, the answer to this question is straightforward. The series (15.16) converges for physical values of E; it also converges when E is continued away from the real axis to some parts of the complex plane. However, the series *diverges* as soon as we move E to the neighborhood of the left-hand cut. As soon as any series diverges, the analytic properties of its

[12] See Newton (1966) Section 10.3.2, or Goldberger and Watson (1964) Section 10.3.

sum cease to bear any resemblance to those of its individual terms.[13] Thus, from the mathematical point of view we can simply say that since the partial-wave series (15.16) diverges for E on the left-hand cut, there is no reason why the left-hand singularities of the $f_i(E)$ should communicate themselves to $f(E, q^2)$—and, in fact, they do not.

Nonetheless, the fact that the left-hand singularities of the partial-wave amplitudes do all cancel out of the full amplitude is an important result. We can look at the result the other way round and examine how it is that a full amplitude $f(E, q^2)$, which has no left-hand cut, can give a partial-wave amplitude that does. To see this we recall that

$$f_l(E) = \tfrac{1}{2} \int_{-1}^{1} dz \, P_l(z) f(E, q^2) \tag{15.17}$$

where $z = \cos \theta = 1 - q^2/4mE$. If we consider, in particular, the s-wave amplitude and suppose that the Yukawa is sufficiently weak to be treated in Born approximation, then

$$f = f_{\text{Born}} = -\frac{2\gamma m}{\mu^2 + q^2} = -\frac{2\gamma m}{\mu^2 + 4mE(1 - z)} \tag{15.18}$$

and hence,

$$f_0(E) = -\int_{-1}^{1} dz \, \frac{\gamma m}{\mu^2 + 4mE(1 - z)} = \frac{\gamma}{4E} \ln\left(\frac{\mu^2}{\mu^2 + 8mE}\right)$$

which has the expected branch point at $E = -\mu^2/8m$. Thus, while the full amplitude (15.18) has no branch points, the partial-wave amplitude does. This illustrates (what is, in fact, true of all partial waves) that the left-hand branch cut is introduced into the partial-wave amplitude in the process of making the projection (15.17).

15-d.   The Mandelstam Representation

For the reasons already mentioned, the nonforward dispersion relation is not enough to set up a calculational scheme based on the analytic properties of $f(E, q^2)$. To this end one is led to examine the properties of $f(E, q^2)$ as an analytic function of the *two* complex variables E and q^2. The first step in doing this is to examine $f(E, q^2)$ as a function of q^2 with E fixed and real. This problem was first considered by Lehmann, who established analyticity for q^2 inside a certain ellipse, now called the Lehmann ellipse. This result was subsequently extended to show that for a Yukawa potential $\tilde{f}(E, q^2) = f - f_{\text{Born}}$ is

---

[13] Consider, for example, the function $z/(z - 1)$, which can be expanded in the series $\sum_0^\infty z^{-n}$. This series is convergent for $|z| > 1$ but divergent for $|z| < 1$. Thus, the fact that every term in the series has a singularity at $z = 0$ implies nothing about the original function, which is in fact analytic at $z = 0$.

analytic in the whole plane of q^2 except for a cut for $-\infty < q^2 \leqslant -4\mu^2$. (Notice that in this case the cut $\{-\infty < q^2 \leqslant -4\mu^2\}$ is completely disjoint from the physical region $\{0 \leqslant q^2 < \infty\}$.) This result can be proved in various ways, of which perhaps the most satisfactory is that of Regge—which we shall describe in Section 15-g. The attractive feature of Regge's method is that it also determines the behavior of $f(E, q^2)$ as $q^2 \to \infty$. Specifically, we shall see that $f(E, q^2)$ is bounded by some power of q^2

$$f(E, q^2) = O(q^2)^\alpha \qquad (15.19)$$

as $q^2 \to \infty$ in any direction, and for all values of E. In particular, for certain potentials (a necessary condition is that the potential have no bound states) the power α is negative; in this case $f(E, q^2)$ *goes to zero* as $q^2 \to \infty$ in any direction. Anticipating these results, we can now establish the properties of $f(E, q^2)$ as a function of E and q^2.

We already know that for fixed real q^2 the amplitude $f(E, q^2)$ satisfies the nonforward dispersion relation (15.15), which we write, *for the case of no bound states*, as:

$$f(E, q^2) = -\frac{2\gamma m}{q^2 + \mu^2} + \frac{1}{2\pi i} \int_0^\infty \frac{dE'}{E' - E}$$
$$\times [f(E' + i0, q^2) - f(E' - i0, q^2)] \quad (15.20)$$

We recall that the methods described so far prove this relation only for $0 \leqslant q^2 < 4\mu^2$. However, we shall now see how it can be extended to all q^2.

Examining the integrand of (15.20) we note that for fixed E' each of the terms is analytic in the plane of q^2, cut from $-\infty$ to $-4\mu^2$.[14] This suggests that we could now write the integrand itself as a dispersion integral in q^2. In particular, if we consider the case that the power α in (15.19) is negative and $f(E, q^2) \to 0$ as $q^2 \to \infty$, we can immediately write down the required dispersion relation:

$$\frac{1}{2i} [f(E' + i0, q^2) - f(E' - i0, q^2)] = \frac{1}{\pi} \int_{-\infty}^{-4\mu^2} \frac{dq'^2}{q'^2 - q^2} \rho(E', q'^2) \quad (15.21)$$

Here the function $\rho(E', q'^2)$—called the *double spectral function*—is just the discontinuity across the q^2 cut of the term in brackets on the left. Clearly, $\rho(E', q'^2)$ is defined only for $0 \leqslant E' < \infty$ and $-\infty < q'^2 \leqslant -4\mu^2$ and is real.

---

[14] There is also a *pole* at $q^2 = -\mu^2$ coming from the Born term $-2\gamma m/(q^2 + \mu^2)$. However, this cancels out of the integrand and so need not concern us.

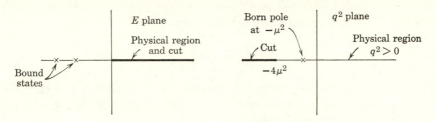

FIGURE 15.5. The amplitude $f(E, q^2)$ for a single Yukawa is analytic in the product of the cut planes of E and q^2.

Combining (15.20) and (15.21) we immediately arrive at the *double dispersion relation* or *Mandelstam representation*

$$f(E, q^2) = -\frac{2\gamma m}{q^2 + \mu^2} + \frac{1}{\pi^2}\int_0^\infty \frac{dE'}{E' - E}\int_{-\infty}^{-4\mu^2} \frac{dq'^2}{q'^2 - q^2}\,\rho(E', q'^2) \qquad (15.22)$$

Since the nonforward dispersion relation (15.20) was originally proved only for $0 \leqslant q^2 < 4\mu^2$ the relation (15.22) is *a priori* true only for the same values of q^2. However, a moment's inspection should convince the reader that the integral (15.22) is analytic in q^2 for all q^2 (except on the cut $-\infty < q^2 \leqslant -4\mu^2$), and consequently it defines the unique analytic continuation of $f(E, q^2)$ for all values of q^2.[15] Therefore, we see that the physical amplitude $f(E, q^2)$ can be analytically continued throughout the complex planes of E and q^2 except for the cuts $\{0 \leqslant E < \infty\}$ and $\{-\infty < q^2 \leqslant -4\mu^2\}$ and the Born pole at $q^2 = -\mu^2$ (see Fig. 15.5). Its value $f(E, q^2)$ for arbitrary E and q^2 (real or complex) is given by the Mandelstam representation in terms of the one real function $\rho(E', q'^2)$ [plus the Born term $-2\gamma m/(q^2 + \mu^2)$].

The Mandelstam representation as given in (15.22) applies to a Yukawa potential that has no bound states and for which $f(E, q^2) = O(q^2)^\alpha$ as $q^2 \to \infty$ with α negative. It develops that if there are bound states, then the number α is certainly greater than zero, and the Mandelstam representation is therefore modified in two ways. First, there are bound state poles. Second, $f(E, q^2)$ does not go to zero as $q^2 \to \infty$ and our usual method of deriving a dispersion relation does not work. What we can do instead is to apply our usual method to the function $f(E, q^2)/(q^2)^m$ with m any integer greater than α. This function goes to zero as $q^2 \to \infty$ and its only disagreeable feature is an m-fold pole at

---

[15] If E is physical and we continue in q^2 to physical values $q^2 > 4\mu^2$, then we do recover the physical amplitude, since we already know that $f(E, q^2)$ is analytic for E physical and all q^2, and the process of analytic continuation is unique.

the origin. This introduces extra terms into the dispersion relation, which can easily be evaluated using the standard result:

$$\oint dz \, \frac{g(z)}{z^{p+1}} = \frac{2\pi i}{p!} \left(\frac{d^p g}{dz^p} \right)_{z=0}$$

In this way we can write a dispersion relation in q^2, which is then substituted into the nonforward dispersion relation to give what is called the *Mandelstam representation with* m *subtractions*[16]

$$\begin{aligned}
f(E, q^2) = &- \frac{2\gamma m}{q^2 + \mu^2} + \sum \frac{\Gamma_n P_{l_n}(\cos \theta)}{E - E_n} \\
&+ \sum_{v=0}^{m-1} (q^2)^v \frac{1}{\pi} \int_0^\infty \frac{dE'}{E' - E} \tau_v(E') \\
&+ \frac{(q^2)^m}{\pi^2} \int_0^\infty \frac{dE'}{E' - E} \int_{-\infty}^{-4\mu^2} \frac{dq'^2}{q'^2 - q^2} \frac{\rho(E', q'^2)}{(q'^2)^m} \qquad (15.23)
\end{aligned}$$

Here the poles at $E = E_n$ are the familiar bound-state poles and the terms in $(q^2)^v$ come from the multiple pole at $q^2 = 0$ just discussed (see Problem **15.2**).

The importance of the Mandelstam representation in relativistic scattering can be sketched as follows: In nonrelativistic scattering the underlying dynamical equation, from which everything (including the Mandelstam representation) is derived, is of course the Schrödinger equation. However, in relativistic scattering it is not at all clear that the Schrödinger equation has any simple counterpart, and probably the single most important problem is in fact to *discover* the fundamental dynamical principles. Under these circumstances it is very interesting to note that:

(1) In nonrelativistic two-particle scattering the Mandelstam representation can actually replace the Schrödinger equation as the basic calculational tool. Specifically, we shall see that if the potential is known then one can calculate $f(E, q^2)$ using just the fact that it is given by the Mandelstam representation and satisfies unitarity.

(2) There is reason to think that relativistic amplitudes should satisfy the Mandelstam representation or something quite like it.

These two facts suggest that it might be possible to construct a relativistic dynamics (or at least a prescription for calculation) based on analyticity of

---

[16] The origin of this curious name is this: If the "unsubtracted" form (15.22) converges, then we can use it to write down $f(E, q^2) - f(E, 0)$, and the result has exactly the "once-substracted" form (15.23) with $m = 1$; that is, the "subtracted" dispersion relation is obtained from the "unsubtracted" relation by subtraction. This derivation only makes sense if the "unsubtracted" relation converges.

the amplitudes (as expressed by the Mandelstam representation) plus unitarity. During the last 10 or 15 years this possibility has been the subject of extensive, and sometimes fruitful, research under the general heading of analytic S-matrix theory.

The method by which amplitudes can be calculated using the Mandelstam representation and unitarity is very cumbersome and is certainly not to be regarded as a practical technique in the nonrelativistic context. Its importance is to establish that (in principle at least) analyticity and unitarity do determine the amplitudes in terms of the potential. Here we shall just sketch the method for the case that there are no bound states and that no subtractions are needed. First, we note that the Born amplitude is just the Fourier transform of the potential; thus, knowledge of V is equivalent to knowledge of f_{Born}, and we shall take f_{Born} as our given input. Next, we note that unitarity of S implies that (for a spherical potential):

$$\langle \mathbf{p}'|\, R\, |\mathbf{p}\rangle^* + \langle \mathbf{p}'|\, R\, |\mathbf{p}\rangle = -\int d^3 p'' \langle \mathbf{p}''|\, R\, |\mathbf{p}'\rangle^* \langle \mathbf{p}''|\, R\, |\mathbf{p}\rangle$$

and, hence, for the amplitude

$$\operatorname{Im} f(\mathbf{p}' \leftarrow \mathbf{p}) = \frac{p}{4\pi} \int d\Omega_{p''} f(\mathbf{p}'' \leftarrow \mathbf{p}')^* f(\mathbf{p}'' \leftarrow \mathbf{p})$$

—a result sometimes called the *generalized optical theorem*. If this equation is substituted into the Mandelstam representation (15.22), then some lengthy algebra leads to an equation for the double spectral function $\rho(E, q^2)$ as an integral involving the Born term and the double spectral function itself. The important property of this integral is that it gives $\rho(E, q^2)$ for any given E and q^2 exclusively in terms of its values for *smaller* (absolute) values of q^2. Specifically, for $-9\mu^2 < q^2 < -4\mu^2$ it gives $\rho(E, q^2)$ in terms of the known f_{Born} only; this means we can calculate $\rho(E, q^2)$ in this range by a single integration. In the interval $-16\mu^2 < q^2 < -9\mu^2$ the integral gives $\rho(E, q^2)$ in terms of f_{Born} plus the values of ρ in the first interval $-9\mu^2 < q^2 < -4\mu^2$, which we have now calculated. Thus, a second integration determines ρ in the second interval. This procedure continues and lets us calculate $\rho(E, q^2)$ for any finite values of E and q^2 by some finite number of integrations.

We see that the Mandelstam representation and unitarity let us determine the double spectral function starting from the known Born term (or equivalently, the potential). Finally, we can insert the double spectral function into the Mandelstam representation and determine the amplitude itself. Thus, in nonrelativistic scattering at least, the Mandelstam representation and unitarity allow calculation of the scattering amplitude and could therefore replace the Schrödinger equation as our basic dynamical principle.

Relativistic scattering amplitudes certainly satisfy a unitarity equation. Thus, if they also satisfy the Mandelstam representation, and if one could find some way to obtain the Born amplitude as input (which one certainly can in some cases), then the analysis just sketched suggests that analyticity plus unitarity might indeed form the basis of a relativistic dynamics, which could be used to compute scattering amplitudes. This is the hope that has inspired much of the tremendous theoretical effort expended on the study of analytic properties during the past 10 or 15 years.

It must be admitted that the Mandelstam representation has not entirely justified the more optimistic claims made on its behalf in its early days. First, a tremendous theoretical effort has failed to prove the representation within the framework of relativistic quantum field theory. This is perhaps not so serious if one remembers that the representation was intended (perhaps) to replace the more conventional dynamical principles; thus, the representation, or something like it, might be correct even if it cannot be proved in quantum field theory. However, there is also considerable internal evidence that the form of the representation must be much more complicated than either (15.22) or (15.23). In particular, in relativistic problems one must always take into account inelastic processes that lead to the production of more than two particles, and these enormously complicate the problem. While some progress has been made in handling multiparticle processes, there is still much about them that is not well understood. Thus, for all the progress that has been made in this field (much reinforced by the methods of complex angular momenta described in the next three sections), there is still no conclusive evidence that the Mandelstam representation, or some more complicated equivalent, together with unitarity could actually provide a self contained dynamical system.

Nonetheless, it seems clear that much of the physical content of these ideas is correct, and they have undoubtedly provided one of the most powerful approaches to the handling of relativistic scattering amplitudes. Certainly a familiarity with these ideas is a prerequisite for an understanding of most of the current literature of high-energy physics.

15-e. Complex Angular Momenta

Our final example of the use of analytic functions in scattering theory is the method of complex angular momenta. This method was originally developed by Regge, after whom the whole subject is often called Regge theory. The method centers on the analytic properties of the partial-wave amplitude as a function of energy *and angular momentum*.

We first remark that in the context of the nonrelativistic Schrödinger equation it is a simple matter to let l become a complex variable. Recall that the partial-wave amplitude is determined by the radial equation

$$\left[\frac{d^2}{dr^2} - \frac{l(l+1)}{r^2} - U(r) + p^2\right]y(r) = 0$$

Nothing in this equation requires that l be a positive integer (this requirement comes from the properties of the spherical harmonics). We can certainly let l be an arbitrary complex variable and then study the solutions of the equation as analytic functions of l. This procedure is, of course, exactly analogous to what was done with the variable p in Chapter 12 and, in fact, all of the same techniques apply. Thus, it is a simple matter to show that for any potential satisfying our usual conditions there is a regular solution $\phi_{l,p}(r)$ which is analytic for all p and all l in the half plane $\{\text{Re } l > -\frac{1}{2}\}$. Similarly, there is a Jost function $f_l(p)$ that is analytic in the regions $\{\text{Im } p > 0\}$ and $\{\text{Re } l > -\frac{1}{2}\}$. The restriction of l to $\{\text{Re } l > -\frac{1}{2}\}$ is exactly analogous to that of p to $\{\text{Im } p > 0\}$.[17] If we make more restrictive assumptions on the potential it is possible to continue $f_l(p)$ to the left into $\{\text{Re } l \leqslant -\frac{1}{2}\}$, just as we have frequently continued p down into $\{\text{Im } p < 0\}$. However, in the case of l such a continuation is of little interest and we shall not discuss it any further. We *shall* wish to continue p into its lower half plane, and for this reason we shall restrict most of our discussion to Yukawa potentials, for which the Jost function is analytic in $\{\text{Re } l > -\frac{1}{2}\}$ and the *whole* of the p plane except for the usual cut. The proof of these assertions is so similar to the analysis of Chapter 12 that we shall not give it here.[18]

The extension of l to be a complex (or at least a continuous) variable is also a very natural thing to do and would certainly be expected to yield useful information. The point is that in our study of the partial-wave amplitude we have so far always focussed attention on one single value of l at a time. We have nowhere exploited the fact that all partial waves are determined by one and the same potential. In fact, it should be clear that so far we could just as well have used a different potential $V_l(r)$ for each value of l. Obviously, the fact that there is really just one potential that determines all partial waves is likely to have important consequences. For example, if we consider an

---

[17] The value $-\frac{1}{2}$ in the condition Re $l > -\frac{1}{2}$ occurs as follows. The two independent solutions of the radial equation go like r^{l+1} and $1/r^l$ as $r \to 0$ and therefore exchange roles on the line Re $l = -\frac{1}{2}$. It follows that the definition of ϕ as the solution that behaves like r^{l+1} breaks down to the left of this same line. To extend ϕ to the left of this line requires some procedure for analytic continuation; and this in turn requires additional assumptions on the potential.

[18] There is one small complication: Since $\hat{j}_l(pr)$ has a factor of p^{l+1}, it has a *branch point at* $p = 0$ when l is nonintegral. The same is true of $\hat{n}$ and $\hat{h}^{\pm}$ and, hence, of the Jost function [see Newton, 1964, Eq. (5–10)]. This "kinematic" branch point for unphysical l sometimes needs careful handling but will not cause any trouble in the applications discussed here.

s-wave bound state of some potential, and if we imagine l to move continuously away from zero, then we should be able to follow the corresponding
solution of the radial equation. When l reaches the value $l = 1$, this solution
will correspond to an $l = 1$ bound state of the same potential (provided the
energy is still negative). If we then move l on to $l = 2$ we will arrive at the
corresponding $l = 2$ bound state (provided, again, the energy is still negative), and so on. In this way we should be able to group the bound states of
any potential into families, such that all of the states in each family correspond to the "same" solution. We shall see in the next section that such
families—known as Regge trajectories—do indeed exist.

There are several simple (but nonetheless useful) results which we can prove
as soon as we let l become complex. We begin our quantitative discussion
with some examples of these.

Since the Jost function $f_l(p)$ is analytic in l and p, the same is true of the S
matrix $s_l(p) = [f_{l*}(p^*)]^*/f_l(p)$, except at those points where $f_l(p) = 0$. In
particular, if l and p are real then (as before) $f_l(p)$ cannot vanish and
$|s_l(p)| = 1$, which means that we can define a phase shift $\delta_l(p)$ given by

$$s_l(p) = \exp[2i\delta_l(p)]$$

and $\delta_l(p)$ is then analytic (and *a fortiori* continuous) for all positive real l
and p.

Having defined $\delta_l(p)$ as an analytic function of l we can discuss its derivative. Starting with the radial equation for ϕ and differentiating with respect
to l we find that, as the reader should check (Problem **15.3**),

$$\frac{d}{dr} W\left(\phi, \frac{\partial \phi}{\partial l}\right) = (2l + 1)\frac{\phi^2}{r^2}$$

and hence, that

$$\left[W\left(\phi, \frac{\partial \phi}{\partial l}\right)\right]_0^\infty = (2l + 1)\int_0^\infty dr\, \frac{\phi^2}{r^2} > 0 \qquad [l \text{ and } p \text{ real}] \qquad (15.24)$$

The Wronskian in question is zero at $r = 0$ and can be evaluated at $r = \infty$
using the known asymptotic form of ϕ. Remembering that $f = |f|e^{-i\delta}$ we
get the result

$$p\,|f_l(p)|^2\left(\frac{\pi}{2} - \frac{\partial \delta_l}{\partial l}\right) > 0$$

or

$$\boxed{\frac{\partial \delta_l}{\partial l} < \frac{\pi}{2}} \qquad\qquad (15.25)$$

This result shows that no physical phase shift $\delta_{l+1}(p)$ can exceed the next lower phase shift $\delta_l(p)$ by more than $\pi/2$,

$$\delta_{l+1}(p) < \delta_l(p) + \frac{\pi}{2} \tag{15.26}$$

This has several interesting consequences. For example, the reader will recall that according to Levinson's theorem the number of bound states of angular momentum l is given by

$$n_l = \frac{\delta_l(0)}{\pi}$$

Thus, by (15.26),

$$n_{l+1} < n_l + \tfrac{1}{2}$$

or, since both n_{l+1} and n_l are integers,

$$\boxed{n_{l+1} \leqslant n_l}$$

That is, the number of bound states with angular momentum $l + 1$ can never exceed the number with angular momentum l.

Now is also a convenient time to discuss our first rigorous result concerning the phase shifts for large l. It can be shown that for any potential satisfying our usual assumptions (p. 191) the amplitude satisfies[19]

$$|pf_l| = |\sin \delta_l(p)| \leqslant \frac{\alpha}{l} \qquad [l \text{ large, all } p] \tag{15.27}$$

where α is some constant independent of l and p. Thus, as $l \to \infty$, each $\delta_l(p)$ approaches some integral multiple of π, for any fixed p. Furthermore, according to (15.27) we can choose l large enough so that $|\sin \delta_l(p)| < \tfrac{1}{2}$ (say) *for all p*; thus, since $\delta_l(p)$ is continuous it must be close to the *same* integral multiple of π, whatever the value of p. We already know [see (11.38)] that for large p, $\delta_l(p)$ is close to 0 (not just $n\pi$); and it therefore follows that the limit of $\delta_l(p)$ as $l \to \infty$ is also 0 (not just $n\pi$). That is, having adjusted the phase shift to go to zero as $p \to \infty$, we find that it also goes to zero as $l \to \infty$.

[19] I have not found a proof of this result in the published literature. However, the interested reader will have no difficulty constructing a proof from standard results. For example, the bound given at the bottom of p. 85 of Alfaro and Regge (1965), combined with the obvious bound $|U(r)| <$ constant$/r^2$ gives for large l, $|\sin \delta_l| <$ constant $\int dp \, \hat{j}_l(p)/p^2 <$ constant$/l$ where the last inequality is obtained by explicit integration [see Alfaro and Regge, 1965, p. 191, Eq. (D.4)].

15-f.   Regge Poles

The partial-wave amplitude $f_l(E)$ is analytic in l and E except for poles at those points where the Jost function [which we now call $f_l(E)$] vanishes. Since $f_l(E)$ is an analytic function of *two* variables, we can look on its poles as poles in E whose positions depend on the value of l, or *vice versa*. For some purposes it is convenient to take the latter view; that is, to focus attention on a definite value of E and look for the poles of $f_l(E)$ as a function of l. When looked at in this way the poles of $f_l(E)$ are called Regge poles.[20] For a given value of E they will be located at points $l = \alpha_1(E)$, $\alpha_2(E)$, ... and if we then let E vary continuously each $\alpha_i(E)$ will trace out a locus in the plane of l. These loci are called *Regge trajectories*.

Now let us focus attention on negative real values of E. In this case it is easily seen (see Problem **15.4**) that the poles of $f_l(E)$ must occur at real values of l; that is, the Regge poles $\alpha_i(E)$ must be real. In particular, let us suppose that for some $E_0 < 0$ there is an s-wave bound state. This means that the s-wave Jost function $f_0(E_0)$ is zero, and hence, in our new terminology, that $f_l(E_0)$ has a Regge pole at $l = 0$. Let us label this pole $l = \alpha(E)$ and ask what happens as we move E away from E_0. In particular, we shall *increase* E, moving to the right along the real axis, and we now show that the Regge pole at $l = \alpha(E)$ moves to the right along the real l axis. We do this by showing that the derivative $d\alpha/dE$ is always positive.

If the Jost function $f_l(E)$ is zero at $l = \alpha(E)$ with $E < 0$ then the regular solution $\phi_{\alpha, E}(r)$ vanishes as $r \to 0$ and ∞ and is square integrable. Now, we know that ϕ satisfies the radial equation

$$\left[\frac{d^2}{dr^2} - \frac{\alpha(\alpha + 1)}{r^2} - U(r) + 2mE \right] \phi_{\alpha, E}(r) = 0 \qquad (15.28)$$

Differentiating with respect to E we find:

$$[\cdots] \frac{\partial \phi}{\partial E} - \left(\frac{2\alpha + 1}{r^2} \frac{d\alpha}{dE} - 2m \right) \phi = 0$$

We then multiply this equation by ϕ and subtract from it the original equation (15.28) multiplied by $\partial \phi / \partial E$. This gives

$$\frac{d}{dr} W\left(\phi, \frac{\partial \phi}{\partial E} \right) - \left(\frac{2\alpha + 1}{r^2} \frac{d\alpha}{dE} - 2m \right) \phi^2 = 0$$

---

[20] The difference between Regge poles and the poles discussed in Chapter 12 is only a question of point of view. Here we fix E and discuss the poles in l, there we fixed l and discussed poles in E. In either case one is really just seeking values of l *and* E for which $f_l(E) = 0$.

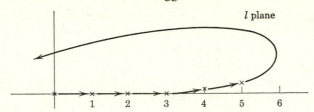

FIGURE 15.6. Typical Regge trajectory for a Yukawa potential.

or, when integrated from 0 to ∞

$$(2\alpha + 1)\frac{d\alpha}{dE}\int_0^\infty dr\,\frac{\phi^2}{r^2} = 2m\int_0^\infty dr\phi^2$$

This establishes the required result that $d\alpha/dE > 0$.

If we return to our Regge pole at $l = \alpha(E_0) = 0$ and start increasing E, then we now know that $\alpha(E)$ starts increasing from $l = 0$ towards $l = 1$. Now for all E (less than 0) the zero of $f_l(E)$ at $l = \alpha(E)$ implies that there is a normalizable solution of the radial equation. As $\alpha(E)$ moves from 0 to 1 this does not have any physical significance. However, when $\alpha(E)$ reaches the point $l = 1$ (provided it does so while E is still negative) the zero of $f_l(E)$ implies a pole of the p-wave amplitude and the solution of the Schrödinger equation is a bona fide p-wave bound state. Thus, the same pole that began at $l = 0$, $E = E_0$ as an s-wave bound state has become a p-wave bound state at $l = 1$ and $E = E_1$, say.

If we continue to move E to the right, we will continue to encounter bound states as $\alpha(E)$ passes through the points $l = 2, 3, \ldots$ until eventually the energy reaches the threshold $E = 0$. If we move E to the right of threshold then the zero of $f_l(E)$ at $l = \alpha(E)$ cannot remain on the real axis. In fact, it is easily shown (see Problem 15.4) that if we move E along the upper edge of the physical cut, then $\alpha(E)$ moves up into the complex plane. An example of a Regge trajectory that defines four bound states ($l = 0, 1, 2, 3$) and then leaves the axis between $l = 3$ and 4 is shown in Fig. 15.6.

Once the Regge trajectory $\alpha(E)$ leaves the real axis it no longer passes through the integers and cannot lead to bound states. However, as long as it remains close to the axis it defines a *resonance* each time its real part, Re $\alpha(E)$, passes through an integer. To see this let us suppose $f_l(E)$ has a pole at $E = E_0$ and $l = \alpha(E_0)$ with $\alpha(E_0)$ close to some integer l_{int}. If we move l the short distance to l_{int} the pole must move to some point E_1 close to E_0. Since l_{int} is real the pole at E_1 must be on the unphysical sheet and is therefore a resonance in the partial wave of angular momentum l_{int}. The trajectory shown in Fig. 15.6 has two such resonances, at $l = 4$ and 5.

As E increases, the Regge trajectories will in general move away from the real axis and cease to be physically observable. As $E \to +\infty$, the behavior of the trajectories depends on the details of the potential. For example, the trajectories of a square well move off to infinity as $E \to +\infty$. For a Yukawa, or suitable superposition of Yukawas, it can be shown (see Newton, 1966, p. 411) that the real part of any trajectory satisfies

$$\operatorname{Re} \alpha(E) \leqslant \frac{\text{constant}}{p} - \tfrac{1}{2} \qquad (15.29)$$

Thus, as $E \to +\infty$ the trajectories have to turn back and eventually disappear into the left half plane. This is the behavior illustrated in Fig. 15.6.

We conclude that all of the bound states and resonances of a potential can be divided into families. Each family contains all of the states on a single Regge trajectory and all members of a family are, in this sense, manifestations of a single physical phenomenon. A familiar example of this is provided by the electron–proton system (i.e., hydrogen), which has infinitely many Regge trajectories, the first of which contains the states $1s$, $2p$, $3d$, . . . , the next $2s$, $3p$, $4d$, . . . , and so on. This example is inevitably atypical since the Coulomb potential has infinitely many bound states, and no resonances. (The trajectories have all reached $l = +\infty$ by the time E gets to threshold.) With short range potentials there are generally only a finite number of trajectories (in the physically relevant region $\operatorname{Re} l \geqslant 0$, at least); and, in general, a trajectory will define some bound states and some resonances.

It is an attractive idea that a similar analysis might be applied in relativistic scattering, and that all of the relativistic "bound states" or "particles" can be classified according to the Regge trajectories on which they lie. In fact, there is reasonably strong evidence that this is the case, and tentative assignments of the particles to trajectories have been made. However, it should also be said that the situation in relativistic scattering is considerably more complicated. In particular, for reasons too involved to discuss here, relativistic Regge trajectories would define physically observable particles only at *alternate* integral values of l.

15-g.   The Watson Transform

The Partial-Wave Series and the Lehmann Ellipses.   One of the most fruitful applications of Regge theory is in the study of analytic and asymptotic properties of the full amplitude $f(E, q^2)$ as a function of the squared momentum transfer q^2. The starting point is, of course, the partial-wave series, which relates the full and partial amplitudes,

$$f(E, q^2) = \sum (2l + 1)f_l(E)P_l(\cos \theta) \qquad (15.30)$$

with $q^2 \equiv 2p^2(1 - \cos \theta)$ and $E = p^2/2m$. It can be shown that for any potential satisfying our usual conditions[21]

$$|f_l(E)| < \frac{\text{constant } p^\epsilon}{l^{2+\epsilon}} \qquad [l \text{ large, all } E > 0]$$

for some $\epsilon > 0$. Since the Legendre polynomials satisfy $|P_l| \leqslant 1$ for all physical θ, it immediately follows that the physical partial-wave series is convergent for all such potentials.

If we now try to continue (15.30) to complex values of q^2 (and hence complex $\cos \theta$) we find that $P_l(\cos \theta)$ grows exponentially as $l \to \infty$. Specifically, the behavior of $P_l(\cos \theta)$ is indicated by the bound[22]

$$|P_l(\cos \theta)| \leqslant \tau(\theta)l^{-\frac{1}{2}}e^{|\text{Im }\theta| \, l} \tag{15.31}$$

where the precise form of the function $\tau(\theta)$ need not concern us here. Since $P_l(\cos \theta)$ grows exponentially as $l \to \infty$, the partial-wave series converges only if the amplitude *diminishes* exponentially. For arbitrary potentials this is unfortunately not the case. However, if we restrict attention to Yukawa potentials [or any potential satisfying $|V| \leqslant \text{constant } \exp(-\mu r)/r$] then it can be shown (see Alfaro and Regge, 1965, Section 8.4) that for physical E

$$|f_l(E)| < \sigma(E)l^{-\frac{1}{2}}e^{-\alpha l} \tag{15.32}$$

where $\sigma(E)$ is some function of E and the number α is given by

$$\cosh \alpha = 1 + \frac{\mu^2}{2p^2} \tag{15.33}$$

Inserting the two bounds (15.31) and (15.32) into the partial-wave series we find that (for physical E)

$$|f(E, q^2)| < \text{constant } \tau(\theta)\sigma(E) \sum_l e^{(|\text{Im }\theta| - \alpha)l}$$

which is obviously convergent whenever $|\text{Im } \theta| < \alpha$. Since the partial-wave series is a series of polynomials in $\cos \theta$, each term is analytic in $\cos \theta$. It follows that the series defines $f(E, q^2)$ as an analytic function of $\cos \theta$ everywhere in the region defined by

$$|\text{Im } \theta| < \alpha$$

It is a straightforward exercise (see Problem **15.5**) to show that this condition confines $\cos \theta$ to an ellipse with foci at ± 1 and semi-major axis $(1 + \mu^2/2p^2)$.

----

[21] The argument is similar to that sketched in footnote[19] in Section 15-e but uses the bound $|U(r)| \leqslant \text{constant}/r^{3+\epsilon}$.

[22] For this and other properties of Legendre functions see Erdélyi (1955) Vol. 1, Chapter 3.

This ellipse is called the *small Lehmann ellipse* and is shown in Fig. 15.7, part (a).

If $f(E, q^2)$ is analytic in cos θ with E fixed, it follows at once that it is analytic in $q^2 = 2p^2(1 - \cos \theta)$, and the region of analyticity in terms of q^2 is the ellipse shown in Fig. 15.7 part (b). Notice that the Born pole $f_{\text{Born}} = -2\gamma m/(q^2 + \mu^2)$ is located at $q^2 = -\mu^2$ on the boundary of this ellipse, and is, in fact, the reason that the ellipse cannot be enlarged.

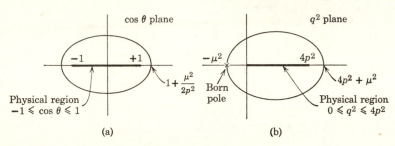

(a)   (b)

FIGURE 15.7. (a) For a Yukawa potential the partial-wave series defines f as an analytic function of cos θ in the small Lehmann Ellipse. (b) The corresponding ellipse when f is considered as a function of $q^2 = 2p^2(1 - \cos \theta)$.

If we consider, instead of f, the difference $\tilde{f} = f - f_{\text{Born}}$ it can be shown that $\tilde{f}_l(E)$ satisfies a bound like (15.32) and (15.33) but with μ replaced by 2μ.[23] This means that $\tilde{f}$ is analytic in larger ellipses—called the large Lehmann ellipses—obtained from those of Fig. 15.7 by replacing μ with 2μ.

The Watson Transform. To continue beyond the Lehmann ellipses we introduce the techniques of complex angular momenta. We first replace the partial-wave series by a cunningly chosen integral:

$$f(E, q^2) = \frac{1}{2i} \oint_C dl \frac{(2l + 1)f_l(E)P_l(-\cos \theta)}{\sin \pi l} \tag{15.34}$$

where the contour C is shown in Fig. 15.8 part (a), and encircles all of the positive integers. In this integral $f_l(E)$ is now the partial-wave amplitude analytically continued to complex l; the function $P_l(z)$ is the Legendre function of degree l, which is entire in l and analytic in z except on a cut $\{-\infty < z \leqslant -1\}$. It will be seen that the denominator sin πl has been ingeniously chosen to give poles at all integers and the argument of P_l has been changed to $-\cos \theta$ to compensate for the alternating sign of sin πl.

[23] The proof is similar to that of (15.32) (see Alfaro and Regge, 1965, Section 8.4). Strictly speaking, one can only replace μ by $2\mu - \epsilon$ with ϵ arbitrarily small; however, this implies the same region of analyticity.

The integral (15.34) is convergent in the same region as the partial-wave series from which it was derived; namely, the small Lehmann ellipse. However, we can now distort the contour of integration to obtain an integral that is convergent in some larger region. Specifically, we can open up the contour as shown in Fig. 15.8, part (b), until it runs down the line Re $l = -\frac{1}{2}$. When we distort the contour in this way two points need attention. First, the integral should include two large quarter circles. However, it can

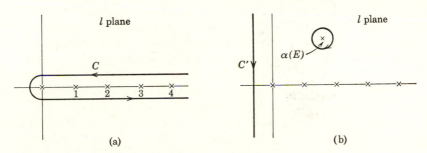

FIGURE 15.8. The two contours used in connection with the Watson transformation.

be shown that, for a Yukawa potential, $f_l(E) \rightarrow 0$ as $|l| \rightarrow \infty$ so fast that these quarter circles do not contribute (see Newton, 1964, Chapter 6). Second, the integrand has poles (Regge poles) at various points $l = \alpha(E)$ in the first quadrant. The fact that $f_l(E) \rightarrow 0$ as $|l| \rightarrow \infty$ guarantees that these poles are finite in number. Thus, as the contour sweeps across them we pick up a finite number of small loops, one around each pole as shown in Fig. 15.8, part (b). The final result is

$$f(E, q^2) = \frac{1}{2i} \int_{C'} dl(2l + 1)f_l(E) \frac{P_l(-\cos\theta)}{\sin\pi l} + \sum_{i=1}^{n} \beta_i(E)P_{\alpha_i(E)}(-\cos\theta) \quad (15.35)$$

where the $\beta_i(E)$ are just the residues of the integrand at the Regge poles (with the P_l factored out explicitly).

The passage from the integral of (15.34) to that of (15.35) was used by Watson and Sommerfeld in electromagnetic scattering and is therefore called the Watson (or Sommerfeld–Watson) transform. Its advantage is that, while the two expressions agree inside the small Lehmann ellipse, the new expression is actually convergent in a much larger region, and therefore allows continuation of $f(E, q^2)$ into this larger region. To see this, we first note that since $P_\alpha(z)$ is analytic in z except on a cut from $-\infty$ to -1, the finite sum of Regge pole terms in (15.35) is certainly analytic except for a cut

$$\{1 \leqslant \cos\theta < \infty\}$$

Concerning the integral we note that the term $P_l/\sin \pi l$ drops exponentially as $\text{Im } l \to \pm\infty$

$$\left|\frac{P_l(-\cos \theta)}{\sin \pi l}\right| \leqslant \rho(\theta) l^{-\frac{1}{2}} e^{-|\text{Re } \theta \text{ Im } l|}$$

while it can be shown for a Yukawa potential (see Newton, 1964, Chapter 6) that

$$|f_l(E)| \leqslant \nu(E) l^{-\frac{1}{2}}$$

(where the exact form of the functions ρ and ν need not concern us). Thus, the integral in (15.35) is convergent for all $\cos \theta$ provided $\text{Re } \theta \neq 0$. The condition $\text{Re } \theta \neq 0$ requires only that $\cos \theta$ avoid the line $\{1 \leqslant \cos \theta < \infty\}$; and the integral of (15.35) is therefore analytic except for $1 \leqslant \cos \theta < \infty$. Since the same is true of the Regge-pole terms we conclude that the Watson transform (15.35) defines $f(E, q^2)$ as an analytic function of $\cos \theta$ in the whole plane except for the cut $\{1 \leqslant \cos \theta < \infty\}$.

In terms of the variable $q^2 = 2p^2(1 - \cos \theta)$, the Watson transform defines $f(E, q^2)$ as an analytic function of q^2 except for a cut from 0 to $-\infty$. Since we already know that $f(E, q^2)$ is analytic in the small Lehmann ellipse of Fig. 15.7, part (b), we conclude that it is actually analytic in the union of these two regions; specifically, the plane of q^2 cut in $\{-\infty < q^2 \leqslant -\mu^2\}$. Finally, we know that the Born amplitude $f_{\text{Born}} = -2\gamma m/(q^2 + \mu^2)$ has a pole at $q^2 = -\mu^2$, while $f - f_{\text{Born}}$ is analytic in the large Lehmann ellipse. This means that $f - f_{\text{Born}}$ is analytic in q^2 with the smaller cut

$$\{-\infty < q^2 \leqslant -4\mu^2\}$$

as anticipated in Section 15-d.

Asymptotic Behavior. A remarkable feature of the Watson transform is that it not only allows continuation of $f(E, q^2)$ in q^2, but also determines its asymptotic form as $q^2 \to \infty$. The essential point is that as $|z| \to \infty$ the Legendre function $P_l(z)$ goes as

$$P_l(z) \xrightarrow[|z| \to \infty]{} \lambda(l) z^l$$

Referring back to the Watson transform (15.35) we see that the integral along the line $\text{Re } l = -\frac{1}{2}$ (often called the "background integral") goes to zero like $(\cos \theta)^{-\frac{1}{2}}$ as $|\cos \theta| \to \infty$. Each of the Regge pole terms behaves like $(\cos \theta)^{\alpha_i}$. Thus, in terms of the variable $q^2 = 2p^2(1 - \cos \theta)$, we find that:

$$f(E, q^2) \xrightarrow[|q^2| \to \infty]{} \gamma(E)(q^2)^{\alpha(E)} \tag{15.36}$$

where $\alpha(E)$ denotes that Regge pole furthest to the right at the energy E. In particular, if there are *no* Regge poles with $\text{Re } \alpha \geqslant 0$, then $f(E, q^2)$ actually goes to zero as $|q^2| \to \infty$.

The importance of the result (15.36) in the proof of q^2 dispersion relations should be clear. If there are no poles with Re $\alpha \geqslant 0$, then we can write an unsubtracted dispersion relation. If the potential has bound states, then we know that (at least for some energies) there *are* trajectories in the right half plane. Specifically, if l_0 is the highest angular momentum for which there is a bound state, we know that there is at least one trajectory with Re $\alpha(E) \geqslant l_0$ for some E. Thus, as $|q^2| \to \infty$, the amplitude $f(E, q^2)$ behaves like $(q^2)^{l_0}$ at least (and possibly much worse). In this case we can write a dispersion relation only for $f/(q^2)^m$ where m is some integer greater than l_0; that is, whenever there are bound states one must write a subtracted dispersion relation, the number of subtractions needed being at least as great as the angular momentum of the bound state with largest l.

To prove the Mandelstam representation it was necessary to write a q^2 dispersion relation that was valid for *all* $E \geqslant 0$. This requires that the power $\alpha(E)$ in (15.36) be less than some fixed α_{max} for all E. For a Yukawa potential this is easily seen to be the case. The bound (15.29) guarantees the result except in the immediate neighborhood of $E = 0$; but the result is certainly true in the neighborhood of $E = 0$, since at any fixed E the poles are finite in number. Thus, for the case of Yukawas we now have all the information needed to establish the Mandelstam representation.

It should be clear that if the ideas of Regge theory carry over into relativistic scattering, they will provide a powerful tool in the exploitation of analytic properties. In fact, their usefulness would be even greater than our discussion so far would suggest. In nonrelativistic scattering the asymptotic form,

$$f(E, q^2) \xrightarrow[|q^2| \to \infty]{} \gamma(E)(q^2)^{\alpha(E)} \tag{15.37}$$

has no direct physical significance since the limit $q^2 \to \infty$ with E fixed is necessarily unphysical. However, in relativistic scattering there is (as we have already mentioned) a close connection between the amplitude f for any process $a + b \to c + d$ and that for the "crossed" process $a + \bar{c} \to \bar{b} + d$. This connection can be approximately described as follows. In relativistic scattering the amplitude is considered as a function $f(s, q^2)$ of the squared total energy s (rather than E) and of q^2. The physical amplitude $f(s, q^2)$ for $a + b \to c + d$ is defined for $s \geqslant 4mc^2$ (assuming all particles have the same mass m) and for $q^2 \geqslant 0$. If we continue $f(s, q^2)$ analytically to the region $s \leqslant 0$ and $q^2 \leqslant -4mc^2$, the resulting function is the physical amplitude for the crossed process with squared energy $-q^2$ and momentum transfer $-s$. Thus, if the limit (15.37) holds for a relativistic process $a + b \to c + d$ then it actually tells us the physical high-energy limit of the crossed process $a + \bar{c} \to \bar{b} + d$.

As we have already indicated, there is fairly good evidence that the ideas of Regge theory *do* carry over into relativistic scattering. However, there is also evidence that the theory is considerably more complicated than its non-relativistic counterpart. (For example, it appears that the relativistic amplitude must have branch cuts, as well as poles, in the plane of l.) The reader who wishes to pursue these matters cannot do better than read the comprehensive monograph of Collins and Squires (1968).

PROBLEMS

15.1.   By taking $E = 0$ show that the forward dispersion relation (15.12) implies the relation (15.13) for the s-wave scattering length.

15.2.   Assuming that $f(E, q^2)$ has all the necessary analytic properties and is $O(q^2)^\alpha$ where α is less than some integer m, derive the "m times subtracted" Mandelstam representation (15.23).

15.3.   Starting from the radial equation for $\phi_{l,p}(r)$ and the known form of ϕ as $r \to 0$ and ∞, prove that $\partial \delta_l / \partial l < \pi/2$ [i.e., fill in the details of the argument leading to (15.25)].

15.4.   (a) Suppose that the amplitude has a pole (or the Jost function a zero) for some $E < 0$ and $l = \alpha(E)$ (with Re $\alpha > -\frac{1}{2}$). Use the radial equation to prove that Im $\alpha(\alpha + 1) = 0$ and hence that α is real.
(b) Show that if the amplitude has a pole for physical E ($E > 0$) and $l = \alpha(E)$ then Im $\alpha > 0$.

15.5.   It was shown in Section 15-g that the partial-wave series defines $f(E, q^2)$ (for a Yukawa) as an analytic function of cos θ provided $|\text{Im } \theta| < \alpha$ where cosh $\alpha = 1 + \mu^2/2p^2$. Show that this condition confines cos θ to the interior of an ellipse (the small Lehmann ellipse) with foci at ± 1 and semi-major axis $1 + \mu^2/2p^2$.

16 The Scattering Operator in Multichannel Scattering

16-a   Channels

16-b   Channel Hamiltonians and Asymptotic
            States

16-c   Orthogonality and Asymptotic
            Completeness

16-d   A Little More Mathematics

16-e   The Scattering Operator

So far in this book we have discussed only processes in which two structure-less particles undergo an elastic collision. It is now time to move on and discuss collisions involving several particles (some of which may be composite like atoms or nuclei) and including inelastic processes such as excitation and disintegration. In fact, almost all processes of experimental interest do include these complications. A typical example from atomic physics is the set of processes:

$$e + H \rightarrow \begin{cases} e + H & \text{(elastic scattering)} \\ e + H^* & \text{(excitation)} \\ e + e + p & \text{(ionization or breakup)} \end{cases} \tag{16.1}$$

(where H^* denotes any of the excited states of hydrogen); or in nuclear physics:

$$p + {}^{12}C \rightarrow \begin{cases} p + {}^{12}C & \\ n + {}^{12}N & \text{(charge exchange)} \\ p + {}^{9}He + {}^{0}Be, & \text{etc.} \end{cases} \tag{16.2}$$

or in particle physics:

$$\pi^- + p \rightarrow \begin{cases} \pi^- + p \\ K^0 + \Lambda^0 \\ \pi^- + \pi^0 + \pi^0 + p, \qquad \text{etc.} \end{cases} \qquad (16.3)$$

Each of the different sets of final particles in each of these three examples is called a *channel* (a term we shall define more precisely later), and for this reason processes of this kind are called *multichannel processes*.

An important simplifying feature common to the examples (16.1) and (16.2) is that it is natural to regard the composite particles (the hydrogen atom, the carbon nucleus, etc.) as being made up of certain elementary constituents (electrons, protons, etc.) and that the numbers of these elementary constituents are conserved.[1] Thus, in the process (16.1) the constituents are two electrons and one proton, and all of the channels contain these same three particles. Similarly in (16.2) the basic constituents, which appear in all channels, are seven protons and six neutrons. This feature—that one can break down a process in terms of its basic, conserved constituents—is normally characteristic of nonrelativistic processes. [It is *not* in general true of relativistic processes like (16.3) or processes involving photons.] It means that in discussing a nonrelativistic process with a given set of constituent particles, we can consistently confine attention to those processes involving just these constituents—which is even so a sufficiently formidable task, as we shall see in the next six chapters.

In this chapter we shall cover that part of multichannel scattering that corresponds to the material of Chapter 2 for the scattering of one particle; that is, we set up a description of the collision process in terms of asymptotic free states and define a unitary S operator that maps each in state onto the corresponding out state.

Since the general discussion of multichannel scattering can become badly obscured behind a cloud of notation, we shall for much of the chapter confine attention to a simple specific three-particle model.

16-a.   Channels

To illustrate the essential features of multichannel scattering we begin by discussing a system of three spinless particles, a, b, c, which interact via

---

[1] I do not wish to suggest by this that there is any fundamental distinction between "elementary" and "composite" particles. On the contrary, it is generally accepted that such a distinction is meaningless. Nonetheless, in any given *nonrelativistic* problem we can usually make a reasonable working distinction: A particle is elementary if it is satisfactorily described by a single coordinate in the multiparticle wave function. Thus, in atomic physics the "elementary" particles are usually electrons and nuclei; in low-energy nuclear physics they are usually electrons, protons, and neutrons.

TABLE 16.1   THE MODEL THREE-PARTICLE SYSTEM

Particles	a, b, c
Bound States	$(bc), (bc)^*, (ac)$
	(and possible bound states of all three particles)

short-range two-body potentials. We shall suppose that the particles b and c have two bound states, a ground state (bc) and an excited state $(bc)^*$; that a and c have one bound state (ac); and that, apart from possible bound states of all three particles, there are no other bound states. There are no real particles with precisely these simple properties. Nonetheless, we can think of a and b as electrons (which are somehow distinguishable) and c as a proton, in which case our system gives a crude model for electron–hydrogen scattering. Or we can imagine a to be a proton, b a neutron, and c some stable nuclear "core" such as ^{16}O; in this case the (bc) bound state is ^{17}O, the (ac) state is ^{17}F, and we have a simple model for the scattering of protons off ^{17}O. We shall later find it convenient to suppose the "core" c to be very heavy and to treat it as fixed. (This is, of course, just a simple way of eliminating the CM motion and considering just the relative motion.) For the moment, however, we make no commitment as to the masses of a, b, c and we consider the complete, momentum-conserving motion of all three particles.

For definiteness we suppose that we are interested in the disintegration process

$$a + (bc) \rightarrow a + b + c \tag{16.4}$$

However, we must immediately recognize that the given initial state $a + (bc)$ will in general lead to several different final states in addition to the particular one of interest. We can enumerate all possible types of final state simply by enumerating all possible groupings of the particles a, b, c into two or more stable subsystems. These groupings are known as *channels* and the possible channels in the present model are shown in Table 16.2. (We do *not* include among the channels the bound states of all three particles. These remain bound at all times and do not communicate with the scattering states, in which we are interested.) We shall say that the process of interest (16.4) leads from an in state in channel 1 to an out state in channel 0. Clearly,

TABLE 16.2   POSSIBLE CHANNELS

Channel number	0	1	2	3
Channel	$a + b + c$	$a + (bc)$	$a + (bc)^*$	$b + (ac)$

however, an in state in channel 1 can generally lead to out states in any of the four channels 0, 1, 2, or 3,

$$a + (bc) \rightarrow \begin{cases} a + b + c & \text{(breakup)} \\ a + (bc) & \text{(elastic scattering)} \\ a + (bc)* & \text{(excitation)} \\ b + (ac) & \text{(rearrangement)} \end{cases}$$

In exactly the same way the final state of interest can arise from several different initial states. In fact, a final state in any one of the four channels 0, 1, 2, 3 can arise from an initial state in any of the same four channels. Therefore, there are 16 qualitatively distinct processes to be considered.

A schematic view of multichannel scattering, which illustrates the name channel, is shown in Fig. 16.1 The various possible in channels are shown as tubes, or channels, through which a fluid of probability can flow into a junction. This junction represents the actual collision, and from it lead the various possible out channels. In practice the in state always lies in a definite channel, which means that all of the fluid enters by one channel. The corresponding out state is usually a superposition of the various possible channels and the fluid therefore leaves through several channels in some definite proportions.

In practice, experiments are performed with initial states whose energy and momentum are rather well defined. Since energy and momentum are conserved, this means that at certain energies some of the channels may not be accessible. For example, the original process of interest,

$$a + (bc) \rightarrow a + b + c$$

cannot occur when the incident kinetic energy is insufficient to overcome the binding energy of (bc). In fact, in our model there are four *threshold energies*, at each of which one of the four channels "opens up." Thus, if we suppose that the energies of the three bound states (bc), $(bc)*$ and (ac) occur in the order,

$$E_{(bc)} < E_{(bc)*} < E_{(ac)} < 0$$

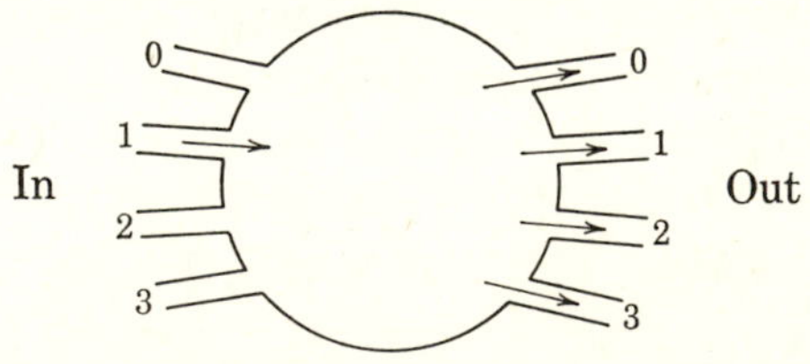

FIGURE 16.1. A schematic view of multichannel scattering as a quantum-mechanical irrigation system.

then for energies below $E_{(bc)}$ there may be bound states of all three particles but there are no scattering states. In the energy range between $E_{(bc)}$ and $E_{(bc)*}$ we can scatter a off (bc) but no inelastic processes can occur; thus, only elastic scattering of $a + (bc)$ is possible and, as we shall see, it is given by an amplitude with all the general properties (invariance properties, partial-wave decomposition, etc.) of the elastic amplitude of Chapters 2–15. For energies between $E_{(bc)*}$ and $E_{(ac)}$ excitation is possible, and there are two open channels allowing four possible processes. At $E_{(ac)}$ the channel $b + (ac)$ opens up and there are then three open channels and nine possible processes. And finally at $E = 0$, the disintegration channel $a + b + c$ opens up, there are four open channels, and all 16 processes are possible.

Our simple model should make clear all of the essential descriptive ideas related to the concept of a channel. In general, a channel is simply a set of particles (elementary or composite) that can enter or leave a collision. In nonrelativistic theory one deals with systems of a fixed number of elementary particles, $i = 1, \ldots, N$. We shall always take as channel 0 that in which all N particles move freely. The remaining channels $\alpha = 1, 2, \ldots$ are groupings of the particles into n_α stable fragments $(2 \leqslant n_\alpha < N)$ each of which is either one of the original particles or some definite bound state of some of them. It should be emphasized that to identify a channel it is necessary to specify both the grouping of the particles and the internal state of each group. Thus, in our three-particle example, the channels

$$a + b + c$$
$$a + (bc)$$
$$b + (ac)$$

are distinguished by the arrangement of the particles into groups, while the channels

$$a + (bc)$$
$$a + (bc)^*$$

have the same grouping and are distinguished by the different internal states of (bc) and $(bc)^*$.[2]

The number of channels for a given system can be either finite or infinite. In our model, which can be regarded as typical of nuclear physics in this respect, the number is finite. But when there are attractive Coulomb forces the number is generally infinite. (For example, the e–H system has an

---

[2] Some authors use the concept of an *arrangement channel*, which is defined as a *set of channels* all with the same arrangement, or grouping, of particles. Thus, in our model the two channels $a + (bc)$ and $a + (bc)^*$ would constitute a single arrangement channel.

infinite number of channels, since the hydrogen atom has infinitely many bound states.) For simplicity of discussion we shall suppose in this chapter that the number of channels, $\alpha = 0, 1, \ldots, n$, is finite, although very little is changed when n becomes infinite.

To conclude this section there are three points that need emphasis. First, it is often a matter of taste whether one chooses to regard two states as belonging to the same channel or not. For example, one could regard two spin-up electrons as belonging to a different channel from that of two spin-down electrons; however, one would usually take the view that they are different spin states within the same channel. Similarly, a state of two neutrons and another of two protons could be regarded as defining different channels or as different isospin states within the same channel. In practice this sort of ambiguity causes no confusion.

A second ambiguity concerns the question of what constitutes a stable fragment. An essential feature of our definition of a channel is that it is a set of *stable* fragments that can enter or leave a collision. The reason that the fragments must be stable is that the channels specify the grouping of particles in the asymptotic free states, which are defined as $t \to \pm\infty$. Since only a stable particle can live an infinitely long time, it is clear that—in principle at least—it makes no sense to speak of a channel containing unstable fragments.

Nonetheless, scattering experiments *are* done with unstable particles. The atomic physicist measures cross sections for excitation of atoms, even though the excited atom will eventually decay back to its ground state. One of the nuclear physicist's most important tools is the unstable neutron; and almost all elementary particles are unstable—many with exceedingly short lifetimes. The point is that whenever one speaks of a scattering experiment involving unstable incident or outgoing fragments, the lifetime of these fragments, however short, is nonetheless much longer than the characteristic time of the actual collision. Thus, one can wait for what is (as regards the collision) an "infinitely" long time and still observe the unstable fragments well before they decay.

In a given context it is usually clear what systems are to be considered as stable. It is also usually possible to treat a given problem in a theoretically consistent way. For example, suppose that we wish to compute the scattering of neutrons off some nucleus. Now, while neutrons are in reality unstable, the forces that cause the neutron to decay are the weak interactions, whereas the collision of the neutron and nucleus will be completely dominated by the strong (and perhaps the electromagnetic) interactions. Accordingly, we can expect to treat the collision by completely neglecting the weak interactions, and once the weak interactions are "switched off" the neutron is stable. In other words, in the theoretical model that we use to describe the collision, the neutron *is* stable and is a legitimate fragment for an in or out channel. Similarly, in the excitation of hydrogen by electrons one would expect to be able to neglect interactions with the radiation field, and in this approximation the excited states of hydrogen are stable.

In the following discussion we shall always assume that for any process under consideration there is a consistent model in which all initial and final fragments of interest are stable.[3]

Finally, it should be emphasized that there are various different possible choices for the zero of energy. In the preceding discussion we have taken the energy when all three particles are well separated and stationary as our zero. Theoretically, this is the most natural choice (in nonrelativistic problems) since the total energy is then just the sum of all kinetic energies plus all potentials (each of which goes to zero for large separations). However, it should be borne in mind that for an experiment starting, for example, in the channel $a + (bc)$, a very natural choice for zero-point would be the energy of a and (bc) when well separated and stationary. This choice differs from the previous one by the amount $E_{(bc)}$.

16-b.   Channel Hamiltonians and Asymptotic States

We now return to our three-particle model and begin our discussion of the quantum-mechanical evolution of a collision experiment. The discussion will proceed as a natural generalization of the one-channel discussion of Chapter 2. In particular, the three essential results that lead up to the introduction of the S operator—the asymptotic condition, the orthogonality theorem and asymptotic completeness—are close analogues of the corresponding results of Chapter 2.

The time development of any state is determined by the Hamiltonian, which we take to have the form[4]:

$$H = \frac{\mathbf{P}_a^2}{2m_a} + \frac{\mathbf{P}_b^2}{2m_b} + \frac{\mathbf{P}_c^2}{2m_c} + V_{ab}(\mathbf{x}_{ab}) + V_{ac}(\mathbf{x}_{ac}) + V_{bc}(\mathbf{x}_{bc})$$
$$\equiv H^0 + V$$

[3] Even when this is not so, it may be possible to compute apparently meaningful cross sections; but in this case the logical consistency of the formalism is not too clear. This whole problem is much more important in relativistic scattering, where very unstable particles are the norm, rather than the exception; in this field a considerable effort has been made to define the notion of an S-matrix element involving unstable fragments.

[4] We choose this form just for simplicity. It is certainly possible to accommodate more general interactions—many-body forces, nonlocal interactions, spin-dependent forces (for particles with spin), etc.

where $\mathbf{x}_{ij} \equiv \mathbf{x}_i - \mathbf{x}_j$ and H^0 is the sum of the three kinetic energies. In terms of this Hamiltonian the general orbit of the system has the usual form

$$U(t) \, |\psi\rangle \equiv e^{-iHt} \, |\psi\rangle$$

where now, of course, $|\psi\rangle$ is any vector in the three-particle Hilbert space $\mathscr{H} = \mathscr{L}^2(\mathbb{R}^9)$ defined by wave functions of the three coordinates,

$$\psi(\mathbf{x}_a, \mathbf{x}_b, \mathbf{x}_c) \equiv \psi(\underline{\underline{x}})$$

Here we have introduced a double underscore to denote the set of all particle coordinates

$$\underline{\underline{x}} \equiv (\mathbf{x}_a, \mathbf{x}_b, \mathbf{x}_c)$$

Let us now consider a scattering orbit originating in channel 0; that is, which originated as three freely moving particles, $a + b + c$. For such an orbit we naturally expect that

$$e^{-iHt} \, |\psi\rangle \xrightarrow[t \to -\infty]{} e^{-iH^0 t} \, |\psi_{\text{in}}\rangle$$

for some in state $|\psi_{\text{in}}\rangle$. We shall see that this result, and a corresponding result for out asymptotes, are indeed true for the appropriate states $|\psi\rangle$, and that they are true for exactly the same reasons as before—as the three particles move apart all of their interactions cease to have any effect.

Suppose however we consider an orbit $U(t) \, |\psi\rangle$ which originated in channel 1, $a + (bc)$. When the two particles a and (bc) move apart (as we follow the orbit back in time) the interactions V_{ab} and V_{ac} between the particle a and the two particles b and c become ineffective. On the other hand the interaction V_{bc} between b and c can never lose its importance. Indeed, it is only because of V_{bc} that the bound state (bc) remains bound. Thus, for an orbit that originated in channel 1, the part of H that is effective long before the collision is not H^0 but, rather, the "*channel 1 Hamiltonian*":

$$H^1 = \frac{\mathbf{P}_a^2}{2m_a} + \frac{\mathbf{P}_b^2}{2m_b} + \frac{\mathbf{P}_c^2}{2m_c} + V_{bc}$$

and for such an orbit we must expect that

$$e^{-iHt} \, |\psi\rangle \xrightarrow[t \to -\infty]{} e^{-iH^1 t} \, |\psi_{\text{in}}\rangle \tag{16.5}$$

for some $|\psi_{\text{in}}\rangle$.

It is important to note that if the orbit $U(t) \, |\psi\rangle$ does originate in channel 1, then the wave function of $|\psi_{\text{in}}\rangle$ must describe a state in which the positions $\mathbf{x}_a$ of particle a and $\bar{\mathbf{x}}_{bc}$ of the center of mass of (bc) move arbitrarily, but *the relative motion of b and c is fixed as that appropriate to the bound state (bc);* that is,

$$\langle \underline{\underline{x}} \, | \, \psi_{\text{in}}\rangle = \chi(\mathbf{x}_a, \bar{\mathbf{x}}_{bc}) \phi_{(bc)}(\mathbf{x}_{bc}) \qquad [\text{in state} = a + (bc)] \tag{16.6}$$

Here $\chi(\mathbf{x}_a, \bar{\mathbf{x}}_{bc})$ describes the motion of the incident particles a and (bc) and is an arbitrary normalizable function of $\mathbf{x}_a$ and the center of mass $\bar{\mathbf{x}}_{bc}$ of b and c. On the other hand $\phi_{(bc)}(\mathbf{x}_{bc})$ is uniquely determined as the wave function for the internal motion of the bound state (bc). Thus, not every vector $|\psi_{\text{in}}\rangle$ in $\mathscr{H}$ can label an in asymptote of channel 1; only those vectors in the subspace $\mathscr{S}^1$ made up from wave functions of the form (16.6) can do so. This subspace $\mathscr{S}^1$ is called the *channel 1 subspace* and consists of *those vectors that can label in or out asymptotes in channel 1.*

The action of the channel Hamiltonian H^1 on the vectors of $\mathscr{S}^1$ is especially simple. We can write

$$H^1 = \frac{\mathbf{P}_a^2}{2m_a} + \frac{\bar{\mathbf{P}}_{bc}^2}{2M_{bc}} + \left(\frac{\mathbf{P}_{bc}^2}{2m_{bc}} + V_{bc}\right)$$

where $\bar{\mathbf{P}}_{bc}$ and $\mathbf{P}_{bc}$ are the total and relative momentum operators for b and c, while M_{bc} and m_{bc} are their total and reduced masses. Then the term in parentheses $(\cdots)$ is the Hamiltonian of the relative motion of b and c, and its action on $\phi_{(bc)}$ is simply to multiply by $E_{(bc)}$. Thus, the asymptotic behavior (16.5) of an orbit that originates in channel 1 can be rewritten as:

$$e^{-iHt}|\psi\rangle \to e^{-iH^1 t}|\psi_{\text{in}}\rangle = \exp\left[-i\left(\frac{\mathbf{P}_a^2}{2m_a} + \frac{\bar{\mathbf{P}}_{bc}^2}{2M_{bc}} + E_{(bc)}\right)t\right]|\psi_{\text{in}}\rangle \quad (16.7)$$

That is, the asymptotic behavior is just that of two freely moving particles—one of mass m_a, the other of mass $M_{bc} = (m_b + m_c)$—except for the additional phase factor $\exp(-iE_{(bc)}t)$.

Exactly parallel considerations apply to all other channels and the whole situation can be summarized as in Table 16.3. In Table 16.3, function χ is

TABLE 16.3.   CHANNEL HAMILTONIANS AND WAVE FUNCTIONS

α	Channel	Channel Hamiltonian H^α	Typical wave function in channel subspace $\mathscr{S}^\alpha$
0	$a + b + c$	$H^0 = \sum \dfrac{\mathbf{P}_i^2}{2m_i}$	$\chi(\mathbf{x}_a, \mathbf{x}_b, \mathbf{x}_c)$
1	$a + (bc)$	$H^1 = \sum \dfrac{\mathbf{P}_i^2}{2m_i} + V_{bc}$	$\chi(\mathbf{x}_a, \bar{\mathbf{x}}_{bc})\phi_{(bc)}(\mathbf{x}_{bc})$
2	$a + (bc)^*$	$H^2 = \sum \dfrac{\mathbf{P}_i^2}{2m_i} + V_{bc}$	$\chi(\mathbf{x}_a, \bar{\mathbf{x}}_{bc})\phi_{(bc)*}(\mathbf{x}_{bc})$
3	$b + (ac)$	$H^3 = \sum \dfrac{\mathbf{P}_i^2}{2m_i} + V_{ac}$	$\chi(\mathbf{x}_b, \bar{\mathbf{x}}_{ac})\phi_{(ac)}(\mathbf{x}_{ac})$

an arbitrary normalizable function of its arguments. For example, the asymptotes of channel 0 have an arbitrary motion of all three particles and the channel subspace $\mathcal{S}^0$ is the whole of $\mathcal{H}$. (That is, *any* vector in $\mathcal{H}$ can label an in or out asymptote of channel 0.) Notice that the channel Hamiltonians H^1 and H^2 for channels 1 and 2 are the same. This is obviously true in general for any two channels that have the same groupings of the particles and differ only in the internal motion of one or more fragments—as with $a + (bc)$ and $a + (bc)*$.

In the general N-particle case, a channel α is specified by the grouping of the N particles into n_α freely moving fragments ($2 \leqslant n_\alpha \leqslant N$), each fragment being either one of the original N particles or a definite bound state of some subset. The corresponding channel Hamiltonian H^α is obtained by deleting from H those potentials that link different fragments,

$$H^\alpha = H - \sum{}' V_{ij}$$

where $\sum{}'$ denotes a sum over all pairs ij for which particles i and j belong to different fragments of channel α. The in and out asymptotes in channel α are identified by wave functions in the subspace $\mathcal{S}^\alpha$ comprising those functions with the form:

$$\chi(\mathbf{y}_1, \dots, \mathbf{y}_{n_\alpha})\phi_1(z_1) \cdots \phi_{n_\alpha}(z_{n_\alpha}) \qquad (16.8)$$

where χ is an arbitrary function of the centers of mass $\mathbf{y}_1, \dots, \mathbf{y}_{n_\alpha}$ of the n_α fragments. The term $\phi_\nu(z_\nu)$ is the bound-state wave function of the νth fragment with internal coordinates z_ν. (If the νth fragment happens to be a single particle, then $\phi_\nu \equiv 1$.)[5]

We are now in a position to state the asymptotic condition, which asserts the existence of an actual orbit with any prescribed in or out asymptote in any channel.

Asymptotic Condition.   If the particle interactions $V_{ij}(\mathbf{x}_{ij})$ all satisfy our usual assumptions, then for every vector $|\psi_{\text{in}}\rangle$ in any channel subspace $\mathcal{S}^\alpha$ there is a vector $|\psi\rangle$ that satisfies

$$\boxed{e^{-iHt}|\psi\rangle \xrightarrow[t \to -\infty]{} e^{-iH^\alpha t}|\psi_{\text{in}}\rangle} \qquad (16.9)$$

---

[5] In general, some of the fragments will have nonzero orbital angular momentum. In this case it is most natural to regard states that differ only in the orientation of a fragment's orbital angular momentum as being different "spin" orientations within the same channel (see Problem **16.1** and Section 17-a).

and which is given in terms of a *channel Møller operator* $\mathbf{\Omega}^{\alpha}_{+}$ as

$$|\psi\rangle = \mathbf{\Omega}^{\alpha}_{+} |\psi_{\text{in}}\rangle = \lim_{t \to -\infty} e^{iHt} e^{-iH^{\alpha}t} |\psi_{\text{in}}\rangle \qquad (16.10)$$

and similarly for every $|\psi_{\text{out}}\rangle$ in $\mathscr{S}^{\alpha}$ as $t \to +\infty$, with $|\psi\rangle = \mathbf{\Omega}^{\alpha}_{-} |\psi_{\text{out}}\rangle$.[6]

It should be emphasized that there are separate Møller operators $\mathbf{\Omega}^{\alpha}_{\pm}$ for each channel α and that they are defined by the limit (16.10) only for those vectors in $\mathscr{S}^{\alpha}$. In fact, we shall find that $\mathscr{S}^{\alpha}$ is the largest space on which we need to define $\mathbf{\Omega}^{\alpha}_{\pm}$.

Outline of Proof: The proof is very similar to that of the single-channel result and we content ourselves with a brief sketch. Just as in the one-channel case, all that is needed is to show that the limit in (16.10) exists. We write the vector concerned as the integral of its derivative and find that the desired limit exists provided the integral

$$\int_{-\infty}^{0} dt \, \|(H - H^{\alpha})e^{-iH^{\alpha}t}\psi_{\text{in}}\|$$

is convergent. Now the difference $(H - H^{\alpha})$ contains just those potentials that link the different fragments of channel α; while the vector $\exp(-iH^{\alpha}t) |\psi_{\text{in}}\rangle$ propagates like a vector representing n_{α} freely moving fragments, which therefore move apart in accordance with the familiar $t^{-3/2}$ law (for the case of a Gaussian). The integrand therefore drops at least like $t^{-3/2}$ (for any Gaussian)[7] and the integral is convergent as required.

QED

The asymptotic condition guarantees that every vector in the channel subspace $\mathscr{S}^{\alpha}$ labels a possible in or out asymptote in channel α. If $|\psi_{\text{in}}\rangle$ is in $\mathscr{S}^{\alpha}$ then the vector

$$|\psi\rangle = \mathbf{\Omega}^{\alpha}_{+} |\psi_{\text{in}}\rangle$$

is that actual state of the system which has developed from the in state labelled by $|\psi_{\text{in}}\rangle$ in channel α. Similarly, if $|\psi_{\text{out}}\rangle$ is in $\mathscr{S}^{\alpha}$, then

$$|\psi\rangle = \mathbf{\Omega}^{\alpha}_{-} |\psi_{\text{out}}\rangle$$

is the state that will develop into the out state $|\psi_{\text{out}}\rangle$ in channel α.

With this result we can already answer the question of greatest practical importance: What is the probability that a system that enters a collision in

---

[6] In accordance with the conventions set up in Chapter 4, we use bold face for $\mathbf{\Omega}^{\alpha}_{\pm}$, since our system is translationally invariant. Just as in the single-channel case we shall find that $\mathbf{\Omega}^{\alpha}_{\pm} = 1_{\text{cm}} \otimes \Omega^{\alpha}_{\pm}$, where $\Omega^{\alpha}_{\pm}$ describes the *relative* motion of the system.

[7] Just as in the one-channel case the mathematically inclined will have no trouble in checking that, if the result is true for any Gaussian wave packet, it is true for any (proper) wave function whatever.

channel α with in asymptote $|\phi\rangle$ (in $\mathcal{S}^\alpha$) be observed to leave in channel α' with out asymptote $|\phi'\rangle$ (in $\mathcal{S}^{\alpha'}$)? If the in state was $|\phi\rangle$ in channel α, then the actual state at $t = 0$ would be $\mathbf{\Omega}_+^\alpha |\phi\rangle$. If the out state were going to be $|\phi'\rangle$ in channel α', then the actual state at $t = 0$ would have to be $\mathbf{\Omega}_-^{\alpha'} |\phi'\rangle$. The required probability amplitude is just the overlap of these two states. Therefore, the required probability is

$$w(\phi', \alpha' \leftarrow \phi, \alpha) = |\langle\phi'| \, \mathbf{\Omega}_-^{\alpha'\dagger}\mathbf{\Omega}_+^\alpha \, |\phi\rangle|^2 \qquad (16.11)$$

However, before we continue our discussion of scattering probabilities it is desirable to establish some more formalism.

16-c.   Orthogonality and Asymptotic Completeness

Just as in single-channel scattering any state that has developed from some in asymptote (or will develop into some out asymptote) should be orthogonal to any bound state of all N particles. In addition, we would expect that any two states that have developed from in asymptotes *in different channels* should be mutually orthogonal. Similarly, any two states that will develop into out asymptotes of different channels should be mutually orthogonal. These results are the content of the orthogonality theorem.

Orthogonality Theorem.   Let $|\phi\rangle$ be any bound state of all N particles and let

$$|\psi\rangle = \mathbf{\Omega}_+^\alpha |\psi_{\text{in}}\rangle$$

and

$$|\psi'\rangle = \mathbf{\Omega}_+^{\alpha'} |\psi'_{\text{in}}\rangle \qquad\qquad (16.12)$$

with $|\psi_{\text{in}}\rangle$ in $\mathcal{S}^\alpha$ and $|\psi'_{\text{in}}\rangle$ in $\mathcal{S}^{\alpha'}$ and $\alpha \neq \alpha'$. Then

$$\langle\phi \,|\, \psi\rangle = \langle\psi' \,|\, \psi\rangle = 0$$

and likewise with $\mathbf{\Omega}_+$ replaced by $\mathbf{\Omega}_-$ (and "in" by "out") in (16.12).

We can give an alternative statement if we define $\mathcal{B}$ to be the subspace spanned by the bound states of all N particles and $\mathcal{R}_\pm^\alpha$ to be the ranges of $\mathbf{\Omega}_\pm^\alpha$. With these definitions $\mathcal{R}_+^\alpha$ is *the subspace of all states that originated in the in channel α*, while $\mathcal{R}_-^\alpha$ is *the subspace of all states that will terminate in the out channel α.* The theorem can be stated as:

$$\mathcal{B} \perp \mathcal{R}_+^\alpha \perp \mathcal{R}_+^{\alpha'} \qquad [\text{all } \alpha, \alpha'; \ \alpha \neq \alpha']$$

and similarly,[8]

$$\mathcal{B} \perp \mathcal{R}_-^\alpha \perp \mathcal{R}_-^{\alpha'} \qquad [\text{all } \alpha, \alpha'; \ \alpha \neq \alpha']$$

---

[8] We do *not* claim that $\mathcal{R}_+^\alpha \perp \mathcal{R}_-^{\alpha'}$. Indeed, if $\mathcal{R}_+^\alpha$ *were* orthogonal to $\mathcal{R}_-^{\alpha'}$, then an orbit that originated in channel α could never terminate in channel α'; that is, the inelastic process $(\alpha' \leftarrow \alpha)$ would be impossible.

Outline of Proof: The orthogonality of the bound states to the scattering states is true for the same reasons as in the one-channel case. Therefore, we discuss the case of the two vectors (16.12) that have come from different in channels α and α'. The scalar product $\langle \psi' \,|\, \psi \rangle$ can be evaluated at any time during the evolution of the corresponding orbits. In particular, it can be evaluated for t large and negative when the two vectors are well approximated by their in asymptotes. Thus,

$$\langle \psi' \,|\, \psi \rangle = \langle \psi' |\, e^{iHt}e^{-iHt} \,|\, \psi \rangle = \lim_{t \to -\infty} \langle \psi'_{\text{in}} |\, e^{iH^{\alpha'}t}e^{-iH^{\alpha}t} \,|\, \psi_{\text{in}} \rangle \qquad (16.13)$$

There are now two possibilities to be considered: First, we discuss the case where channels α and α' have the same arrangement of particles into groups and differ only in the internal state of some fragment; let us consider, for example, the channels $a + (bc)$ and $a + (bc)^*$ in our model. In this case the channel Hamiltonians H^{α} and $H^{\alpha'}$ are the same, and the product (16.13) reduces to $\langle \psi'_{\text{in}} \,|\, \psi_{\text{in}} \rangle$. This product is zero because the bound-state wave functions $\phi_{(bc)}$ and $\phi_{(bc)^*}$ are orthogonal. In the second case the channels α and α' correspond to different arrangements of the particles; e.g., $a + (bc)$ and $b + (ac)$. In this case, as t becomes large, particle a becomes well separated from c in the first channel, but remains bound to c in the second. Thus, the overlap of the two asymptotic states goes to zero, the limit (16.13) vanishes, and the original product must therefore be zero. In either case the physical reason for the orthogonality is clear; for t large and negative $U(t) \,|\, \psi \rangle$ and $U(t) \,|\, \psi' \rangle$ represent entirely distinct states and are therefore orthogonal. But since the scalar product of the two vectors is time-independent it must therefore be zero at all times, in particular at $t = 0$; that is, $\langle \psi' \,|\, \psi \rangle = 0$.

The orthogonality of two states with out asymptotes in different channels can be proved in exactly the same way, and our proof is complete.   **QED**

So far we have discussed only those scattering orbits that have developed from a definite in channel, or will develop into some definite out channel. These are certainly not the most general kind of scattering orbits. For example, suppose that $|\psi^1\rangle$ and $|\psi^2\rangle$ are states that have developed from in asymptotes in two different channels, $\alpha = 1$ and $\alpha = 2$ (say),

$$|\psi^1\rangle = \Omega_+^1 \,|\, \psi_{\text{in}}^1 \rangle$$

and

$$|\psi^2\rangle = \Omega_+^2 \,|\, \psi_{\text{in}}^2 \rangle$$

Then the superposition principle asserts that the vector $|\psi\rangle = |\psi^1\rangle + |\psi^2\rangle$ defines an allowed physical state, and we can ask the question: What is the asymptotic form of the orbit defined by $|\psi\rangle$? Now, we know that as $t \to -\infty$

$$e^{-iHt}|\psi^1\rangle \to e^{-iH^1t}|\psi^1_{\text{in}}\rangle$$

and

$$e^{-iHt}|\psi^2\rangle \to e^{-iH^2t}|\psi^2_{\text{in}}\rangle$$

Adding these two results we immediately see that the orbit defined by $|\psi^1\rangle + |\psi^2\rangle$ has the asymptotic form

$$e^{-iHt}(|\psi^1\rangle + |\psi^2\rangle) \to e^{-iH^1t}|\psi^1_{\text{in}}\rangle + e^{-iH^2t}|\psi^2_{\text{in}}\rangle$$

That is, this orbit originates as a *superposition of states* in the two in channels 1 and 2.

Having once recognized the existence of orbits that originate as a superposition of two in channels, we must obviously expect that the most general scattering orbit would be one which originated as a superposition of *all* possible in channels. That is, the general scattering orbit should have the asymptotic form

$$e^{-iHt}|\psi\rangle \xrightarrow[t \to -\infty]{} e^{-iH^0t}|\psi^0_{\text{in}}\rangle + \cdots + e^{-iH^nt}|\psi^n_{\text{in}}\rangle \qquad (16.14)$$

where each $|\psi^\alpha_{\text{in}}\rangle$ lies in the appropriate channel subspace $\mathscr{S}^\alpha$ and

$$|\psi\rangle = \Omega^0_+|\psi^0_{\text{in}}\rangle + \cdots + \Omega^n_+|\psi^n_{\text{in}}\rangle \qquad (16.15)$$

In the asymptotic form (16.14) it is natural to regard each term $\exp(-iH^\alpha t)|\psi^\alpha_{\text{in}}\rangle$ as *the component of the in asymptote in the channel* α. To identify the general in asymptote it is obviously essential to specify its components in all channels. This is most easily done by specifying all of the vectors $|\psi^\alpha_{\text{in}}\rangle$; that is, we can identify the general in asymptote by giving the sequence

$$\{|\psi^0_{\text{in}}\rangle, \ldots, |\psi^n_{\text{in}}\rangle\} \qquad (16.16)$$

where each $|\psi^\alpha_{\text{in}}\rangle$ identifies the component of the incoming asymptotic behavior in the corresponding channel α.

In the same way we should expect the general scattering orbit to evolve as $t \to \infty$ into a superposition of all possible out channels

$$e^{-iHt}|\psi\rangle \xrightarrow[t \to \infty]{} e^{-iH^0t}|\psi^0_{\text{out}}\rangle + \cdots + e^{-iH^nt}|\psi^n_{\text{out}}\rangle \qquad (16.17)$$

with

$$|\psi\rangle = \Omega^0_-|\psi^0_{\text{out}}\rangle + \cdots + \Omega^n_-|\psi^n_{\text{out}}\rangle \qquad (16.18)$$

This asymptotic form is, of course, identified by the sequence

$$\{|\psi^0_{\text{out}}\rangle, \ldots, |\psi^n_{\text{out}}\rangle\} \qquad (16.19)$$

It should be noted that in practice the in state always lies in one definite channel α and, hence, is given by a sequence

$$\{0, \ldots, 0, |\psi_{in}^{\alpha}\rangle, 0, \ldots, 0\}$$

On the other hand the out state is usually a superposition of several channels and has the general form (16.19) with several nonzero components (except of course in the special case where all inelastic processes are energetically impossible). This asymmetry between the in and out states is a reflection of the type of experiment that can be done in *practice;* in *principle,* the in states can be arbitrary superpositions of all channels just as can the out states.

We would expect to find that every scattering state defines an orbit that behaves as in (16.14) when $t \to -\infty$ *and* as in (16.17) when $t \to \infty$. Of course we would *not* expect that every state of the system should be a scattering state, since in general there will also be the bound states of all N particles. What we do expect is that the scattering states [satisfying (16.14) and (16.17)] *together with* the bound states should span the space $\mathscr{H}$ of *all* states. If this expectation is realized we say that the theory is asymptotically complete.

To make these ideas more precise we write $\mathscr{H}$ as the direct sum

$$\mathscr{H} = \mathscr{B} \oplus \mathscr{R}$$

which simply defines $\mathscr{R}$ as the space of states orthogonal to all bound states. What we expect is that the subspace $\mathscr{R}$ should consist precisely of those vectors $|\psi\rangle$ which behave as in (16.14) when $t \to -\infty$, and that these should be the same as those which behave as in (16.17) when $t \to \infty$. Now any vector $|\psi\rangle$ satisfying (16.14) is a superposition of vectors coming from the in channels $0, \ldots, n$,

$$|\psi\rangle = \Omega_+^0 |\psi_{in}^0\rangle + \cdots + \Omega_+^n |\psi_{in}^n\rangle \tag{16.20}$$

that is, it is a sum of vectors, one from each of the orthogonal subspaces $\mathscr{R}_+^0, \ldots, \mathscr{R}_+^n$. Thus, if every vector of $\mathscr{R}$ satisfies (16.14), and vice versa, then $\mathscr{R}$ has to be the direct sum:

$$\mathscr{R} = \mathscr{R}_+^0 \oplus \cdots \oplus \mathscr{R}_+^n$$

Similarly, if every vector of $\mathscr{R}$ satisfies (16.17), then $\mathscr{R}$ is the direct sum of the spaces $\mathscr{R}_-^0, \ldots, \mathscr{R}_-^n$. Thus, a concise statement of the ideas of the last paragraphs is this:

Asymptotic Completeness.   A multichannel scattering theory is asymptotically complete if:

$$\mathscr{H} = \mathscr{B} \oplus \mathscr{R}$$

where $\mathscr{B}$ is the space spanned by the bound states of all N particles, and

$$\mathscr{R} = \mathscr{R}_+^0 \oplus \cdots \oplus \mathscr{R}_+^n = \mathscr{R}_-^0 \oplus \cdots \oplus \mathscr{R}_-^n$$

where $\mathscr{R}_\pm^\alpha$ is the subspace of all states that originated as (or will develop into) an asymptotic state in channel α.

Asymptotic completeness for a three-body system with suitable potentials was proved by Faddeev (1965) and his proof was extended to the N-body case by Hepp (1969). The proof is extremely complicated and will not be given here. We shall content ourselves with *assuming* (as is certainly reasonable on physical grounds) that the multiparticle systems considered here *are* asymptotically complete.

Our findings so far can now be summarized: The space $\mathscr{H}$ of all possible states of our N-particle system can be decomposed into two orthogonal parts, $\mathscr{H} = \mathscr{B} \oplus \mathscr{R}$. The subspace $\mathscr{B}$ is spanned by the bound states of all N particles; that is, in any state of $\mathscr{B}$ all N particles remain localized together all the time. The subspace $\mathscr{R}$ is the space of scattering states, in which the particles split up into two or more fragments as $t \to \pm\infty$.

The in and out asymptotes of any scattering orbit can lie in the various channels $\alpha = 0, \ldots, n$ or, more generally, can be a superposition of these channels as in (16.14) and (16.17). Those in (or out) asymptotes that lie in a definite channel α are labelled by vectors $|\psi_{\text{in}}\rangle$ (or $|\psi_{\text{out}}\rangle$) in the channel subspace $\mathscr{S}^\alpha$. The corresponding actual states at $t = 0$ are given by the action of the Møller operators,

$$|\psi\rangle = \boldsymbol{\Omega}_+^\alpha |\psi_{\text{in}}\rangle \qquad (\text{or } |\psi\rangle = \boldsymbol{\Omega}_-^\alpha |\psi_{\text{out}}\rangle)$$

The range of $\boldsymbol{\Omega}_\pm^\alpha$ is denoted by $\mathscr{R}_\pm^\alpha$, which is therefore the subspace of all states that originated in (or will develop into) the channel α.

It should be emphasized that the subspaces $\mathscr{R}_+^\alpha$ and $\mathscr{R}_-^\alpha$ *are not* in general the same. Indeed, if it were true that $\mathscr{R}_+^\alpha = \mathscr{R}_-^\alpha$, then every orbit that came from the in channel α must necessarily evolve into the same out channel α; that is, no inelastic processes could occur. The observed fact that inelasticity *does* occur implies that each $\mathscr{R}_+^\alpha$ overlaps several of the spaces $\mathscr{R}_-^0, \ldots, \mathscr{R}_-^n$ and vice versa.

The general in asymptote is a superposition of the various in channels, as in (16.14), and is labelled

$$\{|\psi_{\text{in}}^0\rangle, \ldots, |\psi_{\text{in}}^n\rangle\} \tag{16.21}$$

where $|\psi_{\text{in}}^\alpha\rangle$ identifies the component of the asymptote in channel α. Similarly, the general out asymptote is a superposition of the various out channels and is labelled

$$\{|\psi_{\text{out}}^0\rangle, \ldots, |\psi_{\text{out}}^n\rangle\} \tag{16.22}$$

Finally the two decompositions of $\mathscr{R}$ as the direct sums

$$\mathscr{R} = \mathscr{R}_+^0 \oplus \cdots \oplus \mathscr{R}_+^n = \mathscr{R}_-^0 \oplus \cdots \oplus \mathscr{R}_-^n \qquad (16.23)$$

express the fact that every scattering state can be written as a superposition of states each of which originated in a definite in channel α, and also as a superposition of states each of which will develop into a definite out channel α. This implies that every scattering orbit has in and out asymptotes of the forms (16.21) and (16.22).

So far the multichannel problem is a very straightforward generalization of the single-channel case. Our next step is the definition of the S operator and at this point we require some additional tools.

16-d.   A Little More Mathematics

Direct Sums. So far we have used the *direct sum* as a decomposition of a given space $\mathscr{H}$ into two or more smaller subspaces. The decompositions (16.23) are examples, and a simpler example would be the decomposition of the real plane $\mathbb{R}^2$ into two one-dimensional spaces

$$\mathbb{R}^2 = \mathbb{R}_x^1 \oplus \mathbb{R}_y^1$$

We now wish to introduce a slightly different kind of direct sum, which is used to build up a single larger space from two or more given smaller spaces.

If $\mathscr{H}_1$ and $\mathscr{H}_2$ are two given Hilbert spaces, we can define a third space $\mathscr{H}$ made up of all pairs

$$|\Psi\rangle = \{|\psi_1\rangle, |\psi_2\rangle\}$$

with $|\psi_1\rangle$ in $\mathscr{H}_1$ and $|\psi_2\rangle$ in $\mathscr{H}_2$. It is a simple exercise to check that if we define addition by

$$|\Psi\rangle + |\Phi\rangle = \{|\psi_1\rangle + |\phi_1\rangle, |\psi_2\rangle + |\phi_2\rangle\}$$

and a scalar product

$$\langle \Psi \mid \Phi \rangle = \langle \psi_1 \mid \phi_1 \rangle + \langle \psi_2 \mid \phi_2 \rangle \qquad (16.24)$$

then $\mathscr{H}$ is itself a Hilbert space. The reader will recognize this as precisely the method by which a real two-dimensional space of vectors $\mathbf{x} = (x_1, x_2)$ can be built up from two one-dimensional spaces with addition defined as

$$\mathbf{x} + \mathbf{y} = (x_1 + y_1, x_2 + y_2)$$

and the scalar product

$$\mathbf{x} \cdot \mathbf{y} = x_1 y_1 + x_2 y_2$$

It is usual to call the space $\mathscr{H}$ of pairs $\{|\psi_1\rangle, |\psi_2\rangle\}$ the *direct sum* of $\mathscr{H}_1$ and $\mathscr{H}_2$ and to write:

$$\mathscr{H} = \mathscr{H}_1 \oplus \mathscr{H}_2$$

Clearly this new kind of direct sum, sometimes called an *external* direct sum, is different from the direct sum used up to now (sometimes called an *internal* direct sum). Nonetheless, the two concepts are intimately related and it is really very natural to use the same notation for both. The point is that every pair in the new space $\mathscr{H}$ can be written as the sum of two orthogonal vectors

$$\{|\psi_1\rangle, |\psi_2\rangle\} = \{|\psi_1\rangle, 0\} + \{0, |\psi_2\rangle\}$$

Therefore $\mathscr{H}$ is the direct sum (in the old, internal, sense) of two subspaces, one containing all vectors of the form $\{|\psi_1\rangle, 0\}$, the other made up of all vectors $\{0, |\psi_2\rangle\}$. Now the set of all vectors $\{|\psi_1\rangle, 0\}$ with $|\psi_1\rangle$ in $\mathscr{H}_1$ is obviously closely related to $\mathscr{H}_1$ itself, and one can in fact *identify* these two spaces. Similarly, if we identify the set of vectors $\{0, |\psi_2\rangle\}$ with the space $\mathscr{H}_2$, then what we have done is to *embed* the original spaces $\mathscr{H}_1$ and $\mathscr{H}_2$ in the larger space $\mathscr{H}$ and the relation of $\mathscr{H}$ to $\mathscr{H}_1$ and $\mathscr{H}_2$ is precisely the direct sum in its original (internal) sense.

The relationship of the two kinds of direct sum is perhaps easiest to understand in terms of the example of the real plane $\mathbb{R}^2$. If we start with a given plane we can decompose it as the (internal) direct sum of two lines; if we start with two lines we can form a plane as their (external) direct sum. Whichever procedure is followed, there is an obvious sense in which the relation of the plane to the two lines is really the same in either case.

Whatever the reader may feel as to its merits, we shall follow traditional usage and employ the same notation for both kinds of direct sum. What kind is being used in any given situation should always be clear from the context.

There are two properties of direct sums that we shall need and which the reader can easily check: First, if the vectors $|1\rangle_1, |2\rangle_1, |3\rangle_1, \ldots$ are an orthonormal basis of $\mathscr{H}_1$, and similarly, $|1\rangle_2, |2\rangle_2, |3\rangle_2, \ldots$ of $\mathscr{H}_2$, then the set containing

$$\{|1\rangle_1, 0\}, \{|2\rangle_1, 0\}, \ldots \qquad \text{and} \qquad \{0, |1\rangle_2\}, \{0, |2\rangle_2\}, \ldots$$

is an orthonormal basis of $\mathscr{H} = \mathscr{H}_1 \oplus \mathscr{H}_2$. It follows, in particular, that the dimension of $\mathscr{H}_1 \oplus \mathscr{H}_2$ is the sum of the dimensions of $\mathscr{H}_1$ and $\mathscr{H}_2$.

Second, the method of construction of the direct sum of two spaces can easily be extended to any finite number (and, indeed, a countably infinite number[9]) of given spaces. In particular, the general vector in

$$\mathscr{H} = \mathscr{H}_1 \oplus \cdots \oplus \mathscr{H}_n$$

---

[9] In this case there is one small complication. The definition (16.24) of the scalar product becomes an infinite series and the space $\mathscr{H}_1 \oplus \mathscr{H}_2 \oplus \cdots$ must therefore be restricted to those sequences $\{|\psi_1\rangle, |\psi_2\rangle, \ldots\}$ for which this series converges.

is a sequence

$$|\Psi\rangle = \{|\psi_1\rangle, \ldots, |\psi_n\rangle\}.$$

with each $|\psi_i\rangle$ in the corresponding space $\mathcal{H}_i$.

Linear Operators Between Different Spaces. The linear operators that we have used up to now have been operators that map certain vectors in a given space $\mathcal{H}$ onto certain vectors *in the same space* $\mathcal{H}$. In general, a linear operator can map vectors of one space $\mathcal{H}$ onto those of a second, different, space $\mathcal{H}'$; and we shall now need to discuss such operators. Their definition is a trivial extension of the usual one:

An operator A is a linear operator from $\mathcal{H}$ into $\mathcal{H}'$ if for each of certain vectors $|\psi\rangle$ in $\mathcal{H}$ there is a unique vector $A|\psi\rangle$ in $\mathcal{H}'$ such that

$$A(a|\psi\rangle + b|\phi\rangle) = aA|\psi\rangle + bA|\phi\rangle$$

Clearly the domain $\mathcal{D}(A)$ of A lies in $\mathcal{H}$, while the range $\mathcal{R}(A)$ lies in $\mathcal{H}'$.

In particular we shall be interested in isometric operators between two spaces, which are defined in this more general context as follows:

A linear operator Ω from $\mathcal{H}$ into $\mathcal{H}'$ is *isometric from $\mathcal{H}$ onto $\mathcal{H}'$*, if its domain is $\mathcal{H}$, its range $\mathcal{H}'$, and it preserves the norm.

As with our earlier definition, it is easily checked that an isometric operator from $\mathcal{H}$ onto $\mathcal{H}'$ is a one-to-one map from $\mathcal{H}$ onto $\mathcal{H}'$. Therefore, it has an inverse that is seen to be isometric from $\mathcal{H}'$ onto $\mathcal{H}$. In fact, our new definition can be seen to include our earlier one. Recall that an operator Ω was said to be isometric on $\mathcal{H}$ if its domain was $\mathcal{H}$, its range some subspace $\mathcal{R}$ of $\mathcal{H}$, and it preserved the norm. We can regard $\mathcal{R}$ as a space in its own right and then Ω is clearly isometric from $\mathcal{H}$ onto $\mathcal{R}$ in the new sense. We also remark that a unitary operator can be regarded as a special case of an isometric operator for which the two spaces $\mathcal{H}$ and $\mathcal{H}'$ are one and the same.

Another example is the channel Møller operator $\boldsymbol{\Omega}_+^\alpha$ (or $\boldsymbol{\Omega}_-^\alpha$), for any α. This operator maps the channel subspace $\mathcal{S}^\alpha$ onto the range $\mathcal{R}_+^\alpha$. It is easily checked that it preserves the norm [since it is the limit of the unitary operator $\exp(iHt)\exp(-iH^\alpha t)$] and we can therefore describe it as isometric from the space $\mathcal{S}^\alpha$ onto the space $\mathcal{R}_+^\alpha$.

It is a simple matter to check (exactly as in Chapter 1) that if Ω is isometric from $\mathcal{H}$ onto $\mathcal{H}'$ then $\Omega^\dagger$ is *the inverse* of Ω; that is,[10] $\Omega^\dagger$ maps $\mathcal{H}'$ back onto $\mathcal{H}$ and

$$\Omega^\dagger\Omega|\psi\rangle = |\psi\rangle \qquad [\text{any } |\psi\rangle \text{ in } \mathcal{H}]$$

---

[10] The adjoint of an operator from $\mathcal{H}$ into $\mathcal{H}'$ is defined, as one would expect, by the relation $\langle\phi|A|\psi\rangle = \langle\psi|A^\dagger|\phi\rangle^*$ for any $|\psi\rangle$ in $\mathcal{H}$ and $|\phi\rangle$ in $\mathcal{H}'$. Notice that $A^\dagger$ maps $\mathcal{H}'$ into $\mathcal{H}$.

16-e.   The Scattering Operator

To anyone who has attentively read the preceding two sections, our next move should be clear. In Section 16-c we saw that the general in asymptote of a multichannel system is identified by a sequence

$$\{|\psi_{in}^0\rangle, \ldots, |\psi_{in}^n\rangle\}$$

where $|\psi_{in}^\alpha\rangle$ lies in the channel α subspace $\mathscr{S}^\alpha$ and identifies the component of the in state in that channel. In the terminology of Section 16-d we can therefore say that each in asymptote is given by a vector

$$|\Psi_{in}\rangle = \{|\psi_{in}^0\rangle, \ldots, |\psi_{in}^n\rangle\} \qquad (16.25)$$

in the space

$$\mathscr{H}_{as} = \mathscr{S}^0 \oplus \cdots \oplus \mathscr{S}^n$$

This new space $\mathscr{H}_{as}$, which we call the *space of asymptotic states*, is (in an obvious sense) bigger than $\mathscr{H}$, since $\mathscr{S}^0$ alone is equal to $\mathscr{H}$. It is, however, the natural space for labelling the asymptotic states, each vector in $\mathscr{H}_{as}$ corresponding to a unique in (or, of course, out) asymptote, and vice versa.

We saw in Section 16-c that if the in asymptote is given by the sequence (16.25) then the actual state at $t = 0$ is:

$$|\psi\rangle = \Omega_+^0 |\psi_{in}^0\rangle + \cdots + \Omega_+^n |\psi_{in}^n\rangle$$

Accordingly, we define a linear operator Ω_+ on our new space $\mathscr{H}_{as}$ as

$$\begin{aligned} \Omega_+ |\Psi_{in}\rangle &= \Omega_+\{|\psi_{in}^0\rangle, \ldots, |\psi_{in}^n\rangle\} \\ &= \Omega_+^0 |\psi_{in}^0\rangle + \cdots + \Omega_+^n |\psi_{in}^n\rangle \end{aligned} \qquad (16.26)$$

With this definition Ω_+ maps $\mathscr{H}_{as}$ into $\mathscr{H}$ and for each in asymptote labelled by $|\Psi_{in}\rangle$, the corresponding actual state at $t = 0$ is just $|\psi\rangle = \Omega_+|\Psi_{in}\rangle$. Similarly, we can define Ω_- so that the actual state at $t = 0$ corresponding to any out asymptote $|\Psi_{out}\rangle$ is just $|\psi\rangle = \Omega_-|\Psi_{out}\rangle$.

It is not hard to prove (see Problem **16.3**) that, provided the theory is asymptotically complete, $\Omega_\pm$ are actually isometric from $\mathscr{H}_{as}$ onto $\mathscr{R}$, the space of scattering states. Therefore, the situation is as follows:

$$\mathscr{H}_{as} \xrightarrow[\text{isometric}]{\Omega_+} \mathscr{R} \xleftarrow[\text{isometric}]{\Omega_-} \mathscr{H}_{as}$$

or, in terms of vectors,

$$|\Psi_{in}\rangle \xrightarrow{\Omega_+} |\psi\rangle \xleftarrow{\Omega_-} |\Psi_{out}\rangle$$

$$\underset{\text{asymptote}}{} \quad \underset{\substack{\text{actual state} \\ \text{at } t=0}}{} \quad \underset{\text{out asymptote}}{}$$

The analogy with the single-channel problem is now complete. In particular, we can invert

$$|\psi\rangle = \boldsymbol{\Omega}_- \,|\Psi_{out}\rangle$$

to give

$$|\Psi_{out}\rangle = \boldsymbol{\Omega}_-^\dagger \,|\psi\rangle = \boldsymbol{\Omega}_-^\dagger \boldsymbol{\Omega}_+ \,|\Psi_{in}\rangle$$

Thus, if we define the scattering operator

$$\mathbf{S} = \boldsymbol{\Omega}_-^\dagger \boldsymbol{\Omega}_+$$

then **S** maps each in asymptote onto the corresponding out asymptote

$$|\Psi_{out}\rangle = \mathbf{S}\,|\Psi_{in}\rangle$$

Since **S** is isometric from $\mathcal{H}_{as}$ onto $\mathcal{H}_{as}$, it is actually unitary—a property with the same interpretation as before.

The probability amplitude for a system that enters a collision with in asymptote given by $|\Phi\rangle$ in $\mathcal{H}_{as}$ to be observed leaving with out asymptote $|\Phi'\rangle$ is just the matrix element $\langle\Phi'|\,\mathbf{S}\,|\Phi\rangle$,

$$\begin{aligned}
w(\Phi' \leftarrow \Phi) &= |\langle\Phi'|\,\mathbf{S}\,|\Phi\rangle|^2 \\
&= |\langle\Phi'|\,\boldsymbol{\Omega}_-^\dagger \boldsymbol{\Omega}_+\,|\Phi\rangle|^2
\end{aligned} \qquad (16.27)$$

In practice, the experimental initial state is prepared in a definite channel and the in asymptote $|\Phi\rangle$ has the form

$$|\Phi\rangle = \{0, \ldots, 0, |\phi\rangle, 0, \ldots, 0\}$$

with $|\phi\rangle$ in a definite $\mathcal{S}^\alpha$. The corresponding out state $\mathbf{S}\,|\Phi\rangle$ is naturally not (in general) in any definite channel. However, in practice one always monitors for a final state that *does* lie in a definite channel; that is, in practice we are interested in a $|\Phi'\rangle$ of the form

$$|\Phi'\rangle = \{0 \ldots, 0, |\phi'\rangle, 0, \ldots, 0\}$$

with $|\phi'\rangle$ in a definite $\mathcal{S}^{\alpha'}$. For these states we have:

$$\boldsymbol{\Omega}_+\,|\Phi\rangle = \boldsymbol{\Omega}_+^\alpha \,|\phi\rangle$$

and

$$\boldsymbol{\Omega}_-\,|\Phi'\rangle = \boldsymbol{\Omega}_-^{\alpha'}\,|\phi'\rangle$$

Thus, for such states the probability (16.27) can be written as

$$w\,(\phi', \alpha' \leftarrow \phi, \alpha) = |\langle\phi'|\boldsymbol{\Omega}_-^{\alpha'\dagger}\,\boldsymbol{\Omega}_+^\alpha|\,\phi\rangle|^2$$

This is precisely the result quoted in (16.11) and serves to emphasize that if one is interested solely in the computation of transition probabilities between two definite channels, it is actually unnecessary to introduce $\mathcal{H}_{as}$. It is only if we wish to express the required probability amplitudes as matrix elements of a single unitary operator **S** that this space is needed.

In nonrelativistic quantum mechanics there is no doubt that the fundamental Hilbert space describing the evolution of an N-particle system is $\mathscr{H} = \mathscr{L}^2(\mathbb{R}^{3N})$ (times an appropriate spin space if necessary). The space $\mathscr{H}_{\rm as}$ is an auxiliary space introduced because it provides the most natural labelling of the general asymptotes.

In relativistic quantum mechanics this situation is by no means so clear cut. It is not at all clear what is the fundamental space for describing a system that can undergo processes such as

$$\pi^- + p \to \begin{cases} K^0 + \Lambda \\ \pi^- + \pi^0 + \pi^0 + p \end{cases}$$

On the other hand the construction of the space $\mathscr{H}_{\rm as}$ presents no problems. Each configuration, $\pi^- + p$, $K^0 + \Lambda$, $\pi^- + \pi^0 + \pi^0 + p$, defines a separate channel and a distinct channel space $\mathscr{S}^\alpha$. And $\mathscr{H}_{\rm as}$ is just the direct sum (usually over infinitely many channels) of these spaces.

Perhaps the most extreme view concerning the two spaces $\mathscr{H}_{\rm as}$ and $\mathscr{H}$ is that sometimes taken in analytic S-matrix theory. This is that all physical experiments are in reality scattering experiments and hence that all relevant information is contained in the S operator (which maps the space $\mathscr{H}_{\rm as}$ onto itself). Thus, whether or not one can find a space $\mathscr{H}$ which describes the moment-by-moment evolution of a system and "interpolates" between the in and out asymptotes, the existence of such a space is irrelevant and should play no role in a satisfactory physical theory.

PROBLEMS

16.1.   Even when all particles in a system are spinless it usually happens that certain fragments in some of the channels have nonzero orbital angular momentum; and one then has to decide how to handle the "spin" of these fragments. Suppose that in our three-particle model, a, b, and c are spinless, that their interactions are rotationally invariant, and that the state (bc) has orbital angular momentum $l = 1$. Discuss the form of the wave functions describing asymptotic free motion in the channel $a + (bc)$, treating different m values as different orientations of the (bc) "spin," all belonging to the same channel.

16.2.   Consider the scattering of an electron e' off a hydrogen atom $(e\text{–}p)$, the two electrons e and e' being considered distinguishable. Enumerate all channels and channel Hamiltonians, and write down typical wave functions representing asymptotic states for each channel. For simplicity you may ignore the electron and proton spins, but you should take into account the angular momentum of the bound states. Bound states with different values of the quantum numbers n and l should be considered as defining different channels; different values of m should be treated as in Problem **16.1**.

16.3. We have asserted that the Møller operator defined in (16.26) is isometric from $\mathcal{H}_{as}$ to $\mathcal{R}$. To prove this consider the following: Let Ω^1 and Ω^2 be isometric from certain spaces $\mathcal{S}^1$ and $\mathcal{S}^2$ onto two other spaces $\mathcal{R}^1$ and $\mathcal{R}^2$, where $\mathcal{R}^1$ and $\mathcal{R}^2$ are subspaces of some $\mathcal{R}$ with $\mathcal{R} = \mathcal{R}^1 \oplus \mathcal{R}^2$. Define Ω on $\mathcal{H}_{as} = \mathcal{S}^1 \oplus \mathcal{S}^2$ such that

$$\Omega\{|\psi^1\rangle, |\psi^2\rangle\} = \Omega^1 |\psi^1\rangle + \Omega^2 |\psi^2\rangle$$

Prove that Ω is isometric from $\mathcal{H}_{as}$ onto $\mathcal{R}$. [You must show that Ω is defined everywhere on $\mathcal{H}_{as}$, that every vector in $\mathcal{R}$ is the image under Ω of some vector in $\mathcal{H}_{as}$, and that Ω preserves the norm. Notice that (as is the case with the actual spaces of interest) $\mathcal{R} = \mathcal{R}^1 \oplus \mathcal{R}^2$ is an internal direct sum, while $\mathcal{S} = \mathcal{S}^1 \oplus \mathcal{S}^2$ is external.]

Cross Sections and Invariance Principles in Multichannel Scattering

17

17-a   **The Momentum-Space Basis Vectors**

17-b   **Conservation of Energy and the On-Shell T Matrix**

17-c   **Cross Sections**

17-d   **Rotational Invariance**

17-e   **Time-Reversal Invariance**

In this chapter we continue our general development of multichannel scattering theory. In Chapter 16 we showed that the in and out asymptotes of a multichannel scattering orbit are labelled by sequences of vectors of the form:

$$\{|\psi_{\text{in}}^0\rangle, \ldots, |\psi_{\text{in}}^n\rangle\}$$

(and similarly for the out asymptotes) where each $|\psi_{\text{in}}^\alpha\rangle$ identifies the channel α component of the asymptotic state. These sequences can be viewed as vectors in the space $\mathcal{H}_{\text{as}}$ of asymptotic states, in which case each in asymptote is carried onto the corresponding out asymptote by the unitary operator **S** on $\mathcal{H}_{\text{as}}$. In this chapter we construct a momentum-space basis of $\mathcal{H}_{\text{as}}$ and then define an on-shell T matrix in terms of the elements of the momentum-space S matrix. We then derive expressions for the various cross sections in terms of the on-shell T matrix. Finally, we explore some consequences of the various possible invariance principles.

17-a.   The Momentum-Space Basis Vectors

The operator $\mathbf{S}$ is a unitary operator on the space $\mathscr{H}_{as}$ of asymptotic states. Just as in the one-channel case, we shall find that it is the momentum-space matrix elements of $\mathbf{S}$ in terms of which the experimental cross sections are expressed, and for this reason we must now discuss the momentum-space basis of $\mathscr{H}_{as}$. This basis will be a generalization of the bases of plane-wave states $|\mathbf{p}\rangle$ used in the one-particle case, and of the corresponding states $|\mathbf{p_1}, \mathbf{p_2}\rangle \equiv |\bar{\mathbf{p}}, \mathbf{p}\rangle$ in the two-particle problem. For simplicity we discuss first our simple model of three particles a, b, c introduced in the previous chapter.

Since $\mathscr{H}_{as}$ is the direct sum of the channel spaces $\mathscr{S}^{\alpha}$ we can begin by constructing a basis for each $\mathscr{S}^{\alpha}$. We start with the channel 0, in which a, b, and c all move freely, and for which the general asymptote is labelled by an arbitrary wave function $\chi(\mathbf{x}_a, \mathbf{x}_b, \mathbf{x}_c)$ of all three positions. The appropriate basis is clearly given by the three-particle momentum eigenstates:

$$|\mathbf{p}_a, \mathbf{p}_b, \mathbf{p}_c; 0\rangle \equiv |\underline{p}, 0\rangle$$

with corresponding wave functions

$$\langle \underline{x} \mid \underline{p}, 0\rangle = (2\pi)^{-9/2} \exp[i(\mathbf{p}_a \cdot \mathbf{x}_a + \mathbf{p}_b \cdot \mathbf{x}_b + \mathbf{p}_c \cdot \mathbf{x}_c)]$$

and normalization

$$\langle \underline{p}', 0 \mid \underline{p}, 0\rangle = \delta_3(\mathbf{p}'_a - \mathbf{p}_a)\, \delta_3(\mathbf{p}'_b - \mathbf{p}_b)\, \delta_3(\mathbf{p}'_c - \mathbf{p}_c) \equiv \delta_9(\underline{p}' - \underline{p})$$

[$\underline{p}$ stands for the set $(\mathbf{p}_a, \mathbf{p}_b, \mathbf{p}_c)$ and 0 is the channel number $\alpha = 0$.] These vectors are, of course, eigenvectors of the free Hamiltonian,

$$H^0 \, |\underline{p}, 0\rangle = \left(\frac{\mathbf{p}_a^2}{2m_a} + \frac{\mathbf{p}_b^2}{2m_b} + \frac{\mathbf{p}_c^2}{2m_c} \right) |\underline{p}, 0\rangle \equiv E_{\underline{p}}^0 \, |\underline{p}, 0\rangle$$

In channel 1, $a + (bc)$, the asymptotes are identified by wave functions

$$\chi(\mathbf{x}_a, \bar{\mathbf{x}}_{bc}) \phi_{(bc)}(\mathbf{x}_{bc})$$

where χ is an arbitrary function of the position $\mathbf{x}_a$ and the center of mass $\bar{\mathbf{x}}_{bc}$ of b and c, while $\phi_{(bc)}$ is the bound-state wave function of (bc). Since $\phi_{(bc)}$ is fixed these functions can be spanned by products of plane waves in $\mathbf{x}_a$ and $\bar{\mathbf{x}}_{bc}$ with the fixed function $\phi_{(bc)}(\mathbf{x}_{bc})$. That is, the appropriate basis vectors are:

$$|\mathbf{p}_a, \bar{\mathbf{p}}_{bc}; 1\rangle \equiv |\underline{p}, 1\rangle$$

with wave functions

$$\langle \underline{x} \mid \underline{p}, 1 \rangle = (2\pi)^{-3} \exp[i(\mathbf{p}_a \cdot \mathbf{x}_a + \bar{\mathbf{p}}_{bc} \cdot \bar{\mathbf{x}}_{bc})\phi_{(bc)}(\mathbf{x}_{bc})$$

and normalization

$$\langle \underline{p}', 1 \mid \underline{p}, 1 \rangle = \delta_3(\mathbf{p}'_a - \mathbf{p}_a)\,\delta_3(\bar{\mathbf{p}}'_{bc} - \bar{\mathbf{p}}_{bc}) \equiv \delta_6(\underline{p}' - \underline{p})$$

These vectors represent states in which particle a moves with momentum $\mathbf{p}_a$, while b and c are bound together into the bound state (bc), whose CM moves with momentum $\bar{\mathbf{p}}_{bc}$. In this channel we use the label $\underline{p}$ for the pair $(\mathbf{p}_a, \bar{\mathbf{p}}_{bc})$, and it is unnecessary to specify the wave function $\phi_{(bc)}$ in writing $|\underline{p}, 1\rangle$ since the channel label 1 already includes this information.

The energy operator for asymptotic states in channel 1 is the channel Hamiltonian H^1, and the vectors $|\underline{p}, 1\rangle$ are, of course, eigenvectors of this operator,

$$H^1 |\underline{p}, 1\rangle = \left(\frac{\mathbf{p}_a^2}{2m_a} + \frac{\bar{\mathbf{p}}_{bc}^2}{2M_{bc}} + E_{(bc)}\right) |\underline{p}, 1\rangle \equiv E_{\underline{p}}^1 |\underline{p}, 1\rangle$$

where $E_{(bc)}$ is the internal energy of the bound state (bc). Thus, the improper vectors $|\underline{p}, 1\rangle$ will represent in and out states of definite energy $E_{\underline{p}}^1$ in channel 1.

The momentum basis vectors for the other channels are constructed in exactly the same way and are shown in Table 17.1.

TABLE 17.1.   CHANNEL BASIS FUNCTIONS

α	Channel	Wave function $\langle \underline{x} \mid \underline{p}, \alpha \rangle$	Energy $E_{\underline{p}}^\alpha$
0	$a + b + c$	$(2\pi)^{-9/2} \exp[i(\mathbf{p}_a \cdot \mathbf{x}_a + \mathbf{p}_b \cdot \mathbf{x}_b + \mathbf{p}_c \cdot \mathbf{x}_c)]$	$\dfrac{p_a^2}{2m_a} + \dfrac{p_b^2}{2m_b} + \dfrac{p_c^2}{2m_c}$
1	$a + (bc)$	$(2\pi)^{-3} \exp[i(\mathbf{p}_a \cdot \mathbf{x}_a + \bar{\mathbf{p}}_{bc} \cdot \bar{\mathbf{x}}_{bc})]\phi_{(bc)}(\mathbf{x}_{bc})$	$\dfrac{p_a^2}{2m_a} + \dfrac{\bar{p}_{bc}^2}{2M_{bc}} + E_{(bc)}$
2	$a + (bc)^*$	$(2\pi)^{-3} \exp[i(\mathbf{p}_a \cdot \mathbf{x}_a + \bar{\mathbf{p}}_{bc} \cdot \bar{\mathbf{x}}_{bc})]\phi_{(bc)^*}(\mathbf{x}_{bc})$	$\dfrac{p_a^2}{2m_a} + \dfrac{\bar{p}_{bc}^2}{2M_{bc}} + E_{(bc)^*}$
3	$b + (ac)$	$(2\pi)^{-3} \exp[i(\mathbf{p}_b \cdot \mathbf{x}_b + \bar{\mathbf{p}}_{ac} \cdot \bar{\mathbf{x}}_{ac})]\phi_{(ac)}(\mathbf{x}_{ac})$	$\dfrac{p_b^2}{2m_b} + \dfrac{\bar{p}_{ac}^2}{2M_{ac}} + E_{ac}$

The construction of the corresponding bases of $\mathscr{S}^\alpha$ in the general N-particle problem is entirely straightforward and need not be spelled out here. We mention only that, since the general channel α has n_α freely moving fragments, the corresponding basis vector has the form $|\underline{p}, \alpha\rangle$ where $\underline{p}$ labels the n_α momenta of these fragments.

Since $\mathscr{H}_{as}$ is the direct sum

$$\mathscr{H}_{as} = \mathscr{S}^0 \oplus \cdots \oplus \mathscr{S}^n$$

we can obtain an orthonormal basis of $\mathscr{H}_{as}$ by combining any orthonormal bases of $\mathscr{S}^0, \ldots, \mathscr{S}^n$. In particular, the vectors $|\underline{p}, \alpha\rangle$ (α fixed, all $\underline{p}$) are an (improper) orthonormal basis of $\mathscr{S}^\alpha$ and, hence, the set of vectors

$$\{0, \ldots, 0, |\underline{p}, \alpha\rangle, 0, \ldots, 0\} \qquad [\alpha = 0, \ldots, n; \text{ all } \underline{p}] \qquad (17.1)$$

is the desired momentum basis of $\mathscr{H}_{as}$. It should be noted that to span $\mathscr{H}_{as}$ we must include all momenta *and* all channels.

The basis vector (17.1) represents an asymptote with momenta $\underline{p}$ lying completely in the channel α. Since the symbol (17.1) is very cumbersome we shall abbreviate it to $|\underline{p}, \alpha\rangle$, writing

$$\{0, \ldots, 0, |\underline{p}, \alpha\rangle, 0, \ldots, 0\} \equiv |\underline{p}, \alpha\rangle$$

that is, we shall use the same symbol $|\underline{p}, \alpha\rangle$ both for the basis vectors of $\mathscr{S}^\alpha$ in $\mathscr{H}$, and for the sequence in $\mathscr{H}_{as}$ consisting of $|\underline{p}, \alpha\rangle$ in the αth position and zeros everywhere else. In practice it will always be clear which kind of vector is being used, and this imprecise usage will cause no confusion. With this convention the orthonormality of our basis of $\mathscr{H}_{as}$ is expressed as[1]

$$\langle \underline{p}', \alpha' \,|\, \underline{p}, \alpha\rangle = \delta_{\alpha'\alpha}\,\delta(\underline{p}' - \underline{p})$$

where the factor $\delta_{\alpha'\alpha}$ reflects the orthogonality (as vectors of $\mathscr{H}_{as}$) of any two vectors belonging to different channels. With respect to this basis the S operator becomes a matrix with elements $\langle \underline{p}', \alpha'|\, \mathbf{S}\, |\underline{p}, \alpha\rangle$ and knowledge of all of these elements is equivalent to knowledge of $\mathbf{S}$. Similarly, an operator equation like the unitarity equation $\mathbf{S}^\dagger \mathbf{S} = 1$ becomes the matrix equation:

$$\sum_{\alpha''} \int d\underline{p}'' \langle \underline{p}', \alpha'|\, \mathbf{S}^\dagger\, |\underline{p}'', \alpha''\rangle\langle \underline{p}'', \alpha''|\, \mathbf{S}\, |\underline{p}, \alpha\rangle = \delta_{\alpha'\alpha}\,\delta(\underline{p}' - \underline{p}) \qquad (17.2)$$

(And from this result follows the multichannel optical theorem—See Problem **17.3**.)

---

[1] To avoid too horrid a clutter of subscripts we shall not indicate the dimension of the delta functions $\delta(\underline{p}' - \underline{p}) = \delta_{3n_\alpha}(\underline{p}' - \underline{p})$. [Because of the factor $\delta_{\alpha'\alpha}$, the two variables $\underline{p}'$ and $\underline{p}$ in $\delta(\underline{p}' - \underline{p})$ always have the same number of dimensions.] Similarly, when we integrate over the momenta of channel α we shall write the volume element as just $d\underline{p}$ rather than $d^{3n_\alpha}\underline{p}$.

To conclude this section we remark that so far we have assumed that all particles are spinless and that all bound states have zero orbital angular momentum. The inclusion of either kind of angular momentum into our formalism is completely straightforward. Suppose, for instance, that in our three-particle model the particle a has spin s and that the state (bc) has orbital angular momentum l. [For example, we could imagine a to be an electron and (bc) to be the $2p$ state of hydrogen—though for simplicity we suppose b and c to be spinless.] Let us then consider an asymptotic state (either in or out) in which a and (bc) move freely. There are various possible orientations of the spin of particle a, and similarly, of the orbital momentum of (bc); and it is natural to regard these as being different "spin" orientations *all within the same channel*. Thus, we denote as channel 1 *all* free states of a and (bc), and the corresponding momentum eigenvectors we label

$$|\mathbf{p}_a, \bar{\mathbf{p}}_{bc}, m_a, m_{bc}; \, 1\rangle$$

where $\mathbf{p}_a$ and $\bar{\mathbf{p}}_{bc}$ are the momenta of the two fragments, m_a is the z component of the spin of particle a, and m_{bc} is the z component of the orbital angular momentum of (bc). The corresponding wave function has the form:

$$(2\pi)^{-3} \exp[i(\mathbf{p}_a \cdot \mathbf{x}_a + \bar{\mathbf{p}}_{bc} \cdot \bar{\mathbf{x}}_{bc})]\chi^{ma}\phi(r_{bc})Y_l^{mbc}(\hat{\mathbf{x}}_{bc})$$

where χ^{ma} is the appropriate spinor for particle a, and the last two factors are the radial and angular wave functions of the (bc) bound state.

It should be clear that, at this stage, fragments with spin (either intrinsic or "orbital") are no more than a slight notational complication, and we shall, for the most part, proceed under the simplifying assumption that none of the fragments in any channel have angular momentum.

17-b.   Conservation of Energy and the On-Shell T Matrix

We are now in a position to establish conservation of energy, which follows in almost exactly the same way as in the one-channel case. The first step is to prove the intertwining relations

$$H\Omega_\pm^\alpha = \Omega_\pm^\alpha H^\alpha \qquad [\text{on } \mathscr{S}^\alpha] \qquad\qquad (17.3)$$

which hold for any vector of the channel subspace $\mathscr{S}^\alpha$ (the domain on which $\Omega_\pm^\alpha$ are defined). These follow at once from the definition

$$\Omega_\pm^\alpha = \lim e^{iHt}e^{-iH^\alpha t} \qquad [\text{on } \mathscr{S}^\alpha]$$

exactly as in Section 3-a. We next consider the S-matrix element $\langle p', \alpha' | \, \mathbf{S} \, | p, \alpha \rangle$ between initial and final states of energies E_p^α and $E_{p'}^{\alpha'}$—which we shall abbreviate as E and E' when there is no danger of confusion. This S-matrix element can be rewritten as

$$\langle \underline{p}', \alpha' | \mathbf{S} | \underline{p}, \alpha \rangle = \langle \underline{p}', \alpha' | \Omega_-^{\alpha'\dagger} \Omega_+^{\alpha} | \underline{p}, \alpha \rangle$$

and using this form we can apply the intertwining relations to prove the required identity:

$$
\begin{aligned}
E \langle \underline{p}', \alpha' | \mathbf{S} | \underline{p}, \alpha \rangle &= \langle \cdots | \Omega_-^{\alpha'\dagger} \Omega_+^{\alpha} H^{\alpha} | \cdots \rangle \\
&= \langle \cdots | \Omega_-^{\alpha'\dagger} H \Omega_+^{\alpha} | \cdots \rangle \\
&= \langle \cdots | H^{\alpha'} \Omega_-^{\alpha'\dagger} \Omega_+^{\alpha} | \cdots \rangle \\
&= E' \langle \cdots | \mathbf{S} | \cdots \rangle
\end{aligned}
$$

or

$$(E - E') \langle \underline{p}', \alpha' | \mathbf{S} | \underline{p}, \alpha \rangle = 0$$

Thus, $\langle \underline{p}', \alpha' | \underset{=}{\mathbf{S}} | \underline{p}, \alpha \rangle$ is zero unless the initial and final energies are equal; that is, $\underset{=}{\mathbf{S}}$ conserves energy and its matrix elements contain the expected factor[2] $\delta(E' - E)$.

In addition to conserving energy, $\mathbf{S}$ also conserves total momentum. This is because all potentials involve only the *relative* positions of particles and the system is therefore translationally invariant. This means that $\mathbf{S}$ commutes with the total momentum operator $\bar{\mathbf{P}} = \sum \mathbf{P}_i$ and, hence, that its matrix elements contain a factor $\delta_3(\bar{\mathbf{p}}' - \bar{\mathbf{p}})$. In fact, we can go further: By using as independent variables the overall CM position $\bar{x}$ and any suitable choice of $(N - 1)$ relative coordinates we can factor the space $\mathcal{H}$ (and similarly $\mathcal{H}_{as}$) as

$$\mathcal{H} = \mathcal{H}_{cm} \otimes \mathcal{H}_{rel}$$

Here, just as in the two-particle case, $\mathcal{H}_{cm}$ describes the motion of the overall CM, and $\mathcal{H}_{rel}$ the relative motion of the particles. [Making a suitable choice of the $(N - 1)$ relative coordinates and momenta that define $\mathcal{H}_{rel}$ requires some care—see Problem 17.2.] The Hamiltonian can then be written as

$$H = H_{cm} + H_{rel}$$

---

[2] We can improve the analogy with the one-channel result if we define an operator H_{as} on $\mathcal{H}_{as}$, such that

$$H_{as} | \underline{p}, \alpha \rangle = E_{\underline{p}}^{\alpha} | \underline{p}, \alpha \rangle$$

This operator is the energy operator for *all* asymptotic states, and it is easily seen that $[H_{as}, \mathbf{S}] = 0$.

where $H_{\text{cm}} = \bar{\mathbf{P}}^2/2M$ and M is the total mass of the system; and just as in two-particle scattering, the $\mathbf{S}$ operator factors as

$$\mathbf{S} = 1_{\text{cm}} \otimes \mathsf{S}$$

The operator S describes the relative motion and is the scattering operator one would obtain directly from the Hamiltonian H_{rel}. Just as in the one-channel case all of the physically interesting information is contained in S and much of our subsequent analysis will be in terms of S.

Returning to the matrix elements of $\mathbf{S}$ we see at once that they can be written as:

$$\langle \underline{p}', \alpha' | \mathbf{S} | \underline{p}, \alpha \rangle = \delta_3(\bar{\mathbf{p}}' - \bar{\mathbf{p}}) \times \text{remainder}$$

and that the remainder depends only on the relative momenta.

We can now combine the consequences of energy and momentum conservation. Before we do so it is convenient, as in the one-channel case, to subtract out a no-scattering term by writing

$$\mathbf{S} = 1 + \mathbf{R}$$

Then, since $\mathbf{R}$, like $\mathbf{S}$, conserves total energy and momentum, the matrix elements can be written as:

$$\boxed{\begin{aligned} \langle \underline{p}', \alpha' | \mathbf{S} | \underline{p}, \alpha \rangle = \delta_{\alpha'\alpha}\, \delta(\underline{p}' - \underline{p}) \\ - 2\pi i\, \delta(E' - E)\, \delta_3(\bar{\mathbf{p}}' - \bar{\mathbf{p}}) t(\underline{p}', \alpha' \leftarrow \underline{p}, \alpha) \end{aligned}} \qquad (17.4)$$

Here the *on-shell T matrix* $t(\underline{p}', \alpha' \leftarrow \underline{p}, \alpha)$ depends only on the sets of initial and final *relative* momenta, which we indicate with a single underscore as $\underline{p}$ and $\underline{p}'$. (That is, $\underline{p}$ denotes the set of $n_\alpha - 1$ suitably chosen relative momenta for the fragments of channel α.) It is defined only on the energy shell $E' = E$; that is, it is defined only for those $\underline{p}$ and $\underline{p}'$ consistent with conservation of energy. It can, of course, be computed directly from the operator S of the relative motion, whose matrix elements $\langle \underline{p}', \alpha' | \mathsf{S} | \underline{p}, \alpha \rangle$ have a decomposition similar to (17.4), but without the factor $\delta_3(\bar{\mathbf{p}}' - \bar{\mathbf{p}})$.

The interpretation of the decomposition (17.4) is much the same as in the one-channel case. The first term is the S matrix for no scattering and leaves all momenta and channels unchanged. The second represents the actual scattering; it conserves total energy and total momentum but can in general change the relative momenta and the channel.

The fact that S can only connect states of the same energy leads to important restrictions on what we can call the "channel structure" of the S matrix. To illustrate this, let us return to our three-particle model with its four channels $\alpha = 0, \ldots, 3$, as shown in Table 17.2. We can regard the discrete channel indices α' and α in $\langle \underline{p}', \alpha' | \mathsf{S} | \underline{p}, \alpha \rangle$ or $t(\underline{p}', \alpha' \leftarrow \underline{p}, \alpha)$ as labelling the

TABLE 17.2.   CHANNELS AND THRESHOLDS

α	0	1	2	3
Channel	$a + b + c$	$a + (bc)$	$a + (bc)^*$	$b + (ac)$
Threshold	0	$E_{(bc)}$ $<$	$E_{(bc)*}$ $<$	$E_{(ac)} < 0$

rows and columns of a matrix in "channel space," each element of which is a function of the momenta $\underline{p}'$ and $\underline{p}$. For example, in our model we have a matrix with up to four rows and columns. However, at certain energies the dimension is actually less than four as we now discuss. For definiteness we shall confine our attention to the CM frame (that is, we discuss the S matrix of the relative motion), and we now imagine the energy to be increased from some large negative value.

All states in a given channel have energy at least as great as the channel's threshold energy. Thus, for energies less than the lowest threshold $E_{(bc)}$ there are no states in any channel and, hence, no S matrix. When the energy increases into the range $E_{(bc)} \leqslant E < E_{(bc)*}$ there are states in channel 1 but none in any other. Thus, in this energy range, S is a 1×1 "channel-space" matrix

$$\langle \mathbf{p}', 1|\ \mathsf{S}\ |\mathbf{p}, 1 \rangle$$

which has many of the properties of the one-channel S matrix. For example, if we choose an angular momentum basis of the channel 1 subspace and assume rotational invariance, then exactly the arguments of Chapter 6 show that:

$$\langle E', l', m'; 1|\ \mathsf{S}\ |E, l, m; 1 \rangle$$
$$= \delta(E' - E)\, \delta_{l'l}\delta_{m'm}e^{2i\delta_l(E)} \qquad [E_{(bc)} \leqslant E < E_{(bc)*}]$$

where the phase shift $\delta_l(E)$ is real.

If we now increase the energy into the range $E_{(bc)*} \leqslant E < E_{(ac)}$ there are states in channels 1 and 2 but in no others. Thus, S becomes a 2×2 "channel-space" matrix:

$$\begin{pmatrix} \langle \mathbf{p}', 1|\ \mathsf{S}\ |\mathbf{p}, 1 \rangle & \langle \mathbf{p}', 1|\ \mathsf{S}\ |\mathbf{p}, 2 \rangle \\ \langle \mathbf{p}', 2|\ \mathsf{S}\ |\mathbf{p}, 1 \rangle & \langle \mathbf{p}', 2|\ \mathsf{S}\ |\mathbf{p}, 2 \rangle \end{pmatrix} \qquad (17.5)$$

Clearly, as we increase the energy past each threshold the corresponding channel opens up and the matrix gains one dimension. Finally, when $E \geqslant 0$ all channels are open and the S matrix has its full complement of four rows and four columns.

In conclusion, note that the relative motion in any two-body channel is specified by a single momentum: the relative momentum of the two fragments. Thus, as long as only two-body channels are open, each element $\langle p', \alpha' | S | p, \alpha \rangle$ is in fact labelled by just two momenta $\mathbf{p}'$ and $\mathbf{p}$, as indicated for example in (17.5). In this energy range the multichannel S matrix is more complicated than that of the one-channel problem only inasmuch as it has more than one element. Once the channels with three or more bodies open up, the corresponding matrix elements depend on several variables and the situation is markedly more complicated.

17-c.　Cross Sections

We must now set about computing the observable cross sections in terms of the on-shell T matrix. Since it is an experimental fact that almost all processes of current interest have two-body initial states, we shall confine our discussion to such processes, and consider a process leading from a two-body in channel α to an arbitrary out channel α' (with n' bodies). In our three-particle model we could consider any of the processes:

$$a + (bc) \rightarrow \begin{cases} a + (bc) \\ a + (bc)^* \\ b + (ac) \\ a + b + c \end{cases} \tag{17.6}$$

(the first three having two-body final states, the last having a three-body final state). For simplicity, we work in the CM frame, considering just the relative motion, and take an initial state $|\phi\rangle$ in $\mathscr{S}^\alpha$, prescribed by its momentum-space wave function $\phi(\mathbf{p})$ in the relative momentum $\mathbf{p}$ of the two initial particles.[3] As usual $\phi(\mathbf{p})$ will be well peaked about the mean incident momentum. The corresponding in asymptote must, of course, be written as:

$$|\Psi_{\text{in}}\rangle = |\Phi\rangle = \{0, \ldots, 0, |\phi\rangle, 0, \ldots, 0\}$$

where $|\phi\rangle$ occupies the αth position.

We begin by computing the probability that, with the given in state $|\Psi_{\text{in}}\rangle = |\Phi\rangle$ in channel α, the final particles be observed in channel α' with their momenta in some prescribed volume in the $(3n' - 3)$-dimensional

---

[3] Strictly speaking we should consider an initial two-particle wave packet of the form $\phi_1(\mathbf{p}_1) \phi_2(\mathbf{p}_2)$. However, just as in Chapters 3 and 4, we get the same results by considering a wave packet $\phi(\mathbf{p})$ in the relative motion. In particular, if we think of a process like e-H scattering where one of the target particles is very heavy, then the heavy particle is practically fixed and the relative motion is a close approximation to the actual motion of the light particles. In this case $\phi(\mathbf{p})$ is (to a close approximation) just the wave function of the incident light particle.

space of the relative momenta of this channel. First, if we consider an infinitesimal $d\underline{p}'$ about some $\underline{p}'$ then, since $|\Psi_{\text{out}}\rangle = S\,|\Phi\rangle$, this probability is just[4]

$$w(d\underline{p}',\,\alpha' \leftarrow \phi,\,\alpha) = d\underline{p}'\,|\langle \underline{p}',\,\alpha'\,|\,\Psi_{\text{out}}\rangle|^2$$

$$= d\underline{p}'\,|\langle \underline{p}',\,\alpha'|\,S\,|\Phi\rangle|^2 \qquad (17.7)$$

If we consider instead some finite volume Δ' in the space of the final momenta, then the required probability is obtained by integrating this result over Δ',

$$w(\Delta',\,\alpha' \leftarrow \phi,\,\alpha) = \int_{\Delta'} d\underline{p}'\,|\langle \underline{p}',\,\alpha'|\,S\,|\Phi\rangle|^2 \qquad (17.8)$$

In principle, Δ' can be an arbitrary volume in the space of the final momenta. Later, we shall return to discuss some examples of what are convenient choices for Δ' in practice. Here we mention only that if the final channel is a two-body channel, then $\underline{p}'$ reduces to a single momentum $\mathbf{p}'$ (the relative momentum of the two final particles) and the natural choice for Δ' is the familiar cone defined by an element of solid angle about any fixed direction.

Exactly as in the one-channel case, the precise details of the in wave function $\phi(\mathbf{p})$ are unknown and we must repeat the experiment many times with ϕ suitably randomized with respect to impact parameter. (Just as in the one-channel case we do not need explicitly to randomize with respect to the shape of ϕ since our answer will turn out to be independent of shape—under the appropriate conditions.) When this is done the total number of particles observed to emerge in channel α' with their momenta in Δ' is

$$N_{\text{sc}}(\Delta',\,\alpha') = \sum_i w(\Delta',\,\alpha' \leftarrow \phi_{\boldsymbol{\rho}_i},\,\alpha)$$

$$= n_{\text{inc}} \int d^2\rho\, w(\Delta',\,\alpha' \leftarrow \phi_{\boldsymbol{\rho}},\,\alpha)$$

$$\equiv n_{\text{inc}}\sigma(\Delta',\,\alpha' \leftarrow \phi,\,\alpha)$$

where as usual $\boldsymbol{\rho}$ is perpendicular to the mean incident momentum and $\phi_{\boldsymbol{\rho}}$ denotes the wave packet obtained when ϕ is displaced by $\boldsymbol{\rho}$. The cross section

$$\sigma(\Delta',\,\alpha' \leftarrow \phi,\,\alpha) = \int d^2\rho\, w(\Delta',\,\alpha' \leftarrow \phi_{\boldsymbol{\rho}},\,\alpha) \qquad (17.9)$$

is clearly the effective cross section of the target for scattering of the packet ϕ into the volume Δ' in channel α'.

---

[4]Here $d\underline{p}'$ is an infinitesimal volume element in the $(3n' - 3)$-dimensional space of the relative momenta of channel α'. The result is the appropriate generalization of the one-channel probability of Section 3-d, $w(d^3p' \leftarrow \psi_{\text{in}}) = d^3p'\,|\langle \mathbf{p}'\,|\,\psi_{\text{out}}\rangle|^2$.

To evaluate the cross section we return to (17.8) for the probability w. The matrix element that we have to calculate is

$$\langle \underline{p}', \alpha' | \, \mathsf{S} \, | \Phi_\rho \rangle = \int d^3 p \langle \underline{p}', \alpha' | \, \mathsf{S} \, | \mathbf{p}, \alpha \rangle e^{-i\mathbf{p}\cdot\mathbf{p}} \phi(\mathbf{p})$$

As before, we can replace the S-matrix element by the appropriate multiple of the on-shell T matrix. (For elastic scattering, $\alpha' = \alpha$, this requires that we avoid the forward direction; for inelastic scattering this restriction is unnecessary.) This gives:

$$\langle \underline{p}', \alpha' | \, \mathsf{S} \, | \Phi_\rho \rangle = -2\pi i \int d^3 p \; \delta(E' - E) \, t(\underline{p}', \alpha' \leftarrow \mathbf{p}, \alpha) e^{-i\mathbf{p}\cdot\mathbf{p}} \phi(\mathbf{p})$$

Substituting into (17.8) and then into (17.9) we can do two of the integrations (exactly as in Chapter 3) and obtain:

$$\sigma(\Delta', \alpha' \leftarrow \phi, \alpha) = (2\pi)^4 m \int_{\Delta'} d\underline{p}' \int d^3 p \, \frac{1}{p_\parallel} \, \delta(E' - E) \, |t(\underline{p}', \alpha' \leftarrow \mathbf{p}, \alpha)|^2 \, |\phi(\mathbf{p})|^2$$

where m denotes the reduced mass of the initial two particles in channel α. Provided the in wave packet ϕ is sufficiently well peaked about its mean momentum (which we now call $\mathbf{p}$) the integral over $\mathbf{p}$ disappears as the normalization integral for ϕ and we obtain an answer that is independent of the shape of ϕ and so can be written:

$$\boxed{\sigma(\Delta', \alpha' \leftarrow \mathbf{p}, \alpha) = (2\pi)^4 \, \frac{m}{p} \int_{\Delta'} d\underline{p}' \, \delta(E' - E) \, |t(\underline{p}', \alpha' \leftarrow \mathbf{p}, \alpha)|^2} \qquad (17.10)$$

The simplest application of this result is to the case of a two-body final channel [for example, any of the first three processes in (17.6)]. In this case the final relative momenta $\underline{p}'$ reduce to a single momentum $\mathbf{p}'$ and the volume Δ' is a volume in the corresponding three-dimensional space. Because of energy conservation there is no interest in measurement of the magnitude of $\mathbf{p}'$, and the smallest interesting volume Δ' is the familiar cone defined by the infinitesimal solid angle $d\Omega$. For this case (17.10) becomes:

$$\sigma(d\Omega, \alpha' \leftarrow \mathbf{p}, \alpha) = (2\pi)^4 \, \frac{m}{p} \, d\Omega \int_0^\infty p'^2 \, dp'$$

$$\times \, \delta\!\left(\frac{p'^2}{2m'} + W_{\alpha'} - \frac{p^2}{2m} - W_\alpha\right) |t(\mathbf{p}', \alpha' \leftarrow \mathbf{p}, \alpha)|^2 \quad (17.11)$$

since the energy E in a two body-channel α is $p^2/2m + W_\alpha$, where W_α denotes the channel threshold. Rewriting the left-hand side in the familiar form

$(d\sigma/d\Omega)\,d\Omega$ this gives:

$$\frac{d\sigma}{d\Omega}\,(\mathbf{p}',\alpha'\leftarrow\mathbf{p},\alpha)=(2\pi)^4 mm'\frac{p'}{p}\,|t(\mathbf{p}',\alpha'\leftarrow\mathbf{p},\alpha)|^2$$

[α and α' two-body channels]   (17.12)

In the case of elastic scattering, $\alpha'=\alpha$, this result reduces to

$$\frac{d\sigma}{d\Omega}\,(\mathbf{p}',\alpha\leftarrow\mathbf{p},\alpha)=(2\pi)^4 m^2\,|t(\mathbf{p}',\alpha\leftarrow\mathbf{p},\alpha)|^2 \qquad (17.13)$$

which has exactly the form of the one-channel result and includes the latter as a special case. For inelastic processes, $\alpha'\neq\alpha$, the result (17.12) differs from (17.13) in two respects. The reduced masses m and m' may be different [as, for example, in the rearrangement collision $a+(bc)\to b+(ac)$]. And the initial and final momenta may be different, since p' is fixed by energy conservation to satisfy

$$\frac{p'^2}{2m'}+W_{\alpha'}=\frac{p^2}{2m}+W_{\alpha}$$

For example, in the excitation process $a+(bc)\to a+(bc)^*$,

$$p'=[p^2-2m(E_{(bc)^*}-E_{(bc)})]^{\frac12}$$

It is convenient, as in the one-channel case, to introduce a *scattering amplitude*, defined as[5]

$$f(\mathbf{p}',\alpha'\leftarrow\mathbf{p},\alpha)=-(2\pi)^2(m'm)^{\frac12}\,t(\mathbf{p}',\alpha'\leftarrow\mathbf{p},\alpha) \qquad (17.14)$$

in terms of which the cross section for any process with two-body initial and final states is

$$\boxed{\frac{d\sigma}{d\Omega}\,(\mathbf{p}',\alpha'\leftarrow\mathbf{p},\alpha)=\frac{p'}{p}\,|f(\mathbf{p}',\alpha'\leftarrow\mathbf{p},\alpha)|^2} \qquad (17.15)$$

If we wish to apply the general result (17.10) for processes with more than two final particles we must first decide what is a reasonable choice for the volume Δ' in the final momentum space; and we now briefly discuss this question for the case of a three-body final state. For definiteness we consider a process of the type

$$a+(bc)\to a+b+c$$

---

[5] As usual, definitions differing by assorted factors of m, m', p and p' are found.

and for simplicity we suppose that particle c is very heavy. For example, we could consider the ionization process

$$e + H \rightarrow e + e + p$$

For such a system the two relative momenta can be taken to be simply the *actual* momenta $\mathbf{p}_a$ and $\mathbf{p}_b$ of the two light particles a and b, and we must now decide what properties of these two final momenta it will be useful to observe. For example, we can measure the number of particles b ejected with their momenta in the infinitesimal volume defined by dE_b and $d\Omega_b$, and the number of particles a with their direction in $d\Omega_a$. However, the energy of the particles a is then fixed by energy conservation. Thus, in this case the smallest volume Δ' of interest has $\mathbf{p}_b$ fixed in the infinitesimal volume defined by $d\Omega_b$ and dE_b, but $\mathbf{p}_a$ anywhere in the infinite cone defined by $d\Omega_a$. For such a volume we write

$$\sigma(d\Omega_a, d\Omega_b, dE_b; \alpha' \leftarrow \mathbf{p}, \alpha) = \frac{d^3\sigma}{d\Omega_a \, d\Omega_b \, dE_b} \, d\Omega_a \, d\Omega_b \, dE_b$$

where, as can easily be checked from (17.10),

$$\frac{d^3\sigma}{d\Omega_a \, d\Omega_b \, dE_b} = (2\pi)^4 \frac{m_a^2 m_b p_a p_b}{p} \, |t(\mathbf{p}_a, \mathbf{p}_b; \alpha' \leftarrow \mathbf{p}, \alpha)|^2 \qquad (17.16)$$

The differential cross section $d^3\sigma/d\Omega_a \, d\Omega_b \, dE_b$ represents the most detailed type of measurement possible for a three-body final state. It is often convenient to make a less detailed measurement. For example, one could measure the direction of the scattered particle a, but completely ignore the ejected particle b. This would correspond to taking for Δ' the usual cone $d\Omega_a$ for $\mathbf{p}_a$ times the whole of the sphere of allowed values of $\mathbf{p}_b$. It gives a differential cross section $d\sigma/d\Omega_a$, which is obtained from (17.16) by integrating over Ω_b and E_b. Another possibility is to monitor the energy E_b of the ejected particle but to ignore both directions. The corresponding cross section $d\sigma/dE_b$ is obtained by integrating (17.16) over all Ω_a and Ω_b, and gives the energy spectrum of the ejected particle.

To conclude this section we remark that if we take the whole of the momentum space of channel α' for the volume Δ', then we obtain the *total cross section for scattering into channel* α', which we denote $\sigma(\alpha' \leftarrow \mathbf{p}, \alpha)$. (In particular, for $\alpha' = \alpha$ this is the *total elastic cross section*.) If we then sum this over all final channels α' we obtain the *total cross section*,

$$\sigma(\mathbf{p}, \alpha) = \sum_{\alpha'} \sigma(\alpha' \leftarrow \mathbf{p}, \alpha)$$

$$= (2\pi)^4 \frac{m}{p} \sum_{\alpha'} \int d\underline{p}' \, \delta(E' - E) \, |t(\underline{p}', \alpha' \leftarrow \mathbf{p}, \alpha)|^2 \qquad (17.17)$$

while if we restrict this sum to $\alpha' \neq \alpha$, we obtain the *total inelastic cross section*.

17-d.   Rotational Invariance

If our multichannel system is rotationally invariant, then the rotation operator $R(\alpha)$ commutes with the Hamiltonians H and H^α. It follows that it commutes with $\Omega^\alpha_\pm$ and, hence, with S,

$$S = R^\dagger SR \tag{17.18}$$

If all of the particles and their bound states are spinless, the effect of the rotation R on the channel basis vectors $|\underline{p}, \alpha\rangle$ is just

$$R|\underline{p}, \alpha\rangle = |\underline{p}_R, \alpha\rangle$$

where $\underline{p}_R$ denotes the effect of rotating all of the relative momenta $\underline{p}$. Thus, taking matrix elements of (17.18), we obtain the very natural result:

$$\langle \underline{p}', \alpha'| S |\underline{p}, \alpha\rangle = \langle \underline{p}'_R, \alpha'| S |\underline{p}_R, \alpha\rangle$$

and corresponding equalities for the T matrix $t(\underline{p}', \alpha' \leftarrow \underline{p}, \alpha)$. If any of the fragments of channels α or α' have spins (due either to the intrinsic spins of the original particles, or to the orbital momentum of the bound states), then this result is complicated by the transformation properties of the spin indices. These complications must then be handled by the same methods as described in Chapters 6 and 7.

As in the one-channel case, the simplest way to exploit rotational invariance is to work in an angular-momentum basis. This is especially simple if we consider an energy at which only two-body channels are open and suppose further that none of the bodies have spin. In this case the angular-momentum basis (of the relative motion) in each channel is a simple orbital basis with vectors $|E, l, m; \alpha\rangle$. Rotational invariance implies that the corresponding S matrix has the form:

$$\langle E', l', m'; \alpha'| S |E, l, m; \alpha\rangle = \delta(E' - E) \delta_{l'l}\delta_{m'm}s^l_{\alpha'\alpha}(E) \tag{17.19}$$

That is, the scattering for given E and l is determined by an $n \times n$ matrix $S^l(E)$ in "channel-space," n being the number of open channels at the energy E.

The passage between the momentum and angular-momentum bases proceeds exactly as in the one-channel case. Corresponding to the definition

$$f_l(E) = \frac{s_l(E) - 1}{2ip}$$

of the one-channel amplitude, we define the multichannel partial-wave amplitude as:

$$f^l_{\alpha'\alpha}(E) = \frac{s^l_{\alpha'\alpha}(E) - \delta_{\alpha'\alpha}}{2i(p_{\alpha'}p_\alpha)^{1/2}} \qquad (17.20)$$

Here p_α denotes the momentum in channel α when the total energy is E—that is,

$$p_\alpha = [2m_\alpha(E - W_\alpha)]^{1/2}$$

where W_α is the threshold of channel α—and the Kronecker delta $\delta_{\alpha'\alpha}$ is to ensure that the amplitude is related to the matrix elements of the operator $S - 1$ in the normal way. As usual, definitions differing from (17.20) by assorted factors of $2\pi i$ and the momenta are common; with our definition it is easily checked (Problem **17.4**) that the multichannel partial-wave series has the simple form (for α and α' both two-body channels):

$$f(\mathbf{p}', \alpha' \leftarrow \mathbf{p}, \alpha) = \sum_l (2l + 1)f^l_{\alpha'\alpha}(E)P_l(\cos\theta) \qquad (17.21)$$

The unitarity of S implies that the $n \times n$ matrix $S^l(E)$ of (17.19) is a unitary matrix. Since the dimension of this matrix depends on the energy, it is of some interest to discuss the unitary equation as a function of energy. We start with the energy E just above the threshold of the lowest channel (which we label $\alpha = 1$), and imagine E to be increased steadily past the various thresholds W_α (which we label in order of increasing energy). When $W_1 \leqslant E < W_2$, the matrix $S^l(E)$ is a one-dimensional unitary matrix and hence has a single element of modulus one $|s^l_{11}| = 1$, or

$$s^l_{11}(E) = e^{2i\delta_l(E)} \qquad [W_1 \leqslant E < W_2]$$

with $\delta_l(E)$ real. If we consider the partial-wave amplitude defined in (17.20) this means that the quantity $p_1 f^l_{11}$ lies on the unitary circle (see Fig. 17.1).

When E moves above the first inelastic threshold $S^l(E)$ becomes a (2×2) unitary matrix; and more generally, when E moves above W_n (still assumed to be a two-body threshold) $S^l(E)$ becomes an $n \times n$ matrix satisfying

$$S^l(E)^\dagger S^l(E) = 1$$

If we consider the $(1, 1)$ matrix element of this equation, we see that

$$|s^l_{11}|^2 + |s^l_{21}|^2 + \cdots + |s^l_{n1}|^2 = 1 \qquad (17.22)$$

Clearly, as soon as there is any inelasticity the original elastic element s^l_{11} must have modulus *less than* one, $|s^l_{11}| < 1$ or

$$s^l_{11}(E) = \epsilon_l(E)e^{2i\delta_l(E)} \qquad (17.23)$$

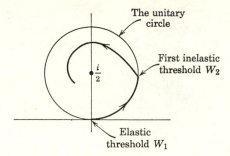

FIGURE 17.1. Typical behavior of the elastic partial-wave amplitude as a function of energy.

where δ_l is still real but ϵ_l, which is called the *inelasticity factor*, satisfies

$$0 \leqslant \epsilon_l(E) < 1$$

Alternatively we can write,

$$s_{11}^l(E) = e^{2i\eta_l(E)}$$

with η_l complex. In this form we can say that inelasticity forces the "phase shift" η_l to have a positive imaginary part. The result means that the elastic partial-wave amplitude moves *inside* the unitary circle as shown in Fig. 17.1.

The result (17.23) is easily understood. Much as in the one-channel case, $|s_{11}^l|^2$ is the ratio of outgoing to ingoing flux in channel 1, when the incident beam is pure l wave in channel 1. As long as only channel 1 is open, this ratio must be one; but as soon as other competing channels open up it must be less than one.

It is clear from (17.22), and its equivalent with 1 replaced by α, that every element $s_{\alpha'\alpha}^l$ must have modulus less than one. In particular, for any elastic element the number $p_\alpha f_{\alpha\alpha}^l$ must lie inside the unitary circle. However only the first amplitude f_{11}^l has a purely elastic interval.

Once a three-body channel opens up the situation becomes more complicated. The basis vectors for the relative motion in a three-body channel are labelled by E, l, m *plus* an internal energy (a continuous variable) and two internal angular momenta. The discussion of such channels is obviously much more difficult. Thus, if we return to (17.22) and suppose that the channel α has three bodies, then the simple term $|s_{\alpha 1}^l|^2$ is replaced by an integral and sum over the additional labels. However, this new term is still positive and our conclusion that $|s_{11}^l| \leqslant 1$ remains valid.

17-e. Time-Reversal Invariance

The discussion of parity invariance for multichannel scattering is little different from that for the one-channel case, and need not be spelled out here (see Problem **17.5**). However, in the case of time-reversal invariance, the multichannel problem offers some striking new possibilities.

If the dynamics are invariant under time reversal then we can show just as before that $S = T^\dagger S^\dagger T$ and, hence, that

$$\langle \chi | \, S \, | \phi \rangle = \langle \phi_T | \, S \, | \chi_T \rangle \tag{17.24}$$

For example, if all particles are spinless, T invariance simply implies that[6]

$$\langle \underline{p}', \alpha' | \, S \, | \underline{p}, \alpha \rangle = \langle -\underline{p}, \alpha | \, S \, | -\underline{p}', \alpha' \rangle \tag{17.25}$$

The interesting new feature of this result is that it relates the amplitudes for two qualitatively different processes, $\alpha' \leftarrow \alpha$ and $\alpha \leftarrow \alpha'$. In recent years there has been considerable interest in testing T invariance, and there have been several experiments to check the result (17.25) for various processes. Here we mention just one example—the experiment of Bodansky, et al. (1966), which used the two processes

$$p + {}^{25}\mathrm{Mg} \rightleftarrows d + {}^{24}\mathrm{Mg} \tag{17.26}$$

Since the particles involved in these reactions have spins ($\frac{1}{2}$, $\frac{5}{2}$, 1, and 0, respectively) we must generalize the result (17.25) accordingly. If we denote by m_1, m_2, m_3, m_4 the third components of the spins of the four particles in (17.26), then (as the reader can readily check) T invariance implies that the amplitudes for the two processes concerned satisfy

$$f_\rightarrow(\mathbf{p}', m_3, m_4 \leftarrow \mathbf{p}, m_1, m_2) = f_\leftarrow(-\mathbf{p}, -m_1, -m_2 \leftarrow -\mathbf{p}', -m_3, -m_4) \tag{17.27}$$

where, to avoid writing channel labels, we have introduced arrow subscripts $\rightleftarrows$ corresponding to the directions of the two reactions (17.26).

Now, according to (17.15), the square of the left-hand side of this result is (p/p') times the corresponding differential cross section. If we then sum this over all possible values of m_1, m_2, m_3, m_4, we obtain $(2s_1 + 1)(2s_2 + 1)$ times the *unpolarized*[7] cross section for momenta ($\mathbf{p}' \leftarrow \mathbf{p}$). Exactly the same

---

[6] The results (17.24) and (17.25) are variously referred to as "the principle of micro-reversability," "the reciprocity theorem," and "principle of detailed balance." Unfortunately, there is considerable confusion as to the precise meaning of these terms. In particular, the use of "detailed balance" seems regrettable since historically this was used for the equality of the probabilities for ($\chi \leftarrow \phi$) and ($\phi \leftarrow \chi$)—in general, a quite different proposition from equality of the probabilities for ($\chi \leftarrow \phi$) and ($\phi_T \leftarrow \chi_T$). (Of course in certain situations the two results are related; for example, if all particles are spinless, invariance under PT implies that $\langle \underline{p}', \alpha' | \, S \, | \underline{p}, \alpha \rangle = \langle \underline{p}, \alpha | \, S \, | \underline{p}', \alpha' \rangle$, which is equivalent to detailed balance.) To avoid getting entangled in this confusing situation we shall refer to (17.24) by the unambiguous title of "time-reversal invariance."

[7] Remember that to obtain the unpolarized cross section, one sums over final spins and *averages* over initial spins. Thus if we simply sum over all spins we obtain $(2s_1 + 1)$ $(2s_2 + 1)$ times the unpolarized cross section.

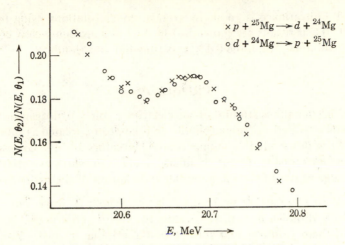

FIGURE 17.2.   Relative cross sections $N(E, \theta_2)/N(E, \theta_1)$ for the processes

$$p + {}^{25}\text{Mg} \rightleftharpoons d + {}^{24}\text{Mg}$$

at fixed angles ($\theta_1 \approx 30°$, $\theta_2 \approx 120°$) and various energies (after Bodansky, et al., 1966).

considerations apply to the right-hand side. Finally, since rotational invariance implies that the unpolarized cross sections depend only on the energy and angle between $\mathbf{p}$ and $\mathbf{p}'$, the result (17.27) means that the total unpolarized counts (normalized to equal incident densities) into any given solid angle for the two processes concerned satisfy

$$(2s_1 + 1)(2s_2 + 1)\frac{p}{p'} N_\rightarrow(E, \theta) = (2s_3 + 1)(2s_4 + 1)\frac{p'}{p} N_\leftarrow(E, \theta) \quad (17.28)$$

That is, apart from a proportionality factor depending on the spins and momenta, the unpolarized angular distributions for the two processes $p + {}^{25}\text{Mg} \rightleftharpoons d + {}^{24}\text{Mg}$ should be the same at all energies and angles— provided the interactions are invariant under rotations and time reversal.

The experiment of Bodansky, et al. used unpolarized beams of protons incident on ${}^{25}\text{Mg}$ and deuterons on ${}^{24}\text{Mg}$, both from a tandem Van der Graaf with energies adjusted to give equal CM energies for the two processes (~15 MeV protons and ~10 MeV deuterons). To avoid the difficulties of measuring absolute cross sections they plotted the ratio $N(E, \theta_2)/N(E, \theta_1)$ for the two processes at two fixed angles ($\theta_1 \approx 30$ deg and $\theta_2 \approx 120$ deg) and at several energies. The various proportionality factors in (17.28) cancel out of this ratio to give

$$\frac{N_\rightarrow(E, \theta_2)}{N_\rightarrow(E, \theta_1)} = \frac{N_\leftarrow(E, \theta_2)}{N_\leftarrow(E, \theta_1)}$$

provided T invariance is good (invariance under rotations being taken for granted). Their results are shown in Fig. 17.2 and are completely consistent with T invariance within the 0.4% experimental uncertainties.

PROBLEMS

17.1. Consider the scattering of an electron e' off a hydrogen atom (ep), with e' and e treated as distinguishable and the proton treated as spinless and fixed. Write down suitable momentum eigenvectors and the corresponding wave functions for the various channels. Include the spins of both electrons and use eigenstates of the total angular momentum of the hydrogen atoms.

17.2. (a) In two-particle problems it is convenient to work with the CM and relative variables $\bar{\mathbf{p}}, \bar{\mathbf{x}}, \mathbf{p}, \mathbf{x}$. Discuss the various criteria that lead to the choice of these variables and show that the criteria are met. You should include at least the following—which are not all independent:

(1) The new momenta must include the total momentum, the generator of displacements of the whole system.
(2) The second of the new position variables must be unchanged by displacements.
(3) The new variables must be canonically conjugate pairs; that is, they must satisfy the right commutation relations.
(4) The new momenta should be given by $-i\,\mathbf{\nabla}_{\bar{x}}$ and $-i\,\mathbf{\nabla}_x$ in the x representation.
(5) The Jacobian of the transformation from $(\mathbf{x}_1, \mathbf{x}_2)$ to $(\bar{\mathbf{x}}, \mathbf{x})$ should be 1.
(6) The kinetic energy should be a sum of squares of the momenta.
(7) The space $\mathscr{H}$ of two-particle states should factor as $\mathscr{H}_{\mathrm{cm}} \otimes \mathscr{H}_{\mathrm{rel}}$, and the Hamiltonian decompose as $H = H_{\mathrm{cm}} + H_{\mathrm{rel}}$, in such a way that $\bar{\mathbf{p}}, \bar{\mathbf{x}}, H_{\mathrm{cm}}$ act only on $\mathscr{H}_{\mathrm{cm}}$ and $\mathbf{p}, \mathbf{x}, H_{\mathrm{rel}}$ on $\mathscr{H}_{\mathrm{rel}}$.)

(b) In our three-particle model, with spinless particles a, b, c, consider the following set of coordinates: the position of b relative to c, the position of a relative to the CM of b and c, and the position of the overall CM. Write down the corresponding momenta and show that these variables satisfy all of your criteria discussed in (a).

17.3. Prove the multichannel optical theorem for any two-body initial channel α,

$$\operatorname{Im} f(\mathbf{p}, \alpha \leftarrow \mathbf{p}, \alpha) = \frac{p}{4\pi}\, \sigma(\mathbf{p}, \alpha)$$

where $f(\mathbf{p}, \alpha \leftarrow \mathbf{p}, \alpha)$ is the forward elastic amplitude, and $\sigma(\mathbf{p}, \alpha)$ is the total cross section for an initial state of relative momentum $\mathbf{p}$ in channel α.

Don't forget that the total cross section generally includes many-body final channels.

17.4. For any two-body channels α and α' derive the multichannel partial-wave series (17.21); (review the one-channel discussion of Section 6-c).

17.5. Consider a multichannel system in which all particles are spinless and all bound states have zero angular momentum.
(a) What does invariance under parity imply for the momentum-space S matrix?
(b) Show that invariance under PT implies that the momentum-space S matrix is symmetric.

17.6. If one has no idea how to calculate the T matrix for a given process, one can sometimes get a rough idea of how the cross section will behave by assuming that the T matrix is *constant* and carrying out the appropriate integral (17.10) over the final momenta. This procedure is known as a *phase space calculation*, since it just takes account of the volume of phase space (or, more correctly, momentum space) available to the final particles. As a simple example of the method, consider the ionization process:

$$a + (bc) \rightarrow a + b + c$$

with c infinitely heavy. Compute the energy spectrum $d\sigma/dE_b$ of the ejected particle b from the general result (17.10) by integrating over the momentum of a and the direction of b, assuming that the T matrix element is constant. Show that, under this assumption,

$$\frac{d\sigma}{dE_b} = \begin{cases} \text{constant} \times \dfrac{(E_b)^{\frac{1}{2}}(E - E_b - W)^{\frac{1}{2}}}{E^{\frac{1}{2}}} & E_b < E - W \\ 0 & E_b > E - W \end{cases}$$

where E is the incident kinetic energy and W is the ionization energy of (bc) (and we assume that $E > W$).

Integrate this result over E_b and show that (on the basis of phase space alone) the total ionization cross section should behave like $(E - W)^2$ just above the threshold $E = W$.

It should be emphasized that the use of phase space calculations requires some skill and luck, since the procedure is really very ambiguous. The point is that the definition of the T matrix is fairly arbitrary and in place of $t(\underline{p}', \alpha' \leftarrow \underline{p}, \alpha)$ we could just as well have introduced $t' = g(\underline{p}', \underline{p})t$ where $g(\underline{p}', \underline{p})$ is almost any function of all the momenta. Clearly it makes a big difference whether we make the approximation $t = $ constant, or $t' = $ constant. In relativistic problems there is a generally accepted definition of a covariant T matrix; but even here it is by no means clear that this is the "correct" T matrix to approximate by a constant.

18 Fundamentals of Time-Independent Multichannel Scattering

18-a   **The Stationary Scattering States**

18-b   **The Lippmann–Schwinger Equations**

18-c   **The T Operators**

18-d   **Born Approximation; Elastic Scattering**

18-e   **Born Approximation; Excitation**

In Chapters 16 and 17 we have set up the time-dependent theory of multichannel scattering. This provides a description of the multichannel collision process, as it actually occurs, in terms of the S operator and on-shell T matrix. We now turn to the corresponding time-independent theory. Its main purpose is to furnish a means for the actual computation of the T matrix for given interactions; and for this reason it is the time-independent theory that is the day to day concern of the practising physicist.

In one-channel scattering we saw that the time-independent formalism centers around the Green's operators $G(z)$ and $G^0(z)$, the T operator $T(z)$, and the stationary scattering states $|\mathbf{p}\pm\rangle$; the on-shell T matrix being given by any of the equivalent expressions

$$t(\mathbf{p}' \leftarrow \mathbf{p}) = \langle\mathbf{p}'|\, T\,(E_p + i0)\,|\mathbf{p}\rangle \tag{18.1}$$

$$= \langle\mathbf{p}'|\, V\,|\mathbf{p}+\rangle = \langle\mathbf{p}'-|\, V\,|\mathbf{p}\rangle \tag{18.2}$$

Both $T(z)$ and $|\mathbf{p}\pm\rangle$ were given in terms of $G^0(z)$ by Lippmann–Schwinger equations, which provided essentially equivalent approaches to finding the quantity of real interest, the on-shell T matrix.

In this chapter we shall see how the one-channel results summarized in (18.1) and (18.2) extend to the multichannel case. Superficially at least, the most important complication is that all quantities acquire additional channel labels. As we have already seen the single free Hamiltonian H^0 of the one-channel case is replaced by a family of "free" channel Hamiltonians H^α, describing the "free" evolution in the various different channels. In place of the single free Green's operator $G^0(z)$, there is a family of channel Green's operators $G^\alpha(z) = (z - H^\alpha)^{-1}$. In place of the stationary states $|\mathbf{p}\pm\rangle$, there is a family of stationary states $|\underline{p}, \alpha\pm\rangle$. In place of the single T operator there is a double family of T operators $T^{\beta\alpha}(z)$, one for each pair of channels α and β.

In the one-channel case we first developed the theory of the T operator and then that of the stationary scattering states. In the present case we reverse this order of presentation, since it is the stationary-state formalism that appears most commonly in practical calculations.[1] Accordingly, in the first two sections we give the definitions and some properties of the stationary scattering states $|\underline{p}, \alpha\pm\rangle$ and derive expressions analogous to (18.2) for the on-shell T matrix. In Section 18-c we set up the corresponding theory of the T operators $T^{\beta\alpha}(z)$ and establish the analogue of (18.1). Finally, in the last two sections we discuss the simplest approximate method for calculation of the on-shell T matrix—the inevitable Born approximation.

18-a.   The Stationary Scattering States

In single-channel scattering the stationary states $|\mathbf{p}\pm\rangle$ were eigenvectors of the full Hamiltonian H defined as $\Omega_{\pm}|\mathbf{p}\rangle$. Loosely speaking this meant that $|\mathbf{p}+\rangle$ (for example) is the actual state at $t = 0$ corresponding to the plane-wave in state $|\mathbf{p}\rangle$. More precisely, the actual state $|\phi+\rangle$ at $t = 0$ corresponding to a normalized in wave packet $|\phi\rangle$ has the same expansion in terms of the $|\mathbf{p}+\rangle$ as does $|\phi\rangle$ in terms of the plane waves $|\mathbf{p}\rangle$.

We now define multichannel stationary scattering states as

$$\boxed{|\underline{p}, \alpha\pm\rangle = \Omega_{\pm}^\alpha |\underline{p}, \alpha\rangle} \qquad (18.3)$$

where as usual $\underline{p}$ denotes a set of $(n_\alpha - 1)$ relative momenta of the n_α bodies in channel α and $|\underline{p}, \alpha\rangle$ is the corresponding "free" plane-wave state.[2]

---

[1] The reader may reasonably ask why we used the reverse order in the one-channel case. The principal reason is that by introducing the operators $G(z)$ and $T(z)$ first (without reference to $|\mathbf{p}\pm\rangle$) one can get a better understanding of the off-shell T operator. Also, perhaps there is some advantage to working things both ways round.

[2] As in Chapter 17, whenever possible we consider just the relative motion. Thus $\underline{p}$ is a set of relative momenta , $|\underline{p}, \alpha\rangle$ a vector in the space of the relative motion, and H and H^α will denote the relative parts of the full and channel Hamiltonians. Alternatively, the reader may prefer to consider a process like electron–atom collisions where one body can be considered infinitely heavy and hence fixed. In this case $\underline{p}$ is just the actual momenta of the $(n_\alpha - 1)$ light bodies and H and H^α are *the* Hamiltonian and channel Hamiltonian.

Using the intertwining relations (17.3) we can immediately see that these states are eigenstates of the full Hamiltonian H with energy $E = E_{\underline{p}}^{\alpha}$. (This means that their wave functions $\langle \underline{x} \mid \underline{p}, \alpha \pm \rangle$ satisfy the time-independent Schrödinger equation, which provides one route to the actual computation of the states.) In the same loose sense as applied to the single-channel case, $|\underline{p}, \alpha + \rangle$ can be regarded as the actual state at $t = 0$ coming from an in asymptote *in channel* α with momenta $\underline{p}$. Similarly $|\underline{p}, \alpha - \rangle$ can be thought of as the actual state at $t = 0$ that would evolve into an out asymptote in channel α with momenta $\underline{p}$. As before, of course, the real significance of the improper vectors is found by smearing them into proper normalizable states.

Just as in the one-channel case we can derive an expression for $|\underline{p}, \alpha \pm \rangle$ in terms of $|\underline{p}, \alpha \rangle$ and the Green's operator. We first note, in the now familiar way, that for a proper normalizable vector $|\phi\rangle$ in the channel α subspace,

$$\Omega_{\pm}^{\alpha} |\phi\rangle = \lim e^{iHt} e^{-iH^{\alpha}t} |\phi\rangle$$

$$= |\phi\rangle + i \int_{0}^{\mp\infty} dt \, e^{iHt} V^{\alpha} e^{-iH^{\alpha}t} |\phi\rangle \qquad (18.4)$$

where we have introduced the *channel* α *scattering potential*

$$\boxed{V^{\alpha} \equiv H - H^{\alpha}}$$

This consists of all potentials that link different freely moving fragments in channel α; that is, V^{α} is precisely that part of the potential which becomes ineffective as the particles move apart in channel α. If V^{α} were zero, an in state in channel α would not be scattered.

The known convergence of the integral (18.4) allows us to insert the familiar damping factor $e^{\pm \epsilon t}$ with the limit $\epsilon \downarrow 0$; and in this form we can apply (18.4) to the improper states $|\underline{p}, \alpha\rangle$ to give:

$$|\underline{p}, \alpha \pm \rangle = |\underline{p}, \alpha\rangle + \lim_{\epsilon \downarrow 0} i \int_{0}^{\mp\infty} dt \, e^{\pm \epsilon t} e^{iHt} V^{\alpha} e^{-iH^{\alpha}t} |\underline{p}, \alpha\rangle$$

Since $|\underline{p}, \alpha\rangle$ is an eigenvector of the channel Hamiltonian H^{α} with energy $E = E_{\underline{p}}^{\alpha}$, this gives:

$$|\underline{p}, \alpha \pm \rangle = |\underline{p}, \alpha\rangle + \lim_{\epsilon \downarrow 0} i \int_{0}^{\mp\infty} dt \, e^{-i(E \pm i\epsilon - H)t} V^{\alpha} |p, \alpha\rangle$$

or

$$\boxed{|\underline{p}, \alpha \pm \rangle = |\underline{p}, \alpha\rangle + G(E \pm i0) V^{\alpha} |\underline{p}, \alpha\rangle} \qquad (18.5)$$

where we have introduced the full Green's operator $G(z) = (z - H)^{-1}$.

The result (18.5) is the analogue of the one-channel result $|\mathbf{p}\pm\rangle = |\mathbf{p}\rangle + G(\pm)V|\mathbf{p}\rangle$, and with its help we can obtain expressions for the on-shell T matrix. The momentum-space S-matrix elements can be written as:

$$\langle \underline{p}', \beta| \mathsf{S} |\underline{p}, \alpha\rangle = \langle \underline{p}', \beta| \Omega_-^{\beta\dagger}\Omega_+^{\alpha} |\underline{p}, \alpha\rangle$$
$$= \langle \underline{p}', \beta- | \underline{p}, \alpha+\rangle \qquad (18.6)$$

Now, using (18.5) we can write $|\underline{p}, \alpha+\rangle$ as

$$|\underline{p}, \alpha+\rangle = |\underline{p}, \alpha-\rangle + [G(E + i0) - G(E - i0)]V^{\alpha} |\underline{p}, \alpha\rangle$$

which can be substituted into (18.6) to give:

$$\langle \underline{p}', \beta| \mathsf{S} |\underline{p}, \alpha\rangle = \delta_{\beta\alpha}\, \delta(\underline{p}' - \underline{p}) + \langle \underline{p}', \beta-| [G(E + i0) - G(E - i0)] V |\underline{p}, \alpha\rangle$$

Finally, since the bra $\langle \underline{p}', \beta-|$ is an eigenvector of H with eigenvalue $E' = E_{p'}^{\beta}$, the factor in brackets comes outside as:

$$\left[\frac{1}{E - E' + i0} - \frac{1}{E - E' - i0}\right] = -2\pi i\, \delta(E' - E)$$

to give

$$\langle \underline{p}', \beta| \mathsf{S} |\underline{p}, \alpha\rangle = \delta_{\beta\alpha}\delta(\underline{p}' - \underline{p}) - 2\pi i\, \delta(E' - E)\langle \underline{p}', \beta-| V^{\alpha} |\underline{p}, \alpha\rangle \quad (18.7)$$

That is, the on-shell T matrix is given by

$$\boxed{\, t(\underline{p}', \beta \leftarrow \underline{p}, \alpha) = \langle \underline{p}', \beta- |V^{\alpha}|\underline{p}, \alpha\rangle \,} \qquad (18.8)$$

Similarly, if we were to rewrite the bra $\langle \underline{p}', \beta-|$ of (18.6) in terms of $\langle \underline{p}', \beta+|$ we would find the alternative result

$$\boxed{\, t(\underline{p}', \beta \leftarrow \underline{p}, \alpha) = \langle \underline{p}', \beta| V^{\beta} |\underline{p}, \alpha+\rangle \,} \qquad (18.9)$$

These two expressions are the multichannel versions of the one-channel results (18.2). The version in (18.8) contains the scattering potential V^{α} of the initial channel α and is therefore referred to as the *"prior"* version. That in (18.9) has the potential V^{β} appropriate to the final channel β and is called the *"post"* version.

18-b.   The Lippmann–Schwinger Equations

It is clear from the results (18.8) and (18.9) that knowledge of the states $|\underline{p}, \alpha+\rangle$ or $|\underline{p}, \alpha-\rangle$ implies knowledge of the on-shell T matrix, and the great bulk of the literature on scattering theory is therefore concerned with the problem of calculating $|\underline{p}, \alpha\pm\rangle$. Just as in the one-channel case the explicit expressions $|\underline{p}, \alpha\pm\rangle = |\underline{p}, \alpha\rangle + G(\pm)V^\alpha|\underline{p}, \alpha\rangle$ are of little direct use since they require knowledge of the full Green's operator $G(z)$. And, again, just as in the one-channel case, they can be converted into implicit Lippmann–Schwinger equations in terms of certain "free" Green's operators, as we now show.

In the one-channel problem there is a single free Hamiltonian H^0 and a single corresponding Green's operator $G^0(z) = (z - H^0)^{-1}$. In the multichannel problem there are various different "free" channel Hamiltonians H^α, describing the "free" motion in the various channels, and we have to define corresponding channel Green's operators as

$$\boxed{G^\alpha(z) = (z - H^\alpha)^{-1}}$$

The first step in establishing the Lippmann–Schwinger equation is to use the familiar identity

$$A^{-1} = B^{-1} + B^{-1}(B - A)A^{-1}$$

with $A = (z - H)$ and $B = (z - H^\alpha)$. This gives the resolvent equations

$$G(z) = G^\alpha(z) + G^\alpha(z)V^\alpha G(z) \tag{18.10}$$

where, as before, $V^\alpha \equiv H - H^\alpha$. Returning to (18.5) we multiply through by $G^\alpha V^\alpha$ to give

$$G^\alpha V^\alpha |\underline{p}, \alpha\pm\rangle = G^\alpha V^\alpha |\underline{p}, \alpha\rangle + G^\alpha V^\alpha G V^\alpha |\underline{p}, \alpha\rangle$$

Since by (18.10) $G^\alpha V^\alpha G = (G - G^\alpha)$, this reduces to the important result:

$$G^\alpha(E \pm i0)V^\alpha |\underline{p}, \alpha\pm\rangle = G(E \pm i0)V^\alpha |\underline{p}, \alpha\rangle \tag{18.11}$$

Finally, substitution into (18.5) gives the desired Lippmann–Schwinger equation

$$\boxed{|\underline{p}, \alpha\pm\rangle = |\underline{p}, \alpha\rangle + G^\alpha(E \pm i0)V^\alpha |\underline{p}, \alpha\pm\rangle} \tag{18.12}$$

This equation, which is an integral equation for the corresponding wave functions, and the Schrödinger equation provide the two principal approaches to the computation of the stationary scattering states.

It should be emphasized that the multichannel Lippmann–Schwinger equation is considerably harder to handle than its single-channel equivalent. Some of the difficulties are readily apparent. For example, the Green's operator on the right is the channel Green's operator $G^\alpha(z) = (z - H^\alpha)^{-1}$. Except for the case $\alpha = 0$ (all particles free), H^α contains some potentials and the corresponding Green's operator G^α cannot, in general, be exactly calculated (in contra-distinction to the one-channel situation where G^0 *is* known exactly). A less obvious, though more profound, difficulty is that where the one-channel equation is (or can be easily converted to) a *non-singular* integral equation, the multichannel equations are in general highly singular. This makes both the theoretical study and practical use of the equations much more difficult. In fact, it is only in the last few years that any significant progress on the rigorous study of these equations has been made (see Faddeev, 1965).

18-c.   The *T* Operators

Just as in the one-channel case, the time-independent multichannel theory has two essentially equivalent formulations, one in terms of the stationary scattering states, the other in terms of the *T* operators. Having established the former, we can now easily set up the latter.

We start with the expression,

$$t(\underline{p}', \beta \leftarrow \underline{p}, \alpha) = \langle \underline{p}', \beta - | V^\alpha | \underline{p}, \alpha \rangle$$

into which we substitute:

$$|\underline{p}, \beta - \rangle = |\underline{p}, \beta \rangle + G(E - i0)V^\beta |\underline{p}, \beta \rangle$$

This gives

$$t(\underline{p}', \beta \leftarrow \underline{p}, \alpha) = \langle \underline{p}', \beta | [V^\alpha + V^\beta G(E + i0)V^\alpha] |\underline{p}, \alpha \rangle$$

or

$$\boxed{t(\underline{p}', \beta \leftarrow \underline{p}, \alpha) = \langle \underline{p}', \beta | T^{\beta\alpha}(E + i0) |\underline{p}, \alpha \rangle} \qquad (18.13)$$

where we have defined the *T* operator,

$$\boxed{T^{\beta\alpha}(z) = V^\alpha + V^\beta G(z)V^\alpha} \qquad (18.14)$$

As anticipated there is a double family of *T* operators $T^{\beta\alpha}(z)$. The on-shell *T* matrix for a transition from channel α to channel β is given by the on-shell matrix element of the *appropriate* $T^{\beta\alpha}(z)$ between the free states $|\underline{p}', \beta \rangle$ and $|\underline{p}, \alpha \rangle$.[3]

One feature of the definition (18.14) of $T^{\beta\alpha}(z)$ that deserves comment is the apparent asymmetry between the indices α and β. There seems no obvious reason why it is the potential V^α and not V^β that appears in the first term. Indeed, if we had started with the "post" form $\langle \underline{p}', \beta | V^\beta |\underline{p}, \alpha + \rangle$ for the

---

[3] Although $T^{\beta\alpha}(z)$ is labelled by two channel indices, there are not actually as many *T* operators as there are pairs of channels. The point is that if two channels α and α' have the same groupings of the particles, then the channel Hamiltonians H^α and $H^{\alpha'}$ are the same, and hence $V^\alpha = V^{\alpha'}$. Thus, $T^{\beta\alpha}(z)$ and $T^{\beta'\alpha'}(z)$ are different only if the channels α and α' or β and β' have different arrangements.

on-shell T matrix, we would have been led to a result similar to (18.13) but with the T operator

$$\tilde{T}^{\beta\alpha}(z) = V^\beta + V^\beta G(z) V^\alpha$$

If the channels α and β have different groupings of the particles, then the channel Hamiltonians H^α and H^β are different and hence, $V^\alpha \neq V^\beta$. Thus, the two T operators $T^{\beta\alpha}(z)$ and $\tilde{T}^{\beta\alpha}(z)$ are generally different. Then which is the correct one? The answer is that both lead to the same on-shell T matrix, since

$$\langle \underline{p}', \beta | (T^{\beta\alpha} - \tilde{T}^{\beta\alpha}) | \underline{p}, \alpha \rangle = \langle \underline{p}', \beta | (V^\alpha - V^\beta) | \underline{p}, \alpha \rangle$$
$$= \langle \underline{p}', \beta | (H^\beta - H^\alpha) | \underline{p}, \alpha \rangle$$
$$= (E' - E) \langle \underline{p}', \beta | \underline{p}, \alpha \rangle \qquad (18.15)$$

which is zero on the energy shell. Since the two T operators give the same on-shell T matrix, and since they differ only by the constant operator $(V^\alpha - V^\beta)$, there is no interest in retaining both. For definiteness we shall use $T^{\beta\alpha}(z)$ as defined in (18.14).

As one would expect, the explicit definition (18.14) of $T^{\beta\alpha}(z)$ can be replaced by an implicit Lippmann–Schwinger equation. To do this we multiply (18.14) on the left by $G^\beta(z)$ to give

$$G^\beta(z) T^{\beta\alpha}(z) = (G^\beta + G^\beta V^\beta G) V^\alpha = G(z) V^\alpha$$

Substitution back into (18.14) gives the desired equation

$$\boxed{T^{\beta\alpha}(z) = V^\alpha + V^\beta G^\beta(z) T^{\beta\alpha}(z)} \qquad (18.16)$$

It is the similar structure of the Lippmann–Schwinger equations (18.16) for the T operators and (18.12) for the stationary states that guarantees that the two approaches to the on-shell T matrix are essentially equivalent. For the most part we shall work with the stationary states since they appear most commonly in the practical literature. It should, however, be mentioned that in rigorous discussions of the subject there are some advantages to working with the T operators.

18-d. The Born Approximation; Elastic Scattering

To conclude this chapter on basic ideas we discuss the most basic of all methods of computation—the Born approximation. One can arrive at this approximation using either the stationary states or the T operators. One assumes that the effect of the scattering potential is small and completely

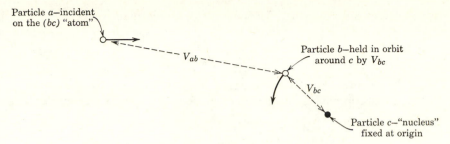

Particle a—incident on the (bc) "atom"

Particle b—held in orbit around c by V_{bc}

V_{ab}

V_{bc}

Particle c—"nucleus" fixed at origin

FIGURE 18.1.   Particle a incident on the bound state (bc).

ignores the second term in the Lippmann–Schwinger equation. For example, this gives

$$|\underline{p}', \beta-\rangle \approx |\underline{p}', \beta\rangle$$

Insertion into the exact T matrix $\langle \underline{p}', \beta-| V^{\alpha} |\underline{p}, \alpha\rangle$ then gives the Born approximation:

$$\boxed{t(\underline{p}', \beta \leftarrow \underline{p}, \alpha) \approx \langle \underline{p}', \beta| V^{\alpha} |\underline{p}, \alpha\rangle} \quad \cdots \text{[Born approximation]}   (18.17)$$

This is the "prior" form of the Born approximation because it involves the potential V^{α} of the incident channel. There is a corresponding "post" form $\langle \underline{p}', \beta| V^{\beta} |\underline{p}, \alpha\rangle$, which is identically equal to (18.17) (on shell) by (18.15).

The Born approximation (18.17) is obviously the first term in an infinite series in powers of the operator $G^{\beta}(E + i0)V^{\beta}$ and we naturally expect that it will be a good approximation either if V^{β} is weak or, since $G^{\beta}(z) = (z - H^{\beta})^{-1}$, if the energy is high. However, remarkably little has been rigorously proved about convergence of the multichannel Born series, and we shall simply accept (what is found to be the case) that under suitable circumstances (mostly in atomic physics) the Born approximation does give satisfactory results.[4]

To understand the interesting general features of the Born approximation it is best to confine attention to the simplest relevant model. We can consider as an example our three-particle model with particles a, b, and c. We suppose that c is infinitely heavy and hence effectively fixed. We imagine a to be incident on the (bc) bound system (channel 1) and we suppose at first that a does not interact with c at all (Fig. 18.1). In this case there are just two potentials: V_{bc}, which holds the (bc) "atom" together, and V_{ab},

---

[4] That the multichannel Born series is much more complicated than its one-channel equivalent is quite apparent when one remembers that it is a series in powers of the *channel potential* V^{β}, *not* the whole potential V. In this connection, if one were to make the *whole* potential weak, then all bound states would cease to exist and the problem would cease to be a multichannel problem.

which causes the scattering. The channel 1 Hamiltonian is

$$H^1 = \frac{\mathbf{P}_a^2}{2m_a} + \frac{\mathbf{P}_b^2}{2m_b} + V_{bc}$$

while the channel 1 scattering potential is $V^1 = V_{ab}$.

Let us first compute the Born approximation for channel 1 elastic scattering

$$a + (bc) \rightarrow a + (bc)$$

The wave function for an asymptotic state $|\mathbf{p}, 1\rangle$ of channel 1 is

$$\langle \mathbf{x}_a, \mathbf{x}_b \,|\, \mathbf{p}, 1 \rangle = (2\pi)^{-3/2} e^{i\mathbf{p} \cdot \mathbf{x}_a} \phi_1(\mathbf{x}_b)$$

where we now denote by ϕ_1 the (bc) bound-state wave function. Thus, the on-shell T matrix for elastic scattering in Born approximation is

$$t(\mathbf{p}', 1 \leftarrow \mathbf{p}, 1) = \langle \mathbf{p}', 1| \, V^1 \,|\mathbf{p}, 1\rangle$$

$$= (2\pi)^{-3} \int d^3x_a \int d^3x_b e^{-i\mathbf{p}' \cdot \mathbf{x}_a} \phi_1(\mathbf{x}_b)^* V_{ab}(\mathbf{x}_a - \mathbf{x}_b) e^{i\mathbf{p} \cdot \mathbf{x}_a} \phi_1(\mathbf{x}_b)$$

[Born approximation]   (18.18)

The result (18.18) can be rewritten in two ways. First, by regrouping we obtain:

$$t(\mathbf{p}', 1 \leftarrow \mathbf{p}, 1) = (2\pi)^{-3} \int d^3x_a e^{-i\mathbf{q} \cdot \mathbf{x}_a} \int d^3x_b V_{ab}(\mathbf{x}_a - \mathbf{x}_b) \, |\phi_1(\mathbf{x}_b)|^2$$

$$\equiv (2\pi)^{-3} \int d^3x_a e^{-i\mathbf{q} \cdot \mathbf{x}_a} \bar{V}(\mathbf{x}_a)$$

where $\mathbf{q} \equiv \mathbf{p}' - \mathbf{p}$. The second expression has exactly the form of the one-channel Born approximation for scattering by the potential $\bar{V}(\mathbf{x}_a)$. Since $\bar{V}(\mathbf{x}_a)$ is the result of averaging the actual potential $V_{ab}(\mathbf{x}_a - \mathbf{x}_b)$ with respect to the distribution $|\phi_1(\mathbf{x}_b)|^2$, this result coincides precisely with the "static" approximation discussed in Section 9-d in connection with scattering of electrons off atoms. That is, if we compute the elastic electron–atom amplitude using the multichannel Born approximation, then we obtain the same result as if we use the one-channel Born approximation, treating the atom as a static charge distribution.

We can rewrite the result (18.18) a little differently by taking as variables of integration $\mathbf{x} = \mathbf{x}_a - \mathbf{x}_b$ and $\mathbf{x}_b$. This gives

$$t(\mathbf{p}', 1 \leftarrow \mathbf{p}, 1) = (2\pi)^{-3} \int d^3x \, e^{-i\mathbf{q} \cdot \mathbf{x}} V_{ab}(\mathbf{x}) \int d^3x_b e^{-i\mathbf{q} \cdot \mathbf{x}_b} \, |\phi_1(\mathbf{x}_b)|^2$$

$$= t(\mathbf{p}' \leftarrow \mathbf{p})|_{b \text{ fixed}} \, g_1(\mathbf{q}) \qquad \text{[Born approximation]} \quad (18.19)$$

That is, in Born approximation the elastic scattering of a off the bound b particle is the same as that off a fixed target b, *times a certain function* $g_1(\mathbf{q})$. This function $g_1(\mathbf{q})$ is just the Fourier transform of the probability distribution $|\phi_1(\mathbf{x}_b)|^2$ and is called the *elastic form factor*.

The concept of the form factor has played a very important role in scattering theory. From the present discussion and that of Section 9-d, it should be clear that the form factor arises whenever the scatterer [the (bc) "atom" in our case] instead of being a fixed point particle, is spread out according to some distribution. Measurement of $g_1(\mathbf{q})$ gives information about this distribution. For example, it develops that the charge of a proton is not concentrated in a point and measurement of the $e–p$ form factor gives information on the charge distribution inside the proton. Since in the forward direction $(\mathbf{q} = 0)$

$$g_1(0) = 1$$

we see that, in the neighborhood of the forward direction, scattering off a bound target is just like scattering off a fixed point target.[5] Thus, to obtain useful information about the target structure, one must make measurements at large momentum transfers—that is, at high energy and large angle (see Problem **18.1**).

Because of the great importance of form factors it should be emphasized that they are defined by (18.19) only in the context of the Born approximation. In general, the corresponding result for the exact amplitudes is simply not true. Both the actual amplitude and that for b fixed depend on $\mathbf{p}$ and $\mathbf{p}'$ separately (not just $\mathbf{q} = \mathbf{p}' - \mathbf{p}$) and their quotient cannot be written as $g_1(\mathbf{q})$.

Before going on to discuss *inelastic* processes we note that we can easily extend the above discussion to the case where the incident projectile a interacts both with the bound particle b and the "nucleus" c. In this case the scattering potential V^1 is

$$V^1 = V_{ab} + V_{ac}$$

Clearly, in Born approximation the elastic amplitude is just the sum of the amplitudes obtained from each potential separately. Since c is fixed this is just:

$$t(\mathbf{p}', 1 \leftarrow \mathbf{p}, 1) = t_b(\mathbf{p}' \leftarrow \mathbf{p})\big|_{b \text{ fixed}} \, g_1(\mathbf{q}) + t_c(\mathbf{p}' \leftarrow \mathbf{p})\big|_{c \text{ fixed}} \quad (18.20)$$

18-e.   The Born Approximation; Excitation

We now go on to apply the Born approximation to the simplest form of inelastic process—excitation of the target by the projectile. For example, we

---

[5] This is to be expected. Roughly speaking, particles scattered near the forward direction have large impact parameter and so are insensitive to the target structure.

can consider the process

$$a + (bc) \rightarrow a + (bc)^*$$

leading from channel 1 to channel 2 of our model. According to (18.17) the Born approximation for this reaction is:

$$t(\mathbf{p'}, 2 \leftarrow \mathbf{p}, 1) = \langle \mathbf{p'}, 2| \, V^1 \, |\mathbf{p}, 1\rangle$$

$$= (2\pi)^{-3} \int d^3x_a \int d^3x_b e^{-i\mathbf{p'} \cdot \mathbf{x}_a} \phi_2(\mathbf{x}_b)^*$$

$$\times \, [V_{ab}(\mathbf{x}_{ab}) + V_{ac}(\mathbf{x}_a)] e^{i\mathbf{p} \cdot \mathbf{x}_a} \phi_1(\mathbf{x}_b) \qquad (18.21)$$

where $\phi_1(\mathbf{x}_b)$ and $\phi_2(\mathbf{x}_b)$ are the wave functions of the initial and final bound states (bc) and $(bc)^*$. First, since ϕ_1 and ϕ_2 are orthogonal the contribution of the potential V_{ac} linking a to the "nucleus" c is zero. (This is to be expected. In the absence of V_{ab}, the passage of the projectile a could obviously not cause excitation at all.) The contribution of the potential V_{ab} can be rewritten as in the elastic case to give:

$$t(\mathbf{p'}, 2 \leftarrow \mathbf{p}, 1) = (2\pi)^{-3} \int d^3x \, e^{-i\mathbf{q} \cdot \mathbf{x}} V_{ab}(\mathbf{x}) \int d^3x_b e^{-i\mathbf{q} \cdot \mathbf{x}_b} \phi_2(\mathbf{x}_b)^* \phi_1(\mathbf{x}_b)$$

$$= t(\mathbf{p'} \leftarrow \mathbf{p})|_{b \text{ fixed}} g_2(\mathbf{q}) \qquad [\text{Born approximation}] \quad (18.22)$$

In this last result $g_2(\mathbf{q})$ is called the *inelastic form factor* and is just the Fourier transform of the product $\phi_2^* \phi_1$.[6] Since the process is inelastic, $p' \neq p$. [Remember that $p'^2 = p^2 - 2m(E_2 - E_1)$ where $E_2 - E_1$ is the excitation energy of the final state $(bc)^*$.] Thus, the first factor in (18.22) is the elastic T matrix (in Born approximation) *off the energy shell*. This is our first example of a result mentioned earlier—that the on-shell amplitude for a many-body process can often be expressed approximately in terms of the *off-shell* amplitudes for certain related two-body processes.

The use of the inelastic Born approximation can be well illustrated by the case of $1s \rightarrow 2p$ excitation of atomic hydrogen by electrons. For the purposes of the illustration we make some simplifying assumptions. We treat the two electrons as distinguishable (that is, we ignore any effects of the exclusion principle), we ignore all spins, and we treat the nucleus as infinitely heavy and, hence, fixed. (All of these approximations prove relatively unimportant in the energy range where the Born approximation is expected to be good.) We label the general state of the target atom by the usual three quantum numbers n, l, m, and first compute the amplitude for excitation from the ground state $(1, 0, 0)$ to an arbitrary level (n, l, m).

---

[6] In general, if the target has several levels there is a separate form factor $g_n = \int e^{-i\mathbf{q} \cdot \mathbf{x}} \phi_n^* \phi_1$ for excitation of each level n. In fact, strictly speaking one should write $g_{n'n}$ as the form factor for the target initially in the nth level to be excited (or de-excited) to the n'th level.

The scattering potential V^1 consists of the Coulomb attraction between the incident electron and the nucleus, plus the repulsion between the two electrons. Of these only the latter contributes to excitation and the relevant amplitude is [recall that the amplitude is $-(2\pi)^2 m$ times the T matrix]:

$$f(\mathbf{p}', nlm \leftarrow \mathbf{p}, 100) = -(2\pi)^2 m \langle \mathbf{p}', nlm| \, V^1 \, |\mathbf{p}, 100 \rangle$$

$$= -\frac{m}{2\pi} \int d^3x_a \int d^3x_b$$

$$\times \, e^{-i\mathbf{p}'\cdot\mathbf{x}_a} \, \phi_{nlm}(\mathbf{x}_b)^* \, \frac{e^2}{|\mathbf{x}_a - \mathbf{x}_b|} \, e^{i\mathbf{p}\cdot\mathbf{x}_a} \phi_{100}(\mathbf{x}_b)$$

This can be manipulated as in (18.22) to give

$$= -\frac{me^2}{2\pi} \int d^3x \, \frac{e^{-i\mathbf{q}\cdot\mathbf{x}}}{r} \int d^3x' e^{-i\mathbf{q}\cdot\mathbf{x}'} \phi_{nlm}(\mathbf{x}')^* \phi_{100}(\mathbf{x}')$$

The first integral suffers the usual diseases associated with the Coulomb potential and is divergent. However, if we follow the correct procedure and use a cut off [for example, $e^{-r/\rho}$ with $\rho \to \infty$] it is easily evaluated as $4\pi/q^2$. This gives

$$f(\mathbf{p}', nlm \leftarrow \mathbf{p}, 100) = \frac{2}{aq^2} \int d^3x \, e^{-i\mathbf{q}\cdot\mathbf{x}} \phi_{nlm}(\mathbf{x})^* \phi_{100}(\mathbf{x})$$

The bound state wave functions $\phi_{nlm}(\mathbf{x})$ are explicitly known and have the form $(1/r)y_{nl}(r) Y_l^m(\hat{\mathbf{x}})$. The integral can therefore be evaluated by expanding the factor $e^{-i\mathbf{q}\cdot\mathbf{x}}$ as:

$$e^{i\mathbf{q}\cdot\mathbf{x}} = \frac{4\pi}{qr} \sum_{l=0}^{\infty} i^l j_l(qr) \sum_{m=-l}^{l} Y_l^m(\hat{\mathbf{q}})^* Y_l^m(\hat{\mathbf{x}}) \tag{18.23}$$

Thus, in particular, the $1s \to 2p$ amplitude is:

$$f(\mathbf{p}', 21m \leftarrow \mathbf{p}, 100) = aI(qa)Y_1^m(\hat{\mathbf{q}})^* \tag{18.24}$$

where $I(qa)$ is a certain dimensionless radial integral, which the reader can easily write down.

The differential cross section corresponding to the amplitude (18.24) is:

$$\frac{d\sigma}{d\Omega}(\mathbf{p}', 21m \leftarrow \mathbf{p}, 100) = |f(\mathbf{p}', 21m \leftarrow \mathbf{p}, 100)|^2$$

In practice, neither the final magnetic state m nor the final direction $\mathbf{p}'$ is measured. (The cross section for the $1s \to 2p$ excitation is measured by

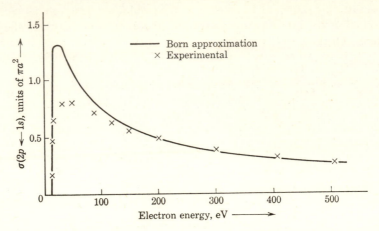

FIGURE 18.2. The cross section $\sigma(2p \leftarrow 1s)$ for excitation of the $2p$ states of hydrogen by an electron. (Courtesy Dr. S. J. Smith.)

counting the photons emitted in the subsequent $2p \rightarrow 1s$ transition.) Thus, the interesting cross section is,

$$\sigma(2p \leftarrow 1s) = \int d\Omega_{\mathbf{p}'} \sum_m \frac{d\sigma}{d\Omega} \, (\mathbf{p}', 21m \leftarrow \mathbf{p}, 100)$$

$$= \frac{3a^2}{4\pi} \int d\Omega_{\mathbf{p}'} \, |I(qa)|^2 \qquad (18.25)$$

where we have used the relation

$$\sum_m |Y_l^m(\hat{\mathbf{q}})|^2 = \frac{2l+1}{4\pi}$$

The angular integral can be evaluated with perseverance and the resulting cross section is plotted as a function of incident momentum in Fig. 18.2. It will be seen that the agreement with experiment is reasonable above about 100 eV.

   This example is typical of the situations in which the multichannel Born approximation is normally useful—namely at reasonably high energies in processes (such as those of atomic physics) where electromagnetic interactions dominate. In nuclear and high-energy physics, where strong interactions are usually important, the Born approximation is almost never reliable. We shall return later to a number of alternative techniques (coupled-channel approximation, distorted-wave Born approximation, etc.) that may be useful in circumstances where the Born approximation is not.

PROBLEMS

18.1. Consider the elastic form factor $g_1(q)$ defined in (18.19) for scattering of a off target (bc) [assuming that c is infinitely heavy and that the (bc) target has $l = 0$]. By expanding $g_1(q)$ as a Taylor series about $q = 0$, show that useful information about the target can only be obtained with momentum transfers satisfying $q \gtrsim 1/r$, where r is the root-mean-square radius of the target. Since $q \leqslant 2p$, this means that the minimum momentum needed to probe a target of size r is $p \sim 1/r$, in agreement with uncertainty principle arguments.

18.2. Consider the scattering of a particle a off a bound state (bc) with neither b nor c infinitely heavy. Show that in Born approximation the elastic T matrix has the form:

$$t_{\text{el}}(\mathbf{p}' \leftarrow \mathbf{p}) = t_{ab}(\mathbf{p}' \leftarrow \mathbf{p})|_{b \text{ fixed}} g\left(\frac{m_c \mathbf{q}}{m_b + m_c}\right) + t_{ac}(\mathbf{p}' \leftarrow \mathbf{p})|_{c \text{ fixed}} g\left(\frac{m_b \mathbf{q}}{m_b + m_c}\right)$$

where $\mathbf{p}$ and $\mathbf{p}'$ are the initial and final momenta of a in the overall CM frame $(\mathbf{q} = \mathbf{p}' - \mathbf{p})$ and the form factor $g(\mathbf{k})$ is defined as the Fourier transform of the (bc) probability density $|\phi_{bc}(\mathbf{x}_{bc})|^2$. (Your only problem is to sort out the various relative and CM coordinates. One way is to write down the Born approximation for the complete motion—including plane waves for a and the (bc) CM—and then factor out the overall momentum δ function.)

18.3. (a) Using the "prior" form of the Born approximation (18.17) show that the T matrix for the ionization process $a + (bc) \rightarrow a + b + c$ (with c infinitely heavy) has the form

$$t_{\text{ion}}(\mathbf{p}_a, \mathbf{p}_b \leftarrow \mathbf{p})$$

$$= t_{ab}(\mathbf{p}_a \leftarrow \mathbf{p})|_{b \text{ fixed}} \, \tilde{\phi}_{bc}(\mathbf{p}_a + \mathbf{p}_b - \mathbf{p}) + t_{ac}(\mathbf{p}_a \leftarrow \mathbf{p})|_{c \text{ fixed}} \, \tilde{\phi}_{bc}(\mathbf{p}_b)$$

where $\tilde{\phi}_{bc}(\mathbf{k})$ is the Fourier transform of the (bc) wave function $\phi_{bc}(\mathbf{x}_{bc})$—times $(2\pi)^{-3/2}$.

(b) The "post" form of the Born approximation would contain an additional term involving V_{bc}. Prove that this term is zero. [The use of the Born approximation for rearrangement collisions is actually a controversial subject. It is generally agreed that a better approximation would be to drop the second (so-called "core") term in part (a), since this corresponds to the interaction V_{ac} of a with the fixed "core" c and V_{ac} alone could obviously not cause ionization at all.]

Properties of the Multichannel Stationary Wave Functions

19

19-a   Asymptotic Form of the Wave Functions; Collisions Without Rearrangement

19-b   Asymptotic Form of the Wave Functions; Rearrangement Collisions

19-c   Expansion in Terms of Target States

19-d   The Optical Potential

In this chapter we continue our general discussion of the time independent theory, concentrating on the properties of the stationary states $|\underline{p}, \alpha+\rangle$ and their wave functions $\langle \underline{x} \mid p, \alpha+ \rangle$. We discuss only those collisions with two-body initial states since these are the only processes of serious practical importance. For simplicity, we suppose that while the target is an arbitrary bound state of several particles, the projectile is just a single particle such as an electron, proton, or neutron.

In Sections 19-a and 19-b we examine the behavior of the wave functions at large distances and establish analogues of the familiar one-channel result:

$$\langle \mathbf{x} \mid \mathbf{p}+ \rangle \xrightarrow[x \to \infty]{} (2\pi)^{-3/2} \left[e^{i\mathbf{p} \cdot \mathbf{x}} + f(p\hat{\mathbf{x}} \leftarrow \mathbf{p}) \frac{e^{ipr}}{r} \right] \qquad (19.1)$$

In establishing these results we are led to an expansion of the wave functions in terms of the stationary states of the target. We show in Section 19-c that this expansion, which we call the *target-state expansion*, reduces the multichannel problem to an infinite set of coupled one-particle Schrödinger equations.   This infinite set of equations leads to the *close-coupling* or *coupled-channel approximation*, which is obtained by truncation to give a finite (and approximate) set of N coupled equations.   Under suitable conditions this truncation can be made so as to account explicitly for all channels that are important or of interest, and to neglect only channels that are unimportant.   Under these conditions the approximation provides a useful method of computation.

Finally, in Section 19-d we show that by introducing a suitable potential V_{opt}, called the *optical potential*, one can reduce the multichannel problem to an *exact* set of N coupled one-particle equations.   We shall see that in these equations N channels are treated explicitly, while the effects of all other channels are taken into account (exactly) by the optical potential.   The optical potential is usually very complicated and cannot be exactly calculated.   Its principal importance is rather that one can often make a good phenomenological guess for V_{opt}.   With this guessed V_{opt} the N coupled equations describe N channels explicitly and account approximately for the effects of all other channels.   In this way it is often possible (especially in nuclear physics) to get extremely good fits to large bodies of experimental data.

19-a.   Asymptotic Form of the Wave Functions;
Collisions Without Rearrangement

The multichannel analogue of the one-channel asymptotic form (19.1) is quite complicated in general, and we shall therefore consider just a few simple examples (for a much more comprehensive and careful discussion see Nuttall and Webb, 1969).   We begin by considering our three-particle model with light particles a and b and a heavy fixed particle c.   We allow for several bound states $(bc)_1, \ldots , (bc)_n$ and $(ac)_1, \ldots , (ac)_{n'}$ and perhaps also some bound states of a and b.   There are then n channels of the form

$$a + (bc)_1, \ldots , a + (bc)_n$$

and n' of the form

$$b + (ac)_1, \ldots , b + (ac)_{n'}$$

In addition, there is the channel 0 $(a + b + c)$ and perhaps further channels of the form $(ab) + c$.

We shall consider a process in which particle a is incident with momentum $\mathbf{p}$ on the ground state $(bc)_1$.[1]   We label this channel as channel 1, and the corresponding stationary state is then $|\mathbf{p}, 1+\rangle$.   This has wave function

---

[1] The target could just as well be some excited state.   However, since in practice it is almost always in the ground state we consider this case.

$\langle \mathbf{x}_a, \mathbf{x}_b \mid \mathbf{p}, 1+ \rangle$ depending on the two variables $\mathbf{x}_a$ and $\mathbf{x}_b$, and the question immediately arises: Do we want the asymptotic behavior as $\mathbf{x}_a \to \infty$, or $\mathbf{x}_b \to \infty$, or perhaps both? In fact, all of these cases are of interest, as can readily be seen. If $\mathbf{x}_a \to \infty$ with $\mathbf{x}_b$ fixed, then the wave function should show the effects of those channels in which particle a moves far away from b and c (and which are open at the energy considered). However, we would not expect to see those channels in which a is captured $[a + (bc)_1 \to b + (ac)_i]$, nor those in which a and b move off together $[a + (bc)_1 \to (ab)_i + c]$. Thus, as $\mathbf{x}_a \to \infty$ with $\mathbf{x}_b$ fixed, the wave function $\langle \mathbf{x}_a, \mathbf{x}_b \mid \mathbf{p}, 1+ \rangle$ should consist of an incident plane wave in channel 1 plus outgoing waves in the channels $a + (bc)_i$ and $a + b + c$. Instead, if we let $\mathbf{x}_b \to \infty$ with $\mathbf{x}_a$ fixed, then we should see the outgoing waves in the channels $b + (ac)_i$ and $a + b + c$, but not the others.

We begin with the case that $\mathbf{x}_a \to \infty$ with $\mathbf{x}_b$ fixed. We proceed in close analogy with the one-channel analysis of Section 10-c, starting from the Lippmann–Schwinger equation

$$|\mathbf{p}, 1+ \rangle = |\mathbf{p}, 1 \rangle + G^1(E + i0)V^1 |\mathbf{p}, 1+ \rangle \qquad (19.2)$$

To get at the wave function $\langle \mathbf{x}_a, \mathbf{x}_b \mid \mathbf{p}, 1+ \rangle$ we need to know the spatial matrix elements $\langle \mathbf{x}_a, \mathbf{x}_b \mid G^1(z) \mid \mathbf{x}_a', \mathbf{x}_b' \rangle$ of the channel 1 Green's operator. In the one-channel case we evaluated the matrix element $\langle \mathbf{x} \mid G^0(z) \mid \mathbf{x}' \rangle$ of the free Green's operator by inserting a complete set of eigenvectors $|\mathbf{p} \rangle$ of the free Hamiltonian H^0. In the present case we seek analogous eigenvectors of the channel 1 Hamiltonian, which we write as:

$$H^1 = \frac{\mathbf{P}_a^2}{2m_a} + \frac{\mathbf{P}_b^2}{2m_b} + V_{bc} \equiv \frac{\mathbf{P}_a^2}{2m_a} + H_{(bc)} \qquad (19.3)$$

Here the second term is the Hamiltonian of particle b in the field of the fixed particle c. From this it is clear that we can take as eigenfunctions of H^1 the products,

$$(2\pi)^{-3/2} e^{i\mathbf{p} \cdot \mathbf{x}_a} \phi_\alpha(\mathbf{x}_b) \qquad (19.4)$$

where $\phi_\alpha(\mathbf{x}_b)$ denotes any of the eigenfunctions of $H_{(bc)}$ with energy E_α. These are of two types; there are the n bound states $(bc)_1, \ldots, (bc)_n$ with wave functions $\phi_1(\mathbf{x}_b), \ldots, \phi_n(\mathbf{x}_b)$. Also, there are the continuum states, which for example we could choose to be the outgoing wave scattering states of b in the field of c. In either case the functions (19.4) are eigenfunctions of H^1 with energy $(p^2/2m_a) + E_\alpha$.

The reader will notice that when $\phi_\alpha(\mathbf{x}_b)$ is one of the bound-state wave functions $\phi_1, \ldots, \phi_n$, the function (19.4) is precisely the free wave function of the channel $a + (bc)_\alpha$. The same is *not* true of the continuum functions;

when $\phi_\alpha(\mathbf{x}_b)$ is one of these, the function (19.4) is not the free wave function of any channel. Therefore, the basis of eigenfunctions (19.4) of the channel Hamiltonian H^1 is, thus, a curious hybrid. The discrete part consists of free channel wave functions $\langle \mathbf{x}_a, \mathbf{x}_b \mid \mathbf{p}, \alpha \rangle$ where α runs over all channels with the arrangement $a + (bc)_\alpha$. The continuous part consists of functions that have no simple relation to any channels. In fact, we shall be mainly interested in the discrete part of the basis, and it is for this reason that we identify the functions $\phi_\alpha(\mathbf{x}_b)$ with the same label α as used to identify channels. However, it must be remembered that this α runs over *some of the channels* (specifically those with the same arrangement as channel 1), and it also runs over a continuous range corresponding to the continuum states of b in the potential V_{bc}. To emphasize this point we shall write "sums" over α with the symbol $\displaystyle\int\!\!\!\!\!\!\sum$.

Now we can insert a complete set of the states (19.4) into the required Green's function, to give:

$$\langle \mathbf{x}_a, \mathbf{x}_b| \, G^1(z) \, |\mathbf{x}_a', \mathbf{x}_b' \rangle = (2\pi)^{-3} \int\!\!\!\!\!\!\sum d^3p \int_\alpha \frac{e^{i\mathbf{p}\cdot(\mathbf{x}_a - \mathbf{x}_a')}\phi_\alpha(\mathbf{x}_b)\phi_\alpha(\mathbf{x}_b')^*}{z - (p^2/2m_a) - E_\alpha} \qquad (19.5)$$

The integral over $\mathbf{p}$ can be performed, exactly as in the one-channel case. Each term in the sum over α has a pole at $p = [2m_a(z - E_\alpha)]^{1/2}$; and for the case $z = E + i0$ the result is

$$\langle \mathbf{x}_a, \mathbf{x}_b| \, G^1(E + i0) \, |\mathbf{x}_a', \mathbf{x}_b' \rangle = -\frac{m_a}{2\pi} \int\!\!\!\!\!\!\sum_\alpha \frac{e^{ip_\alpha|\mathbf{x}_a - \mathbf{x}_a'|}}{|\mathbf{x}_a - \mathbf{x}_a'|} \, \phi_\alpha(\mathbf{x}_b)\phi_\alpha(\mathbf{x}_b')^*$$

where $p_\alpha \equiv [2m(E - E_\alpha)]^{1/2}$ is the momentum of particle a if the target (bc) is left with energy E_α. In particular, we shall want the Green's function for $r_a \gg r_a'$, in which case,

$$\langle \mathbf{x}_a, \mathbf{x}_b| \, G^1(E + i0) \, |\mathbf{x}_a', \mathbf{x}_b' \rangle \xrightarrow[\mathbf{x}_a \to \infty]{} -\frac{m_a}{2\pi} \int\!\!\!\!\!\!\sum_\alpha \frac{e^{ip_\alpha r_a}}{r_a} \, e^{-ip_\alpha \hat{\mathbf{x}}_a \cdot \mathbf{x}_a'}\phi_\alpha(\mathbf{x}_b)\phi_\alpha(\mathbf{x}_b')^* \qquad (19.6)$$

We are now ready to establish the asymptotic form of the stationary wave functions. From the Lippmann–Schwinger equation (19.2) it follows that:

$$\langle \mathbf{x}_a, \mathbf{x}_b \mid \mathbf{p}, 1+ \rangle = \langle \mathbf{x}_a, \mathbf{x}_b \mid \mathbf{p}, 1 \rangle$$
$$+ \int d^3x_a' \int d^3x_b' \langle \mathbf{x}_a, \mathbf{x}_b| \, G^1(E + i0) \, |\mathbf{x}_a', \mathbf{x}_b' \rangle$$
$$\times \, V^1(\mathbf{x}_a', \mathbf{x}_b')\langle \mathbf{x}_a', \mathbf{x}_b' \mid \mathbf{p}, 1+ \rangle$$

Substituting (19.6) for the Green's function we obtain:

$$\xrightarrow[x_a \to \infty]{} (2\pi)^{-3/2} \left[e^{i\mathbf{p} \cdot \mathbf{x}_a} \phi_1(\mathbf{x}_b) - (2\pi)^2 m_a \int_\alpha \frac{e^{ip_\alpha r_a}}{r_a} \phi_\alpha(\mathbf{x}_b) \langle p_\alpha \hat{\mathbf{x}}, \alpha | V | \mathbf{p}, 1 \rangle \right] \qquad (19.7)$$

or[2]

$$
\boxed{
\begin{aligned}
\langle \mathbf{x}_a, \mathbf{x}_b \mid \mathbf{p}, 1+\rangle \xrightarrow[x_a \to \infty]{} (2\pi)^{-3/2} \Big[& e^{i\mathbf{p} \cdot \mathbf{x}_a} \phi_1(\mathbf{x}_b) \\
& + \int_\alpha f(p_\alpha \hat{\mathbf{x}}_a, \alpha \leftarrow \mathbf{p}, 1) \frac{e^{ip_\alpha r_a}}{r_a} \phi_\alpha(\mathbf{x}_b) \Big]
\end{aligned}
}
\qquad (19.8)
$$

This result is a natural generalization of the well-known one-channel result (19.1). The first term is the expected incident plane wave in channel 1; the second term has a sum over α. This sum includes n discrete values, corresponding to the bound states $(bc)_1, \ldots, (bc)_n$, plus continuous values corresponding to the continuum states of the target (bc). For the moment we focus attention on the n discrete terms. Each of these is the product of three factors: an amplitude, a spherically spreading wave for particle a, and the target function $\phi_\alpha(\mathbf{x}_b)$. Since the momentum of the spherical wave is:

$$p_\alpha = [2m_a(E - E_\alpha)]^{1/2} \qquad (19.9)$$

we see that each term in the sum represents the particle a travelling outwards with energy $E - E_\alpha$ having excited the target to the state $(bc)_\alpha$ with energy E_α.

If the total energy E is less than E_α, then the momentum (19.9) in channel α is pure imaginary. In this case the contribution of channel α to the asymptotic form (19.8) vanishes exponentially as $r_a \to \infty$. Thus, the sum in (19.8) can be taken to include only those channels that are open at energy E. In particular, if the energy is below that necessary to disintegrate the target, the "sum" in (19.8) is a genuine sum with no contribution from the continuum.

If the incident energy is sufficient to break up the target, then the "sum" in the asymptotic form (19.8) includes an integral over the continuum states. It can be shown that these terms represent outgoing waves in the breakup channel $a + b + c$, but we shall not go into this here.

It is not hard to extend these results to more general systems, in which both projectile and target are composites made up of several particles (atom–atom or nucleus–nucleus collisions, for example). Here we consider just an intermediate case in which a single particle is incident on an arbitrary target.

---

[2] The notation in both of these equations is really appropriate only to the discrete terms in the sum. Only for these terms is the wave function $\exp(i\mathbf{p}' \cdot \mathbf{x}_a)\phi_\alpha(\mathbf{x}_b)$ correctly identified by the bra $\langle \mathbf{p}', \alpha |$ as in (19.7). Hence, only for these terms is the use of the notation $f(\mathbf{p}', \alpha \leftarrow \mathbf{p}, 1)$ in (19.8) appropriate.

We denote by $\mathbf{x}$, $\mathbf{p}$ and m the position, momentum, and mass of the projectile and by $\underline{x}_{\mathrm{tar}}$ the set of coordinates of the particles inside the target. The channel Hamiltonian of the incident channel (number 1) has the form:

$$H^1 = \frac{\mathbf{P}^2}{2m} + H_{\mathrm{tar}}$$

where H_{tar} contains all kinetic energies and interactions within the target. For simplicity we consider a target with a heavy fixed core (such as an atom with its heavy nucleus) although as usual our discussion applies equally to the relative motion of a translationally invariant system.

The analysis goes through just as above. The Green's operator $G^1(z)$ is evaluated in terms of a complete set of eigenfunctions $\phi_\alpha(\underline{x}_{\mathrm{tar}})$ of the target Hamiltonian. These consist of the bound states $\phi_1, \ldots, \phi_n$ of the target plus a complicated set of continuum states. An analysis parallel to that leading from (19.2) to (19.8) then gives the natural result,

$$\langle \mathbf{x}, \underline{x}_{\mathrm{tar}} \mid \mathbf{p}, 1+\rangle \xrightarrow[\mathbf{x}\to\infty]{} (2\pi)^{-3/2}\left[e^{i\mathbf{p}\cdot\mathbf{x}}\,\phi_1(\underline{x}_{\mathrm{tar}}) \right.$$
$$\left. + \int_\alpha f(p_\alpha\hat{\mathbf{x}}, \alpha \leftarrow \mathbf{p}, 1)\,\frac{e^{ip_\alpha r}}{r}\,\phi_\alpha(\underline{x}_{\mathrm{tar}}) \right] \tag{19.10}$$

Here the first term is the incident wave in channel 1, the discrete terms of the sum represent the outgoing projectile with the target left in one of the states $\phi_1, \ldots, \phi_n$, and the continuum terms represent the various possible breakup processes as before.

The cross sections for elastic scattering and for excitation can easily be read off from the asymptotic form (19.10). The first term represents a steady incident flux,

$$\text{incident flux} = (2\pi)^{-3}\frac{p}{m}$$

in channel 1. The αth term in the sum represents a scattered flux per unit solid angle

$$\text{scattered flux} = (2\pi)^{-3}\,|f(\mathbf{p}_\alpha, \alpha \leftarrow \mathbf{p}, 1)|^2 \frac{p_\alpha}{m}$$

in channel α. Thus (with all the reservations that apply to the corresponding derivation in the one-channel case), we arrive at the answer:

$$\frac{d\sigma}{d\Omega}(\mathbf{p}_\alpha, \alpha \leftarrow \mathbf{p}, 1) = \frac{\text{scattered flux/solid angle}}{\text{incident flux/area}} = \frac{p_\alpha}{p}\,|f(\mathbf{p}_\alpha, \alpha \leftarrow \mathbf{p}, 1)|^2$$

as expected.

19-b.   Asymptotic Form of the Wave Functions; Rearrangement Collisions

So far we have considered the behavior of the stationary wave function $\langle x \mid \mathbf{p}, 1+ \rangle$ as the coordinate of the original projectile approaches infinity, with all target coordinates fixed. As expected this displayed the effects of those channels with the same arrangement as that of the incident channel 1. If we wish to see the effects of rearrangement collisions, then we must let some of the target coordinates go to infinity. Here we shall consider the same three-particle model as above with a incident on the ground state $(bc)_1$, and examine the wave function as $\mathbf{x}_b$ goes to infinity with $\mathbf{x}_a$ fixed. This should display the effects of the rearrangement collisions

$$a + (bc)_1 \to b + (ac)_\beta$$

The analysis of this case is quite similar to the previous one. The first step is to obtain some analogue of the Lippmann–Schwinger equation (19.2) in terms of the final channel Green's operator $G^\beta(z) = (z - H^\beta)^{-1}$, where H^β is the Hamiltonian of the final channels under consideration,

$$H^\beta = \frac{\mathbf{P}_b^2}{2m_b} + \frac{\mathbf{P}_a^2}{2m_a} + V_{ac} \tag{19.11}$$

We start with the original Lippmann–Schwinger equation

$$|\mathbf{p}, 1+\rangle = |\mathbf{p}, 1\rangle + G^1(E + i0)V^1 |\mathbf{p}, 1+\rangle$$

and rewrite G^1 in terms of G^β as

$$G^1 = G^\beta + G^\beta(V^\beta - V^1)G^1$$

This gives

$$|\mathbf{p}, 1+\rangle = |\mathbf{p}, 1\rangle + G^\beta V^1 |\mathbf{p}, 1+\rangle + G^\beta(V^\beta - V^1)G^1 V^1 |\mathbf{p}, 1+\rangle$$

Since $G^1 V^1 |\mathbf{p}, 1+\rangle$ in the last term is the same as $|\mathbf{p}, 1+\rangle - |\mathbf{p}, 1\rangle$, this gives

$$|\mathbf{p}, 1+\rangle = G^\beta V^\beta |\mathbf{p}, 1+\rangle + |o\rangle \tag{19.12}$$

where

$$|o\rangle = |\mathbf{p}, 1\rangle - G^\beta(V^\beta - V^1) |\mathbf{p}, 1\rangle$$

It can be shown that the vector $|o\rangle$ is orthogonal to all vectors in the channels of interest—that is, vectors in the subspaces of the channels $b + (ac)_\beta$. Therefore it need not concern us here and we shall omit it from the next few equations. The first term in (19.12) can be treated by the techniques of Section 19-a.

To evaluate the spatial matrix elements of $G^\beta(z)$ we need a set of eigenfunctions for the channel Hamiltonian H^β of (19.11). These have the form:

$$(2\pi)^{-3/2} e^{i\mathbf{p}\cdot\mathbf{x}_b} \chi_\beta(\mathbf{x}_a)$$

where $\chi_\beta(\mathbf{x}_a)$ stands for any of the (ac) bound-state wave functions or the corresponding continuum functions. This gives an expression for the matrix elements of G^β analogous to (19.6) for G^1 (but with $\mathbf{x}_a$, m_a, and ϕ_α replaced by $\mathbf{x}_b$, m_b, and χ_β). Substitution into the expression (19.12) for $|\mathbf{p}, 1+\rangle$ gives:

$$\langle \mathbf{x}_a, \mathbf{x}_b \mid \mathbf{p}, 1+\rangle \xrightarrow[\mathbf{x}_b \to \infty]{} (2\pi)^{-3/2}\left[-(2\pi)^2 m_b \int_\beta \frac{e^{ip_\beta r_b}}{r_b} \chi_\beta(\mathbf{x}_a)\langle p_\beta \hat{\mathbf{x}}_b, \beta| V^\beta |\mathbf{p}, 1+\rangle \right]$$

(19.13)

Bearing in mind that for the rearrangement

$$a + (bc)_1 \to b + (ac)_\beta$$

the amplitude is:

$$f(\mathbf{p}_\beta, \beta \leftarrow \mathbf{p}, 1) = -(2\pi)^2 (m_a m_b)^{1/2}\, t(\mathbf{p}_\beta, \beta \leftarrow \mathbf{p}, 1)$$

we can rewrite (19.13) in the final form:

$$\langle \mathbf{x}_a, \mathbf{x}_b \mid \mathbf{p}, 1+\rangle \xrightarrow[\mathbf{x}_b \to \infty]{} (2\pi)^{-3/2}\left(\frac{m_b}{m_a}\right)^{1/2} \int_\beta f(p_\beta \hat{\mathbf{x}}_b, \beta \leftarrow \mathbf{p}, 1) \frac{e^{ip_\beta r_b}}{r_b} \chi_\beta(\mathbf{x}_a)$$

(19.14)

where $p_\beta = [2m(E - E_\beta)]^{1/2}$.

This result has exactly the expected form; there is no incident wave. The discrete part of the sum runs over those open channels with the arrangement $b + (ac)$ (the closed-channel terms go to zero exponentially and can be dropped) and each term represents the particle b moving out in a spherical wave leaving particle a in the bound state $\chi_\beta(\mathbf{x}_a)$. Finally, it can be shown that the continuum terms represent the channel $a + b + c$ as before.

The relevant cross sections can be read off from (19.14). The scattered flux per unit solid angle in any of the discrete channels β is

$$(2\pi)^{-3} \frac{m_b}{m_a} |f(\mathbf{p}_\beta, \beta \leftarrow \mathbf{p}, 1)|^2 \frac{p_\beta}{m_b}$$

while the incident flux in channel 1 is $(2\pi)^{-3} p/m_a$ as before. Thus,

$$\frac{d\sigma}{d\Omega}(\mathbf{p}_\beta, \beta \leftarrow \mathbf{p}, 1) = \frac{p_\beta}{p} |f(\mathbf{p}_\beta, \beta \leftarrow \mathbf{p}, 1)|^2$$

as expected.

19-c.   Expansion in Terms of Target States

In the last two sections we examined the wave function $\langle x \mid \mathbf{p}, 1+ \rangle$ for large values of its arguments. In particular, for the case that a single projectile (coordinate $\mathbf{x}$) is incident on an arbitrary target (coordinates x_{tar}), we saw that when $\mathbf{x} \to \infty$ the wave function has the natural expansion:

$$\langle \mathbf{x}, x_{\text{tar}} \mid \mathbf{p}, 1+ \rangle \xrightarrow[\mathbf{x}\to\infty]{} (2\pi)^{-3/2} \left[e^{i\mathbf{p}\cdot\mathbf{x}} \phi_1(x_{\text{tar}}) + \int_{\alpha} f(p_\alpha \hat{\mathbf{x}}, \alpha \leftarrow \mathbf{p}, 1) \frac{e^{ip_\alpha r}}{r} \phi_\alpha(x_{\text{tar}}) \right]$$

$$(19.15)$$

Here, the amplitudes $f(\mathbf{p}_\alpha, \alpha \leftarrow \mathbf{p}, 1)$ are the amplitudes for elastic scattering and excitation (for the discrete values of α—the continuous terms are related to the breakup process). This suggests that if one is principally interested in processes that do not involve rearrangements it might be useful to make an exact expansion (not just a large-argument expansion) of the wave function in terms of the stationary states $\phi_\alpha(x_{\text{tar}})$ of the target system. We shall call such an expansion an *expansion in terms of target states*, and we shall see that it does indeed provide a useful way of approaching the multichannel problem.

The wave function $\langle \mathbf{x}, x_{\text{tar}} \mid \mathbf{p}, 1+ \rangle$ depends on $\mathbf{x}$ and x_{tar} while the target eigenstates $\phi_\alpha(x_{\text{tar}})$ depend only on x_{tar}. Thus, the coefficients η_α in an expansion of the former in terms of the latter must be functions of $\mathbf{x}$,

$$\langle \mathbf{x}, x_{\text{tar}} \mid \mathbf{p}, 1+ \rangle = \int_{\alpha} \eta_\alpha(\mathbf{x}) \phi_\alpha(x_{\text{tar}}) \qquad (19.16)$$

Here we remind the reader that the wave function on the left is an eigenfunction of the full Hamiltonian H with energy E and describes a collision in which the projectile is incident with momentum $\mathbf{p}$ on the target ground state ϕ_1. The functions $\phi_\alpha(x_{\text{tar}})$ on the right are a complete set of eigenfunctions of the target Hamiltonian H_{tar} with energy E_α and include the bound and continuum states of the target system. The functions $\eta_\alpha(\mathbf{x})$ are, for the moment, just the coefficients in the expansion of the former in terms of the latter. However, we shall see that they have an extremely useful interpretation.

We first note that if we choose $\mathbf{x}$ large and compare the expansion (19.16) with the asymptotic form (19.15) we find that:

$$\eta_\alpha(\mathbf{x}) \xrightarrow[\mathbf{x}\to\infty]{} (2\pi)^{-3/2} \left[e^{i\mathbf{p}\cdot\mathbf{x}} \delta_{\alpha 1} + f(p_\alpha \hat{\mathbf{x}}, \alpha \leftarrow \mathbf{p}, 1) \frac{e^{ip_\alpha r}}{r} \right] \qquad (19.17)$$

In particular, the coefficient $\eta_1(\mathbf{x})$ behaves just like a one-channel scattering wave function, with an incident plane wave plus a spherically spreading scattered wave times the elastic amplitude. The coefficients $\eta_\alpha(\mathbf{x})(\alpha \neq 1)$ behave similarly except that in their case there is no incident wave.

The form (19.17) suggests that we can view the functions $\eta_\alpha(\mathbf{x})$ as one-particle wave functions describing the motion of the projectile. The existence of a whole family of these functions reflects the existence of many channels, each $\eta_\alpha(\mathbf{x})$ describing that part of the motion that is in the channel labelled by α.

It should be emphasized that, useful as this point of view is, it is, nonetheless, somewhat misleading. Thus, if we view (19.16) as a decomposition of the full wave function into its "channel components," then the discrete terms correspond to the elastic and excitation channels, while the continuum terms describe the breakup channels. However, all rearrangement channels appear to be missing. The fact is that the view of (19.16) as a decomposition into channels is an oversimplification. The effects of *all* channels are, of course, included in the original wave function and, hence, in its full expansion. However, the effect of each "nonrearrangement" channel is explicitly displayed by the separate terms of the expansion as $\mathbf{x} \to \infty$; while that of the rearrangement channels (which show up only when some of the target coordinates go to infinity) is concealed in the whole infinite sum.

With this warning we return to the properties of the "wave functions" $\eta_\alpha(\mathbf{x})$. Thus far we know that their asymptotic form generalizes that of a single-channel wave function. We now show that they satisfy a natural generalization of the single-channel Schrödinger equation, namely an infinite set of coupled one-particle equations.

The full state $|\mathbf{p}, 1+\rangle$ is an eigenstate of the Hamiltonian H, which we can write as:

$$H = H^1 + V^1 = \frac{\mathbf{P}^2}{2m} + H_{\mathrm{tar}} + V^1$$

(where as usual V^1 consists of those potentials linking the projectile to the target). If we now substitute the expansion

$$\sum_\alpha \eta_\alpha(\mathbf{x})\phi_\alpha(\underline{x}_{\mathrm{tar}})$$

of the full wave function into the Schrödinger equation and remember that $\phi_\alpha(\underline{x}_{\mathrm{tar}})$ is an eigenfunction of H_{tar} with energy E_α, this gives

$$\sum_\alpha \left[-\frac{\nabla^2}{2m} + E_\alpha + V^1(\mathbf{x}, \underline{x}_{\mathrm{tar}})\right]\eta_\alpha(\mathbf{x})\phi_\alpha(\underline{x}_{\mathrm{tar}}) = E\sum_\alpha \eta_\alpha(\mathbf{x})\phi_\alpha(\underline{x}_{\mathrm{tar}})$$

We now multiply through by $\phi_\alpha(\underline{x}_{\text{tar}})^*$ and integrate over the target coordinates $\underline{x}_{\text{tar}}$. Because the functions ϕ_α are orthonormal, this gives

$$-\frac{\nabla^2}{2m}\eta_\alpha(\mathbf{x}) + \sum_{\alpha'}\bar{V}_{\alpha\alpha'}(\mathbf{x})\eta_{\alpha'}(\mathbf{x}) = (E - E_\alpha)\eta_\alpha(\mathbf{x}) \qquad (19.18)$$

where we have introduced the *potential matrix* $\bar{V}$ given by

$$\bar{V}_{\alpha\alpha'}(\mathbf{x}) = \int d\underline{x}_{\text{tar}}\phi_\alpha(\underline{x}_{\text{tar}})^* V^1(\mathbf{x}, \underline{x}_{\text{tar}})\phi_{\alpha'}(\underline{x}_{\text{tar}}) \qquad (19.19)$$

Thus, each of the wave functions $\eta_\alpha(\mathbf{x})$ satisfies a one-particle Schrödinger equation, in which, however, the potential term couples all of the $\eta_\alpha(\mathbf{x})$ together.

The equations (19.18) are a continuously infinite set of coupled equations and are of little practical value as they stand. Their usefulness lies in the possibility that, under certain conditions, it may be a good approximation to retain only a small number N of the discrete terms and none of the continuum. This approximation is known variously as the *coupled-channel approximation*, the *close-coupling approximation*, or the *N-state approximation*.[3] It reduces (19.18) to a finite set of coupled equations, which can be put into a compact matrix form. For example, the *two-state* approximation can be written as:

$$-\frac{\nabla^2}{2m}\begin{pmatrix}\eta_1\\\eta_2\end{pmatrix} + \begin{pmatrix}\bar{V}_{11} & \bar{V}_{12}\\\bar{V}_{21} & \bar{V}_{22}\end{pmatrix}\begin{pmatrix}\eta_1\\\eta_2\end{pmatrix} = \begin{pmatrix}E - E_1 & 0\\0 & E - E_2\end{pmatrix}\begin{pmatrix}\eta_1\\\eta_2\end{pmatrix} \qquad (19.20)$$

which, together with the boundary conditions:

$$\eta_1 \to (2\pi)^{-3/2}\left(e^{i\mathbf{p}\cdot\mathbf{x}} + f_{11}\frac{e^{ipr}}{r}\right)$$

$$\eta_2 \to (2\pi)^{-3/2}\left(f_{21}\frac{e^{ip_2r}}{r}\right)$$

can be solved to give η_1 and η_2.

It is interesting that the diagonal elements of the potential matrix

$$\bar{V}_{\alpha\alpha}(\mathbf{x}) = \int d\underline{x}_{\text{tar}} |\phi_\alpha(\underline{x}_{\text{tar}})|^2\, V^1(\mathbf{x}, \underline{x}_{\text{tar}})$$

are just the averaged potential seen by the projectile in the static field of the target state ϕ_α. In particular, therefore, the *one-state approximation*

$$-\frac{\nabla^2}{2m}\eta_1(\mathbf{x}) + \bar{V}_{11}\eta_1(\mathbf{x}) = (E - E_1)\eta_1(\mathbf{x})$$

---

[3] Also, sometimes, the Tamm–Dancoff approximation. However, this last is usually reserved for bound-state problems.

precisely reproduces the elastic scattering in the *static approximation* discussed in Section 9-d. More generally, if we write the N-state approximation as:

$$\left[-\frac{\nabla^2}{2m} + \bar{V}_{\alpha\alpha} - (E - E_\alpha)\right]\eta_\alpha(\mathbf{x}) = -\sum_{\alpha' \neq \alpha} \bar{V}_{\alpha\alpha'}\eta_{\alpha'}(\mathbf{x})$$

then the left-hand side alone would describe elastic scattering within channel α in the static approximation, while the inhomogeneous terms on the right describe the coupling to the other $N - 1$ channels.

In a general way, it is easy to understand the conditions under which the coupled-channel approximation should be useful. For example, if the incident energy is below the threshold for excitation of levels $N + 1$ and higher, then one can reasonably hope that an N-state approximation using the lowest N levels would be fairly realistic. (For example e–H scattering in the neighborhood of the $n = 2$ threshold has been treated with some success by a three-state approximation including just the $1s$, $2s$, and $2p$ states.) The reason is that each term $\eta_\alpha(\mathbf{x})\phi_\alpha(x_{\text{tar}})$ in the full wave function measures the probability of excitation of the level ϕ_α of the target. If the energy is less than E_α this probability—even for "virtual" excitation during the collision—can be expected to be small. Thus, $\eta_{(N+1)} \cdots$ should be small and, hence, can reasonably be neglected.

More generally, even when many channels are open, it may be that neglect of all but a small number of the most important channels will give reliable answers for the scattering in these channels. For example, many nuclei have a single strongly excitable low-lying collective state (such as the 2^+ state in ^{12}C or ^{24}Mg). This means that one can treat the scattering of nucleons or alpha particles off such nuclei in a two-state approximation, including just the ground state and this one collective state.

Finally, we remark that the coupled-channel equations can often be further simplified. For example, a partial-wave analysis retaining just a few angular momenta may be satisfactory at low energies (see Problem **19.5**), or, if the off-diagonal elements $\bar{V}_{\alpha\alpha'}$ are small, it may be sufficient to solve first for η_1 neglecting all coupling and then to use this η_1 to find the remaining η_α, retaining just the coupling to channel 1.

19-d.   The Optical Potential

In Section 19-c we saw that by expanding the full wave function that describes a projectile incident on a composite target in terms of target eigenstates as

$$\sum_\alpha \eta_\alpha(\mathbf{x})\phi_\alpha(x_{\text{tar}})$$

we can convert the original many-body problem into the infinite set of coupled one-body equations (19.18). If for some reason we are only interested in a certain subset of N channels, then all the information we require is contained in the corresponding N functions $\eta_\alpha(\mathbf{x})$. Unfortunately, the equations (19.18) couple all of the η_α together, and it is generally impossible to solve just for the η_α of interest. Of course, we can simply ignore all of the other η_α, in which case we obtain the finite N-state equations for the η_α of interest. However, these equations are only an approximation and are frequently a very bad approximation since they make no allowance for the loss of particles into those channels that are ignored.

In this section we shall show that for any given choice of N channels, it is possible to define an operator V_{opt}, called the optical potential, such that the N wave functions $\eta_\alpha(\mathbf{x})$ *exactly* satisfy N coupled equations with a potential matrix given by V_{opt}.[4] To begin, we shall consider the case $N = 1$; that is, we shall consider just the wave function $\eta_1(\mathbf{x})$ describing the elastic scattering, and we shall show that one can define a one-particle operator V_{opt} such that $\eta_1(\mathbf{x})$ exactly satisfies the one-particle equation:

$$\left(-\frac{\nabla^2}{2m} + V_{\text{opt}}\right)\eta_1(\mathbf{x}) = (E - E_1)\eta_1(\mathbf{x}) \tag{19.21}$$

Since $\eta_1(\mathbf{x})$ also satisfies the one-channel boundary condition,

$$\eta_1(\mathbf{x}) \to (2\pi)^{-\frac{3}{2}}\left(e^{i\mathbf{p}\cdot\mathbf{x}} + f_{11}\frac{e^{ipr}}{r}\right) \tag{19.22}$$

this exactly reduces the computation of the elastic scattering to an equivalent one-channel problem.

Obviously, there is a price to pay for this sensational simplification of the multichannel problem. The optical potential V_{opt} is generally an extremely complicated operator. In particular, it is nonlocal, energy-dependent, and, for energies above the first inelastic threshold, it is not Hermitian. This situation is exactly what one should expect; if the Hamiltonian $(\mathbf{P}^2/2m) + V_{\text{opt}}$ were Hermitian, the corresponding evolution operator would be unitary. This would mean that solutions of the corresponding time-dependent Schrödinger equation would have constant norm or, equivalently, that the stationary wave function $\eta_1(\mathbf{x})$ would have equal incoming and outgoing fluxes. This situation would allow for no loss of flux due to inelastic processes. Accordingly, if inelastic processes *do* occur—that is, if E is above the inelastic threshold—and if we wish to describe the elastic scattering by an

---

[4] For $N > 1$ the potential V_{opt} is sometimes called the *generalized* optical potential.

equivalent one-particle Hamiltonian, then that Hamiltonian has to be non-Hermitian.

In practice, the optical potential V_{opt} is far too complicated for exact use in actual computations. Rather, its importance lies in the mere knowledge of its existence. On the one hand this makes possible a number of theoretical developments, such as the so-called formal theory of resonances. (see Newton, 1966, Section 16.5). On the other, it justifies attempts to fit elastic scattering data (or more generally, the data in any N chosen channels) by means of simple phenomenological optical potentials—a technique that has proved extremely profitable, especially in nuclear physics.[5]

To prove that the elastic wave function $\eta_1(\mathbf{x})$ does satisfy the one-particle equation (19.21), we must define the optical potential V_{opt}. Before doing this it is convenient to introduce some more notation. First, we note that the equation (19.21), together with the boundary condition (19.22), equivalent to the one-particle equation:

$$|\eta_1\rangle = |\mathbf{p}\rangle + G_{\mathrm{opt}}(E + i0)V_{\mathrm{opt}}|\mathbf{p}\rangle \qquad (19.23)$$

Here $|\eta_1\rangle$ denotes the one-particle state vector corresponding to the wave function $\eta_1(\mathbf{x})$, the vector $|\mathbf{p}\rangle$ is, as usual, a one-particle plane wave of momentum $\mathbf{p}$, and

$$G_{\mathrm{opt}}(z) = (z - H_{\mathrm{opt}})^{-1}$$

is the one-particle Green's operator corresponding to the one-particle Hamiltonian[6]

$$H_{\mathrm{opt}} = \frac{\mathbf{P}^2}{2m} + E_1 + V_{\mathrm{opt}}$$

We shall establish that $|\eta_1\rangle$ is given by (19.23) by showing that (19.23) follows from the exact expression

$$|\mathbf{p}, 1+\rangle = |\mathbf{p}, 1\rangle + G(E + i0)V^1|\mathbf{p}, 1\rangle \qquad (19.24)$$

for the full multiparticle stationary state $|\mathbf{p}, 1+\rangle$.

To make the connection between the desired equation (19.23) and the known equation (19.24), we note that the wave function $\eta_1(\mathbf{x})\phi_1(\underline{x}_{\mathrm{tar}})$ is the component of the full wave function

$$\langle \mathbf{x}, \underline{x}_{\mathrm{tar}} \mid \mathbf{p}, 1+\rangle = \sum_\alpha \eta_\alpha(\mathbf{x})\phi_\alpha(\underline{x}_{\mathrm{tar}})$$

---

[5] It is from here that the name "optical potential" is derived. Quite good fits can be obtained with simple phenomenological optical potentials of the form $U(r) + iW(r)$. The Schrödinger equation for such a potential exactly reproduces the scattering of light by a medium of complex refractive index with real and imaginary parts related to $U(r)$ and $W(r)$.

[6] The term E_1 is added to ensure that the zero of energy coincides with that used in the full many-body system.

in the channel 1 subspace $\mathscr{S}^1$. [This latter being just the subspace of functions $\chi(\mathbf{x})\,\phi_1(x_{\text{tar}})$ with χ arbitrary.] This suggests that we introduce the projection operator Λ onto the channel 1 subspace as[7]

$$\Lambda\,|\psi\rangle = |\psi\rangle \qquad [\,|\psi\rangle \text{ in } \mathscr{S}^1]$$
$$= 0 \qquad [\,|\psi\rangle \text{ orthogonal to } \mathscr{S}^1]$$

in terms of which we can write

$$\Lambda\,|\mathbf{p}, 1+\rangle = |\eta_1\rangle \otimes |\phi_1\rangle$$

Thus, by multiplying (19.23) by the target state $|\phi_1\rangle$ we see that the equation (19.23) (that we wish to prove) is exactly equivalent to the multiparticle equation

$$\Lambda\,|\mathbf{p}, 1\rangle = |\mathbf{p}, 1\rangle + G_{\text{opt}}(E + i0)V_{\text{opt}}\,|\mathbf{p}, 1\rangle \qquad (19.25)$$

Here we have used the fact that the channel 1 free state $|\mathbf{p}, 1\rangle$ is just the product $|\mathbf{p}\rangle \otimes |\phi_1\rangle$, and, for convenience, we shall define the operators G_{opt} and V_{opt} on the multiparticle states to coincide with the one-particle G_{opt} and V_{opt} *on the channel* 1 *subspace* $\mathscr{S}^1$ *but to be zero on vectors orthogonal to* $\mathscr{S}^1$ (this last statement makes no difference in (19.25) since $|\mathbf{p}, 1\rangle$ lies completely in $\mathscr{S}^1$).

Our next step is to multiply the full equation (19.24) by Λ to give

$$\Lambda\,|\mathbf{p}, 1+\rangle = |\mathbf{p}, 1\rangle + \Lambda G(E + i0)V^1\Lambda\,|\mathbf{p}, 1\rangle \qquad (19.26)$$

(The extra Λ on the right makes no difference since $|\mathbf{p}, 1\rangle$ lies in $\mathscr{S}^1$ and so is equal to $\Lambda\,|\mathbf{p}, 1\rangle$.) Equation (19.26) is known to be true, while (19.25) is equivalent to the result that we wish to prove. Thus, our proof will be complete if we can define V_{opt} and show that it satisfies

$$\boxed{G_{\text{opt}}(E + i0)V_{\text{opt}} = \Lambda G(E + i0)V^1\Lambda} \qquad (19.27)$$

To define V_{opt} we introduce a second projector

$$M = 1 - \Lambda$$

which projects onto the orthogonal compliment $\mathscr{S}^{1\perp}$ of the channel 1 subspace $\mathscr{S}^1$.[8] For any operator A we now introduce four new operators

$$A_{\Lambda\Lambda} = \Lambda A\Lambda, \qquad A_{\Lambda M} = \Lambda A M,$$
$$A_{M\Lambda} = M A\Lambda, \qquad A_{MM} = M A M$$

---

[7] We remind the reader that the projection operator on $\mathscr{S}$ acts as the identity on $\mathscr{S}$, and as the zero operator on vectors orthogonal to $\mathscr{S}$. It is easily seen that the characteristic properties of a projector are $\Lambda = \Lambda^2 = \Lambda^\dagger$.

[8] This means that $\Lambda M = 0$, $\Lambda + M = 1$, and $M^2 = M = M^\dagger$.

The operators $A_{\Lambda\Lambda}$ and A_{MM} are *restrictions* of A to the subspaces $\mathscr{S}^1$ and $\mathscr{S}^{1\perp}$, while $A_{\Lambda M}$ and $A_{M\Lambda}$ are the parts of A that link the two subspaces. Clearly, the sum of the four operators is the original operator A itself. Finally, we introduce two partial Green's operators G_Λ and G_M such that

$$(z - H_{\Lambda\Lambda})G_\Lambda(z) = \Lambda$$
$$(z - H_{MM})G_M(z) = M$$

that is, $G_\Lambda(z)$ is the Green's operator for $H_{\Lambda\Lambda}$ on the subspace $\mathscr{S}^1$ and similarly for $G_M(z)$ on $\mathscr{S}^{1\perp}$.

We are now ready to define the optical Hamiltonian as

$$H_{\mathrm{opt}}(E) = H_{\Lambda\Lambda} + V_{\Lambda M}^1 G_M(E + i0)V_{M\Lambda}^1 \qquad (19.28)$$

That is, H_{opt} is the restriction of H to the channel 1 subspace, *plus* a second term. The significance of the first term is easily seen. We can write

$$H = H^1 + V^1 = \frac{\mathbf{P}^2}{2m} + H_{\mathrm{tar}} + V^1$$

Then, since the action of H_{tar} on any vector of $\mathscr{S}^1$ is to give E_1, it follows that on the channel 1 subspace,

$$H_{\Lambda\Lambda} = \frac{\mathbf{P}^2}{2m} + E_1 + \bar{V}_{11} \qquad [\mathrm{on}\ \mathscr{S}^1]$$

where $\bar{V}_{11}$ is just the static field of the target state ϕ_1,

$$\bar{V}_{11}(\mathbf{x}) = \int d\underline{x}_{\mathrm{tar}}\ |\phi_1(\underline{x}_{\mathrm{tar}})|^2\ V^1(\mathbf{x}, \underline{x}_{\mathrm{tar}})$$

Thus, we can write the optical Hamiltonian (on $\mathscr{S}^1$) in the form

$$\frac{\mathbf{P}^2}{2m} + E_1 + V_{\mathrm{opt}}$$

where

$$V_{\mathrm{opt}}(E) = \bar{V}_{11} + V_{\Lambda M}^1 G_M(E + i0)V_{M\Lambda}^1 \qquad (19.29)$$

Neglect of the second term would lead us back to the one-state approximation of the last section. In particular, since the potential $\bar{V}_{11}$ is Hermitian, it alone would never produce inelasticity. Meanwhile, the second term contains $V_{\Lambda M}^1$ and $V_{M\Lambda}^1$ and clearly describes the effects of coupling to all other channels. The Green's operator $G_M(E + i0)$, and, hence, $V_{\mathrm{opt}}(E)$, is Hermitian for energies less than E_2 (the lowest threshold in the subspace $\mathscr{S}^{1\perp}$) but is non-Hermitian above E_2, just as anticipated.[9]

[9] The skew-Hermitian part of $V_{\mathrm{opt}}(E)$ can be shown to be negative-definite. This guarantees that it describes absorption (not creation) of particles from channel 1.

With V_{opt} defined, we now have only to establish that it satisfies the identity (19.27). To do this we rewrite G in terms of G_Λ and G_M as

$$G = (G_\Lambda + G_M) + (G_\Lambda + G_M)(V^1_{\Lambda M} + V^1_{M\Lambda})G$$

Multiplying from the left by Λ or M gives equations for ΛG and MG,

$$\Lambda G = G_\Lambda + G_\Lambda V^1_{\Lambda M}G$$
$$MG = G_M + G_M V^1_{M\Lambda}G$$

Inserting the second of these into the first we find:

$$\Lambda G = G_\Lambda + G_\Lambda V^1_{\Lambda M}(G_M + G_M V^1_{M\Lambda}G)$$

We now multiply on the left by $(z - H_{\Lambda\Lambda})$ and, remembering that

$$(z - H_{\Lambda\Lambda})G_\Lambda = \Lambda$$

obtain:

$$(z - H_{\Lambda\Lambda})\Lambda G = \Lambda + V^1_{\Lambda M}(G_M + G_M V^1_{M\Lambda}G)$$

or, bringing the last term on the right over to the left,

$$(z - H_{opt})\Lambda G = \Lambda + V^1_{\Lambda M}G_M$$

Finally, multiplication on the right by $V^1\Lambda$ gives

$$(z - H_{opt})\Lambda G V^1\Lambda = V^1_{\Lambda\Lambda} + V^1_{\Lambda M}G_M V^1_{M\Lambda}$$

or, with $z = E + i0$,

$$\Lambda G(E + i0)V^1\Lambda = G_{opt}(E + i0)V_{opt}(E)$$

as required.

This completes the proof that the elastic wave function $\eta_1(\mathbf{x})$ is completely determined by the one-particle Schrödinger equation with the energy dependent one-particle potential $V_{opt}(E)$. It should be emphasized that while the optical potential (in principle at least) determines the elastic scattering exactly, it gives no information on any of the inelastic processes (except for the total cross section via the optical theorem). However, as already noted. the method of the optical potential can be generalized to handle any chosen finite set of channels. Thus, for a given set of N channels one can introduce a generalized optical potential V_{opt} (depending of course on the choice of channels) such that the corresponding N wave functions $\eta_\alpha(\mathbf{x})$ exactly satisfy a set of N coupled equations with the form of the coupled-channel equations of Section 19-c, but with a potential matrix derived from V_{opt}. The analysis of this general case is parallel to that given above, except that Λ becomes the projector onto the subspace spanned by the N channel spaces $\mathscr{S}^\alpha$ of interest (see Problem **19.7**).

PROBLEMS

19.1. The asymptotic form of any multichannel Green's function has many terms corresponding to the many different channels, as in (19.6). Go over the analysis between (19.5) and (19.6) and show that the terms corresponding to *closed channels* ($E_\alpha > E$) in (19.6) go to zero exponentially as $x_a \to \infty$. [This will establish that the same is true of the closed channel terms in the wave function (19.8).]

19.2. For the three-particle model of Sections 19-a and 19-b, write down (without proof) the asymptotic form of the wave function $\langle \mathbf{x}_a, \mathbf{x}_b \mid \mathbf{p}, 1+ \rangle$ as $\mathbf{x}_a$ and $\mathbf{x}_b$ go to infinity with $\mathbf{x}_a - \mathbf{x}_b$ fixed. [This will be analogous to (19.14), but should display the pickup collisions $a + (bc) \to (ab) + c$.]

19.3. Consider a collision between two composite particles A and B (e.g. nucleus–nucleus scattering). Show that the full wave function $\langle \mathbf{x} \mid \mathbf{p}, 1+ \rangle$ can be expanded in terms of products $\phi_\alpha(\underline{x}_A)\chi_\beta(\underline{x}_B)$ where $\phi_\alpha(\underline{x}_A)$ and $\chi_\beta(\underline{x}_B)$ are eigenfunctions of the internal Hamiltonians of the composites A and B; and that the expansion coefficients $\eta_{\alpha\beta}(\mathbf{x})$ satisfy a set of coupled one-particle equations of exactly the form (19.18) (notice that each channel is identfiied by a pair of indices α, β here).

19.4. (a) Show that the potential matrix (19.19), which enters the coupled-channel equations (19.18), is Hermitian

$$\bar{V}_{\alpha\alpha'}(\mathbf{x}) = \bar{V}_{\alpha'\alpha}(\mathbf{x})^*$$

(b) For the case that all interactions are rotationally invariant and that α and α' label target states of zero "spin" (i.e., zero total angular momentum), show that $\bar{V}_{\alpha\alpha'}(\mathbf{x})$ is rotationally invariant,

$$\bar{V}_{\alpha\alpha'}(\mathbf{x}) = \bar{V}_{\alpha\alpha'}(\mathbf{x}_R) \qquad \text{[any R]}$$

and can therefore be written as $\bar{V}_{\alpha\alpha'}(r)$. (Convince yourself that this is *not* true if α and α' label states of nonzero spin.)

19.5. Consider the coupled-channel equations (19.18) in the simple case where all interactions are rotationally invariant and where the equations are truncated to retain only target states of zero spin.
(a) Show that (in this case) the wave functions $\eta_\alpha(\mathbf{x})$ can be expanded in a partial-wave series

$$\eta_\alpha(\mathbf{x}) = (2\pi)^{-3/2} \frac{1}{pr} \sum (2l + 1)i^l \psi_\alpha^l(r)P_l(\cos\theta)$$

and that the radial functions satisfy the coupled radial equations,

$$\frac{d^2}{dr^2}\psi_\alpha^l(r) - \frac{l(l+1)}{r^2}\psi_\alpha^l(r) - \sum_{\alpha'} 2m\bar{V}_{\alpha\alpha'}(r)\psi_{\alpha'}^l(r) - p_\alpha^2\psi_\alpha^l(r) = 0$$

where p is the incident momentum in channel 1 and p_α is the corresponding momentum in channel α $(p_1 \equiv p)$. Write this equation in matrix form using $\mathbf{\Psi}^l$ as the column made up of the ψ_α^l and P the diagonal matrix made up of the p_α.

(b) Use the asymptotic form (19.17) of $\eta_\alpha(\mathbf{x})$ to show that the radial wave functions satisfy

$$\psi_\alpha^l(r) \xrightarrow[r \to \infty]{} \hat{\jmath}_l(pr)\, \delta_{\alpha 1} + pf_{\alpha 1}^l \hat{h}_l^+(p_\alpha r)$$

$$= \frac{i}{2}\left[\hat{h}_l^-(pr)\, \delta_{\alpha 1} - \left(\frac{p_1}{p_\alpha}\right)^{\!\frac{1}{2}} s_{\alpha 1}^l \hat{h}_l^+(p_\alpha r) \right]$$

where the partial-wave S matrix $s_{\alpha\alpha'}^l$ and amplitude $f_{\alpha\alpha'}^l$ are defined in Section 17-d.

19.6. Discuss the form of the potential matrix and coupled-channel equations if one includes channels in which the target has nonzero angular momentum.

19.7. Establish the existence of a generalized optical potential V_{opt} that exactly determines the wave functions $\eta_1(\mathbf{x})$ and $\eta_2(\mathbf{x})$ describing the lowest *two* channels of the system of Section 19-d. [The analysis will closely parallel that of Section 19-d. In fact, $H_{\text{opt}}(E)$ is defined by precisely (19.28) if Λ is taken to be the projector onto the subspace $\mathscr{S}^1 \oplus \mathscr{S}^2$ defined by the two channels of interest. With this definition one can establish an analogue of (19.25) and then show that this equation is equivalent to two coupled Schrödinger equations for $\eta_1(\mathbf{x})$ and $\eta_2(\mathbf{x})$.]

20 Analytic Properties and Multichannel Resonances

20-a   **Analytic Properties**

20-b   **Proof of Analytic Properties**

20-c   **Bound States**

20-d   **Resonances**

20-e   **Decay of a Multichannel Resonance**

In this chapter we discuss the analytic properties of multichannel amplitudes. In fact, we shall only scratch the surface of this extensive and important subject. We shall limit our discussion to the partial-wave amplitudes as functions of energy, and consider just two-body processes that involve only spinless particles. Nonetheless, in this way we can introduce several important features of the multichannel theory, while using only the simplest extensions of the one-channel techniques developed in Chapter 12.

In Section 20-a we discuss qualitatively the multichannel partial-wave amplitudes and argue that they must have branch point singularities at all of the channel thresholds $E = W_1, W_2, \ldots$. In Section 20-b we prove some of the assertions of Section 20-a within the framework of one simple multichannel model (specifically, the N-channel system defined by the N-state equations of Section 19-c). In Section 20-c we establish the expected connection between bound states and poles of the partial-wave amplitudes.

Perhaps the principal usefulness of the formalism developed in the first three sections is that it provides a basis for discussion of multichannel resonances in Sections 20-d and 20-e. In Section 20-d we show how poles of the amplitude on unphysical sheets can lead to resonances; and in Section

20-e we examine (in close analogy with the one-channel analysis of Section 13-d) the time development of scattering in the neighborhood of a very narrow resonance.

20-a.   Analytic Properties

We begin our discussion of multichannel partial-wave amplitudes with a quick review of the corresponding one-channel results.

The One-Channel Case.   The reader will recall (from Chapter 12) that for a large class of potentials both the partial-wave S matrix $s_l(p)$ and the amplitude

$$f_l(p) = \frac{s_l(p) - 1}{2ip}$$

are analytic functions of the incident (relative) momentum p. For example, the Yukawa potential (which can be regarded as typical of the potentials of nuclear and particle physics) has an amplitude that is analytic in p except at certain poles and at two branch cuts, one running from $i\mu/2$ upwards, the other from $-i\mu/2$ downward. For simplicity, in this chapter we shall consider the more restricted (and less realistic) class of *potentials with finite range*—that is, potentials that are identically zero beyond some finite radius $r = a$. For these, $f_l(p)$ has no branch cuts and its only singularities are poles. In the upper half plane these poles are confined to the imaginary axis and are in one-to-one correspondence with the bound states of angular momentum l. In the lower half plane there is no such restriction on the pole positions and all that one can say is that those poles that are sufficiently close to the positive real axis correspond to resonances. This situation is summarized in Fig. 20.1, part (a).

It is often more convenient (especially in multichannel scattering) to think of the amplitude as a function of the CM energy E rather than the momentum

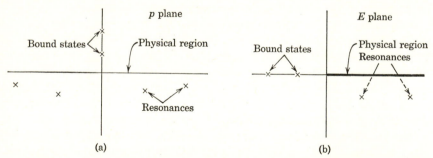

FIGURE 20.1.   Regions of analyticity of the single-channel partial-wave amplitude for a finite-range potential: (a) as a function of p, (b) as a function of E.

$p = (2mE)^{1/2}$. Since p is an analytic function of E except at $E = 0$, and since the amplitude f_l is analytic in p, it follows at once that f_l is analytic in E except at $E = 0$. The singularity at $E = 0$ is a branch point and simply reflects the fact that the mapping from p to E is two-to-one. As the reader will recall, the correspondence of the planes of p and E is traditionally chosen so that the upper half plane $\{\text{Im } p > 0\}$ maps onto one sheet of E (the "physical sheet") with a cut along the positive real axis; the lower half plane $\{\text{Im } p < 0\}$ maps onto the "unphysical sheet" below. The amplitude is analytic on the physical sheet except at the bound-state poles. (There are, of course, further singularities, such as the left-hand cut of a Yukawa potential, if we make less restrictive assumptions about the potential.) The resonance poles lie on the unphysical sheet as shown in Fig. 20.1, part (b). Finally, we recall that, since the amplitude f_l is real when p is imaginary, f_l (or s_l) is real on the negative real axis of E (see Section 12-c). For E positive real, s_l is, of course, unitary $s_l = \exp(2i\delta_l)$.

The Multichannel Case. If we consider a multichannel system, the situation becomes much more complicated. To illustrate this we consider the simplest possible case, a rotationally invariant system all of whose fragments have zero spin, and an energy range where only two-body channels are open. In this case there is a sequence of thresholds $W_1 < W_2 < \cdots$. At each threshold a new channel opens up; and in each interval $\{W_n < E < W_{n+1}\}$ the scattering for angular momentum l is given by an $n \times n$ unitary matrix $S^l(E)$.

In the single-channel case we saw that as a function of energy the one S-matrix element has a branch point at the one threshold $E = 0$; we shall now show that in the multichannel case at least some of the S-matrix elements must have singularities at *every* threshold W_n. To this end we consider the unitarity equation, starting in the lowest interval $\{W_1 < E < W_2\}$ where we have

$$s_{11}(E)^* s_{11}(E) = 1 \qquad [W_1 < E < W_2] \qquad (20.1)$$

(Wherever possible we omit the superscipt l.) In the next interval

$$\{W_2 < E < W_3\}$$

the S matrix is a 2×2 unitary matrix and the 1,1 element of the equation $S^\dagger S = 1$ reads

$$s_{11}(E)^* s_{11}(E) + s_{21}(E)^* s_{21}(E) = 1 \qquad [W_2 < E < W_3] \qquad (20.2)$$

Now, if the S-matrix elements are analytic, then each of the above equations is an analytic identity (provided we replace E by E^* in the factors that carry a star). Then, if we suppose that $s_{11}(E)$ is analytic in a neighborhood of W_2, (20.1) can be continued into the interval $\{W_2 < E < W_3\}$ where (20.2) also holds. If both equations hold, then it immediately follows by subtraction that $s_{21}(E)$ is identically zero—which in general it is not. The only way to avoid this contradiction is for $s_{11}(E)$ to have a singularity at W_2, so that (20.1) cannot be continued into the region where (20.2) holds. Accordingly, $s_{11}(E)$ must be singular at the threshold W_2.

Obviously a similar argument can be applied at W_3, W_4, and so on. At each threshold W_n the dimension of the physical S matrix increases by one, and so does the number of terms in each element of the unitarity equation. If these extra terms are to be nonzero, then at least some of the S-matrix elements must be singular at the threshold. In fact, we shall see in Section 20-b that in general *every* S-matrix element has a singularity at each threshold, and that these singularities are branch points. It is usual to handle these singularities by drawing a single branch cut starting from W_1 and passing through all of the branch points $W_2, \ldots$ and on to $+\infty$, as shown in Fig. 20.2. For obvious reasons this cut is often called the *unitarity cut*.

Since the S-matrix elements have singularities at each threshold, the question naturally arises whether there is any simple relation between the values of a given S-matrix element above and below any given threshold W_n. In fact, we shall see that the physical S-matrix elements on the left of W_n can be related to those on the right by an analytic continuation passing *above* W_n, as shown in Fig. 20.2. Thus, if we start in the lowest interval

$$\{W_1 < E < W_2\}$$

there is a single physical S-matrix element $s_{11}(E)$, which is analytic in this interval. Therefore, we can move E to the right towards W_2 retaining the same analytic function. Because of the singularity at W_2 we cannot continue through the point W_2; however, we shall see that if we continue up into the complex plane and loop above W_2, then when we reach the real axis in $\{W_2 < E < W_3\}$ the resulting function is precisely the physical S-matrix element $s_{11}(E)$ in this new interval.

Once in the interval $\{W_2 < E < W_3\}$ we have three additional elements, s_{12}, s_{21}, s_{22}, which, together with s_{11}, make up the complete 2×2 physical S matrix. All four elements are analytic and can be continued up to W_3, where a similar continuation passing above the branch point yields the corresponding four elements of the physical S matrix in $\{W_3 < E < W_4\}$. In this interval we pick up five more elements to make up the full 3×3 physical S matrix and we can then continue on to W_4, and so on.

We shall see that we can also start in any interval $\{W_n < E < W_{n+1}\}$ with the $n \times n$ physical S matrix and, reversing the above procedure, continue to the left passing above W_n into the interval $\{W_{n-1} < E < W_n\}$. In this way we get an $n \times n$ analytic matrix of which only the $(n-1) \times (n-1)$ leading submatrix is the actual physical S matrix, however. This suggests that the most general procedure would be to start above the *highest* threshold W_N

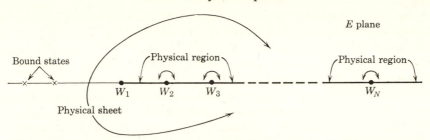

FIGURE 20.2.   Physical sheet of the multichannel S matrix. Physical values on either side of any threshold are connected by continuing above the threshold.

where all channels are open and the S matrix has its full compliment of N^2 elements. By continuing to the left passing above each threshold we obtain an $N \times N$ matrix of analytic functions $s_{\alpha'\alpha}(E)$ with the property that in each interval $\{W_n < E < W_{n+1}\}$ its leading $n \times n$ submatrix is the physical unitary S matrix; the remaining N–n "unphysical" rows and columns, although well defined analytic functions, are not of direct physical significance.[1]

The Physical Sheet. Since the physical values of the S matrix are all linked up by continuing above the threshold branch points, we can draw the branch cut through $W_1, W_2, \ldots$ in such a way that the physical values of S are all found on the upper edge of the cut, as shown in Fig. 20.2. The plane of E cut in this way is called the *physical sheet*. We shall find that the whole $N \times N$ matrix $\mathsf{S}(E)$ is analytic on the whole physical sheet except on the real axis below W_1, where it has poles at the energies of the bound states of the projectile–target system (this is for finite-range potentials; in general there are further singularities, like the left-hand cut of the Yukawa). Corresponding to the result that the one-channel S matrix is *real* below threshold, we shall find that the multichannel matrix $\mathsf{S}(E)$ is *Hermitian* for $E < W_1$. [This is the general result. If time-reversal is good (as it usually is in scattering problems) then $\mathsf{S}(E)$ is a symmetric matrix, and, hence, is also *real* for $E < W_1$.] We shall also find that we can continue $\mathsf{S}(E)$ downwards through any of the physical intervals $\{W_n < E < W_{n+1}\}$ onto "unphysical sheets"; it may happen that $\mathsf{S}(E)$ has poles on these unphysical sheets close to the real axis, and we shall see in Section 20-d that such poles correspond to resonances.

Finally, we remark that the physical sheet can be characterized in terms of the channel momenta[2]

$$p_\alpha = [2m_\alpha(E - W_\alpha)]^{1/2} \qquad (20.3)$$

---

[1] In relativistic problems there are infinitely many channels and there is no highest threshold. In this case there are infinitely many elements $s_{\alpha'\alpha}(E)$, all of them (presumably) analytic. In practice, one is always forced to ignore all but some finite number of them.
[2] For simplicity, we continue to ignore all channels with more than two bodies. Thus, there is just one (relative) momentum in each channel.

In the one-channel case the physical sheet was characterized as having the momentum p in its upper half plane; we can now see that the physical sheet of the multichannel problem is characterized by having *all* of the momenta $p_1, p_2, \ldots$ in their upper half planes. When E is above the highest threshold W_N, all of the channel momenta are positive real. If we move to the left of W_N passing above W_N, then the momentum p_N becomes positive imaginary. In general, if we move down to the interval $\{W_n < E < W_{n+1}\}$, always passing above all thresholds, the momenta $p_{n+1}, \ldots, p_N$ all become positive imaginary, while $p_1, \ldots, p_n$ remain positive real. Thus, everywhere in the physical region all channel momenta are either positive real (for the open channels) or positive imaginary (for the closed channels). More generally, if we move up from the physical region into the physical sheet all of the p_α move into their upper half planes $\{\mathrm{Im}\, p_\alpha > 0\}$. If we swing around in a counter-clockwise direction passing to the left of W_1, and finally coming up underneath the cut to the right of W_N, then all of the p_α sweep (counterclockwise also) across their upper half planes, to their negative real axes. Thus, on the whole of the physical sheet, with its cut running from W_1 to $+\infty$, all of the channel momenta are in their upper half planes, $\{\mathrm{Im}\, p_\alpha > 0\}$. In fact, we shall see that this is precisely what gives the physical sheet its special role.

20-b.   Proof of Analytic Properties

The One-Channel Case.   Before attempting to prove some of the properties discussed in the last section, let us briefly recall how the corresponding one-channel properties were proved in Chapter 12. The reader will remember that our main tools for the study of the partial-wave amplitude were the various radial wave functions. The normalized radial function (obtained by resolving the stationary state $|\mathbf{p}+\rangle$ into partial waves) was characterized as the solution of the radial Schrödinger equation, which is zero at the origin and has the asymptotic form as $r \to \infty$

$$\psi(r) \xrightarrow[r \to \infty]{} \hat{j}(pr) + pf\hat{h}^+(pr)$$

$$= \frac{i}{2}\,[\hat{h}^-(pr) - s\hat{h}^+(pr)] \tag{20.4}$$

where we omit all unnecessary subscripts l and p. For the study of analytic properties it was more convenient to use the so-called regular solution $\phi(r)$ defined so that as $r \to 0$,

$$\phi(r) \xrightarrow[r \to 0]{} \hat{j}(pr)$$

This has the asymptotic form as $r \to \infty$

$$\phi(r) \xrightarrow[r \to \infty]{} \frac{i}{2}\,[\hat{h}^-(pr)f(p) - \hat{h}^+(pr)f(-p)] \tag{20.5}$$

where $f(p)$ is the so-called Jost function. Comparison of (20.4) and (20.5) shows that:

$$\psi(r) = \frac{\phi(r)}{f(p)} \tag{20.6}$$

(that is, ψ and ϕ differ just in their normalization, their ratio being the Jost function) and,

$$s = \frac{f(-p)}{f(p)} \tag{20.7}$$

Finally, the regular solution $\phi(r)$ satisfies an integral equation of the form:

$$\phi(r) = j(pr) + \int_0^r dr' g(r, r')U(r')\phi(r') \tag{20.8}$$

examination of which gives the important expression for the Jost function:

$$f(p) = 1 + \frac{1}{p} \int_0^\infty dr \hat{h}^+(pr)U(r)\phi(r) \tag{20.9}$$

We saw in Section 12-b that for any potential satisfying our usual assumptions (p. 191), the integral equation (20.8) for $\phi(r)$ can be solved by iteration and defines $\phi(r)$ as an entire function of p (i.e., analytic for all p). This in turn meant that the integral (20.9) defines $f(p)$ as an analytic function of p everywhere in the upper half plane {Im $p > 0$}. [Recall that in the upper half plane the exponential decrease of $\hat{h}^+$ as $r \to \infty$ cancels the increase of $\phi(r)$; in the lower half plane $\hat{h}^+$ increases and the integral will generally diverge.] With more restrictive assumptions on the potential, $f(p)$ can be continued into the lower half plane {Im $p < 0$}. In particular, in this chapter we are assuming that the potential $U(r)$ is *identically zero* beyond some finite radius a, in which case the integral (20.9) converges for *all* p and defines $f(p)$ as an entire function of p. This leads to the analytic properties for s or f discussed at the beginning of Section 20-a.

A Multichannel Model.   A corresponding analysis of the analytic properties of any real multichannel system is extremely complicated and well beyond the scope of this book.[3] Here we shall consider just the simple model system provided by the coupled-channel equations of the N-state approximation discussed in Section 19-c. The reader will recall from that discussion that the

---

[3] In fact, not very much has been rigorously proved about the analytic properties of multi-particle amplitudes in general. For some results in three-particle scattering see Rubin, Sugar, and Tiktopoulos (1967).

starting point was an expansion of the stationary state $|\mathbf{p}, 1+\rangle$, which describes a projectile of momentum $\mathbf{p}$ incident in channel 1 on a complex target, in terms of a basis of eigenstates of the target,

$$\langle \mathbf{x}, \underline{x}_{\text{tar}} \mid \mathbf{p}, 1+\rangle = \sum_{\alpha}^{\text{\it f}} \eta_{\alpha}(\mathbf{x})\phi_{\alpha}(\underline{x}_{\text{tar}})$$

The full Schrödinger equation reduces to an infinite set of coupled one-particle equations for the functions $\eta_{\alpha}(\mathbf{x})$. If we then approximate these equations by truncation, we obtain the N-state coupled equations

$$\nabla^2 \eta_{\alpha}(\mathbf{x}) - \sum_{\alpha'=1}^{N} U_{\alpha\alpha'}(\mathbf{x})\eta_{\alpha'}(\mathbf{x}) + p_{\alpha}^2 \eta_{\alpha}(\mathbf{x}) = 0 \qquad [\alpha = 1, \ldots, N] \quad (20.10)$$

where $U_{\alpha\alpha'}(\mathbf{x})$ is $2m$ times the element $\bar{V}_{\alpha\alpha'}(\mathbf{x})$ of the potential matrix defined in (19.19) and p_{α} is the channel momentum as before. The equation (20.10), together with the boundary conditions

$$\eta_{\alpha}(\mathbf{x}) \xrightarrow[r \to \infty]{} (2\pi)^{-\frac{3}{2}}\left[\delta_{\alpha 1}e^{i\mathbf{p}\cdot\mathbf{x}} + f(p_{\alpha}\hat{\mathbf{x}}, \alpha \leftarrow \mathbf{p}, 1)\frac{e^{ip_{\alpha}r}}{r}\right] \quad (20.11)$$

provide a self-contained (though approximate) framework for computation of the wave functions and amplitudes.

We shall adopt the N-state equations (20.10) and (20.11) as the starting point for our discussion of analytic properties. Although these equations are admittedly only an approximation to the multiparticle system for which they were originally set up, it should be emphasized that they are nonetheless a self-consistent set of equations, describing a very reasonable model system—specifically a system in which a projectile is incident on a target with just N states (all bound); and which therefore has exactly N channels each with just two bodies. The great advantages of this model are that it allows an analysis exactly parallel to the one-channel analysis just sketched, and that within the model one can give a very clear discussion of the correspondence between bound states and poles of $S(E)$.[4]

Within our theoretical model every state is uniquely specified by the set of N wave functions $\eta_1(\mathbf{x}), \ldots, \eta_N(\mathbf{x})$, where $\eta_{\alpha}(\mathbf{x})$ describes that part of the

---

[4] The only inelastic processes that can occur in our model are excitations. This is inherent in the coupled-channel approximation, which is not well suited to discussion of rearrangements. (In particular, it means that the reduced masses of all channels are the same.) One could adopt a set of equations of the form (20.10) as a model to describe (two-body) rearrangement collisions. (In this case there would be different reduced masses for the different channels.) However, the theoretical basis for such a model is not clear and we shall not discuss it further here.

motion pertaining to channel α. It is natural and convenient to group these functions into a column

$$\boldsymbol{\eta}(\mathbf{x}) = \begin{bmatrix} \eta_1(\mathbf{x}) \\ \cdot \\ \cdot \\ \cdot \\ \eta_N(\mathbf{x}) \end{bmatrix}$$

In this case each state of the system is given by a unique column function $\boldsymbol{\eta}(\mathbf{x})$ and the Schrödinger equation (20.10) can be written in the compact matrix form:

$$\nabla^2\boldsymbol{\eta}(\mathbf{x}) - U(\mathbf{x})\boldsymbol{\eta}(\mathbf{x}) + P^2\boldsymbol{\eta}(\mathbf{x}) = 0$$

where we now use P to denote the diagonal matrix of channel momenta

$$P \equiv \begin{bmatrix} p_1 & & 0 \\ & \cdot & \\ & & \cdot \\ & & & \cdot \\ 0 & & p_N \end{bmatrix} \qquad [p_\alpha \equiv [2m(E - W_\alpha)]^{1/2}]$$

and where $U(\mathbf{x})$ is the $N \times N$ potential matrix made up of the elements $U_{\alpha\alpha'} = 2m V_{\alpha\alpha'}$. We recall (from Problem **19.4**) that the matrix $U(\mathbf{x})$ is always Hermitian. [If time-reversal invariance holds, then $U(\mathbf{x})$ is real and also symmetric, but we shall not need to assume this.]

We are assuming that our system is rotationally invariant, and further, that all of its fragments have zero spin. This being the case, the wave function $\boldsymbol{\eta}(\mathbf{x})$ can be resolved into partial waves and the radial wave functions $\boldsymbol{\psi}^l(r)$ (from which we shall omit the superscript l) satisfy the matrix radial equation

$$\left[\frac{d^2}{dr^2} - \frac{l(l+1)}{r^2} - U(r) + P^2\right]\boldsymbol{\psi}(r) = 0 \tag{20.12}$$

So far we have considered the case that $\boldsymbol{\eta}(\mathbf{x})$ represents a collision whose incident state is in channel 1. [This is reflected in the factor $\delta_{\alpha 1}$ on the incident plane wave in (20.11).] In general, the incident wave can lie in any of the channels $\alpha = 1, \ldots, N$ and there are therefore N distinct solutions $\boldsymbol{\eta}_1(\mathbf{x}), \ldots, \boldsymbol{\eta}_N(\mathbf{x})$ to be considered, where each $\boldsymbol{\eta}_\alpha(\mathbf{x})$ describes a collision beginning in channel α. This means that for each angular momentum l there are N distinct radial functions $\boldsymbol{\psi}_\alpha(r)$, each having N components $\psi_{\alpha'\alpha}(r)$ with the asymptotic form (see Problem **19.5**)

$$\psi_{\alpha'\alpha}(r) \xrightarrow[r \to \infty]{} \hat{\jmath}(p_\alpha r)\,\delta_{\alpha'\alpha} + p_\alpha f_{\alpha'\alpha}\hat{h}^+(p_{\alpha'}r)$$

$$= \frac{i}{2}\left[\hat{h}^-(p_\alpha r)\,\delta_{\alpha'\alpha} - \left(\frac{p_\alpha}{p_{\alpha'}}\right)^{1/2} s_{\alpha'\alpha}\hat{h}^+(p_{\alpha'}r)\right] \tag{20.13}$$

where

$$s_{\alpha'\alpha} = \delta_{\alpha'\alpha} + 2i(p_{\alpha'}p_\alpha)^{\frac{1}{2}}f_{\alpha'\alpha} \qquad (20.14)$$

(We suppose for the moment that $E > W_N$ so that all channels are open, all channel momenta are real, and, hence, the functions $\hat{h}^\pm(p_\alpha r)$ are all ingoing or outgoing waves.)

As a final matter of notation we shall assemble the functions $\psi_{\alpha'\alpha}(r)$ into an $N \times N$ matrix $\Psi(r)$. Each of the N columns of this matrix is a solution of the radial equation (20.12) [which means that the matrix $\Psi(r)$ itself satisfies the equation] the αth column being the solution for which the incident wave is in channel α. The asymptotic forms (20.13) can then be rewritten in matrix form as

$$\Psi(r) \xrightarrow[r\to\infty]{} \hat{j}(Pr) + \hat{h}^+(Pr)FP$$

$$= \frac{i}{2}\left\{\hat{h}^-(Pr) - \hat{h}^+(Pr)P^{-\frac{1}{2}}SP^{\frac{1}{2}}\right\} \qquad (20.15)$$

where we have introduced the diagonal matrices

$$\hat{j}(Pr) \equiv \begin{bmatrix} \hat{j}(p_1r) & & & 0 \\ & \cdot & & \\ & & \cdot & \\ & & & \cdot \\ 0 & & & \hat{j}(p_Nr) \end{bmatrix}$$

and

$$\hat{h}^\pm(Pr) \equiv \begin{bmatrix} \hat{h}^\pm(p_1r) & & & 0 \\ & \cdot & & \\ & & \cdot & \\ & & & \cdot \\ 0 & & & \hat{h}^\pm(p_Nr) \end{bmatrix}$$

and F and S are as usual the matrices made up of $f_{\alpha'\alpha}$ and $s_{\alpha'\alpha}$ with

$$S = 1 + 2iP^{\frac{1}{2}}FP^{\frac{1}{2}} \qquad (20.16)$$

The Regular Solution and Jost Matrix.   Just as in the one-channel case it is convenient to introduce a different set of N "regular solutions," which we group into a matrix $\Phi(r)$ defined to satisfy

$$\Phi(r) \xrightarrow[r\to 0]{} \hat{j}(Pr)$$

that is, the αth column of $\Phi(r)$ is a solution of the radial equation that, as $r \to 0$, behaves like $\hat{j}(p_\alpha r)$ in channel α and is smaller in all other channels.

As $r \to \infty$ any column of $\Phi(r)$ must behave like some combination of in-going and outgoing waves in all N channels. (We continue to suppose $E > W_N$ so that all channels are open.) Thus, we can write [compare (20.5)]

$$\Phi(r) \xrightarrow[r \to \infty]{} \frac{i}{2} \left\{ \hat{h}^-(Pr)\mathscr{F} - \hat{h}^+(Pr)\mathscr{F}' \right\} \qquad (20.17)$$

for certain square matrices $\mathscr{F}$ and $\mathscr{F}'$, which we call the *Jost matrices*. (We shall see shortly that $\mathscr{F}'$ can be simply expressed in terms of $\mathscr{F}$ so there is really only one Jost matrix to worry about.)

The radial equation (20.12) is a second-order N-component differential equation and, therefore, has $2N$ independent solutions, of which only N are zero at $r = 0$. It follows that any of the columns of $\Psi(r)$ (the normalized solutions) can be expressed as some combination of the N columns of the regular solution $\Phi(r)$. In fact comparison of (20.15) and (20.17) shows that the required combination is just

$$\Psi(r) = \Phi(r)\mathscr{F}^{-1}$$

where $\mathscr{F}^{-1}$ denotes the matrix inverse of $\mathscr{F}$ [compare (20.6)]. Clearly,

$$S = P^{\frac{1}{2}}\mathscr{F}'\mathscr{F}^{-1}P^{-\frac{1}{2}} \qquad (20.18)$$

Just as in the one-channel case, the radial equation plus boundary conditions is equivalent to the (matrix) integral equation

$$\Phi(r) = \hat{j}(Pr) + \int_0^r dr' G(r, r')U(r')\Phi(r') \qquad (20.19)$$

where $G(r, r')$ is the diagonal matrix made up of the Green's functions $g_{l, p_\alpha}(r, r')$ defined in (11.39). If every element of the potential matrix satisfies our usual assumptions, then it is easily checked that $\Phi(r)$ can always be found by iteration. Concerning the dependence of $\Phi(r)$ on the energy or momenta, it is convenient to note that nothing in the integral equation (20.19) (or the radial equation) requires that the channel momenta are related by energy conservation,

$$\frac{p_1^2}{2m} + W_1 = \cdots = \frac{p_N^2}{2m} + W_N = E$$

Accordingly, we can temporarily consider the variables $p_1, \ldots, p_N$ in (20.19) as *independent variables*; and it is easily seen that all elements of $\Phi(r)$ are then entire functions of all N variables $p_1, \ldots, p_N$. Now, if $f(p_1, \ldots, p_N)$ is an analytic function of $p_1, \ldots, p_N$ and if we then set $p_\alpha = [2m(E - W_\alpha)]^{\frac{1}{2}}$, then the resulting function of E,

$$f\left([2m(E - W_1)]^{\frac{1}{2}}, \ldots, [2m(E - W_N)]^{\frac{1}{2}}\right)$$

is an analytic function of N analytic functions of E, and is therefore itself an analytic function of E, except at each threshold W_α where the corresponding p_α has a branch point. It immediately follows that the regular solution $\Phi(r)$ is an analytic function of the energy E except at the N thresholds[5] $W_1, \ldots, W_N$.

If we examine the integral equation (20.19) for large r and compare with the asymptotic form (20.17) of $\Phi(r)$, we see that the Jost matrix can be written as[6]

$$\mathscr{F} = \mathscr{F}(p_1, \ldots, p_N) \equiv \mathscr{F}(P) = 1 + P^{-1} \int_0^\infty dr\, \hat{h}^+(Pr)U(r)\Phi(r) \quad (20.20)$$

with a similar expression, with $\hat{h}^+$ replaced by $\hat{h}^-$, for the second Jost matrix $\mathscr{F}'$. It would be very reasonable to expect that, like its one-channel analogue, the integral (20.20) would lead us to the desired analytic properties of $\mathscr{F}$ and, hence, S. Unfortunately, the convergence of this integral at its upper limit is generally a very troublesome problem to analyze, even when all momenta p_α are in their upper half planes.[7] It is to avoid this difficulty that we are assuming that all elements of the potential matrix $U(r)$ are *identically zero for r greater than some finite radius* a. In this case the integral in (20.20) extends only to $r = a$. It is then seen to be convergent for arbitrary momenta, and to define a Jost matrix all of whose elements are entire functions of the N variables $p_1, \ldots, p_N$.

The S Matrix. We have seen that (for finite-range potentials) the Jost matrix $\mathscr{F}(P)$ is an entire function of $p_1, \ldots, p_N$. It is a simple matter to check [using (20.20) and its counterpart, with $\hat{h}^+$ replaced by $\hat{h}^-$, for $\mathscr{F}'$] that the second Jost matrix $\mathscr{F}'$ is given by

$$\mathscr{F}'(P) = \mathscr{F}(-P)$$

that is, $\mathscr{F}'(P)$ is obtained by continuing $\mathscr{F}(P)$ in all of the momenta from p_α to $-p_\alpha$. This lets us rewrite the asymptotic form of $\Phi(r)$ as

$$\Phi(r) \xrightarrow[r \to \infty]{} \frac{i}{2} [\hat{h}^-(Pr)\mathscr{F}(P) - \hat{h}^+(Pr)\mathscr{F}(-P)] \quad (20.21)$$

---

[5] The singularities of Φ at the thresholds are trivial branch points, which could even be eliminated by a slight redefinition of Φ. However, this is of no especial interest since our real interest is in S, whose branch points are not so trivial.

[6] For the moment we think of $\mathscr{F}$ as a function of the N momenta written $\mathscr{F}(p_1, \ldots, p_N)$. The abbreviation $\mathscr{F}(P)$ is not strictly accurate, since P also denotes the *matrix* with $p_1, \ldots, p_N$ down its diagonal. In practice this useful shorthand is unlikely to cause confusion.

[7] The reason is not hard to see. Because of its matrix character the integrand contains terms like $h^+(p_\alpha r)\phi_{\alpha'\alpha''}(r)$, whose convergence as $r \to \infty$ depends on the imaginary parts of the two different momenta p_α and $p_{\alpha'}$. Clearly, the region of convergence for any given element of $\mathscr{F}$ will be complicated and will depend on which element is considered.

FIGURE 20.3.   Starting at the right of the highest threshold W_N, the physical S matrix for lower energies is obtained by continuing to the left and passing above all thresholds.

and the expression (20.18) for the S matrix as

$$S = P^{1/2}\mathscr{F}(-P)\mathscr{F}(P)^{-1}P^{-1/2} \tag{20.22}$$

We now recall that the inverse of any $N \times N$ matrix A is given by

$$A^{-1} = \frac{\operatorname{cof} A}{\det A}$$

where $\det A$ denotes the determinant of A and $\operatorname{cof} A$ is the $N \times N$ matrix made up of the cofactors of A transposed. We can therefore rewrite S as

$$S = \frac{P^{1/2}\mathscr{F}(-P)\operatorname{cof}\mathscr{F}(P)P^{-1/2}}{\det \mathscr{F}(P)} \tag{20.23}$$

Since both $\operatorname{cof} \mathscr{F}(P)$ and $\det \mathscr{F}(P)$ are obviously analytic wherever $\mathscr{F}(P)$ is, we conclude that all elements of the matrix S are analytic functions of the N variables $p_1, \dots, p_N$ except at those points where $\det \mathscr{F}(P) = 0$. Finally, if we re-impose the conditions $p_\alpha = [2m(E - W_\alpha)]^{1/2}$ and regard S as a function of E, then we find that the $N \times N$ matrix $S(E)$ is an analytic function of E for all E, except for branch points at the thresholds $W_1, \dots, W_N$, and poles at those points where $\det \mathscr{F}(P) = 0$.

The matrix $S(E)$ whose analytic properties we have established is the $N \times N$ matrix, which is the physical S matrix when E is above the highest threshold W_N. We must now prove that if we continue the matrix $S(E)$ down to energies below W_N, then its appropriate leading submatrix is the physical S matrix. This is in fact easily seen if we examine carefully the asymptotic form of Ψ

$$\Psi(r) \xrightarrow[r \to \infty]{} \frac{i}{2} [\hat{h}^-(Pr) - \hat{h}^+(Pr)\mathscr{F}(-P)\mathscr{F}(P)^{-1}] \tag{20.24}$$

When $E > W_N$, all momenta $p_\alpha = [2m(E - W_\alpha)]^{1/2}$ are positive real, the functions $\hat{h}^\pm(p_\alpha r)$ are ingoing and outgoing waves, and $\mathscr{F}(-P)\mathscr{F}(P)^{-1}$ is indeed the $N \times N$ physical S matrix (apart from trivial factors of $P^{1/2}$).

Let us now continue leftward in energy, passing above the thresholds $W_N, W_{N-1}, \dots$ to an interval $\{W_n < E < W_{n+1}\}$ as shown in Fig. 20.3. As discussed in Section 20-a, when we do this the momenta $p_{n+1}, \dots, p_N$ all become positive imaginary, while $p_1, \dots, p_n$ remain real. (If we were to pass *below* any threshold W_α, the corresponding momentum p_α would become negative imaginary and we would get a different result.) Since $p_{n+1}, \ldots, p_N$ are now positive imaginary the functions $\hat{h}^-(p_\alpha r)$ for $\alpha = n + 1, \ldots, N$ are now exponentially increasing, and the corresponding $\hat{h}^+(p_\alpha r)$ exponentially decreasing, as $r \to \infty$. If we now examine the asymptotic form (20.24) of $\Psi(r)$ we see that, while the first n columns of $\Psi(r)$ are still finite as $r \to \infty$, the last $N\text{-}n$ columns are exponentially exploding and are therefore physically unacceptable.

This situation is exactly what we should expect. In the interval

$$\{W_n < E < W_{n+1}\}$$

only the first n channels are open and there are only n physically relevant solutions, one for each permissible in channel. These are precisely the first n columns of $\Psi(r)$. If we look more closely at these n columns, we see from (20.24) that their first n elements consist of ingoing and outgoing waves, while their last $N\text{-}n$ elements fall off exponentially. Thus, it is the first n elements of the first n columns of $\Psi(r)$ that determine the actual scattered flux in the n open channels. According to (20.24) these elements have the asymptotic form[8]

$$\psi_{\alpha'\alpha}(r) \xrightarrow[r \to \infty]{} \frac{i}{2} \left\{ \hat{h}^-(p_\alpha r)\, \delta_{\alpha'\alpha} - \hat{h}^+(p_{\alpha'}r)[\mathscr{F}(-P)\mathscr{F}(P)^{-1}]_{\alpha'\alpha} \right\} \qquad [\alpha', \alpha = 1, \ldots, n]$$

Thus, the elements of the $n \times n$ physical S matrix are:

$$s_{\alpha'\alpha} = p_{\alpha'}^{\frac{1}{2}}[\mathscr{F}(-P)\mathscr{F}(P)^{-1}]_{\alpha'\alpha}p_\alpha^{-\frac{1}{2}} \qquad [\alpha', \alpha = 1, \ldots, n]$$

that is, the $n \times n$ physical S matrix is precisely the $n \times n$ leading submatrix of the $N \times N$ analytic matrix $S(E) = P^{\frac{1}{2}}\mathscr{F}(-P)\mathscr{F}(P)^{-1}P^{-\frac{1}{2}}$.

Finally, we note that it is easy to prove (see Problem **20.1**) that since U is Hermitian, the Jost matrix satisfies the important identity

$$\mathscr{F}(P^*)^\dagger P\mathscr{F}(P) = \mathscr{F}(-P^*)^\dagger P\mathscr{F}(-P) \qquad (20.25)$$

In particular, for $E < W_1$ (all p_α imaginary) this implies that

$$S(E)^\dagger = S(E) \qquad [E < W_1]$$

That is, when continued to energies below the lowest threshold, $S(E)$ is Hermitian.

---

[8] Of course all elements of $\Psi(r)$ have this form. The point is that only for $\alpha, \alpha' = 1, \ldots, n$ are the functions $\hat{h}^\pm$ the required ingoing and outgoing waves.

20-c.   Bound States

Having established that the analytic properties of the partial-wave S matrix are as described in Section 20-a, we can now discuss the connection between the bound states and the poles of $S(E)$ (still in the framework of our N-channel model). In the one-channel case we saw that the existence of a bound state of energy $\bar{E} < 0$ means that the analytically continued S matrix $s(E)$ has a pole at $E = \bar{E}$ on the physical sheet. To understand this we have only to recall the asymptotic form

$$\phi(r) \xrightarrow[r \to \infty]{} \frac{i}{2} [\hat{h}^-(pr)f(p) - \hat{h}^+(pr)f(-p)] \qquad (20.26)$$

The solution $\phi(r)$ is zero at $r = 0$, while the functions $\hat{h}^+$ and $\hat{h}^-$ are exponentially decreasing and increasing as $r \to \infty$ when p is positive imaginary (corresponding to $E < 0$ on the physical sheet). If there is a bound state at $E = \bar{E}$, then the solution that is zero at $r = 0$ (namely ϕ) must be exponentially decreasing as $r \to \infty$; that is, at the energy of the bound state the coefficient $f(p)$ of the exponentially increasing term in (20.26) has to be zero. It follows that $s = f(-p)/f(p)$ has a pole.[9]

The analysis of the multichannel case is very similar. We expect to find bound states of the whole system at energies $\bar{E}$ below the lowest threshold,[10] $\bar{E} < W_1$, and, accordingly, we continue into this region (where, of course, all elements of S are unphysical). We now consider the asymptotic form:

$$\Phi(r) \xrightarrow[r \to \infty]{} \frac{i}{2} [\hat{h}^-(Pr)\mathscr{F}(P) - \hat{h}^+(Pr)\mathscr{F}(-P)] \qquad (20.27)$$

The existence of a bound state at $\bar{E} < W_1$ means that there is a column solution to the radial equation (of the appropriate l), which is zero both at $r = 0$ and as $r \to \infty$. Since this solution is zero at $r = 0$ it can be expressed as some linear combination of the columns of $\Phi(r)$. If we form this particular combination of the columns of $\Phi(r)$ and then let $r \to \infty$ the result must therefore be zero; which [according to (20.27)] means that this combination of the columns of $\mathscr{F}(\bar{P})$ must be zero (that is, the coefficient of $\hat{h}^-$ must vanish). But this means that the columns of $\mathscr{F}(\bar{P})$ are linearly dependent and, hence, that $\det \mathscr{F}(\bar{P}) = 0$. Finally, since

$$S = \frac{P^{1/2}\mathscr{F}(-P)\operatorname{cof}\mathscr{F}(P)P^{-1/2}}{\det\mathscr{F}(P)} \qquad (20.28)$$

[9] Remember that we are considering potentials of finite range so there is no problem with continuing the numerator $f(-p)$ to the position of the bound state.
[10] We shall discuss the possibility of bound states "embedded in the continuum" above W_1 at the end of this section.

it follows that the elements of S all have poles at the bound-state energy.[11] Since the same reasoning can be reversed, we conclude that there is a one-to-one correspondence between the bound states of the system and the poles of S on the physical sheet. In particular, since all bound states have real energies, the poles on the physical sheet are confined to the real axis of E.

If the bound state of energy E and angular momentum l is nondegenerate [apart of course from the inevitable $(2l + 1)$-fold degeneracy], then there is more that can be said about the corresponding pole. The fact that the bound state is nondegenerate means that the corresponding solution of the Schrö-dinger equation is unique; in other words, that there is just one combination of the columns of $\Phi(r)$ that vanishes as $r \to \infty$. This means that there is a unique combination of the columns of $\mathscr{F}(\bar{P})$ that is zero; which is to say that $\mathscr{F}(\bar{P})$ is a matrix of rank $N - 1$, or that it has exactly $N - 1$ independent columns.

With the knowledge that $\mathscr{F}(\bar{P})$ has rank $N - 1$ we can now examine the behavior of S as given by (20.28). It is easily seen that if $\mathscr{F}(\bar{P})$ has rank $N - 1$, then all the rows of cof $\mathscr{F}(\bar{P})$ are multiples of a single row [that is, that cof $\mathscr{F}(\bar{P})$ has rank 1] and the same is therefore true of the whole numerator[12] of S. Much as in the one channel case (see Problem **12.3**) it can be shown that, at a nondegenerate bound state, the zero of det $\mathscr{F}(P)$ is always simple. This means that the pole of S is a simple pole, and we can therefore write

$$S(E) = \frac{A}{E - \bar{E}} \qquad [E \text{ near } \bar{E}]$$

where A is a matrix of rank 1. Finally, we recall that $S(E)$ is Hermitian for $E < W_1$; so the same must be true of the residue matrix A. Thus, we conclude that at the energy of a nondegenerate bound state, $S(E)$ has a simple pole whose residue matrix A is Hermitian and of rank 1.

The general form of the matrix A is easily found. That A has rank 1 means that all of its columns are multiples of one single column. This means that its elements can be written in the form

$$a_{\alpha'\alpha} = \gamma_{\alpha'}\delta_\alpha$$

(The numbers $\gamma_{\alpha'}$, δ_α are obviously not unique.) The fact that A is Hermitian implies that

$$\gamma_{\alpha'}\delta_\alpha = \gamma_\alpha^*\delta_{\alpha'}^*$$

---

[11] It can happen that some (but never all) of the elements of the numerator of S happen to be zero at the bound-state energy, and in this case the corresponding elements of S do not have the pole.

[12] For any two matrices A and B the rank of AB is always less than or equal to that of A or B. In the present case cof $\mathscr{F}$ has rank 1, so the whole numerator must have rank 1 or rank 0. In the latter case, it would be identically zero, which (it is fairly easy to show) it cannot be.

from which it is clear (see Problem **20.2**) that we can adjust γ_α and δ_α so that

$$\delta_\alpha = \pm \gamma_\alpha^*$$

Therefore, our final conclusion is that in the neighborhood of a nondegenerate bound state the analytically continued S matrix has a simple pole of the form:

$$s_{\alpha'\alpha}(E) = \pm \frac{\gamma_{\alpha'}\gamma_\alpha^*}{E - \bar{E}} \qquad (20.29)$$

The result that the residue of $s_{\alpha'\alpha}(E)$ has the form $\gamma_{\alpha'}\,\gamma_\alpha^*$ is sometimes referred to as *factorization of the residue*. It should be emphasized that it depends on the assumption that there is only one state of the energy and angular momentum in question. If there are several such states, then the radial Schrödinger equation has several independent solutions that vanish as $r \to 0$ and ∞. This means that there are several distinct combinations of the columns of $\mathscr{F}(\bar{P})$ that are zero, and, hence, that $\mathscr{F}(\bar{P})$ has rank less than $N - 1$. Therefore the reasoning leading to (20.29) breaks down. In fact, it can be shown that if there are r independent bound states of energy $\bar{E}$ and angular momentum l, then $S(E)$ still has a simple pole but that its residue is an Hermitian matrix of rank r. Since such a matrix can be written as a sum of r Hermitian matrices of rank 1 it follows that in this case $S(E)$ can be written as a sum of r terms each of the form (20.29).

Finally, we note that we have so far discussed bound states with energy below the lowest threshold W_1. In multichannel problems it is in fact possible (even with a perfectly reasonable potential) to have bound states "embedded in the continuum" with energy above W_1. When this happens the S matrix has a pole of the same form (20.29) at the appropriate energy, $W_n < \bar{E} < W_{n+1}$, say. However, it is clear that the physical, open-channel S matrix, being unitary, cannot have a pole. In fact, it is easily shown that the factors $\gamma_1, \ldots, \gamma_n$ are all zero and that the residue matrix is zero in all open channels. In practice it is most unlikely that states of this type will be exactly bound. However, it is entirely possible that there occur *very nearly* bound states of this type. In this case there is a pole close to the interval $\{W_n < E < W_{n+1}\}$ on an unphysical sheet and the scattering has a resonance, as we shall discuss in the next section.

20-d. Resonances

Having established that poles of the S matrix on the physical sheet correspond to bound states, we can now go on to show (as in the one-channel case) that poles on the unphysical sheets can correspond to resonances. We saw in Section 13-a that in the one-channel case $s(E)$ may have a pole below the real axis at $E = E_R - i\Gamma/2$ on the unphysical sheet. If this pole is close enough to the axis and if there are no other nearby singularities, then, bearing in mind that $s(E)$ is unitary, we could write for the most general form of $s(E)$ near E_R

$$s(E) = e^{2i\delta_{bg}} \frac{E - E_R - i\Gamma/2}{E - E_R + i\Gamma/2}$$

where the background phase δ_{bg} is some real number. In the simple case that $\delta_{bg} = 0$ the pole is called a pure Breit–Wigner pole and we find for the corresponding amplitude

$$f(E) = \frac{s-1}{2ip} = -\frac{1}{2p}\frac{\Gamma}{E - E_R + i\Gamma/2} \qquad \text{[Breit–Wigner]} \quad (20.30)$$

Clearly, at a pure Breit–Wigner resonance $|f(E)|$ peaks to its maximum $1/p$ at E_R, and is small where E is far from E_R. In general, when $\delta_{bg} \neq 0$ we can write,

$$f(E) = \frac{1}{2ip}[(s - e^{2i\delta_{bg}}) + (e^{2i\delta_{bg}} - 1)]$$

$$= -\frac{e^{2i\delta_{bg}}}{2p}\frac{\Gamma}{E - E_R + i\Gamma/2} + \frac{e^{2i\delta_{bg}} - 1}{2ip}$$

$$\equiv f_{\text{res}}(E) + f_{bg} \qquad (20.31)$$

where f_{res} is the same as the Breit–Wigner amplitude (20.30) apart from the phase factor $\exp(2i\delta_{bg})$, while f_{bg} is the constant background, corresponding to the phase shift δ_{bg}.

We saw in Sections 13-c and 13-d that the behavior of the scattering near to a resonance depends on the relative widths of the incident wave packet (ΔE) and of the resonance itself (Γ). If $\Delta E \ll \Gamma$, cross sections proportional to $|f|^2$ can be measured in the usual way; if the resonance is pure Breit–Wigner, then the cross section shows a pure bump, while if $\delta_{bg} \neq 0$ it may bump, dip, or both, as discussed in Section 13-a. When the resonance is very narrow, $\Gamma \ll \Delta E$, one sees a very different phenomenon associated with temporary capture into a nearly stable state that subsequently decays with mean life $1/\Gamma$.

We now return to the multichannel S matrix and suppose that it has a pole on the unphysical sheet reached by continuing downwards between W_n and W_{n+1} as shown in Fig. 20.4. As in the one-channel case, there is no general principle that guarantees that all such poles must be simple; however, in practice such poles usually come about as would-be bound-state poles. Since bound-state poles always *are* simple, this type of pole would normally be expected to be simple. In any case we shall consider the case of a

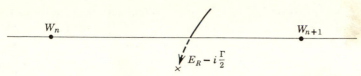

FIGURE 20.4. A resonance pole on the unphysical sheet reached between the thresholds W_n and W_{n+1}.

simple pole. By the same token we shall suppose that the pole is "nondegenerate"; that is, its residue matrix has rank 1, as it would do if it were a nondegenerate bound state. Finally, we suppose that the pole is isolated, in the sense that it is well separated from any other poles and from the nearest thresholds.

We assume first a behavior of the amplitude analogous to the Breit–Wigner form (20.30)[13]

$$f_{\alpha'\alpha} = -\frac{1}{2(p_{\alpha'}p_\alpha)^{1/2}} \frac{a_{\alpha'\alpha}}{E - E_R + i\Gamma/2} \qquad [E \text{ near } E_R] \qquad (20.32)$$

By analogy with the single-channel case we can expect that the form of the residue $a_{\alpha'\alpha}$ is constrained by the requirement that S be unitary, and this is the case: If $f_{\alpha'\alpha}$ is given by (20.32), then

$$S = 1 - \frac{iA}{E - E_R + i\Gamma/2} \qquad (20.33)$$

where we shall now use S *to denote the* $n \times n$ *open-channel* S *matrix.* Now, for E real (in the interval concerned) S satisfies $S^\dagger S = 1$. Substituting the form (20.33) we find, after a little algebra, that

$$i(A^\dagger - A)(E - E_R) + A^\dagger A - \frac{(A + A^\dagger)\Gamma}{2} = 0$$

The unitarity of S requires that this identity hold for all real E (in the neighborhood of E_R). This implies first that $(A^\dagger - A) = 0$, and that A is Hermitian; and then that

$$A^2 = \Gamma A \qquad (20.34)$$

Since A is Hermitian and is assumed to have rank 1 we know from the arguments of the last section that its elements have the form

$$a_{\alpha'\alpha} = \pm\gamma_{\alpha'}\gamma_\alpha^*$$

[13] The factors of p_α and $p_{\alpha'}$ are put in only for convenience; they could, of course, be absorbed in $a_{\alpha'\alpha}$, but it is more convenient to keep them out.

Substituting this form into (20.34) we see first that only the positive sign is possible (Γ cannot be negative or the pole would be on the physical sheet) and second that

$$\sum_{\alpha} |\gamma_\alpha|^2 = \Gamma \tag{20.35}$$

(where the sum is over open channels only since we are discussing only the open-channel S matrix). Thus, the most general form of a simple, non-degenerate Breit–Wigner resonance pole consistent with unitarity is

$$f_{\alpha'\alpha}(E) = -\frac{1}{2(p_{\alpha'}p_\alpha)^{1/2}} \frac{\gamma_{\alpha'}\gamma_\alpha^*}{E - E_R + i\Gamma/2} \tag{20.36}$$

where the γ_α satisfy (20.35). For reasons which will emerge shortly, the number

$$\Gamma_\alpha = |\gamma_\alpha|^2$$

is called the *partial width* of the resonance for channel α. Thus, the requirement (20.35) can be rephrased by saying that the sum of the partial widths is equal to the total width, $\sum \Gamma_\alpha = \Gamma$.

As we shall discuss directly, the Breit–Wigner (20.36) is not the most general possible form of a resonance. However, as in the single-channel case (and for essentially the same reasons), resonances that are reasonably close to the lowest threshold *are* always expected to be more or less pure Breit–Wigner. This is well borne out by experiment, the best known examples being the famous resonances of low-energy neutrons scattering off nuclei. An important feature of the Breit–Wigner (20.36) is that it predicts similar bumps in the cross sections leading from *every* in channel α to *all* out channels α' (unless any of the γ_α happen to be zero). While the various bumps will have different heights (determined by the product of the relevant partial widths $\Gamma_{\alpha'}\Gamma_\alpha$), all bumps should occur at the same energy E_R and have the same width Γ. This striking property can be clearly seen in Fig. 20.5, which shows a resonance in the scattering of slow neutrons off ^{123}Te; the Breit–Wigner bump is observed at the same energy (2.3 eV) and with the same width (0.11 eV) in both of the processes

$$n + {}^{123}\text{Te} \rightarrow \begin{cases} n + {}^{123}\text{Te} \\ \gamma + {}^{124}\text{Te} \end{cases}$$

Of course, the Breit–Wigner of (20.33) and (20.36) is not the most general form consistent with unitarity and with the assumption that S is dominated by an isolated, simple, "nondegenerate" pole. The general form consistent with the latter assumption is clearly

$$\mathsf{S} = B - \frac{iC}{E - E_R + i\Gamma/2}$$

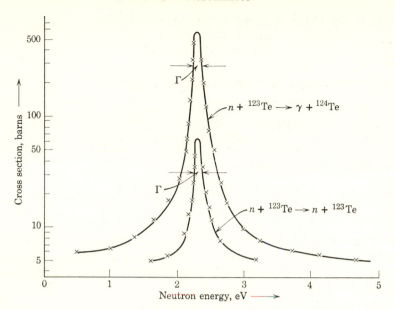

FIGURE 20.5.   An isolated Breit–Wigner resonance in the scattering of slow neutrons off ^{123}Te.   (Data from Hughes and Harvey, 1955.)

In this case we can explore the consequences of unitarity in two steps: First, as E moves away from E_R, we note that S approaches B; thus since S is unitary for all E it follows that B must be. Therefore, it is natural to relabel B as S^{bg}, and write

$$S = S^{bg}\left(1 - \frac{iA}{E - E_R + i\Gamma/2}\right)$$

Since both S and S^{bg} are unitary, so is the factor in parenthesis. Exactly our previous argument then shows that the elements of A must have the form:

$$a_{\alpha'\alpha} = \delta_{\alpha'}\delta_{\alpha}^* \qquad [\text{with } \sum |\delta_\alpha|^2 = \Gamma]$$

The residue of S at the pole is given by the matrix $S^{bg}A$ whose elements can therefore be written

$$\gamma_{\alpha'}\delta_\alpha^*$$

where the "vector" $\{\gamma_\alpha\}$ is just $S^{bg}\{\delta_\alpha\}$. Hence, we arrive at the final form for the general simple "nondegenerate" resonance pole

$$s_{\alpha'\alpha}(E) = s_{\alpha'\alpha}^{bg} - i\,\frac{\gamma_{\alpha'}\delta_\alpha^*}{E - E_R + i\Gamma/2}$$

In terms of the amplitude this gives [compare (20.31)]

$$\boxed{F(E) = F^{bg} + F^{res}(E)}$$

(20.37)

where F^{bg} is the amplitude corresponding to S^{bg} and

$$\boxed{f^{res}_{\alpha'\alpha}(E) = -\frac{1}{2(p_{\alpha'}p_\alpha)^{1/2}} \frac{\gamma_{\alpha'}\delta^*_\alpha}{E - E_R + i\Gamma/2}}$$

(20.38)

The numbers γ_α and δ_α satisfy

$$\{\gamma_\alpha\} = S^{bg}\{\delta_\alpha\}$$

and

$$\sum |\gamma_\alpha|^2 = \sum |\delta_\alpha|^2 = \Gamma$$

The quantity $|\delta_\alpha|^2$ is called the partial width for *entering* the resonance from channel α, while $|\gamma_{\alpha'}|^2$ is the partial width for *leaving* the resonance and going into channel α'. If the system is time-reversal invariant, then the matrix $S(E)$, and, therefore, also its residue, is symmetric. In this case we can adjust the phases of γ_α and δ_α such that $\gamma_\alpha = \delta^*_\alpha$ and

$$f^{res}_{\alpha'\alpha}(E) = -\frac{1}{2(p_{\alpha'}p_\alpha)^{1/2}} \frac{\gamma_{\alpha'}\gamma_\alpha}{E - E_R + i\Gamma/2} \qquad \text{[T invariance]}$$

In particular, if the resonance is a pure Breit–Wigner and time reversal is good, then the residue is both symmetric and Hermitian, and hence also real. In this case, γ_α can also be chosen real.

If the resonance is wide compared to the width of the incident packet ($\Delta E \ll \Gamma$), then just as in the one-channel case we can compute and observe cross sections in the ordinary way. In particular, the contribution of any partial wave to a process ($\alpha' \leftarrow \alpha$) has the form

$$\sigma^l_{\alpha'\alpha}(E) = 4\pi(2l + 1)\,|f^l_{\alpha'\alpha}(E)|^2$$

To see how this behaves as a function of energy near a resonance one has simply to analyze the behavior of the various possibilities implied by (20.37) and (20.38). A few general statements can be made: When E is well below or above E_R the amplitude is equal to the constant F^{bg}. As E traverses an interval of the order of Γ about E_R, each element $f^{res}_{\alpha'\alpha}(E)$ describes a circle counterclockwise in the complex plane. Since each amplitude must lie in the appropriate unitary circle, it follows that as E passes E_R each amplitude starts at some fixed point (inside the unitary circle), rapidly describes a complete circle entirely within the unitary circle, and returns to its starting point. Four possible loci for an elastic amplitude $f^l_{\alpha\alpha}$ (times p_α) are shown in Fig. 20.6. Below each locus is shown the behavior of the corresponding elastic

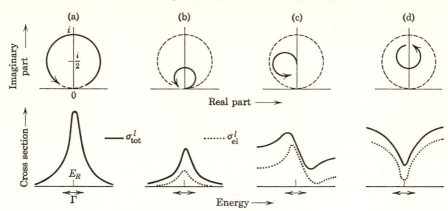

FIGURE 20.6. **Four possible resonances all corresponding to an isolated pole at** $E_R - i\Gamma/2$. **Top line shows behavior of elastic amplitudes; bottom line the corresponding cross sections.**

and total cross sections (σ_{el}^l and σ_{tot}^l) for the incident channel α. [The total cross section can be computed using the optical theorem, which for given l reads $\sigma_{tot}^l = 4\pi(2l + 1)\mathrm{Im}\,f_{el}/p$. Notice that the inelastic cross section can be read off as $\sigma_{tot} - \sigma_{el}$.] Diagram (a) shows a pure Breit–Wigner, which is completely elastic in channel α ($\Gamma_\alpha = \Gamma$; all other partial widths zero). Diagram (b) shows a Breit–Wigner with plenty of inelasticity (Γ_α a little less than half of Γ). (An extreme example of this type of resonance is the neutron resonance of Fig. 20.5 where the inelastic peak is ten times the elastic.) Diagrams (c) and (d) show resonances with substantial inelasticity and background.

It should be remembered that the form (20.37) of a resonance is an idealization and the precise curves shown are seldom seen in practice. A convenient way to express this is to say that the background amplitude F^{bg} is not really a constant but varies (slowly) with E. This has the effect of smudging out the edges of the resonance curves.

20-e. Decay of a Multichannel Resonance

Finally, we discuss the effects of a resonance that is narrow compared to the incident wave packet ($\Gamma \ll \Delta E$). Just as in the one-channel case we need to examine the behavior of the actual time-dependent wave packet describing the collision. The analysis so closely parallels that of the one-channel case (Section 13-d) that we shall give only the briefest sketch here.

The time-dependent state of our N-channel system is given by an N-component wave function

$$\Psi(\mathbf{x}, t) = \begin{bmatrix} \psi_1(\mathbf{x}, t) \\ \cdot \\ \cdot \\ \cdot \\ \psi_N(\mathbf{x}, t) \end{bmatrix}$$

where each $\psi_\alpha(\mathbf{x}, t)$ describes that part of the motion in channel α. This time-dependent wave packet can be expanded in terms of the stationary scattering states, and the behavior of the former for large r deduced from the known behavior of the latter. This leads us to an expression of the form

$$\Psi(\mathbf{x}, t) = \Psi^{\text{in}}(\mathbf{x}, t) + \Psi^{\text{sc}}(\mathbf{x}, t)$$

where the column Ψ^{in} consists at all times of a freely moving packet in the incident channel (α say) and is zero in all other channels; the column Ψ^{sc} is zero until the collision occurs and then contains the familiar outgoing waves in all open channels.

We now suppose that the incident energy coincides with the energy E_R of a sharp isolated resonance of angular momentum l. For simplicity we suppose that the resonance is pure Breit–Wigner (no background),

$$f^l_{\alpha'\alpha} = -\frac{1}{2(p_{\alpha'}p_\alpha)^{1/2}}\frac{\gamma_{\alpha'}\gamma_\alpha^*}{E - E_R + i\Gamma/2}$$

and that all other partial-wave amplitudes are negligible. (It is a straightforward matter to include a background and other partial waves; but, just as in the one-channel case, they do not affect our main conclusion.) With these assumptions it is simple to write down $\Psi^{\text{sc}}(\mathbf{x}, t)$. In particular, for large r the α'th component of Ψ^{sc} turns out to satisfy [Compare (13.14) for the one-channel case]

$$|\psi^{\text{sc}}_{\alpha'}(\mathbf{x}, t)|^2 \underset{r\text{ large}}{=} 2\pi m\Gamma_{\alpha'}\Gamma_\alpha |Y_l^0(\hat{\mathbf{x}})|^2 |\phi_l(E_R)|^2$$

$$\times \frac{e^{-\Gamma(t-r/v_{\alpha'})}}{p_{\alpha'}r^2}\,\theta\left(t - \frac{r}{v_{\alpha'}}\right) \quad (20.39)$$

where $p_{\alpha'}$ is the momentum in channel α' at the resonant energy E_R and $v_{\alpha'} = p_{\alpha'}/m$ is the corresponding velocity; $\phi_l(E)$ is the wave function of the incident packet in the angular momentum representation and $\Gamma_\alpha = |\gamma_\alpha|^2$ as before.

We see from (20.39) that the scattered wave in channel α' has a spherical front that moves outwards with speed $v_{\alpha'}$ and has the angular dependence appropriate to the angular momentum l. The intensity is proportional to

$\Gamma_{\alpha'}\Gamma_{\alpha}$ (α being the initial channel) and, once the wave has arrived at any given point, the intensity falls off like $e^{-\Gamma t}$.

To better understand the role of the partial widths it is useful to compute the total probability of seeing the particle emerge in a given direction in channel α'. This is obtained by integrating (20.39) over all r. Because of the term $\exp(\Gamma r/v_{\alpha'})$ the integration gives a factor $v_{\alpha'}/\Gamma$ and the $v_{\alpha'}$ then cancels the $p_{\alpha'}$ in the denominator to give

$$w(d\Omega', \alpha' \leftarrow \phi, \alpha) = \{2\pi\Gamma_{\alpha}\,|\phi_l(E_R)|^2\}\left\{\frac{\Gamma_{\alpha'}}{\Gamma}\,|Y_l^0(\hat{\mathbf{p}}')|^2\,d\Omega'\right\} \qquad (20.40)$$

The interpretation of the process described by (20.39) and (20.40) is now clear. Equation (20.39) describes the exponential decay of a nearly stable state formed at $t = 0$, the time of the original collision. The probability (20.40) is the product of two factors, one depending only on the initial packet ϕ and channel α, the other depending only on the final direction and channel α'. It follows that the first factor is the probability for capture of the projectile into the metastable state, and the second that of its subsequent decay,

$$\begin{aligned} w_{\text{cap}}(\phi, \alpha) &= 2\pi\Gamma_{\alpha}\,|\phi_l(E_R)|^2 \\ w_{\text{decay}}(d\Omega', \alpha') &= \frac{\Gamma_{\alpha'}}{\Gamma}\,|Y_l^0(\hat{\mathbf{p}}')|^2\,d\Omega' \end{aligned} \qquad (20.41)$$

(The normalization of these is fixed by the requirement that once the state is formed, the total probability of its decay must be one. Since $\sum \Gamma_{\alpha'}/\Gamma = 1$ and $\int d\Omega\,|Y|^2 = 1$ our normalization of w_{decay}—and, hence, w_{cap}—is obviously correct.) In the literature of nuclear physics, where the nearly stable resonant state is known as a *compound nucleus*, the independence of the two probabilities (20.41) is often interpreted by saying that the compound nucleus "forgets" its mode of formation before it decays.

The two probabilities (20.41) clearly show the two roles of the partial widths. The width Γ_{α} enters the capture probability in the product

$$\Gamma_{\alpha}\,|\phi_l(E_R)|^2$$

Since $|\phi_l(E_R)|^2$ is the probability per unit energy that the incident packet has the right energy and angular momentum, it is natural to view Γ_{α} as the *energy width* of the opening from channel α into the resonant state. The product $\Gamma_{\alpha}\,|\phi_l(E_R)|^2$ is then the probability that the incident packet has its energy in the appropriate interval for acceptance.

The role of the partial width $\Gamma_{\alpha'}$ in the decay can be viewed as follows: The exponential decay factor $e^{-\Gamma t}$ means that the total rate of decay of the state is just Γ. Clearly from (20.41) $\Gamma_{\alpha'}$ is just the rate of decay into the particular channel α. Therefore, the result that $\sum \Gamma_{\alpha'} = \Gamma$ simply expresses the obvious fact that the total decay rate is the sum of the rates into the various possible channels.[14]

To conclude we remark that there are certainly more general types of resonant behavior, and other theoretical approaches to the same phenomenon. Concerning the first point, we have discussed an amplitude that is dominated in some energy range by an isolated, simple pole with residue matrix of rank 1 (a "nondegenerate" resonance) in a single angular momentum state l. Much of the discussion was further limited to the case that the pole is pure Breit–Wigner (no background) and that all other partial waves are negligible. It is certainly possible to analyze more general situations: As we have already mentioned, it is easy to add in the effects of a background and of other partial waves. A "degenerate" pole whose residue has rank $r > 1$ can be treated as the sum of r nondegenerate poles. A multiple pole can be treated along similar lines, as too can several nearby poles. (If there are many poles some statistical method of summing may be necessary.) However, it should also be mentioned that it is perfectly possible that the S matrix could show some rapid resonance-like behavior that is too complicated to be analyzed in terms of dominance by some number of poles in one partial-wave amplitude. In particular, there can be effects that involve several different angular momenta at the same time.

Most alternative approaches to the resonance phenomenon (the Wigner–Weisskopf model, "formal" resonance theory, etc.)[15] start from the idea that a resonance is an almost bound state. The simplest such is the Wigner–Weisskopf model, a model designed specifically to let one see the effects of a nearly bound state embedded in the continuum. The model assumes a Hamiltonian H^0 with a continuum of eigenstates plus a single discrete state whose energy lies in the continuum; and a perturbation V, which connects the discrete state with the continuum. A relatively straightforward calculation shows that if the system occupies the unperturbed discrete state at $t = 0$, then the perturbation causes it to leak out exponentially as $t \to \infty$; and further that the S matrix has a simple pole at the expected location. It should be clear that such models can throw considerable light on the dynamical origin of certain resonances. It should also be clear that they are

---

[14] If the resonance is not pure Breit–Wigner and time-reversal invariance does not hold, then the partial widths $|\delta_\alpha|^2$ and $|\gamma_\alpha|^2$ for entering and leaving the resonance may be different. Of the two interpretations just given the first then applies to $|\delta_\alpha|^2$ and the second to $|\gamma_\alpha|^2$.

[15] A clear discussion of the Wigner–Weisskopf model can be found in the recent paper of Dothan and Horn (1970). For "formal" resonance theory see Newton (1966) Section 16.5. Various other approaches can be found in almost any text on nuclear physics—for example, McCarthy (1968) Chapters 8 and 9.

less general than the approach described above, which depends only on the assumed analyticity of the S matrix in neighborhood of the real axis and is quite independent of any specific underlying dynamics.

PROBLEMS

20.1. Use the fact that the potential matrix $U(r)$ is Hermitian to prove that the analytically continued S matrix of Section 20-b is Hermitian below the lowest threshold. (Use the radial equations for $\Phi_P(r)$ and $\Phi_{P*}(r)^\dagger$ to prove that $W[\Phi_{P*}^\dagger, \Phi_P] = 0$ where the Wronskian of two matrices is

$$W[A, B] = A\left(\frac{dB}{dr}\right) - \left(\frac{dA}{dr}\right)B$$

Using the known asymptotic form of Φ deduce the identity (20.25) for the Jost matrices, and hence prove that $S^\dagger = S$ when $E < W_1$, or P is pure imaginary.)

20.2. (a) Show that the elements of a matrix A of rank one can be written as $a_{ij} = \gamma_i \delta_j$.
(b) If A is Hermitian, show that γ_i, δ_j can be chosen so that $\delta_i = \pm\gamma_i^*$.
(c) If A also satisfies $A^2 = \Gamma A$ where $\Gamma > 0$, show that only the plus sign in (b) can occur.
(d) If A is also symmetric, show that we can choose $\delta_i = \gamma_i$.

21

Two More Topics in Multichannel Scattering

21-a   **The Distorted-Wave Born Approximation**

21-b   **Final-State Interactions**

In this chapter we introduce two more facets of multichannel scattering theory: The distorted-wave Born approximation and the theory of final-state interactions. The selection of these two topics is rather arbitrary; and a glance through the literature will suggest numerous other subjects which could almost equally well have been included.[1] As regards the exclusion of so many topics, it can only be said that every book must stop somewhere. As regards the choice of subjects included, it can fairly be said that the two techniques discussed here are not only important as calculational tools, but have had an unusually great impact on the way in which physicists think about multichannel scattering problems.

21-a. The Distorted-Wave Born Approximation

In Section 14-c we discussed the distorted-wave Born approximation for single-channel scattering. We found that if the potential V could be split as

$$V = V_{\mathrm{I}} + V_{\mathrm{II}}$$

with V_{II} small, then the T matrix for scattering by V could be approximated as

$$t(\mathbf{p}' \leftarrow \mathbf{p}) \approx t_{\mathrm{I}}(\mathbf{p}' \leftarrow \mathbf{p}) + \langle \mathbf{p}' -_{\mathrm{I}}| \, V_{\mathrm{II}} \, |\mathbf{p}+_{\mathrm{I}}\rangle$$

---

[1] For example, the tables of contents of Mott and Massey (1965), Goldberger and Watson (1964), and Newton (1966) list many topics all eminently worthy of discussion.

Here $t_I(\mathbf{p}' \leftarrow \mathbf{p})$ is the exact T matrix for scattering by V_I alone. The second term is the matrix element of V_{II} between "distorted waves"—namely, the scattering states appropriate to V_I—and is interpreted as the (first order) amplitude for the scattering by V_{II} in the presence of V_I.

Almost all applications of the DWBA are in multichannel problems, where we must distinguish two distinct types of process: Collisions that do not involve rearrangement, and those that do.

Collisions Without Rearrangement.   If the initial and final channels, α and α', of interest have the same arrangements (that is, we are interested in elastic scattering or excitation only) then the channel potentials V^α and $V^{\alpha'}$ are the same. In this case we can derive the multichannel DWBA exactly as in the one-channel case. We suppose that V^α can be split as

$$V^\alpha = V^\alpha_I + V^\alpha_{II}$$

where the scattering for V^α_I alone is known exactly (or in some good approximation) while the effects of V^α_{II} are small. In this case exactly the arguments of Section 14-c lead us to the DWBA (for collisions *without* rearrangement)

$$t(\underline{p}', \alpha' \leftarrow \underline{p}, \alpha) \approx t_I(\underline{p}', \alpha' \leftarrow \underline{p}, \alpha) + \langle \underline{p}', \alpha' -_I | V^\alpha_{II} | \underline{p}, \alpha +_I \rangle \quad (21.1)$$

with all symbols defined as before.

An interesting feature of the multichannel DWBA (21.1) is that it is often possible and convenient to split the potential V^α in such a way that the first term t_I is exactly zero. For example, let us consider an excitation of the form:

$$a + (bc) \rightarrow a + (bc)^*$$

where c is an infinitely heavy fixed core. We could take for V^α_I the interaction of a with the core c

$$V^\alpha_I = V_{ac}$$

and V^α_{II} would then be the interaction V_{ab} of a with b. Now, if a interacts only with the core c then it cannot excite the bc target, and it follows that, with our choice of V^α_I, the matrix element t_I is exactly zero (see Problem **21.1**). The result (21.1) then reduces to

$$t(\underline{p}', \alpha' \leftarrow \underline{p}, \alpha) \approx \langle \underline{p}', \alpha' -_I | V^\alpha_{II} | \underline{p}, \alpha +_I \rangle \quad (21.2)$$

Another possibility would be to take for V^α_I just the Coulomb interaction of a and c. In this case the result (21.2) still holds and the distorted waves are just Coulomb scattering functions, which are exactly known. In this case the use of the DWBA is especially straightforward.[2]

Rearrangement Collisions. For the case of rearrangement collisions the derivation of the DWBA is a little more complicated. Since the initial and final channels α and α' have different arrangements, the channel potentials V^α and $V^{\alpha'}$ are different. To get a DWBA we must decide first on useful splittings of both V^α and $V^{\alpha'}$ as

and

$$V^\alpha = V_I^\alpha + V_{II}^\alpha$$

$$V^{\alpha'} = V_I^{\alpha'} + V_{II}^{\alpha'}$$

We assume (as is usually the case) that these splittings can be made so that the potentials I alone can not cause any rearrangement; in this case we proceed as follows.

The exact amplitude of interest has the form

$$t(\underline{p}', \alpha' \leftarrow \underline{p}, \alpha) = \langle \underline{p}', \alpha' | V^{\alpha'} | \underline{p}, \alpha+ \rangle$$

As in the one-channel case we rewrite the bra $\langle \underline{p}', \alpha' |$ in terms of $\langle \underline{p}', \alpha' -_I |$ to give

$$t(\underline{p}', \alpha' \leftarrow \underline{p}, \alpha) = \langle \underline{p}', \alpha' -_I | V^{\alpha'} | \underline{p}, \alpha+ \rangle$$
$$- \langle \underline{p}', \alpha' -_I | V_I^{\alpha'} G^{\alpha'}(E + i0) V^{\alpha'} | \underline{p}, \alpha+ \rangle \quad (21.3)$$

Now, the reader will recall that in Section 19-b we saw that the state $|\underline{p}, \alpha+\rangle$ can be written in terms of the channel α' Green's operator as

$$|\underline{p}, \alpha+\rangle = G^{\alpha'} V^{\alpha'} |\underline{p}, \alpha+\rangle + |o\rangle$$

where $|o\rangle$ is orthogonal to all vectors in the channel α' subspace. Inserting this into the last term of (21.3) we find

$$t(\underline{p}', \alpha' \leftarrow \underline{p}, \alpha) = \langle \underline{p}', \alpha' -_I | V^{\alpha'} | \underline{p}, \alpha+ \rangle - \langle \underline{p}', \alpha' -_I | V_I^{\alpha'} | \underline{p}, \alpha+ \rangle$$
$$- \langle \underline{p}', \alpha' -_I | V_I^{\alpha'} | o \rangle$$

Since the potential $V_I^{\alpha'}$ was chosen to cause no rearrangements, the last term is zero; combining the first two terms we arrive at the exact result

$$t(\underline{p}', \alpha' \leftarrow \underline{p}, \alpha) = \langle \underline{p}', \alpha' -_I | V_{II}^{\alpha'} | \underline{p}, \alpha+ \rangle \quad (21.4)$$

[2] It is essential to the present simple discussion that the core c is infinitely heavy. If this is not the case, then the potential V_{ac} can cause excitation and t_I is nonzero. However, it is still possible to get a result similar to (21.2) by choosing for V_I^α some average interaction between a and the CM of b and c.

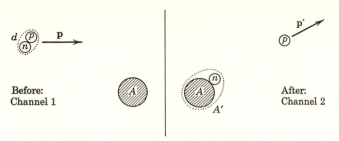

FIGURE 21.1.   The deuteron stripping reaction.

Finally, we approximate the full state $|\underline{p}, \alpha+\rangle$ by $|\underline{p}, \alpha+_{\mathrm{I}}\rangle$ and arrive at the DWBA for rearrangement collisions

$$t(\underline{p}', \alpha' \leftarrow \underline{p}, \alpha) \approx \langle \underline{p}', \alpha'-_{\mathrm{I}}| V_{\mathrm{II}}^{\alpha'} |\underline{p}, \alpha+_{\mathrm{I}}\rangle \qquad (21.5)$$

This is the so-called "post" form of the DWBA since it involves explicitly the potential of the final channel α'. There is, of course, a corresponding "prior" version involving the potential of the initial channel α.

To illustrate the use of the DWBA in rearrangement collisions, we consider an example of considerable historical importance in nuclear physics, the *deuteron stripping* reaction

$$d + A \rightarrow p + A'$$

[usually written $A(d, p)A'$ in nuclear physics] where d denotes a deuteron, A any nucleus with A nucleons, and A' the nucleus obtained by adding a neutron to A. This reaction is shown schematically in Fig. 21.1.

We treat the nucleus A as an infinitely heavy structureless particle, in which case the potential of the initial channel 1 is just

$$V^1 = V_{pA} + V_{nA}$$

while that of the final channel 2 is

$$V^2 = V_{pA} + V_{np}$$

To set up our DWBA we split V^2 so that V_{pA} is the potential I (to be treated "exactly") and V_{np} is potential II (to be treated in first order). With this choice the potential I cannot cause transitions out of channel 2, as required. The potential V^1 we split so that the potential I is some suitable average interaction between A and the deuteron's CM and potential II makes up the difference. (The potential I can be some suitable optical potential to describe absorption effects.) With these choices, the distorted waves that go into the DWBA have an especially simple form. The final wave is just the product of the bound-state wave function $\phi_{A'}(\mathbf{x}_n)$ of the neutron in the nucleus A' with the scattering wave function of the proton in the field of the nucleus A (which we call χ_2^-)

$$\phi_{A'}(\mathbf{x}_n)\,\chi_2^-(\mathbf{x}_p)$$

The initial distorted wave is the product of the deuteron wave function and the scattering state of the deuteron's CM in the field of the nucleus A (denoted χ_1^+)

$$\phi_d(\mathbf{x}_n - \mathbf{x}_p)\,\chi_1^+\left(\frac{\mathbf{x}_n + \mathbf{x}_p}{2}\right)$$

Thus the DWBA (21.5) gives for the required amplitude

$$t(\mathbf{p}', 2 \leftarrow \mathbf{p}, 1) = \int d^3x_n \int d^3x_p \,\phi_{A'}(\mathbf{x}_n)^*\chi_2^-(\mathbf{x}_p)^*$$

$$\times\ V_{np}(\mathbf{x}_n - \mathbf{x}_p)\phi_d(\mathbf{x}_n - \mathbf{x}_p)\chi_1^+\left(\frac{\mathbf{x}_n + \mathbf{x}_p}{2}\right) \quad (21.6)$$

To simplify this integral it is usual to make the so-called zero-range approximation, that the range of V_{np} is very short compared to the deuteron size. Now, the deuteron wave function outside the potential has the form $e^{-\alpha r}/r$ where

$$\alpha^2 = 2m\ [\text{deuteron binding energy}]$$

Thus, in the limit that the potential has negligible range, the normalized deuteron wave function is

$$\phi_d(\mathbf{x}) = \left(\frac{\alpha}{2\pi}\right)^{1/2}\frac{e^{-\alpha r}}{r}$$

everywhere. Using the Schrödinger equation we can write

$$V_{np}(\mathbf{x})\phi_d(\mathbf{x}) = \frac{1}{2m}(\nabla^2 + \alpha^2)\phi_d(\mathbf{x}) = -\frac{(2\pi\alpha)^{1/2}}{m}\,\delta_3(\mathbf{x})$$

With this approximation (21.6) becomes

$$t(\mathbf{p}', 2 \leftarrow \mathbf{p}, 1) = \text{constant}\int d^3x\,\phi_{A'}(\mathbf{x})^*\chi_2^-(\mathbf{x})^*\chi_1^+(\mathbf{x}) \quad (21.7)$$

To evaluate the one remaining integral we need to know the scattering functions χ_2^- and χ_1^+ for the proton and deuteron in the field of the nucleus A. To get a rough idea of the behavior of the amplitude we can make the very radical approximation that these scattering functions are given by undistorted plane waves outside A with total absorption inside,

$$\chi_2^-(\mathbf{x}) = \begin{cases}(2\pi)^{-3/2}e^{i\mathbf{p}'\cdot\mathbf{x}} & r > a \\ 0 & r < a\end{cases}$$

$$\chi_1^+(\mathbf{x}) = \begin{cases}(2\pi)^{-3/2}e^{i\mathbf{p}\cdot\mathbf{x}} & r > a \\ 0 & r < a\end{cases}$$

where a denotes the radius of the nucleus A. With this approximation we obtain

$$t(\mathbf{p}', 2 \leftarrow \mathbf{p}, 1) \approx \text{constant} \int_{r>a} d^3x \, \phi_{A'}(\mathbf{x})^* e^{-i\mathbf{q}\cdot\mathbf{x}} \tag{21.8}$$

where $\mathbf{q} = \mathbf{p}' - \mathbf{p}$ as usual.

We now recall that $\phi_{A'}(\mathbf{x})$ is the bound-state wave function of the captured neutron in the nucleus A'. Assuming that this is a state of definite orbital angular momentum l we can write this as

$$\phi_{A'}(\mathbf{x}) = y(r) Y_l^m(\hat{\mathbf{x}})$$

[Notice that the T matrix should really be written $t(\mathbf{p}', m, 2 \leftarrow \mathbf{p}, 1)$ to identify the m value of the captured neutron.] Expanding the plane wave in (21.8) we can immediately do the angular integration to give

$$t(\mathbf{p}', m, 2 \leftarrow \mathbf{p}, 1) \approx \text{constant } Y_l^m(\hat{\mathbf{q}}) \int_a^\infty dr \, r^2 y(r) j_l(qr)$$

where $j_l(qr)$ is the usual spherical Bessel function. Finally, since $y(r)$ falls off exponentially outside a, the principal contribution to the radial integral is from r close to a, and this suggests our final approximation[3]

$$t(\mathbf{p}', m, 2 \leftarrow \mathbf{p}, 1) \approx \text{constant } Y_l^m(\hat{\mathbf{q}}) j_l(qa) \tag{21.9}$$

In a typical deuteron stripping experiment one does not observe the m value of the captured neutron. Thus, the appropriate cross section is obtained by squaring (21.9) and summing over m. This gets rid of the factor Y_l^m and leaves the result

$$\frac{d\sigma}{d\Omega}(\mathbf{p}', 2 \leftarrow \mathbf{p}, 1) \approx \text{constant } \frac{p'}{p} [j_l(qa)]^2 \tag{21.10}$$

where

$$q = |\mathbf{p}' - \mathbf{p}| = (p'^2 + p^2 - 2p'p \cos\theta)^{1/2}$$

The striking feature of this result is that the angular distribution of the scattered proton is completely determined by the factor $[j_l(qa)]^2$ where l is the orbital momentum of the captured neutron. In spite of all our approximations this result is often surprisingly reliable; in particular, it often correctly predicts the location of the first maximum in the angular distribution of the outgoing protons. This means that the angular momentum l of the captured neutron can be *measured* by comparing the observed angular distribution with the functions $[j_l(qa)]^2$. This procedure is illustrated in

---

[3] Since the integral runs from the nuclear radius outwards, this approximation would be expected to be best if we take a slightly larger than the nuclear radius.

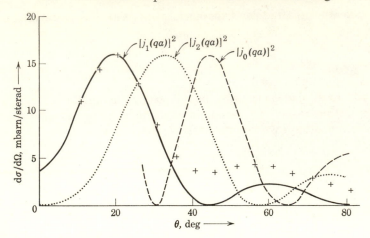

FIGURE 21.2. **Angular distribution of outgoing protons in the reaction**

$$d + {}^{40}Ca \to p + {}^{41}Ca^*$$

at 7 MeV. (Experimental data from Lee et al., 1964.)

Fig. 21.2, which shows the angular distribution of protons in the reaction

$$d + {}^{40}Ca \to p + {}^{41}Ca^*$$

at 7 MeV, where ${}^{41}Ca^*$ is the $\frac{3}{2}^-$ first excited state of ${}^{41}Ca$. It will be seen that the location of the first maximum clearly suggests the correct value $l = 1$ of the captured neutron.[4] Needless to say, markedly better fits to the data are possible if one uses the original DWBA (21.6) [or its zero-range version (21.7)] with more realistic distorted waves calculated by a computer using suitable phenomenological optical potentials.

21-b. Final-State Interactions

The theory of final-state interactions, like the DWBA, applies to reactions where the underlying forces can usefully be split into two pieces, $V = V_{\mathrm{I}} + V_{\mathrm{II}}$. It is usually further restricted to situations (similar to the example at

---

[4] It must be admitted that this is a little fraudulent. Within certain reasonable limits, the radius a (which was taken to be 7.6 fm here) is a parameter that can be adjusted to match the first maximum. However, to bring the next best candidate $[j_2(qa)]^2$ into agreement would require an unreasonable 60% increase in a. Furthermore, the cross sections for other final states can be fitted with their correct l values and with the same radius a. In accordance with footnote 3, the value $a = 7.6$ fm is already rather large for the nuclear radius.

the end of the last section) where the reaction of interest could not occur if the interaction V_{II} were absent. When this is the case, it is reasonable to think of the reaction as being caused by V_{II}, which is therefore referred to as the *primary interaction*. The potential V_I (which for reasons we shall discuss shortly, is called the *final-state interaction*) is then looked on as modifying the reaction caused by V_{II}. It is the purpose of final-state interaction theory to predict these modifications and, by comparison with experiment, to obtain information concerning the "final-state interaction" V_I.

The general conditions for application of final-state interaction theory are much the same as those of the DWBA and, as we shall see, the starting point for the theory is usually precisely the DWBA (21.2). However, it should be emphasized that many results of the theory can be derived by a more general approach based on dispersion relations. This means that the results are more general than the nonrelativistic potential framework within which we shall establish them. In particular the theory can be, and has been, applied to the study of the relativistic interactions of elementary particles.[5]

A classic example of the theory of final-state interactions is given by the photoelectric effect (which we discuss in terms of hydrogen for simplicity)

$$\gamma + H \rightarrow e + p \qquad (21.11)$$

Here the primary interaction which makes the effect possible is the interaction of the electron with the electromagnetic radiation field. This is what destroys the incident photon and knocks out the bound electron. However, one must also take into account the Coulomb attraction of the e and p in the final channel. It is natural to view the whole reaction as being first caused by the primary interaction and then modified by the Coulomb attraction between the two final particles. It is then natural to refer to this Coulomb attraction as the *final-state interaction*.

It should be emphasized that the theoretical tools developed for final-state interactions can equally be applied to processes involving "initial-state interactions," such as the inverse photoelectric effect

$$e + p \rightarrow \gamma + H \qquad (21.12)$$

Here again the primary interaction is obviously that of the electron with the radiation field, while the modifications again come from the Coulomb, e–p attraction, which in this case is an *initial*-state interaction. In practice, it develops that useful applications of the theory usually involve *final*-state interactions, from which circumstance comes the name of the theory. However, before we discuss why this is so, we can usefully look a little further at

---

[5] For this, and several further aspects of final-state interactions see the monograph of Gillespie (1964).

the "initial-state process" (21.12). In this reaction it seems clear the Coulomb attraction will pull the electron into the proton and increase the probability of its capture. Thus, while the Coulomb attraction alone cannot *cause* the process, it is expected to enhance the probability of the process occurring. The interesting thing about this result is that by time-reversal invariance it must also apply to the original photoelectric effect (21.11); that is, we have the quite surprising result that the attraction between the final e and p actually increases the likelihood that the photon can dissociate them.

In the examples just discussed the final-state (or initial state) interactions were the Coulomb force, which is already well understood and can certainly be studied by more direct methods. In most real examples, the method of final-state interactions is used to study interactions that cannot be studied directly. For example, a process like

$$\pi^- + d \rightarrow \gamma + n + n \tag{21.13}$$

can be used to study the *n–n* forces; while processes such as

$$\pi + N \rightarrow \pi + \pi + N \tag{21.14}$$

have been used to study the interactions between two pions. These examples make clear why it is usually *final*-state interactions with which one is concered. If one could set up direct *n–n* or π–π collisions, one would certainly not resort to processes as devious as (21.13) and (21.14) to study their interactions. However, such direct collisions are *not* yet experimentally feasible; and since an initial state with two incident neutrons or pions is experimentally inaccessible, it is clear that the only resort is to processes in which the two particles of interest appear, and hence interact, in the *final* state.

The examples (21.13) and (21.14) illustrate how the method of final-state interactions can be used in practice. In the process $\pi^- + d \rightarrow \gamma + n + n$ it develops that the *n–n* forces produce a characteristic distortion of the energy spectrum of the photon; thus, measurement of this energy spectrum lets one determine certain of the *n–n* scattering parameters. Similarly, in the reaction $\pi + N \rightarrow \pi + \pi + N$ certain features of the π–π interaction can be extracted by studying the distribution of the two pions with respect to energy and angle.

The process $\pi + N \rightarrow \pi + \pi + N$ also illustrates well some of the difficulties of the method. First, it is often unclear how to make a clean separation of the primary and final-state interactions. Much worse, there will obviously be final-state interactions between either pion and the nucleon, in addition to the interaction of interest between the two pions. Obviously there will be problems in sorting out the effects of all these different interactions. We shall return to this point at the end of this section.

To begin our quantitative discussion of final-state interactions we return to the simple photoelectric effect

$$\gamma + H \to e + p$$

assuming the proton to be fixed at the origin. The primary interaction that causes the effect is the interaction of the electron with the radiation field[6]

$$V_{\text{II}} = -\frac{e}{mc} \mathbf{P} \cdot \mathbf{A(x)}$$

where $\mathbf{P}$ is the momentum operator of the electron and $\mathbf{A(x)}$ is the operator of the quantized electromagnetic radiation field. The final-state interaction is the e–p Coulomb attraction

$$V_{\text{I}} = -\frac{e^2}{r}$$

The initial state consists of a photon with momentum $\mathbf{k}$ and the electron bound to the fixed proton. The final state contains no photons, but has a free electron of momentum $\mathbf{p}$. To lowest order in the primary interaction the relevant T-matrix element is given by DWBA as[7]

$$t(\mathbf{p}, e + p \leftarrow \mathbf{k}, \gamma + H) = -\frac{e}{mc} \langle \mathbf{p}, (e + p) - | \mathbf{P} \cdot \mathbf{A(x)} | \mathbf{k}, (\gamma + H) \rangle$$

where $|\mathbf{p}, (e + p) - \rangle$ is the appropriate distorted wave, namely the "minus" scattering state for the electron in the field of the proton. The initial state can be written as the tensor product

$$|\mathbf{k}, (\gamma + H)\rangle = |\mathbf{k}\rangle \otimes |\phi_1\rangle$$

where $|\mathbf{k}\rangle$ is the one-photon state (which should also carry a polarization label $\boldsymbol{\epsilon}$) and $|\phi_1\rangle$ is the state of the electron bound in the target.

If we decompose the field $\mathbf{A(x)}$ in terms of the photon creation and annihilation operators, then only the annihilation operator with the appropriate momentum and polarization contributes to our matrix element, which reduces to

$$t(\mathbf{p} \leftarrow \mathbf{k}) = -\frac{e}{2\pi m(\omega)^{1/2}} \langle \mathbf{p} - | \boldsymbol{\epsilon} \cdot \mathbf{P} e^{i\mathbf{k} \cdot \mathbf{x}} | \phi_1 \rangle \qquad (21.15)$$

---

[6] We assume in this discussion that the reader is familiar with the elementary theory of the quantized electromagnetic field (see, for example, Gottfried, 1966, Chapter VIII or Messiah, 1961, Chapter XXI). Notice that we are treating the electron nonrelativistically and ignoring the A^2 interaction, which does not contribute to the photoelectric effect.

[7] We are assuming that our usual scattering formulas apply to interactions involving quantized fields. While this assumption is correct in the present first-order calculation, it is well known that, in general, there are difficulties. These difficulties arise because the particles continue to interact with the field even when they move far apart; this necessitates the whole machinery of renormalization.

(where we have omitted some unnecessary channel labels). It should be noted that all effects of the final-state interaction are contained in the final bra $\langle \mathbf{p}-|$; if we were to replace the scattering state $\langle \mathbf{p}-|$ by the plane wave $\langle \mathbf{p}|$ we would have the matrix element for the same process in the absence of any final-state interaction.

The matrix element (21.15) is easily evaluated (see Gottfried, 1966, Section 58) and the effects of the final-state interaction examined. However, the results are a little atypical of normal problems for two reasons: First, the photon has spin one; this is reflected in the vector character of the operator $\mathbf{P}$ in (21.15) and means that the matrix element is dominated by p waves at low energies, rather than the s waves of a scalar interaction. Second, the final-state interaction in question is the long range Coulomb force. In order that our example be less exceptional, we now leave the real photoelectric effect and discuss instead a hypothetical "scalar photoelectric effect" in which a scalar particle of momentum $\mathbf{k}$ ejects an "electron" that is subject to a short range, spherical, final-state interaction $V(r)$. In this case the relevant matrix element will have the general form

$$t(\mathbf{p} \leftarrow \mathbf{k}) = \langle \mathbf{p}-| \, B_{\mathbf{k}} \, | \phi_1 \rangle \tag{21.16}$$

where $B_{\mathbf{k}}$ is some operator, which may depend on the incident "photon" momentum $\mathbf{k}$ [compare (21.15)].

In terms of wave functions, the matrix element (21.16) can be written

$$t(\mathbf{p} \leftarrow \mathbf{k}) = \int d^3x \, \langle \mathbf{x} \mid \mathbf{p}- \rangle^* B_{\mathbf{k}}(\mathbf{x}) \phi_1(\mathbf{x}) \tag{21.17}$$

(where we have assumed that $B_{\mathbf{k}}$ is a function of $\mathbf{x}$ only). We now make the important assumption that only the immediate neighborhood of the origin contributes significantly to this integral. This can be either because the primary interaction $B_{\mathbf{k}}(\mathbf{x})$ is confined to a small region, or because the bound-state wave function $\phi_1(\mathbf{x})$ is well localized. In either case we can approximate the distorted wave $\langle \mathbf{x} \mid \mathbf{p}- \rangle$ by its value near $\mathbf{x} = 0$. Since all partial waves except the s wave vanish at $\mathbf{x} = 0$, this gives

$$\langle \mathbf{x} \mid \mathbf{p}- \rangle \approx (2\pi)^{-3/2} \frac{1}{pr} \, \psi_{0,p}(r)^* \tag{21.18}$$

where $\psi_{0,p}(r)$ is the s wave radial wave function of the "electron" in the field of the final state interaction $V(r)$. Now, the normalized radial function $\psi_{0,p}$ is equal to the regular solution $\phi_{0,p}$ divided by the Jost function $f(p)$ [see (11.50)]; while for small r the regular solution is approximately $\sin pr$,

$$\psi_{0,p}(r) = \frac{\phi_{0,p}(r)}{f(p)} \approx \frac{\sin pr}{f(p)}$$

Substituting into (21.18) we find

$$\langle \mathbf{x} \mid \mathbf{p}- \rangle \approx \frac{1}{f(p)^*} (2\pi)^{-\frac{3}{2}} \frac{\sin pr}{pr}$$

and, hence, for the matrix element

$$t(\mathbf{p} \leftarrow \mathbf{k}) \approx \frac{1}{f(p)} (2\pi)^{-\frac{3}{2}} \int d^3x \frac{\sin pr}{pr} B_\mathbf{k}(\mathbf{x}) \phi_1(\mathbf{x}) \qquad (21.19)$$

As we have already noted the effects of the final-state interaction are completely contained in the distorted wave $\langle \mathbf{x} \mid \mathbf{p}- \rangle$. If the final-state interaction were completely absent, then $\langle \mathbf{x} \mid \mathbf{p}- \rangle$ would be replaced by $\langle \mathbf{x} \mid \mathbf{p} \rangle$, which would be approximated close to the origin by $(2\pi)^{-\frac{3}{2}} (\sin pr)/pr$. This would give the same answer as (21.19) except that the denominator $f(p)$ would be absent; that is, the actual amplitude (21.19), including final-state interactions, can be written as

$$\boxed{t(\mathbf{p} \leftarrow \mathbf{k}) = \frac{t_0(\mathbf{p} \leftarrow \mathbf{k})}{f(p)}} \qquad (21.20)$$

where $t_0(\mathbf{p} \leftarrow \mathbf{k})$ denotes the T matrix for the same process in the absence of all final-state interaction and $f(p)$ is the s wave Jost function[8] appropriate to the actual final-state interaction $V(r)$. In other words, the sole effect of the final-state interaction is to enhance the amplitude by the factor $1/f(p)$. Physically, of course, this "enhancement factor" represents the increased (or decreased) amplitude for finding the particle in the region where the primary interaction can take effect.[9]

Using the result (21.20) one can prove that, as anticipated earlier, an attractive final-state interaction will enhance the probability of a reaction, while a repulsive interaction will diminish it. The point is that it can be shown that for any purely attractive force $(\partial V/\partial r \geqslant 0$ for all $r)$ the corresponding Jost function has modulus less than one, $|f(p)| \leqslant 1$; while for any

---

[8] If the s-wave amplitude happens to be zero, then we could repeat our argument using the first nonvanishing l value. In this case (21.20) would hold with $f(p)$ being the Jost function of angular momentum l.

[9] Watson, in this original paper on final-state interactions (1952), derived the somewhat different result $t(\mathbf{p} \leftarrow \mathbf{k}) \approx$ constant $f(p) \, t_0(\mathbf{p} \leftarrow \mathbf{k})$ where $f(p)$ is the s-wave *amplitude* for the final-state interaction. For small p the two results agree since $f = [f(-p) - f(p)]/2ipf(p)$, which, if we retain only terms of order p becomes $f = $ constant$/f(p)$. However, in general, the two results are different and it seems clear that (21.20) is to be preferred. In particular, (21.20) is certainly more reasonable for large p where $f \rightarrow 1$ and, hence, $t \rightarrow t_0$ as one would expect.

purely repulsive force ($\partial V/\partial r \leqslant 0$ everywhere) $|f(p)| \geqslant 1$. We leave this as an exercise for the interested reader; for the special case of a square well see Problem **21.2**.

If the amplitude $t_0(\mathbf{p} \leftarrow \mathbf{k})$ can be calculated, then measurement of the cross section lets one determine $|f(p)|$. Usually, of course, $t_0(\mathbf{p} \leftarrow \mathbf{k})$ cannot be calculated exactly and one must have recourse to further approximations. For example, if we measure the cross section close to threshold (p small) then we can replace the wave function ($\sin pr)/pr$ by one. In this case the only energy dependence in the integral (21.19) is in the primary-interaction term $B_{\mathbf{k}}(\mathbf{x})$. However, close to threshold k is a slowly varying function of p and we can simply replace k by its value k_0 at threshold to give

$$t(\mathbf{p} \leftarrow \mathbf{k}) \approx \frac{\text{constant}}{f(p)}$$

That is, the complete energy dependence of the amplitude close to threshold is given by the Jost function $f(p)$ in the denominator. For the cross section [proportional to $(p/k)\,|t|^2$] this gives

$$\frac{d\sigma}{d\Omega} \approx \text{constant}\, \frac{p}{|f(p)|^2} \qquad [p \text{ near threshold}] \qquad (21.21)$$

Since the Jost function is analytic in p the effect of the enhancement $1/|f(p)|^2$ in (21.21) will be most striking if $f(p)$ vanishes somewhere near threshold, as happens for example if the two final particles are nucleons. We saw in Section 13-b that if the s-wave Jost function vanishes near threshold, then it must do so either on the positive or the negative imaginary axes. (In the former case there is a weakly bound state, like the deuteron; in the latter, there is a "virtual" state, as in the nucleon–nucleon singlet system.) In either case the Jost function can be written (see Problem **13.1**) as

$$f(p) \approx \text{constant}\,(1 + iap) \qquad [p \text{ small}]$$

where a is the s-wave scattering length and is large. The cross section (21.21) then becomes

$$\frac{d\sigma}{d\Omega} \approx \text{constant}\, \frac{p}{1 + a^2 p^2}$$

which shows a sharp peak at $p = 1/|a|$ and allows one to measure the scattering length of the final-state interaction $V(r)$ within a sign.

In most applications of final-state interaction theory one has to handle final states with more than two bodies. While there is no complete theory for this case, there are certain situations where a straightforward generalization of the preceding analysis is possible. Of these the most important are:

(1) Situations where the two final particles of interest have high momenta relative to all other particles, but low momentum relative to one another (giving them plenty of time for interaction with one another).
(2) Situations where the relative energy of the two final particles of interest is at a resonance of these two particles.

In either case it is reasonable to expect that the dominant final-state interaction affecting the two particles of interest will be their mutual interaction. In this case the preceding two-body arguments generalize as follows:
Suppose, for instance, that we observe a reaction with N final particles

$$a + b \rightarrow c + d + e + \cdots + f$$

where c and d are the particles of interest. If, whatever the reason, the only final-state interaction affecting the particles c and d is the c–d potential V_{cd}, then the distorted wave appropriate to the final state will factor as

$$\langle \underline{x} \,|\, \underline{p} - \rangle = \langle \mathbf{x}_{cd} \,|\, \mathbf{p}_{cd} - \rangle \exp\left(i \bar{\mathbf{p}}_{cd} \cdot \bar{\mathbf{x}}_{cd} \right) \times \text{(function of the other } N - 2$$
$$\text{variables)}$$

where $\langle \mathbf{x}_{cd} \,|\, \mathbf{p}_{cd} - \rangle$ is the scattering state for the relative motion of c and d under the potential V_{cd}. Almost exactly the preceding argument then shows that

$$t(\underline{p} \leftarrow \mathbf{p}_a, \mathbf{p}_b) \approx \frac{1}{f(p_{cd})} \times \text{(function independent of } p_{cd}) \qquad (21.22)$$

Here $f(p_{cd})$ is the Jost function of the dominant angular momentum l in c–d scattering. (For low c–d relative energy this will be s-wave; near a c–d resonance, l will be the resonant angular momentum.)
Obviously the result (21.22), when it holds, will allow measurement of certain of the c–d scattering parameters. Probably its most interesting application is when the c–d relative energy (or "subenergy") is close to a c–d resonance. In this case the Jost function in the denominator of (21.22) vanishes just below the real axis and has the approximate form

$$f(p_{cd}) \approx \text{constant} \times \left(E_{cd} - E_R + i \frac{\Gamma}{2} \right)$$

Thus, as a function of the c–d subenergy E_{cd} the cross section for the process

$$a + b \rightarrow c + d + e + \cdots + f$$

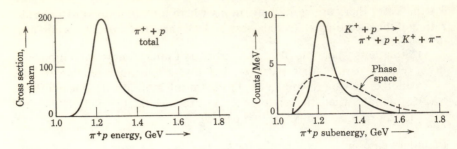

FIGURE 21.3.   (a) The $\pi^+ + p$ total cross section as a function of the total CM energy.   (b) The process $K^+ + p \to \pi^+ + p + K^+ + \pi^-$ as a function of the $\pi^+ p$ subenergy at fixed incident momentum 2.3 GeV/c.

should show a peak whose position and width are characteristic of the c–d resonance. This important property of resonances is in fact true much more generally than our approximate derivation suggests. Unfortunately, to pursue this further would take us beyond the scope of this book.[10] Therefore, we content ourselves with mentioning that the property is certainly well established experimentally. For example, the well known N*(1236) resonance in π–N scattering shows up clearly in the process

$$K^+ + p \to \pi^+ + p + K^+ + \pi^-$$

when plotted as a function of the final $\pi^+ p$ subenergy as shown in Fig. 21.3. (The broken curve is a phase-space calculation of the type described in Problem 17.6 and represents the best guess for the process *if there were no resonance*. Data from Bomse, et al., 1967.)

This property—that resonances of a subsystem appear in the study of the whole system—is, in fact, frequently used to *identify* resonances of particles that cannot be directly studied. For example, the well known ρ resonance in π–π scattering, which has never been directly observed (since no π–π scattering experiment has ever been performed) is nonetheless firmly established on the basis of its appearance in processes such as $\pi + N \to 2\pi + N$ when considered as functions of the π–π subenergy.

[10] If we think of a resonance as a nearly bound state, then the property is easy to understand physically: If the resonance were actually stable with energy E_R, then some number of events (n say) of the type $a + b \to (cd) + e + \cdots + f$ would occur, in all of which the c–d relative energy E_{cd} would be exactly E_R. In addition, n' events with final state $c + d + \cdots + f$ might occur, but with the c–d relative energy more or less smoothly distributed. Now, since the state (cd) is really unstable, it eventually decays. Thus, the original n events would be registered as having final state $c + d + \cdots + f$ but all with E_{cd} in an interval of order Γ about E_R. Accordingly, a plot of this process against E_{cd} should show a bump about $E_{cd} = E_R$.

PROBLEMS

21.1. Prove (the obvious result) that if a impinges on a bound state (bc) with c infinitely heavy, and if a interacts only with c, then the probability for excitation $a + (bc) \rightarrow a + (bc)^*$ is zero. Why is this not so if c has finite mass?

21.2. Compute the s-wave Jost function for a square well of depth V_0 and show that $|f(p)| \lessgtr 1$ according as $V_0 \gtrless 0$.

22 Identical Particles

22-a   The Formalism of Identical Particles

22-b   Scattering of Two Identical Particles

22-c   Multichannel Scattering with Identical Particles

22-d   Transition Probabilities and Cross Sections

22-e   Electron–Hydrogen Scattering

Until now we have restricted our attention to collisions in which all of the particles involved are distinct. Since the majority of interesting experiments do not satisfy this condition—electron–atom collisions involve several identical electrons, nucleon–nucleus collisions involve several identical nucleons, and so on—it is high time we discuss how our scattering theory can be modified to handle identical particles.

There are several ways to set up a scattering theory of identical particles. Perhaps the most completely satisfactory is to use a second-quantized formalism, in which the symmetry requirements of the identical particles are incorporated from the outset (see Hepp, 1965). However, this approach requires the development of an elaborate new machinery, and we shall follow a more pedestrian approach based directly on the formalism already developed for distinct particles. Roughly speaking, our procedure will be to begin by treating the identical particles *as if* they were distinct—for which case we already have the required collision theory. We shall then find that the actual scattering states, properly symmetrized for the identical particles, can be obtained from those of the distinguishable case by using the appropriate symmetrizing projection operators. This in turn will mean that the scattering amplitudes for identical particles (identified by hats as $\hat{f}$) can be expressed as sums or differences of certain related amplitudes for distinct particles. Thus, it follows that all of the computational techniques developed for distinct particles can be immediately applied to the identical particle problem.

Our results can be given a very persuasive interpretation: For example, we shall find that the amplitude for elastic scattering of an electron off a hydrogen atom has the form

$$\hat{f}(\mathbf{p}' \leftarrow \mathbf{p}) = f_{\mathrm{di}}(\mathbf{p}' \leftarrow \mathbf{p}) - f_{\mathrm{ex}}(\mathbf{p}' \leftarrow \mathbf{p}) \qquad (22.1)$$

(We omit spin and channel labels for simplicity.) Here the "direct" amplitude $f_{\mathrm{di}}(\mathbf{p}' \leftarrow \mathbf{p})$ is the amplitude for the incident electron—*treated as distinct from that in the target atom*—to scatter elastically with momentum $\mathbf{p}'$. The "exchange" amplitude $f_{\mathrm{ex}}(\mathbf{p}' \leftarrow \mathbf{p})$ is the amplitude for the process in which the target electron is ejected with momentum $\mathbf{p}'$, while the incident electron is captured. Since the electrons are indistinguishable in reality, these two processes cannot be told apart; and since the electrons are fermions, the appropriate observed amplitude is the difference of the two amplitudes as in (22.1).

We begin the discussion in Section 22-a with a brief survey of the quantum mechanics of identical particles, mainly intended to introduce our notation. In Section 22-b we treat the simple case of the single-channel scattering of just two identical particles. In Sections 22-c and 22-d we treat the general multichannel problem and in Section 22-e we illustrate our results with a discussion of electron–hydrogen scattering.

We shall find it convenient in this chapter to deviate slightly from our earlier notation. We shall identify all states by their wave functions, without reference to a corresponding abstract vector. The action of a linear operator A on a wave function ψ will be written as $A\psi$, and the scalar product of two wave functions ψ and ϕ as $\langle \psi, \phi \rangle$.

22-a.   The Formalism of Identical Particles

We begin with a review of the well-known modifications of the ordinary formalism of quantum mechanics that are needed when a system contains several particles of the same type. For simplicity we first discuss a system containing n particles *all* of the same type.

Symmetric and Anti-Symmetric Wave Functions.   If our n identical particles were actually distinguishable (by having slightly different masses, for example) then the Hilbert space $\mathscr{H}$ of state vectors would be simply the space of all square-integrable functions of their n coordinates,

$$\psi(x_1, \ldots, x_n)$$

Here x_i denotes any suitable choice of coordinates for the ith particle. If the particles are spinless, x_i could be just the position $\mathbf{x}_i$ or the momentum $\mathbf{p}_i$. If the particles have spin, x_i could denote the pair $\mathbf{x}_i$, m_i, where m_i is the z component of spin. If the system is made up of neutrons and protons, which we wish to regard as identical, then x_i could stand for $\mathbf{x}_i$, m_i and the third component of isotopic spin. Throughout this chapter we shall use $\mathscr{H}$ to denote the space that contains *all* functions $\psi(x_1, \ldots, x_n)$ and is appropriate to n *distinguishable* particles.

As is well known, when the particles are indistinguishable, the space $\mathscr{H}$ is no longer the appropriate state space for our system. Instead, the states of our system are given by functions in one of two subspaces, which we shall denote by $\mathscr{H}_s$ and $\mathscr{H}_a$. To characterize these subspaces we must introduce the well-known permutation operators on the functions $\psi(x_1, \ldots, x_n)$. Accordingly, for each permutation Π of the numbers $1, \ldots, n$ we define a unitary operator—also denoted Π—such that

$$\Pi\psi(x_1, \ldots, x_n) = \psi(x_{\Pi 1}, \ldots, x_{\Pi n})$$

In terms of these operators we can then state what appear to be the restrictions on the observables and wave functions of a system of n identical particles:[1]

First, all observables A for a system of n identical particles commute with all permutations,

$$[A, \Pi] = 0 \qquad \text{[any observable } A \text{, all } \Pi]$$

Second, not every wave function $\psi(x_1, \ldots, x_n)$ in $\mathscr{H}$ can represent a state of the n-particle system; rather for a given kind of particle (electron, proton, neutron, pion) just one of two possibilities occurs. *Either* the multi-particle states are represented only by *symmetric wave functions*—that is, functions satisfying

$$\Pi\psi = \psi \qquad \text{[all } \Pi]$$

—in which case the particle is called a *boson; or* the states are represented only by *anti-symmetric functions*—satisfying

$$\Pi\psi = \epsilon_\Pi \psi \qquad \text{[all } \Pi]$$

---

[1] The formalism described here is the conventional one in which all particles are either bosons or fermions. In fact, it is only for the case of the stable particles (electron, proton, neutron, and photon) that the evidence for this is completely beyond question. Discussion of the possibility that some of the unstable particles might be neither bosons nor fermions can be found in Stolt and Taylor (1970) and references given there. Since our main concern is with nonrelativistic problems, and hence with stable particles, these exotic possibilities (called para-statistics) need not concern us here.

where ϵ_Π is $+1$ if the permutation Π is even and -1 if Π is odd—in which case the particle is called a fermion. The subspaces containing just the symmetric or anti-symmetric wave functions we shall denote by $\mathcal{H}_s$ and $\mathcal{H}_a$, respectively.

The Spin-Statistics Theorem.   All particles which are known to be bosons have integral spin, while all known fermions have half-odd integral spin. In fact, within the framework of relativistic quantum field theory it can be proved that this connection must always hold: a result known as the spin-statistics theorem.[2]

Composite Particles.   In situations where a composite particle like an atom or nucleus can be treated as a structureless "elementary" particle, it must be treated as a boson or fermion according as the number of fermions among its constituents is even or odd. Thus, in atomic and molecular physics it is usually possible to treat nuclei as structureless and "elementary." When this is so the wave function for a system containing several ^{4}He nuclei, for example, must be symmetric in the ^{4}He coordinates since ^{4}He contains two protons and two neutrons. On the other hand the wave functions for a similar system made up of ^{3}He nuclei must be anti-symmetric in their co-ordinates since they contain two protons and one neutron.

The Symmetrizing Projection Operators.   To construct symmetric and anti-symmetric wave functions it is convenient to introduce projection operators Λ_s and Λ_a that project an arbitrary wave function onto the subspaces $\mathcal{H}_s$ and $\mathcal{H}_a$. These are defined as

$$\Lambda_s = \frac{1}{n!} \sum_\Pi \Pi$$

$$\Lambda_a = \frac{1}{n!} \sum_\Pi \epsilon_\Pi \Pi$$

(22.2)

When we do not wish to specify whether the particles under consideration are bosons or fermions, we shall write either of these as

$$\Lambda = \frac{1}{n!} \sum_\Pi \eta_\Pi \Pi$$

where Λ denotes either Λ_s or Λ_a and η_Π is $+1$ for bosons but $\eta_\Pi \equiv \epsilon_\Pi$ for fermions.

It is easily checked that Λ_s and Λ_a satisfy the characteristic properties of projectors,

$$\Lambda = \Lambda^\dagger = \Lambda^2$$

---

[2] See Streater and Wightman (1964). For discussion of a proof of the same result in analytic S-matrix theory see Froissart and Taylor (1967).

To see that they are Hermitian one has only to note that in the sums (22.2) over all permutations each Π can be paired with its inverse $\Pi^{-1} = \Pi^\dagger$. To check that $\Lambda^2 = \Lambda$ we proceed thus:

$$\Lambda_s^2 = \left(\frac{1}{n!}\right)^2 \sum_\Pi \sum_{\Pi'} \Pi\Pi'$$

If we write $\Pi\Pi' = \Pi''$, then as Π' runs over all permutations so does Π''. Thus,

$$\Lambda_s^2 = \left(\frac{1}{n!}\right)^2 \sum_\Pi \left(\sum_{\Pi''} \Pi''\right) = \frac{1}{n!} \sum_{\Pi''} \Pi'' = \Lambda_s$$

with a similar argument for Λ_a.

It should be noted that we have normalized Λ so that $\Lambda^2 = \Lambda$. If ψ already happens to be symmetric (or anti-symmetric), then $\Lambda_s\psi = \psi$ (or $\Lambda_a\psi = \psi$) and if ψ is normalized then so obviously is $\Lambda_s\psi$ (or $\Lambda_a\psi$). However, in general $\Lambda\psi$ is not normalized even if ψ is. For example, in the important case where ψ is the product

$$\psi(x_1, \ldots, x_n) = \psi_1(x_1) \cdots \psi_n(x_n)$$

of n orthonormal one-particle functions,

$$\begin{aligned}
\|\Lambda_s\psi\|^2 &= \langle \Lambda_s\psi, \Lambda_s\psi \rangle \\
&= \langle \Lambda_s^2\psi, \psi \rangle \qquad [\text{since } \Lambda = \Lambda^\dagger] \\
&= \langle \Lambda_s\psi, \psi \rangle \qquad [\text{since } \Lambda^2 = \Lambda] \\
&= \frac{1}{n!} \sum_\Pi \langle \Pi\psi, \psi \rangle
\end{aligned}$$

It is easily seen that $\langle \Pi\psi, \psi \rangle = 0$ for any permutation except $\Pi = 1$, and this sum therefore reduces to the single term

$$\|\Lambda_s\psi\|^2 = \frac{1}{n!} \langle \psi, \psi \rangle = \frac{1}{n!}$$

The same result holds in the anti-symmetric case and we conclude that for n orthonormal one-particle functions $\psi_1, \ldots, \psi_n$ the correctly symmetrized and normalized product is

$$(n!)^{1/2}\Lambda(\psi_1 \cdots \psi_n) \tag{22.3}$$

for $\Lambda = \Lambda_s$ or Λ_a.

The great usefulness of the projectors Λ_s and Λ_a in scattering theory is that one can (as we shall see) obtain the scattering states for a system of identical particles simply by acting with the appropriate projector on the corresponding states for a system of distinct particles.

Several Kinds of Particles. If there are different kinds of particles present, then the wave functions must be symmetric or anti-symmetric for permutations *within each separate type of particle*. For instance, suppose we have a system of n electrons, m protons, l alpha particles, etc. Then we denote by $\mathscr{H}$ the space of *all* functions

$$\psi(x_{e1}, \ldots, x_{en}; x_{p1}, \ldots, x_{pm}; x_{\alpha 1}, \ldots, x_{\alpha l}; \ldots)$$

and the subspace of functions that actually represent states of the system contains just those functions with the symmetry

$$\psi(\underbrace{x_{e1}, \ldots, x_{en}}_{\text{anti-symmetric}}; \underbrace{x_{p1}, \ldots, x_{pm}}_{\text{anti-symmetric}}; \underbrace{x_{\alpha 1}, \ldots, x_{\alpha l}}_{\text{symmetric}}; \ldots)$$

(There are no restrictions on—nor even any meaning attached to—permutations of the coordinates referring to different particles.) The projector that produces the correct symmetries is just the product

$$\Lambda = \Lambda^{(e)}\Lambda^{(p)}\Lambda^{(\alpha)} \cdots$$

where $\Lambda^{(e)}$ is the projector appropriate to the n electrons, $\Lambda^{(p)}$ for the m protons, and so on.

The Cluster Law. Before taking up the scattering theory of identical particles we discuss one final and important question of principle. The wave function representing any state of the whole universe should in principle be properly symmetrized with respect to *all* particles of every species. However, in practice, when we compute the energy spectrum of gaseous helium, for example, we anti-symmetrize with respect to the two electrons of any single atom, but completely ignore all other electrons in the gas—not to mention those on the moon! The question is: Is this procedure legitimate?

This question is obviously more subtle than the corresponding question for distinguishable particles. In the distinguishable case it is clear that if two particles are outside of each other's force field then their motions are independent, and in discussing either we can completely ignore the other. On the other hand we know that however far apart two *identical* particles may be, we should, in principle, symmetrize; which simply means that we should *not* consider them as independent. However, it is a remarkable fact, sometimes known as the *cluster law*, that provided two (or more) groups of identical particles are confined to well separated regions of space, and so long as one only makes measurements on each separate group, then it makes no difference whether one symmetrizes with respect to the particles of different groups or not.

We shall not discuss the general proof of the cluster law here (see Messiah, 1961, p. 600). Instead, we shall illustrate it with just one simple example.

Suppose that the world consists of just two electrons, one known to be inside our laboratory with state given by the function $\psi_1(x)$, and one known to be outside the lab with state given by $\psi_2(x)$; and suppose that we are interested in the probability distribution of the two particles. The correctly symmetrized and normalized wave function for the two-electron system is

$$\psi(x_1, x_2) = \frac{1}{\sqrt{2}} [\psi_1(x_1)\psi_2(x_2) - \psi_1(x_2)\psi_2(x_1)]$$

and the corresponding probability density is[3]

$$\begin{aligned} w(x_1, x_2) &= 2 |\psi(x_1, x_2)|^2 \\ &= |\psi_1(x_1)\psi_2(x_2)|^2 + |\psi_1(x_2)\psi_2(x_1)|^2 \\ &\quad - 2 \operatorname{Re} \psi_1(x_1)\psi_2(x_2)\psi_1(x_2)^*\psi_2(x_1)^* \end{aligned}$$

The function $\psi_1(x)$ is known to be zero unless its argument is somewhere inside the lab, while $\psi_2(x)$ is known to be zero unless its argument is somewhere outside the lab. Thus, the last term, with the product $\psi_1(x_1)\psi_2(x_1)^*$, is identically zero; and

$$w(x_1, x_2) = |\psi_1(x_1)\psi_2(x_2)|^2 + |\psi_1(x_2)\psi_2(x_1)|^2 \tag{22.4}$$

That is, as long as the two particles are known to be in different regions of space, the characteristic interference due to the symmetry of the wave function vanishes. The remaining two terms are just the density one would get classically. The fact that there are still two terms merely reflects the fact that we have not specified which particle is inside, and which outside, the lab. In fact, we can now identify that particle inside the lab as particle number 1, and that outside as number 2. With the variable x_1 fixed inside the lab and x_2 outside, the second term in (22.4) is zero; and it is clear that the probabilities of locating our two particles as computed from a properly symmetrized wave function are exactly the same as we would obtain if we had ignored the identity of the particles and used the unsymmetrized wave function $\psi_1(x_1)\psi_2(x_2)$.

This result is of some practical importance in scattering theory. There the states that are measured are the asymptotic free states, in which the various freely moving fragments are all well separated. It is very natural to handle these states as if the particles in different fragments are distinct. For example, the in state of a collision between an electron and a hydrogen atom is naturally identified by a wave function of the form

$$\chi(x_1)\phi(x_2) \tag{22.5}$$

---

[3] A simple way to remember the factor 2 on the first line of this equation is to remember that $\int |\psi|^2 = 1$ (since ψ is normalized) while $\int w = 2$, since each pair of points gets counted twice as one integrates over all space.

where $\chi(x_1)$ is the wave packet of the incident electron (labelled 1) as it emerges from the accelerator, and $\phi(x_2)$ is the wave function of the target electron (labelled 2) bound to its nucleus. The result of the previous paragraph means that the function (22.5) is not only a natural way to identify an asymptotic free state, but is also a completely consistent way, just as long as the particles remain well separated. Of course, as soon as the collision begins, it is essential to use properly symmetrized wave functions; and the relation between the in and out states (namely the scattering amplitude) *is* therefore affected by the symmetrization required for the identical particles. However, the point that we wish to emphasize here is that in the labelling of the in and out asymptotes it is both natural and consistent to ignore the identity of particles in different freely moving fragments.

22-b. Scattering of Two Identical Particles

First, we consider the elastic scattering of just two identical particles, which may or may not have spin. For example, we can consider the scattering of electrons off electrons, of nucleons off nucleons, of alphas off alphas, or several other experimentally interesting processes.

The scattering is determined by the Hamiltonian

$$H = \frac{\mathbf{P}_1^2}{2m} + \frac{\mathbf{P}_2^2}{2m} + V \equiv H^0 + V$$

Since the particles are identical, both H and H^0 commute with the one non-trivial permutation Π, the interchange of particles 1 and 2. As discussed previously, we shall let $\psi(x_1, x_2)$ denote an arbitrary wave function and the general state of the system is then denoted by $\Lambda\psi$, where Λ is either Λ_s or Λ_a according as the particles are bosons or fermions. The general orbit of our system has the form

general orbit: $e^{-iHt}\Lambda\psi = \Lambda e^{-iHt}\psi$ [any ψ]

the two forms being equivalent since $[H, \Lambda] = 0$.

If, for a moment, we disregard the identity of the two particles, then their motion under the Hamiltonian H presents a well-defined mathematical problem, and a problem which we have already solved. In particular, we know that there are just two kinds of orbits, bound orbits and scattering orbits; and for any ψ in the subspace of scattering states we know that[4]

$$e^{-iHt}\psi \xrightarrow[t\to-\infty]{} e^{-iH^0t}\psi_{\text{in}} \qquad [\psi = \mathbf{\Omega}_+\psi_{\text{in}}]$$
$$\xrightarrow[t\to\infty]{} e^{-iH^0t}\psi_{\text{out}} \qquad [\psi = \mathbf{\Omega}_-\psi_{\text{out}}] \qquad (22.6)$$

---

[4] Since we discuss the translationally invariant total motion, we use bold face for the Møller operators.

Now, for the general ψ, the orbit (22.6) does not represent an allowed orbit of the actual system (which consists of two *identical* particles) since it is not properly symmetrized. However we have only to multiply by the projector Λ to obtain an orbit that *is* properly symmetrized. Thus, we have only to multiply the results (22.6) by Λ to obtain the appropriate form of the asymptotic condition for two identical particles:

$$\Lambda e^{-iHt}\psi \xrightarrow[t\to-\infty]{} \Lambda e^{-iH^0 t}\psi_{\text{in}} \qquad [\psi = \Omega_+\psi_{\text{in}}]$$

$$\xrightarrow[t\to\infty]{} \Lambda e^{-iH^0 t}\psi_{\text{out}} \qquad [\psi = \Omega_-\psi_{\text{out}}] \qquad (22.7)$$

Since the asymptotic orbits can be written as

$$\Lambda e^{-iH^0 t}\psi_{\text{in/out}} = e^{-iH^0 t}\Lambda\psi_{\text{in/out}}$$

we arrive at the expected conclusion that every properly symmetrized scattering orbit has in and out free asymptotes and that the general asymptote is uniquely identified by some symmetric or anti-symmetric wave function $\Lambda\psi_{\text{in}}$ or $\Lambda\psi_{\text{out}}$.

It should be noted that a given orbit uniquely determines its symmetrized in and out asymptotes. If, however, we write the asymptotes as $\Lambda\psi_{\text{in}}$ (or $\Lambda\psi_{\text{out}}$) then the function ψ_{in} (or ψ_{out}) is *not* unique, since the symmetrizing operator Λ can project two different functions onto one single function. In particular, we shall choose to label our asymptotes by unsymmetrized functions. For example, a typical in asymptote will be identified by a function $\psi_{\text{in}} = \phi_1(x_1)\phi_2(x_2)$ where ϕ_1 is the wave packet characteristic of the accelerator and ϕ_2 the target wave function. That is, we shall label the asymptotes as if the particles were distinct and identified as 1 (the projectile) and 2 (the target). This being the case, it is important to note that it makes no difference which way round we label the particles; whether we use $\phi_1(x_1)\phi_2(x_2)$ or $\phi_1(x_2)\phi_2(x_1)$, we obtain the same physical state once we multiply by the projector Λ.[5]

Our time-dependent description of the collision process is now complete: The system enters the collision along the asymptote $\Lambda\exp(-iH^0 t)\psi_{\text{in}}$; the actual orbit at all times is $\Lambda\exp(-iHt)\psi$; and as $t \to +\infty$ this approaches the out asymptote $\Lambda\exp(-iH^0 t)\psi_{\text{out}}$. And we can now compute the probability of any prescribed transition. We consider an in asymptote given by a wave function

$$\psi_{\text{in}}(x_1, x_2) = \phi_1(x_1)\phi_2(x_2) \equiv \phi(x_1, x_2) \qquad (22.8)$$

where ϕ_1 and ϕ_2 are orthogonal one-particle functions. (The functions are orthogonal since in practice the particles have well defined and different

---

[5] If the particles are fermions, the two results do differ, but only by a factor of -1, which does not change the physical state.

momenta.) If ϕ_1 and ϕ_2 are normalized, then the correctly symmetrized and normalized in asymptote is

$$\sqrt{2}\Lambda e^{-iH^0 t}\phi$$

It follows from (22.7) that the correctly symmetrized and normalized actual state is

$$\sqrt{2}\Lambda e^{-iHt}\Omega_+\phi$$

and, in particular, at $t = 0$ it is

$$\sqrt{2}\Lambda\Omega_+\phi \qquad (22.9)$$

Similarly, the properly symmetrized and normalized state at $t = 0$ that would evolve into an out asymptote ϕ' [with the same form as (22.8)] is

$$\sqrt{2}\Lambda\Omega_-\phi' \qquad (22.10)$$

The probability amplitude that the in state labelled by ϕ be observed to lead to the out state given by ϕ' is just the overlap of the two states (22.9) and (22.10); that is

$$
\begin{aligned}
w(\phi' \leftarrow \phi) &= |\langle\sqrt{2}\Lambda\Omega_-\phi', \sqrt{2}\Lambda\Omega_+\phi\rangle|^2 \\
&= |2\langle\Lambda\Omega_-\phi', \Omega_+\phi\rangle|^2 \qquad \text{(since } \Lambda = \Lambda^\dagger = \Lambda^2) \\
&= \left|\sum_\Pi \eta_\Pi\langle\Pi\Omega_-\phi', \Omega_+\phi\rangle\right|^2 \\
&= |\langle\Omega_-\phi', \Omega_+\phi\rangle + \eta\langle\Pi\Omega_-\phi', \Omega_+\phi\rangle|^2 \qquad (22.11)
\end{aligned}
$$

where in the last line Π denotes the only nontrivial permutation and $\eta = \pm 1$ according as the particles are bosons or fermions.

The interpretation of this important result is straightforward. The first term $\langle\Omega_-\phi', \Omega_+\phi\rangle$ is exactly the amplitude one would get for the process

$$\phi_1'\phi_2' \leftarrow \phi_1\phi_2 \qquad (22.12)$$

if the particles were distinguishable. To interpret the second term we recall that $\Omega_-\phi'$ is the state at $t = 0$ which would evolve to the out asymptote $\phi' = \phi_1'(x_1)\phi_2'(x_2)$. Thus, the state $\Pi\Omega_-\phi'$ in which the roles of the two particles are exchanged, is the state that would evolve to the out state $\phi_2'(x_1)\phi_1'(x_2)$ with the two particles interchanged. Accordingly the term $\langle\Pi\Omega_-\phi', \Omega_+\phi\rangle$ is the amplitude for the "exchanged" process

$$\phi_2'\phi_1' \leftarrow \phi_1\phi_2 \qquad (22.13)$$

Because the two particles are indistinguishable in reality, the two processes (22.12) and (22.13) cannot be told apart and must both be counted. In classical mechanics the measured probability would be the sum of the separate probabilities for these two processes. The essential feature of the quantum result is that one adds (or subtracts, in the case of fermions) the *amplitudes*, and the probability (22.11) can therefore exhibit interference between the two terms.

We are now ready to compute the scattering cross sections and we begin with the case of two identical spinless bosons. The first step in obtaining the cross section is to replace the probability $w(\phi' \leftarrow \phi)$ of (22.11) by the probability $w(d^3p_1 d^3p_2 \leftarrow \phi)$ that, for given in packets ϕ, the particles emerge in a prescribed volume of the final momentum space. This probability is obtained if we replace the state $\phi' = \phi'_1 \phi'_2$ of (22.11) by the momentum eigenstate $|\mathbf{p}_1, \mathbf{p}_2\rangle$ (and multiply by the volume element $d^3p_1 d^3p_2$). The result is identical to the corresponding probability for distinct particles, except for the presence of the "exchange" term, in which $\mathbf{p}_1$, $\mathbf{p}_2$ are replaced by $\mathbf{p}_2$, $\mathbf{p}_1$. We can then go through precisely the calculations of Section 4-e. As usual we use the variables

$$\bar{\mathbf{p}} = \mathbf{p}_1 + \mathbf{p}_2 \quad \text{and} \quad \mathbf{p} = \frac{\mathbf{p}_1 - \mathbf{p}_2}{2}$$

and since the interchange of $\mathbf{p}_1$ and $\mathbf{p}_2$ sends $\mathbf{p}$ to $-\mathbf{p}$ the final result for the CM cross section is

$$\frac{d\sigma}{d\Omega}(\mathbf{p} \leftarrow \mathbf{p}_0) = |f(\mathbf{p} \leftarrow \mathbf{p}_0) + f(-\mathbf{p} \leftarrow \mathbf{p}_0)|^2$$

$$= |\hat{f}(\mathbf{p} \leftarrow \mathbf{p}_0)|^2 \qquad (22.14)$$

where

$$\hat{f}(\mathbf{p} \leftarrow \mathbf{p}_0) = f(\mathbf{p} \leftarrow \mathbf{p}_0) + f(-\mathbf{p} \leftarrow \mathbf{p}_0)$$

The interpretation of this result is illustrated in Fig. 22.1. The first term is the amplitude one would get for the process $(\mathbf{p} \leftarrow \mathbf{p}_0)$ if the two particles were distinct (Fig. 22.1, part a). The second term $f(-\mathbf{p} \leftarrow \mathbf{p}_0)$ is the amplitude for the experimentally indistinguishable process in which the target particle recoils with momentum $\mathbf{p}$, and the original incident particle therefore emerges with momentum $-\mathbf{p}$ (Fig. 22.1, part b).

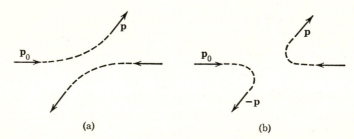

(a)   (b)

FIGURE 22.1.   Two processes that can be thought of as contributing to the cross section $d\sigma/d\Omega(\mathbf{p} \leftarrow \mathbf{p}_0)$ **for two identical particles.**

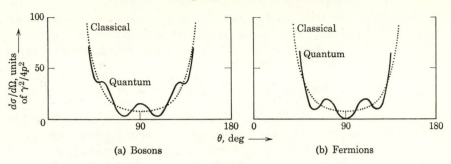

FIGURE 22.2.   **The quantum-mechanical cross sections for two identical particles interacting by a pure Coulomb potential Z^2e^2/r: (a) for two spinless bosons, and (b) for two spin-half fermions with spins parallel; both with strength parameter $\gamma = 5$.**

   In the case where the interactions are spherically symmetric we can rewrite $f(\mathbf{p'} \leftarrow \mathbf{p})$ as $f(E, \theta)$, and the cross section (22.14) becomes

$$\frac{d\sigma}{d\Omega} = |f(E, \theta) + f(E, \pi-\theta)|^2$$

$$= |f(E, \theta)|^2 + |f(E, \pi-\theta)|^2 + 2\,\mathrm{Re}\,f(E, \theta)f(E, \pi-\theta)^* \quad (22.15)$$

The last term is, of course, the characteristic quantum-mechanical interference effect.

   The result (22.15) can be illustrated by the elastic scattering of any two identical spinless nuclei. (A spinless nucleus is of course a boson.) In particular, at low energies only the Coulomb interaction is effective and the amplitude f in (22.15) is the Coulomb amplitude

$$f_C(E, \theta) = -\frac{2mZ^2e^2}{q^2} \exp\left[2i\left(\sigma_0 - \gamma \ln \sin \frac{\theta}{2}\right)\right]$$

where $\gamma = mZ^2e^2/p$ and $q = 2p \sin \theta/2$. Thus, the observed cross section should be

$$\frac{d\sigma}{d\Omega} = \frac{m^2Z^4e^4}{4p^4}\left[\frac{1}{\sin^4 \theta/2} + \frac{1}{\cos^4 \theta/2} + \frac{2\cos(2\gamma \ln \cot \theta/2)}{\sin^2 \theta/2 \cos^2 \theta/2}\right] \quad (22.16)$$

This cross section is shown in Fig. 22.2, part (a) (for one particular choice of the parameters involved), and is compared with the classical result, which contains just the first two terms. It will be seen that the effect of the interference is to superpose an oscillatory term on the classical result. (In particular, at 90 deg the cross section for spinless bosons is always twice the classical result.) The fact that this interference is observed in the scattering of two identical bosons (e.g., in α–α scattering), but is completely absent from the scattering of two distinct particles, is striking evidence in support of the conventional view of identical particles in quantum mechanics.[6]

Returning to the general result (22.14) it should perhaps be emphasized that only the symmetrized amplitude

$$\hat{f}(\mathbf{p}' \leftarrow \mathbf{p}) = f(\mathbf{p}' \leftarrow \mathbf{p}) + f(-\mathbf{p}' \leftarrow \mathbf{p})$$

is experimentally measurable. The amplitude f is the amplitude that would be obtained for the actual interactions if the particles were distinguishable, and, therefore, is a mathematically well-defined quantity. However, since the particles are in fact identical, it is only the symmetrical combination $\hat{f}(\mathbf{p}' \leftarrow \mathbf{p})$ that enters into the observable cross section, and the separate pieces $f(\mathbf{p}' \leftarrow \mathbf{p})$ and $f(-\mathbf{p}' \leftarrow \mathbf{p})$ are experimentally inaccessible. This means that in a pure S-matrix theory, formulated in terms of the experimentally measurable amplitudes only, one would directly postulate the existence of the amplitude $\hat{f}$ satisfying (for the case of rotational invariance)

$$\hat{f}(E, \theta) = \hat{f}(E, \pi - \theta)$$

without any reference to the "fictitious" $f(\mathbf{p}' \leftarrow \mathbf{p})$.

A similar analysis can be applied to particles with arbitrary spin. For simplicity we consider just the case of two identical spin-half fermions which interact via a spin-independent potential. (For example, electron–electron scattering in the approximation where all spin-dependent interactions are neglected.) In this case the spin state is unchanged in any collision and the observed cross section depends only on the properties of the spatial part of the wave function. If we take as a spin basis the eigenfunctions of the total spin, then the corresponding wave functions of the complete two-particle system have the form

$$\psi(\mathbf{x}_1, \mathbf{x}_2)\chi_{\text{trip}}$$

and

$$\psi(\mathbf{x}_1, \mathbf{x}_2)\chi_{\text{sing}}$$

where χ_{trip} denotes any of the triplet spinors with $s = 1$, $m = +1, 0, -1$, and χ_{sing} is the singlet spinor with $s = 0$. Now, the reader will recall that the spinors χ_{trip} are symmetric under permutation of the particles, while χ_{sing} is anti-symmetric. Thus, in order that the total wave function be anti-symmetric, the spatial part $\psi(\mathbf{x}_1, \mathbf{x}_2)$ associated with χ_{trip} must be anti-symmetric, while that associated with χ_{sing} must be symmetric.

---

[6] The presence of the interference term for identical particles was predicted by Mott. One of the first experimental tests was due to Chadwick (1930) who showed that the α–α cross section at 90 deg (in the CM frame of course) was twice the classical result, as shown in Fig. 22.2, part (a).

We suppose first that our two fermions are initially in a triplet state (e.g., both particles polarized in the same direction). The spatial part of the wave function is then anti-symmetric and a calculation similar to that for two bosons gives

$$\left(\frac{d\sigma}{d\Omega}\right)_{\text{trip}} = |f(E, \theta) - f(E, \pi-\theta)|^2 \equiv |\hat{f}_{\text{trip}}(E, \theta)|^2 \qquad (22.17)$$

(We suppose that the potential is rotationally invariant.) That is, the triplet amplitude is the *difference* of the amplitude $f(E, \theta)$ for distinguishable particles and $f(E, \pi-\theta)$. This result is illustrated for the case of a pure Coulomb interaction in Fig. 22.2, part (b). In particular, notice that the triplet cross section is always zero at 90 deg.

Instead, if we consider an initial singlet state, then the spatial part of the wave function has to be symmetric, and we get the result

$$\left(\frac{d\sigma}{d\Omega}\right)_{\text{sing}} = |f(E, \theta) + f(E, \pi-\theta)|^2 \equiv |\hat{f}_{\text{sing}}(E, \theta)|^2$$

That is, two identical *singlet* spin-half fermions would scatter just like two identical spinless bosons. This result and (22.17) illustrate the fact (well-known in the study of bound states) that with systems of identical particles even spin-independent interactions can lead to spin-dependent results. The point is, of course, that different spin states correspond to different permutation symmetries, and the allowed symmetries of the spatial wave function (and hence the observed results) therefore depend on the spin.

In practice, of course, experiments are usually done with unpolarized beams and targets. And even when the particles are polarized their spin state is not necessarily pure singlet or pure triplet. However, any desired spin state can be expressed as an appropriate superposition of singlet and triplet states and the corresponding amplitude computed. In particular, the *unpolarized* cross section is obtained by the usual procedure of summing over final and averaging over initial spins as

$$\left(\frac{d\sigma}{d\Omega}\right)_{\text{unpol}} = \frac{1}{4}\left(\frac{d\sigma}{d\Omega}\right)_{\text{sing}} + \frac{3}{4}\left(\frac{d\sigma}{d\Omega}\right)_{\text{trip}}$$

$$= |f(E, \theta)|^2 + |f(E, \pi-\theta)|^2 - \text{Re}\, f(E, \theta) f(E, \pi-\theta)^*$$

That is, even in a completely unpolarized experiment the characteristic interference terms show up. The fact that these effects are observed (e.g., in *e–e* or *p–p* experiments) confirms both that the particles involved are fermions and that their spin is a half.

Before we go on to the general multichannel problem, one word about total cross sections: In the scattering of two distinct particles the total cross section was defined as

$$\sigma(\mathbf{p}) = \int d\Omega_{p'} \frac{d\sigma}{d\Omega} \, (\mathbf{p}' \leftarrow \mathbf{p}) \qquad \text{[distinct particles]}$$

(with a sum over final spins if the particles have spin). With this definition we can say either that $n_{\text{inc}}\sigma(\mathbf{p})$ is the total number of counts obtained if we completely surround the target with counters, or equivalently that $n_{\text{inc}}\sigma(\mathbf{p})$ is the total number of particles *lost from the incident beam*. In the case of two identical particles these two numbers are not the same: If we completely surround the target with counters then each collision event will be registered twice (since we necessarily count both the scattered and recoil particles) and the number of counts obtained in this way is therefore *twice* the number of particles lost from the incident beam. It is usually more natural to define the total cross section in terms of particles lost from the incident beam, and we therefore adopt the definition

$$\sigma(\mathbf{p}) = \tfrac{1}{2} \int d\Omega_{p'} \frac{d\sigma}{d\Omega} \, (\mathbf{p}' \leftarrow \mathbf{p}) \qquad \text{[identical particles]} \qquad (22.18)$$

22-c.   Multichannel Scattering with Identical Particles

We now proceed to the general multichannel problem and shall find that the scattering theory for systems of identical particles can be obtained from that for distinct particles by the use of symmetrizing projectors, much as in the single-channel case.

We consider an arbitrary system of n particles, at least some of which are identical, and which has Hamiltonian H. We can, of course, imagine H to be the Hamiltonian of n *distinct* particles (the fact that H commutes with the relevant permutation operators does not prevent this) in which case we already understand the corresponding scattering theory as follows.

The general state, which is labelled by an arbitrary function of the n variables x_i, can be written as a superposition of the bound states (of all n particles) and scattering states. The latter are the states of interest and can themselves be classified according to the channels from which they have evolved as $t \to -\infty$ (or to which they will evolve as $t \to +\infty$). A channel α is a grouping of the n particles into n_α stable fragments ($2 \leqslant n_\alpha \leqslant n$) and corresponding to each channel there is a subspace $\mathcal{S}^\alpha$ of wave functions describing free motion of these n_α fragments. If the state ψ of the n particles originated in channel α, then there is a function ψ_{in} in $\mathscr{S}^\alpha$ such that[7]

$$e^{-iHt}\psi \xrightarrow[t \to -\infty]{} e^{-iH^\alpha t}\psi_{\text{in}} \qquad [\psi = \Omega_+^\alpha \psi_{\text{in}}] \qquad (22.19)$$

where H^α is the channel Hamiltonian appropriate to the free motion of the n_α fragments. Similarly if ψ is a state that will evolve into channel α as $t \to +\infty$, then there is a ψ_{out} in $\mathscr{S}^\alpha$ such that

$$e^{-iHt}\psi \xrightarrow[t \to +\infty]{} e^{-iH^\alpha t}\psi_{\text{out}} \qquad [\psi = \Omega_-^\alpha \psi_{\text{out}}] \qquad (22.20)$$

The general scattering state can be expressed as a superposition of states each coming from some definite in channel, and similarly of states each going into some definite out channel.

For an arbitrary wave function ψ the orbit $e^{-iHt}\psi$ does not represent an allowed orbit of the actual system (with identical particles) since ψ does not have the required symmetry. However, we have only to multiply by the appropriate projector Λ to obtain an orbit that does. Under the action of Λ the situation described in the previous paragraph changes. In particular, the number of channels is reduced in two ways.

The first change can be easily understood by considering the example of electron–hydrogen scattering. Here, for every channel with the grouping

$$e_1 + (e_2p) \qquad (22.21)$$

there is a second channel of the form

$$e_2 + (e_1p) \qquad (22.22)$$

obtained by permuting the two electrons. If the electrons are distinguishable these two channels are physically distinct. In reality, however, the electrons are indistinguishable and the distinction between the two channels is without physical meaning. Nonetheless, in the formalism that we choose to adopt, the two channels are mathematically distinct, wave functions of the former having the form

$$\chi(x_1)\phi(x_2)$$

(where χ is an arbitrary wave packet, and ϕ is the hydrogen bound-state wave function) while those of the latter have the form

$$\chi(x_2)\phi(x_1)$$

---

[7] For simplicity we consider a process (such as the collision of an electron with an atom, or nucleon with a nucleus that has a heavy core) in which the target has a heavy core which we can take to be fixed. (Or equivalently, we consider just the relative motion of a translationally invariant system.) Accordingly, we use light face for the Møller operators, which do not conserve momentum (or out of which the momentum-conserving part has been factored).

Needless to say, the answers we obtain will be independent of the labelling
we use to identify the asymptotic states. Still, it is important that we understand the situation and are prepared with some suitable notation. Thus, in
the example just discussed we shall label the channel of (22.21), with the
hydrogen atom in its ground state, by $\alpha = 1$ and the corresponding "exchange" channel of (22.22) by $\alpha = \tilde{1}$.

In general, whenever a channel has two or more identical particles distributed among its freely moving fragments there is a whole family of channels
obtained simply by permuting identical particles among the different fragments. These channels, while mathematically distinct, are physically
indistinguishable, and any one of them can be used to identify the same
physical situation.

To illustrate the second reduction in the number of channels we consider
the slightly more complicated example of electron–helium scattering. The
point is that if the three electrons are distinguishable then there are channels
in which the electrons of the helium atom have a symmetric bound-state
wave function. As soon as we act with the anti-symmetrizing projector these
wave functions are annihilated, corresponding to the obvious fact that the
real helium atom has only anti-symmetric states. It should be clear that this
point is completely trivial and is easily handled. In enumerating the channels
of a system containing identical particles we shall count only those channels
in which the particles of each fragment have the correct allowed symmetry.
By the same token we shall always use channel wave functions which are
properly symmetrized with respect to the internal coordinates of each fragment. Thus, the wave function describing an electron incident on a helium
atom will have the form

$$\chi(x_1)\phi(x_2, x_3)$$

where $\chi(x_1)$ is the incident wave packet and $\phi(x_2, x_3)$ is the properly anti-
symmetric helium wave function. Just as in the one-channel case we shall *not*
bother to symmetrize between the incident electron and those of the target
when labelling the asymptotic states.

22-d. Transition Probabilities and Cross Sections

Having classified the channels of our n-particle system we can now set up
a time-dependent description of the collision process and compute the
various scattering probabilities. If we multiply the appropriate results for
distinct particles [namely (22.19) and (22.20)] by the symmetrizer Λ we
immediately obtain the asymptotic condition for identical particles. An
orbit that originates in channel α has the form

$$\Lambda e^{-iHt}\psi \xrightarrow[t \to -\infty]{} \Lambda e^{-iH^{\alpha}t}\psi_{\text{in}} \qquad [\psi = \Omega_+^{\alpha}\psi_{\text{in}}] \qquad (22.23)$$

while an orbit which is going to terminate in channel α has the form

$$\Lambda e^{-iHt}\psi \xrightarrow[t \to \infty]{} \Lambda e^{-iH^\alpha t}\psi_{\text{out}} \qquad [\psi = \Omega^\alpha_- \psi_{\text{out}}] \qquad (22.24)$$

(It should be noted that because of the projector Λ these orbits are not yet properly normalized.)

Suppose now that we wish for the probability that an in state ϕ in channel α lead to the out state ϕ' in channel α'. If the in state was given by $\psi_{\text{in}} = \phi$ in $\mathscr{S}^\alpha$, then according to (22.23) the actual state at $t = 0$ is

$$a\Lambda\Omega^\alpha_+ \phi \qquad (22.25)$$

where a is some normalization factor yet to be determined. If the out asymptote were going to be ϕ' in $\mathscr{S}^{\alpha'}$, then according to (22.24) the actual state at $t = 0$ would have to be

$$a'\Lambda\Omega^{\alpha'}_- \phi' \qquad (22.26)$$

with a' the appropriate normalization factor. The required transition amplitude is just the overlap of these two functions, and hence,

$$
\begin{aligned}
w(\phi', \alpha' \leftarrow \phi, \alpha) &= (a'a)^2 \,|\langle \Lambda\Omega^{\alpha'}_- \phi', \Lambda\Omega^\alpha_+ \phi\rangle|^2 \\
&= (a'a)^2 \,|\langle \Lambda\Omega^{\alpha'}_- \phi', \Omega^\alpha_+ \phi\rangle|^2 \\
&\propto \left| \sum_\Pi \eta_\Pi \langle \Pi\Omega^{\alpha'}_- \phi', \Omega^\alpha_+ \phi\rangle \right|^2 \qquad (22.27)
\end{aligned}
$$

The interpretation of this result is completely natural. The first term inside the sum (that coming from $\Pi = 1$) is the amplitude one would obtain for the process if the particles were all distinct; the other terms are the amplitudes for the various "exchange" processes differing from the original one by permutations of the final (identical) particles. Since the particles are really indistinguishable, the actual amplitude is obtained by summing all of these amplitudes, multiplied by the appropriate factors ± 1 if the particles are fermions.

To determine the constant of proportionality in (22.27) we must determine the normalization factors a and a' in the states (22.25) and (22.26). The calculation of these factors in general (i.e., with arbitrary numbers of particles of arbitrarily many different types and with arbitrary channels α and α') is straightforward but tedious. For this reason we consider here just one important special case, a collision between a single particle and a target containing n particles of the same type as the projectile (plus perhaps other distinct particles). This includes the important examples of collisions between an electron and an n-electron atom, and between a nucleon and an n-body nucleus (with protons and neutrons treated as identical using the isospin formalism). We shall also confine attention to processes in which the in and out channels have the same arrangements; that is, we consider only elastic scattering and excitation, and do not treat rearrangements like $p + {}^4\text{He} \rightarrow d + {}^3\text{He}$. The interested reader should have no difficulty in extending our considerations to more general situations.

Since the channel α consists of one free particle plus n identical particles bound in the target the initial wave function ϕ has the form

$$\phi(x_0, x_1, \ldots, x_n) = \theta(x_0)\zeta(x_1, \ldots, x_n)$$

where $\theta(x_0)$ is the incident wave packet, $\zeta(x_1, \ldots, x_n)$ is the properly symmetrized wave function describing the motion of the n target particles, and we take both θ and ζ to be normalized. With this form for ϕ we have to determine the proper normalization of the corresponding actual state (22.25) at $t = 0$. Now, we know that the actual orbit has the form

$$\Lambda e^{-iHt}(\Omega_+^\alpha \phi) \xrightarrow[t \to -\infty]{} \Lambda e^{-iH^\alpha t}\phi \tag{22.28}$$

and that the limit is reached as particle 0 moves away from the remaining n particles. When this happens, the wave functions of particle 0 and the remaining n particles cease to overlap; and when this is the case the correctly normalized asymptote is easily seen to be

$$(n + 1)^{\frac{1}{2}}\Lambda e^{-iH^\alpha t}\phi$$

Thus, from (22.28) it follows that the correctly normalized actual state at such times, and hence at all times, is

$$(n + 1)^{\frac{1}{2}}\Lambda e^{-iHt}(\Omega_+^\alpha \phi)$$

That is, the normalization factor in the state (22.25) is $(n + 1)^{\frac{1}{2}}$. Since the same considerations apply to the out channel α', we conclude that

$$w(\phi', \alpha' \leftarrow \phi, \alpha) = (n + 1)^2 \,|\langle \Lambda \Omega_-^{\alpha'} \phi', \Omega_+^\alpha \phi \rangle|^2$$

$$= \left(\frac{1}{n!}\right)^2 \Big| \sum_\Pi \eta_\Pi \langle \Pi \Omega_-^{\alpha'} \phi', \Omega_+^\alpha \phi \rangle \Big|^2 \tag{22.29}$$

To simplify the sum in (22.29) it is convenient to rewrite each of the permutations Π of the $n + 1$ particles $0, 1, \ldots, n$ in the form

$$\Pi = \Pi'\Pi''$$

where Π' exchanges 0 with one of the variables $0, \ldots, n$ leaving all others alone, while Π'' is a permutation of $1, \ldots, n$ only. The sum over the $(n + 1)!$ permutations Π can then be replaced by a sum over the $(n + 1)$ permutations Π' and the $n!$ permutations Π''. Now, the wave function on which Π operates is already symmetrized with respect to particles $1, \ldots, n$ and so each of the $n!$ different permutations Π'' produces the same result.

Summing over these exactly cancels the factor $1/n!$ outside to give

$$w(\phi', \alpha' \leftarrow \phi, \alpha) = \left| {\sum_{\Pi}}' \eta_\Pi \langle \Pi \Omega_-^{\alpha'} \phi', \Omega_+^{\alpha} \phi \rangle \right|^2$$

where now $\sum'$ denotes a sum over the $n + 1$ permutations that exchange the original projectile 0 with one of the particles $0, \ldots, n$ leaving all other particles undisturbed.

Finally we can divide the sum $\sum'$ into one term corresponding to $\Pi = 1$ and the remaining n terms in which particle 0 changes place with one of the target particles $1, \ldots, n$. Since the wave function on which Π acts is already symmetrized with respect to particles $1, \ldots, n$ these last n terms are all the same and we arrive at the final result

$$w(\phi', \alpha' \leftarrow \phi, \alpha) = |\langle \Omega_-^{\alpha'} \phi', \Omega_+^{\alpha} \phi \rangle + \eta n \langle \Pi_{01} \Omega_-^{\alpha'} \phi', \Omega_+^{\alpha} \phi \rangle|^2 \quad (22.30)$$

where as usual $\eta = \pm 1$ depending on whether the particles are bosons or fermions and Π_{01} denotes the permutation which just exchanges particles 0 and 1.

From this result we can immediately proceed to the calculation of the observed cross section. The process under consideration has two-body initial and final states, being the excitation of the n-body target in the state α to a final state α', and the cross section follows from (22.30), exactly as in Section 17-c (we assume spinless particles and target for simplicity):

$$\frac{d\sigma}{d\Omega}(\mathbf{p}', \alpha' \leftarrow \mathbf{p}, \alpha) = \frac{p'}{p} |\hat{f}(\mathbf{p}', \alpha' \leftarrow \mathbf{p}, \alpha)|^2$$

where

$$\boxed{\hat{f}(\mathbf{p}', \alpha' \leftarrow \mathbf{p}, \alpha) = f_{di}(\mathbf{p}', \alpha' \leftarrow \mathbf{p}, \alpha) + \eta n f_{ex}(\mathbf{p}', \alpha' \leftarrow \mathbf{p}, \alpha)} \quad (22.31)$$

Here f_{di} is the amplitude one would calculate for the "direct" process:

direct: $0 + (1\,2 \cdots n)_\alpha \to 0 + (1\,2 \cdots n)_{\alpha'}$

on the assumption that the particles are distinct; that is,

$$f_{di}(\mathbf{p}', \alpha' \leftarrow \mathbf{p}, \alpha) = f(\mathbf{p}', \alpha' \leftarrow \mathbf{p}, \alpha)$$

if, as usual, we use f to denote the amplitude calculated for distinct particles. The amplitude f_{ex} is the corresponding "exchange" amplitude for the ejection of particle 1 with momentum $\mathbf{p}'$ and the simultaneous capture of particle 0:

exchange: $0 + (1\,2 \cdots n)_\alpha \to 1 + (0\,2 \cdots n)_{\alpha'}$ \quad (22.32)

That is,

$$f_{ex}(\mathbf{p}', \alpha' \leftarrow \mathbf{p}, \alpha) = f(\mathbf{p}', \tilde{\alpha}' \leftarrow \mathbf{p}, \alpha)$$

where $\tilde{\alpha}'$ denotes the exchange channel of (22.32). Since the original particle 0 can exchange with any one of the n equivalent particles in the target, the amplitude f_{ex} appears multiplied by n in the expression (22.31) for $\hat{f}$.

With the result (22.31) we have expressed the amplitude $\hat{f}$ for identical particles in terms of two amplitudes, f_{di} and f_{ex}, for processes involving distinct particles. Since we already know how to compute these two amplitudes, our identical-particle problem is now (in principle at least) completely solved.

To conclude this section we should recall that we have considered just the special case that a single projectile is incident on a target containing n particles of the same type as the projectile. It is a straightforward, though tedious, matter to extend our results to more general processes. In particular, the case of a two-body projectile incident on an n-body target (e.g., deuteron–nucleus scattering) is discussed in Problem **22.5**.

22-e. Electron–Hydrogen Scattering

To illustrate the formalism of the last two sections we conclude this chapter with a discussion of the elastic scattering of electrons off atomic hydrogen. The Hamiltonian for the system of two electrons plus a proton (which we take to be infinitely heavy) we take to be

$$H = \frac{\mathbf{P}_1^2}{2m} + \frac{\mathbf{P}_2^2}{2m} - \frac{e^2}{r_1} - \frac{e^2}{r_2} + \frac{e^2}{r_{12}}$$

that is, we neglect all spin-dependent terms. Apart from the break-up channel zero ($e_1 + e_2 + p$) there are just two kinds of channel, both consisting of an electron and a hydrogen atom. There are the channels

$$e_1 + (e_2 p) \tag{22.33}$$

and for each of these there is a (physically indistinguishable) exchange channel

$$e_2 + (e_1 p) \tag{22.34}$$

In particular, the only channels with which we shall be concerned are the incident channel 1, in which electron 1 is incident on the ground state hydrogen atom as in (22.33), and the exchange channel $\tilde{1}$ in which electron 1 is bound in a ground state while electron 2 moves freely, as in (22.34).

The differential cross section for the process in which an electron with momentum $\mathbf{p}$ and spin m_1 is incident on an atom whose electron has spin m_2, and scatters elastically with momentum $\mathbf{p}'$ and spin m_1' leaving the target in its ground state with spin m_2' is

$$\frac{d\sigma}{d\Omega}(\mathbf{p}', m_1', m_2', 1 \leftarrow \mathbf{p}, m_1, m_2, 1) = |\hat{f}(\mathbf{p}', m_1', m_2', 1 \leftarrow \mathbf{p}, m_1, m_2, 1)|^2$$

According to (22.31) the amplitude $\hat{f}$ has the form

$$\hat{f} = f_{\text{di}} - f_{\text{ex}}$$

Here f_{di} is the direct amplitude and can be written as

$$f_{\text{di}}(\mathbf{p}', m_1', m_2', 1 \leftarrow \mathbf{p}, m_1, m_2, 1) = -(2\pi)^2 m \langle \mathbf{p}', m_1', m_2', 1| \, V^1 \, |\mathbf{p}, m_1, m_2, 1+\rangle$$

where V^1 is the channel 1 potential

$$V^1 = -\frac{e^2}{r_1} + \frac{e^2}{r_{12}}$$

and f_{ex} is the amplitude for the rearrangement process

$$e_1 + (e_2 p) \rightarrow e_2 + (e_1 p)$$

which can be written as

$$f_{\text{ex}}(\mathbf{p}', m_1', m_2', 1 \leftarrow \mathbf{p}, m_1, m_2, 1) = -(2\pi)^2 m \langle \mathbf{p}', m_2', m_1', \tilde{1}| \, V^{\tilde{1}} \, |\mathbf{p}, m_1, m_2, 1+\rangle$$

Here $V^{\tilde{1}}$ is the potential of channel $\tilde{1}$,

$$V^{\tilde{1}} = -\frac{e^2}{r_2} + \frac{e^2}{r_{12}}$$

and the bra represents a state in which electron 2 moves freely with momentum $\mathbf{p}'$ and spin m_1' and electron 1 is bound with spin m_2'.

To illustrate the use of these results we can compute the amplitudes concerned in Born approximation. As we have already seen, the Born approximation for the direct amplitude coincides with the Born approximation that is obtained by treating the target atom as a static charge distribution and was given in (9.21) as

$$f_{\text{di}}(\mathbf{p}', m_1', m_2', 1 \leftarrow \mathbf{p}, m_1, m_2, 1)$$

$$= 2a \frac{8 + q^2 a^2}{(4 + q^2 a^2)^2} \langle m_1', m_2' \mid m_1, m_2 \rangle \qquad (22.35)$$

The last factor here is the scalar product of the initial and final spin states; this appears as a simple factor because the interactions are spin-independent.

The evaluation of the exchange amplitude f_{ex} involves the use of the Born approximation for a rearrangement collision—a procedure we have not yet considered in detail. Straightforward application of the Born approximation discussed in Section 18-d would give

$$f_{\text{ex}} = -(2\pi)^2 m \, \langle \mathbf{p}', \tilde{1}| \, V^{\tilde{1}} \, |\mathbf{p}, 1+\rangle \approx -(2\pi)^2 m \, \langle \mathbf{p}', \tilde{1}| \, V^{\tilde{1}} \, |\mathbf{p}, 1 \rangle$$

$$= -(2\pi)^2 m \, \langle \mathbf{p}', \tilde{1}| \left(-\frac{e^2}{r_2} + \frac{e^2}{r_{12}} \right) |\mathbf{p}, 1 \rangle \qquad (22.36)$$

(We have temporarily dropped spin labels.) This approximation is presumably the first term in some sort of expansion of f_{ex} in powers of the potential

$$V^1 = -\frac{e^2}{r_1} + \frac{e^2}{r_{12}}$$

However, the justification for such an expansion is at best very questionable. The potential V^1 contains two terms, the "core interaction" $-e^2/r_1$, and the electron–electron repulsion e^2/r_{12}. Now, it is the core interaction $-e^2/r_1$ that holds the final hydrogen atom together, and this interaction must therefore be (and is) taken into account to all orders in the final-state wave function of (22.36). Therefore, the approximation (22.36) is not the first term in a consistent expansion of f_{ex} in powers of V^1.

It is generally agreed that a more satisfactory procedure is first to use the DWBA for f_{ex}, treating the core interaction exactly and expanding in powers of the electron–electron interaction only. This procedure is obviously not subject to the objections just described. The first term in this power series is exactly zero (since it is the exact amplitude for rearrangement with no electron–electron interaction present—which is zero) and so we find

$$f_{ex} = -(2\pi)^2 m \, \langle \mathbf{p'}, \tilde{1} -_I | \frac{e^2}{r_{12}} | \mathbf{p}, 1+_I \rangle + O\left(\frac{e^2}{r_{12}}\right)^2 \qquad (22.37)$$

Here the subscripts I indicate the scattering states in the absence of the electron–electron interaction and can, in this case, be exactly computed.[8]

The DWBA is often further approximated by expanding the distorted waves in powers of the core interactions. To lowest order this gives

$$f_{ex} \approx -(2\pi)^2 m \, \langle \mathbf{p'}, \tilde{1} | \frac{e^2}{r_{12}} | \mathbf{p}, 1 \rangle \qquad (22.38)$$

It will be observed that this result is the same as the Born approximation (22.36) except that the term involving the core interaction in the Born approximation is completely absent.

It is usually accepted that the approximation (22.38) is more "correct" than the Born approximation (22.36).[9] However, it should be emphasized that, while the superiority of the DWBA (22.37) over the Born approximation (22.36) is reasonably clear, the additional approximations leading from the DWBA to the final result (22.38) are open to the same objections as apply to the original Born approximation. Thus, there is no compelling *a priori*

---

[8] In fact, the wave function for the state $|\mathbf{p}, 1+_1\rangle$ is easily seen to be the product of the Coulomb scattering state of momentum $\mathbf{p}$ for electron 1 times the hydrogen ground-state wave function for electron 2.

[9] See Kang and Sucher (1966).

reason to prefer (22.38) over the simple Born approximation. Nonetheless the arguments leading to (22.38) do at least have the advantage of appearing more systematic. More important, in those cases where it can be tested the approximation (22.38), without any core term, has been found to be in better agreement with experiment. For these reasons we shall use the form (22.38).

According to the approximation (22.38) the exchange amplitude (with spin indices restored) is:

$$f_{ex}(\mathbf{p}', m_1', m_2', 1 \leftarrow \mathbf{p}, m_1, m_2, 1)$$

$$= -\frac{m}{2\pi} \int d^3x_1 \int d^3x_2 e^{-i\mathbf{p}'\cdot\mathbf{x}_2} \phi_{100}(\mathbf{x}_1) \frac{e^2}{r_{12}} e^{i\mathbf{p}\cdot\mathbf{x}_1} \phi_{100}(\mathbf{x}_2)$$

$$\times \langle m_2', m_1' \mid m_1, m_2 \rangle$$

This integral can be evaluated in the limit of large momentum, $pa \gg 1$ (the limit in which the Born approximation should be good) to give

$$_{ex}(\mathbf{p}', m_1', m_2', 1 \leftarrow \mathbf{p}, m_1, m_2, 1)$$

$$= \frac{-32a}{p^2a^2(4 + q^2a^2)^2} \langle m_2', m_1' \mid m_1, m_2 \rangle \quad (22.39)$$

The results (22.35) and (22.39) for f_{di} and f_{ex} can now be substituted into $\hat{f} = f_{di} - f_{ex}$ to give the various cross sections:

$$\frac{d\sigma}{d\Omega}(+ + \leftarrow + +) = \frac{4a^2}{(4 + q^2a^2)^4}\left[(8 + q^2a^2) + \frac{16}{p^2a^2}\right]^2 \quad (22.40)$$

$$\frac{d\sigma}{d\Omega}(+ - \leftarrow + -) = \frac{4a^2}{(4 + q^2a^2)^4}[(8 + q^2a^2)]^2 \quad (22.41)$$

$$\frac{d\sigma}{d\Omega}(- + \leftarrow + -) = \frac{4a^2}{(4 + q^2a^2)^4}\left[\frac{16}{p^2a^2}\right]^2 \quad (22.42)$$

All other cross sections $(m_1', m_2' \leftarrow m_1, m_2)$ are the same as one of these or are zero.[10]

These cross sections are illustrated for an energy of about 100 eV in Fig. 22.3 below. However, before discussing these plots we examine some general features of the results (22.40) to (22.42). We first note that, whereas both the direct and exchange amplitudes contribute to the process $(+ + \leftarrow + +)$, only f_{di} contributes to $(+ - \leftarrow + -)$ and only f_{ex} contributes to the spin–flip process $(- + \leftarrow + -)$. In particular, if f_{ex} were zero then the first

---

[10] We could equally analyze the electron–hydrogen cross section in terms of the singlet and triplet states used in the discussion of electron–electron scattering in Section 22-b. Similarly, the latter discussion could have been presented in terms of the (m_1, m_2) spin states.

two processes would have equal cross sections, while the spin–flip process could not occur at all. These features are true of the exact results as well as the Born approximation and are easy to understand. The underlying interactions are spin-independent and, therefore, the spins of the two electrons cannot change. Thus, the only way for the spin–flip process $(-+\leftarrow+-)$ to occur is for the spin–down target electron to be ejected and the spin–up incident electron to be captured; that is, for the two electrons to exchange. In the same way, since the exchange process necessarily reverses the spins of the initial state $(+-)$, it cannot contribute to the process $(+-\leftarrow+-)$. On the other hand, it obviously can contribute to $(++\leftarrow++)$.

An important feature of the Born results (22.40) to (22.42) is that the ratio of the exchange to the direct amplitude drops like $1/p^2$ at high energies. Since the Born approximation is almost certainly reliable at sufficiently high energies, this result is probably correct. It is also easy to understand, since one would naturally expect the probability of the simultaneous capture of the incident projectile and ejection of the target particle to drop more quickly than that of direct elastic scattering as $p \to \infty$. Since the exchange term reflects the identity of two electrons, this means that the neglect of particle

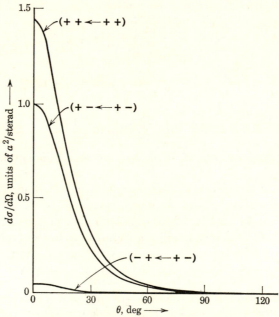

FIGURE 22.3.   **Electron–hydrogen   elastic   cross   sections   at   136 eV   in   Born approximation.**

identity, as practiced on several occasions in the course of this book, is justified at high energies.

Concerning the specific magnitudes of the cross sections (22.40) to (22.42), the plots of Fig. 22.3 show the cross sections at an energy of 136 eV [$(pa)^2 = 10$, to be precise]—an energy at which the Born approximation is probably useful, at least as an order of magnitude estimate.[11] It will be seen that, at this energy, the exchange term is small compared to the direct, but not so small that any precise calculation could afford to neglect it. In particular, the forward cross sections $(++ \leftarrow ++)$ and $(+- \leftarrow +-)$, which are equal if $f_{ex} = 0$, in fact differ by 50%. As the energy increases, of course, this difference will drop like $1/E$.

To conclude, we remark that this example, of electron–hydrogen scattering, shows well that there are certain general conditions under which one can confidently ignore the fact that some of the particles in a collision are identical. Thus, on the basis of this example we can conclude that in the elastic scattering of electrons off atoms (or nucleons off nuclei) at sufficiently high energies the exchange amplitude will be negligible compared to the direct. It also seems clear that in scattering off very stable targets at low energies (e.g., proton–alpha scattering) one can expect the likelihood of ejecting a target particle to be small; and, again, f_{ex} will be negligible compared to f_{di}. Whenever this is the case one can simply ignore the identity of the particles and the whole machinery of this chapter becomes unnecessary.

PROBLEMS

22.1. (a) Suppose there are two different kinds of spinless boson a and b, with the same mass and such that the interaction between two a particles is exactly the same as that of an a with a b, $V_{aa} = V_{ab}$. A beam of low-energy b particles is fired through a target of a particles and is found to be 1% attenuated (i.e., 99% of the beam gets through unscattered). A beam of a particles is then fired with the same energy through the same target of a particles. Assuming that the energy is so low that the scattering is pure s-wave, show that the beam of a particle will be 2% attenuated.

(b) Instead, suppose that the particles a and b are spin-half fermions and that their interactions are spin-independent. If the incident beam and target are completely polarized with their spins parallel, show that under the conditions of part (a) the a target becomes completely transparent to the beam of a particles (i.e., there is no a–a scattering at all).

22.2. Prove the optical theorem, $\text{Im} \hat{f}(\mathbf{p} \leftarrow \mathbf{p}) = (p/4\pi)\sigma(\mathbf{p})$, for the scattering of two identical spinless bosons. (This naturally reflects that the

[11] Unfortunately there are no experimental data to check this.

symmetrized in states are mapped onto the symmetrized out states by a unitary operator. However, we have not discussed this point and the result can be easily proved by writing $\hat{f}(\mathbf{p}' \leftarrow \mathbf{p}) = f(\mathbf{p}' \leftarrow \mathbf{p}) + f(-\mathbf{p}' \leftarrow \mathbf{p})$ and using the known result (3.26) for f. Note that for two identical spinless particles the potential $V(\mathbf{x})$ is automatically invariant under parity (Why?) and so you can freely replace $f(\mathbf{p}' \leftarrow \mathbf{p})$ by $f(-\mathbf{p}' \leftarrow -\mathbf{p})$.

22.3. It is important to know the most general form of the amplitude for two identical particles consistent with the particles' identity and the various possible invariance principles. This is the subject of this problem (which is trivial) and the next (which is not).

The properly symmetrized amplitude for two identical particles to undergo any process $(\phi' \leftarrow \phi)$ must equal ± 1 times that for the process $(\Pi\phi' \leftarrow \phi)$ *and* ± 1 times that for $(\phi' \leftarrow \Pi\phi)$. Equivalently, we can say that the amplitude for $(\phi' \leftarrow \phi)$ must equal ± 1 times that for $(\Pi\phi' \leftarrow \phi)$ and must be *equal* to that for $(\Pi\phi' \leftarrow \Pi\phi)$. For two spinless bosons these two conditions are just

$$\hat{f}(\mathbf{p}' \leftarrow \mathbf{p}) = \hat{f}(-\mathbf{p}' \leftarrow \mathbf{p}) \tag{22.43}$$

and

$$\hat{f}(\mathbf{p}' \leftarrow \mathbf{p}) = \hat{f}(-\mathbf{p}' \leftarrow -\mathbf{p}) \tag{22.44}$$

In this case, the condition (22.44) is precisely the condition of invariance under parity. This means that the scattering of two identical spinless bosons is automatically invariant under parity. Conversely, if we already know that parity is good, then (22.44) contains no new information. In particular, if rotational invariance holds, then we can write $\hat{f}$ as $\hat{f}(E, \theta)$ and (22.44) is automatic; in this case the requirements of particle identity are completely incorporated in (22.43) which becomes

$$\hat{f}(E, \theta) = \hat{f}(E, \pi-\theta)$$

Show that the partial-wave series for scattering of two identical spinless bosons contains only even values of l. (This just reflects the fact that there are no states of odd l.)

22.4. Consider the scattering of two identical spin-half fermions (e.g., proton–proton scattering). The amplitude for $(\mathbf{p}', \chi' \leftarrow \mathbf{p}, \chi)$ can be written as $\chi'^{\dagger} \hat{F}(\mathbf{p}' \leftarrow \mathbf{p})\chi$ and the results analogous to (22.43) and (22.44) are easily seen to be

$$\hat{F}(\mathbf{p}' \leftarrow \mathbf{p}) = -\Pi \hat{F}(-\mathbf{p}' \leftarrow \mathbf{p}) \tag{22.45}$$

and

$$\hat{F}(\mathbf{p}' \leftarrow \mathbf{p}) = \Pi \hat{F}(-\mathbf{p}' \leftarrow -\mathbf{p})\Pi \tag{22.46}$$

where Π is the permutation operator on the four-dimensional spin space. These two conditions can now be applied to the analysis of Problem **6.5,** where (the reader will recall) the matrix $\hat{F}$ was expanded as

$$\hat{F} = \alpha I + \beta_i \sigma_i^1 + \gamma_i \sigma_i^2 + \epsilon_{ij} \sigma_i^1 \sigma_j^2$$

and, assuming rotational invariance, the quantities $\alpha, \beta_i, \gamma_i, \epsilon_{ij}$ were expanded in terms of the vectors **n**, **q**, **k** and 16 independent scalar amplitudes, $a(\theta), b(\theta), \ldots$ (We suppress the energy variable E.)

The condition (22.46) is the more easily applied, since $\Pi \sigma^1 \Pi = \sigma^2$ and vice-versa. Thus, (22.46) implies simply that $\hat{F}$ must be unchanged if we exchange σ^1 and σ^2 and simultaneously replace **p** and **p'** by $-$**p** and $-$**p'**. This gives certain identities among the quantities $\alpha, \ldots, \epsilon_{ij}$ and, hence, reduces the number of independent amplitudes $a(\theta), \ldots$. The condition (22.45) is much harder to apply (see part b). It relates the amplitudes $a(\theta), \ldots$ to their values at $\pi{-}\theta$. Thus, it does not reduce the number of independent amplitudes, but instead requires that they have certain simple properties under the replacement $\theta \to \pi - \theta$.

(a) Using only (22.46), show that the scattering of two identical spin-half particles with rotationally invariant interactions is given by nine independent scalar amplitudes. Show that P invariance reduces this number to six, T reduces it to seven, and P *and* T to five. In particular, for the case of maximum symmetry (R, P, and T) show that $\hat{F}$ can be written

$$\hat{F} = aI + b\,\mathbf{n}{\cdot}(\boldsymbol{\sigma}^1 + \boldsymbol{\sigma}^2) + c\,\mathbf{n}{\cdot}\boldsymbol{\sigma}^1\,\mathbf{n}{\cdot}\boldsymbol{\sigma}^2 + d\,\mathbf{q}{\cdot}\boldsymbol{\sigma}^1\,\mathbf{q}{\cdot}\boldsymbol{\sigma}^2 + e\,\mathbf{k}{\cdot}\boldsymbol{\sigma}^1\,\mathbf{k}{\cdot}\boldsymbol{\sigma}^2$$

(b) Apply the condition (22.45) to this last result to show that the five amplitudes $a, \ldots, e$ can be written as linear combinations of five other amplitudes $A, \ldots, E$ three of which are unchanged when $\theta \to \pi{-}\theta$ and two of which change sign. [The difficulty in applying (22.45) is that it permutes the two final particles, but not the two initial ones. The simplest way to cope with this is to think in terms of spin states with a definite symmetry under permutations; namely the singlet and triplet states. Thus, if you introduce singlet and triplet projectors,

$$\Lambda_0 = \tfrac{1}{4}(1 - \boldsymbol{\sigma}^1 \cdot \boldsymbol{\sigma}^2)$$

and

$$\Lambda_1 = \tfrac{1}{4}(3 + \boldsymbol{\sigma}^1 \cdot \boldsymbol{\sigma}^2)$$

and write

$$\hat{F} = (\Lambda_0 + \Lambda_1)\hat{F} \equiv \hat{F}_0 + \hat{F}_1$$

then (22.45) becomes

$$\hat{F}_0(\mathbf{p'} \leftarrow \mathbf{p}) = +\hat{F}_0(-\mathbf{p'} \leftarrow \mathbf{p})$$

and

$$\hat{F}_1(\mathbf{p'} \leftarrow \mathbf{p}) = -\hat{F}_1(-\mathbf{p'} \leftarrow \mathbf{p})$$

The rest is algebra.]

(c) Review the angular-momentum analysis of (spin ½)–(spin ½) scattering in Problem **6.8**. For the case that the particles are identical and R, P, and T are all good, show that the matrix $S^j(E)$ has only one independent element for j odd, and four for j even (except $j = 0$). (Do not try to write down the partial-wave series; just enumerate the possible states for a given j.) Note that these numbers add up to the five amplitudes of the previous part.

22.5. Consider the elastic scattering of a two-body projectile off an n-body target, all $n + 2$ particles being spinless and identical. (This gives a schematic model for deuteron–nucleus scattering.) Using an analysis similar to that of Section 22-d, show that the properly symmetrized amplitude is the sum of three terms: one representing elastic scattering without exchange, one the exchange of one particle from the projectile with one from the target (this term multiplied by $2n$), and one the exchange of both projectile particles with two from the target [this term multiplied by $n(n - 1)/2$].

References

Alfaro, V. de and Regge, T. (1965), *Potential Scattering*. North-Holland, Amsterdam.

Amrein, W. O., Martin, P. A., and Misra, B. (1970), *Helv. Phys. Acta*, Vol. 43, p. 313.

Bodansky, D., et al. (1966), *Phys. Rev. Letters*, Vol. 17, p. 589.

Bomse, F., et al. (1967), *Phys. Rev.*, Vol. 158, p. 1281.

Bromberg, J. P. (1969), *J. Chem. Phys.*, Vol. 50, p. 3906.

Calogero, F. (1967), *The Variable Phase Approach to Potential Scattering*. Academic Press, New York.

Calucci, G., and Ghirardi, G. C. (1968), *Phys. Rev.*, Vol. 169, p. 1339.

Chadwick, J. (1930), *Proc. Roy. Soc.*, Vol. A128, p. 114.

Chamberlain, O., et al. (1956), *Phys. Rev.*, Vol. 102, p. 1659.

Chew, G. F. (1962), *S-Matrix Theory of Strong Interactions*. W. A. Benjamin, Inc., New York.

Collins, P. D. B., and Squires, E. J. (1968), *Regge Poles in Particle Physics*. Springer–Verlag, Berlin.

Condon, E. U., and Shortley, G. H. (1935), *Theory of Atomic Spectra*. Cambridge University Press, Cambridge.

Courant, R., and Hilbert D. (1953), *Methods of Mathematical Physics*. Interscience, New York.

Demkov, Y. I. (1963), *Variational Principles in the Theory of Collisions*. Pergamon Press, London.

Dollard, J. D. (1964), *J. Math. Phys.*, Vol. 5, p. 729.

Dothan, Y., and Horn, D. (1970), *Phys. Rev.*, Vol. D1, p. 916.

Eden, R. J., et al. (1966), *The Analytic S Matrix*. Cambridge University Press, Cambridge.

Edmonds, A. R. (1957), *Angular Momentum in Quantum Mechanics*. Princeton University Press, Princeton.

Erdélyi, A. (1953), *Higher Transcendental Functions*. McGraw–Hill Book Co., New York.

Faddeev, L. D. (1957), *Vestn. Lenin. Univ.*, Vol. 7, p. 164.

Faddeev, L. D. (1958), *Sov. Phys.—JETP*, Vol. 35, p. 433 (8, 299 in translation).

Faddeev, L. D. (1965), *Mathematical Aspects of the Three-Body Problem in Quantum Scattering Theory*. Israel Program for Scientific Translations, Jerusalem.

Frautschi, S. C. (1963), *Regge Poles and S-Matrix Theory*. W. A. Benjamin, Inc., New York.

Friedman, B. (1956), *Principles and Techniques of Applied Mathematics*. John Wiley & Sons, Inc., New York.

Froissart, M., and Taylor, J. R. (1967), *Phys. Rev.*, Vol. 153, p. 1636.

Geltman, S. (1969), *Topics in Atomic Collision Theory*. Academic Press, Inc., New York.

Gillespie, J. (1964), *Final-State Interactions*. Holden–Day, Inc., San Francisco.

Goldberger, M. L., and Watson, K. M. (1964), *Collision Theory*. John Wiley & Sons, Inc., New York.

Gottfried, K. (1966), *Quantum Mechanics*. W. A. Benjamin, Inc., New York.

Greenberg, J. S., et al. (1960), *Phys. Rev.*, Vol. 120, p. 1393.

Hack, M. N. (1959), *Nuovo Cimento*, Vol. 13, p. 231.

Hepp, K. (1965), "On the Connection Between Wightman and L.S.Z. Quantum Field Theory," in *Axiomatic Field Theory*. Edited by Chretien and Deser. Gordon and Breach, New York.

Hepp, K. (1969), *Helv. Phys. Acta*, Vol. 42, p. 425.

Hughes, D. J., and Harvey, J. A. (1955), *Neutron Cross Sections*. McGraw–Hill Book Co., New York.

Hughes, A. L., McMillen, J. H., and Webb, G. M. (1932), *Phys. Rev.*, Vol. 41, p. 154.

Hunziker, W. (1961), *Helv. Phys. Acta*, Vol. 34, p. 593.

Hunziker, W. (1968), "Mathematical Theory of Multiparticle Systems," in *Lectures in Theoretical Physics: Volume X-A*. Edited by Barut and Brittin. Gordon and Breach, New York.

Ikebe, T. (1960), *Arch Ration. Mech. Anal.*, Vol. 5, p. 1.

Jackson, J. D. (1970), *Rev. Mod. Phys.*, Vol. 42, p. 12.

Jacob, M., and Chew, G. F. (1964), *Strong Interaction Physics.* W. A. Benjamin, Inc., New York.

Jacob, M., and Wick, G. C. (1959), *Ann. Phys.*, Vol. 7, p. 404.

Jordan, T. F. (1969), *Linear Operators for Quantum Mechanics.* John Wiley & Sons, Inc., New York.

Kang, I-J, and Sucher, J. (1966), *Phys. Lett.*, Vol. 20, p. 22.

Kato, T. (1959), *Commun. Pure Appl. Math.*, Vol. 12, p. 402.

Klein, A., and Zemach, C. (1958), *Nuovo Cimento*, Vol. 10, p. 1078.

Kral, N. A., and Gerjuoy, E. (1960), *Phys. Rev.*, Vol. 120, p. 143.

Lee, L. I., et al. (1964), *Phys. Rev.*, Vol. 136B, p. 971.

McCarthy, I. E. (1968), *Introduction to Nuclear Theory.* John Wiley & Sons, Inc., New York.

MacGregor, M. H., Moravcsik, M., and Stapp, H. P. (1960), *Annu. Rev. Nucl. Sci.*, Vol. 10, p. 291.

Messiah, A. M. L. (1961), *Quantum Mechanics.* John Wiley & Sons, Inc., New York.

Mott, N. F., and Massey, H. S. W. (1933 and 1965), *Theory of Atomic Collisions*, First Edition 1933, Third Edition 1965. Oxford University Press, London.

Newton, R. G. (1960), *J. Math. Phys.*, Vol. 1, p. 319.

Newton, R. G. (1964), *The Complex j-Plane.* W. A. Benjamin, Inc., New York.

Newton, R. G. (1966), *Scattering Theory of Waves and Particles.* McGraw-Hill Book Co., New York.

Nuttall, J., and Webb, J. G. (1969), *Phys. Rev.*, Vol. 178, p. 2226.

O'Malley, T. F., Spruch, L., and Rosenberg, L. (1961), *J. Math. Phys.*, Vol. 2, p. 491.

Omnes, R., and Froissart, M. (1963), *Mandelstam Theory and Regge Poles.* W. A. Benjamin, Inc., New York.

Rose, M. E. (1957), *Elementary Theory of Angular Momentum.* John Wiley & Sons, Inc., New York.

Rubin, M., Sugar, R., and Tiktopoulos, G. (1967), *Phys. Rev.*, Vol. 162, p. 1555.

Stolt, R. H., and Taylor, J. R. (1970), *Nucl. Phys.*, Vol. B19, p. 1.

Streater, R. F., and Wightman, A. S. (1964), *PCT, Spin and Statistics and all that.* W. A. Benjamin, Inc., New York.

Squires, E. J. (1963), *Complex Angular Momenta and Particles Physics*. W. A. Benjamin, Inc., New York.

Titchmarsh, E. C. (1939), *The Theory of Functions*, Second Edition. Oxford University Press, London.

Watson, G. N. (1958), *Theory of Bessel Functions*. Cambridge University Press, Cambridge.

Watson, K. M. (1952), *Phys. Rev.*, Vol. 88, p. 1153.

Watson, K. M., and Nuttall, J. (1967), *Topics in Several Particle Dynamics*. Holden–Day, Inc., San Francisco.

Wigner, E. P. (1959), *Group Theory*. Academic Press, Inc., New York.

Wolfenstein, L. (1956), *Annu. Rev. Nucl. Sci.*, Vol. 6, p. 43.

INDEX

Addition theorem, 88
Adiabatic theorem, 136-137
Amplitude, *see* Partial-wave amplitude,
    Scattering amplitude
Amplitude matrix, 75
  measurement of, 80
Analytic continuation, 213
Analytic functions, 213
  integrals of, 215
  series of, 214
Analytic operators, defined, 130
Analytic potentials, 220-221
Analytic S-matrix theory, 286, 301, 336
Analyticity, in cos θ of partial-wave series,
    308-310
  in E and q^2 of the amplitude, 297-300
  in E of the forward amplitude, 291-293
  in l, 302-304
  in strength parameter, 207-208
  of f_l in p or E, 220-223, 287-290
  of Green's operator, 130-132
  of Jost function, 217-223
  of multichannel amplitudes, 392-404
  of regular wave function, 216-217
  of s_l in p, 220-222
Angular momentum, as total plus
    internal, 84
  and helicity eigenstates, 122
  complex, 302-314
  conservation of, and rotational
    invariance, 83-84, 86
Angular momentum eigenstates, 86,
    103, 185-188
  for three-body states, 353
  free, 86, 103
  multichannel, 351
  relation to momentum eigenstates, 87
Anti-bound states, *see* Virtual state
Anti-linear operators, 92-93
Antisymmetric wave functions, 436
Anti-unitary operators, 92-93, 106
Arrangement channel, 319

Asymmetry ratio, 117
Asymptotes, 26
  classical, 23-25
  multichannel, 323-324, 328
Asymptotic completeness, 33
  multichannel, 329-330
Asymptotic condition, 28-30
  multichannel, 324-325
Asymptotic form of stationary wave
    functions, 172
  multichannel, 376-379, 389
  partial-wave, 187-188
Averages over spins, 77

$\mathcal{B}$, space of bound states, 32, 326
Barycentric frame, *see* CM frame
Basis, for multichannel free states, 339-342
  for spin states, 70-72
  of helicity states, 120-123
  of plane waves, 40-41
  of stationary scattering states, 167-168
Bessel functions, 182-184
  bounds on, 211, 218, 235
  orthogonality of, 183
Born approximation, 147-150, 365
  and unitarity, 149-150, 190-191
  conditions for reliability of, 146, 178
  for Coulomb potential, 262-264
  for e-H scattering, 154-155, 368-370,
    455-459
  for e-He scattering, 155-156
  for electron-atom scattering, 153-156
  for excitation, 367-370
  for high l, 193
  for ionization, 371
  for rearrangement collisions, 455-457
  for spherical potential, 148
  for square well, 163
  for Yukawa potential, 150-153, 178
  multichannel, 364-370
  multichannel elastic, 366-367, 371
  partial-wave, 190

Born approximation (*continued*)
  post and prior forms, 365, 371
  second, 151-152
  see also DWBA
Born series, 143-163
  conditions for convergence of, 146
  convergence for weak potentials, 207-208
  for s_l, 207-208
  for stationary scattering states, 170
  for the radial wave function, 207-208
  partial-wave, 190
Boson, defined, 436
Bound states, and poles of S matrix,
      223-226, 405-407
  and resonances, 245-249
  and zeros of det $\mathcal{T}$, 405-406
  and zeros of Jost function, 223-226
  effect on unitarity of $\Omega_{\pm}$, 34, 40
  embedded in the continuum, 407
  of zero energy, 232
Bounded orbits, classical, 23-24
  quantum, 31
Breit-Wigner resonance, 242, 244,
      247-248
  multichannel, 409-410, 412-413

Caley transform, 280-281
Cauchy principal value integral, 289
Cauchy test for convergence, 17-18
Center of mass, *see* CM
Centrifugal barrier, 192
Channel, 317, 319
  basis functions, 339-342
  closed, 389
  Green's function, 375
  Green's operator, 362
  Hamiltonian H^α, 322-324
  Hamiltonian, eigenfunctions of,
      374-375, 379
  Møller operators, 325
  open, 319
  potential V^α, 360
  subspaces, 323-324
  wave functions, 322-324
$\chi_{l,p}^{\pm}(r)$, defined, 224
Classical cross section, 44-46
Classical scattering, 22-25
Clebsch-Gordan coefficient, 71, 108
Close-coupling approximation, 382-383
Closed channels, 389

Cluster law, 439-441
CM coordinate, 58, 356
CM cross section, in terms of
      amplitude, 66-68
  relation to lab cross section, 66
  see also Cross section
CM frame, 64
Complex angular momenta, 302-314
Compound nucleus, 415
Confluent hypergeometric function, 265
Conservation of angular momentum, and
      rotational invariance, 83-84, 86
  for helicity states, 123
  in multichannel scattering, 351
Conservation of energy, 39-41, 55
  in multichannel scattering, 343-345
  in two-particle scattering, 62-63
Conservation of momentum, 62-63
  and translational invariance, 82-83
  in multichannel scattering, 343-344
Conservation of parity, *see* Parity invariance
Convergence of operators, 19
Convergence of vectors, 16-19
  Cauchy test for, 17-18
  weak and strong, 19
Core term, 371, 456-457
Coulomb amplitude f_C, 265
  interference with short-range
      amplitude, 269
Coulomb phase shift σ_l, 261
Coulomb potential, 259-269
  Born approximation for, 262-264
  cross section for, 266
  plus short-range potential, 266-269
  radial functions F_l, G_l, 261, 267
  stationary scattering functions for, 265
  strength parameter γ, 261
  with cutoff, 261-264, 266
Coupled-channel approximation, 382-383
Coupled-channel equations, 382-383,
      398-399
  partial-wave analysis of, 389-390, 399-400
Coupling constant λ, 191
  analyticity in, 207-208
  complex values of, 207
Cross section, at a resonance, 241-244,
      412-413
  at a very narrow resonance, 255, 258
  classical, 44-46
  CM, 66-68

Cross section (*continued*)
  conditions for observation of, 52, 177-178
  definition of, 46-48
  derivation in momentum space, 49-53
  derivation using spatial wave
      functions, 173-178
  derivation using stationary states, 4,
      377, 379
  differential, 51
  elastic, 350
  for Coulomb potential, 266
  for cutoff Coulomb potential,
      263-264, 266
  for particles with spin, 74-75
  for square well, 195
  for three-body final states, 349-350
  for two-body final states, 348-349
  for two identical fermions, 447
  for two identical bosons, 444-445
  in terms of amplitude, 49-51
  independent of incident packet shape, 48
  inelastic, 351
  involving several identical particles, 453
  isotropic at low energy, 194
  large at low energy, 246
  multichannel, 346-351
  partial-wave, 89
  total, 54, 350
  total, for identical particles, 448
  transformation between different
      frames, 64-66
  unpolarized, 77-79, 114-115
Crossed process, 290, 313

Damping equation, 282
Damping factor $\exp(-\epsilon\tau)$, 136-137
Delta function, 55
δ_l, *see* Phase shift
Density matrix, 109-113
  in and out, 113-114
Density operator, 110
Detailed balance, 354
  see also Time-reversal invariance
Deuteron stripping reactions, 421-424
Differential cross section, 51
  transformation from lab to CM, 66
  see also Cross section
Dirac delta function, 55
Dirac picture, 157
Direct amplitude, 435, 453, 455

Direct sums, 10, 331-333
Dispersion relations, double, 297-302
  forward, 291-294
  in relativistic scattering, 290, 300-302
  Kramers-Kronig, 289
  nonforward, 294-296
  partial-wave, 288-290
  subtracted, 299-300, 313
Displacement operator, 82
Distorted-wave Born approximation,
      see DWBA
Domain of an operator, 11
Double dispersion relations, 297-302
Double scattering experiments, 118-119
Double spectral function, 298
DWBA, 270-273, 283-284, 418-424
  for deuteron stripping, 421-424

Effective range expansion, 231, 236-237
Eigenstates, angular momentum and
      helicity, 122
  Coulomb, 265
  momentum, 41
  momentum and helicity, 120-122
  multichannel angular momentum, 351
  multichannel momentum, 339-342
  multichannel scattering, 359-360
  of H and angular momentum, 185-188
  of H^0 and angular momentum, 86, 103
  of total spin, 71
  scattering, 165-166
Electron-helium scattering, 155-156
Electron-hydrogen scattering, 154-155,
      294, 368-370, 454-459
Enhancement factors, 428-430
Entire functions, 207
Euler angles, 120
Evolution operator, 13-14
  for CM and relative motion, 60
  free, 26
  in interaction picture, 158
Exchange amplitude, 435, 453-457
Exchange channel, 450
Exchange process, 435
Excitation in collisions, 367-370
Exponential decay of resonance,
      253-254
  multichannel, 413-416
Exponential potential, 236
Exponentially bounded potentials, 220

f, see Scattering amplitude

f_l, *see* Partial-wave amplitude

$\mathcal{L}_l$, *see* Jost function

Factorable potentials, 37, 141

Factorization of the residue, 407, 417

Fermion, defined, 437

Feynman diagrams (or graphs), 162

Final-state interactions, 424-432

Finite-range potentials, 392

Form factors, 366-368, 371

Formal resonance theory, 416

Forward dispersion relations, 291-294

Free evolution operator, 26

Free Green's function, 170-172

Free Green's operator, 131-133

Free propagator, 161

Free radial functions, 182-185

Free wave packet, 174-175

Galilean invariance, 60

Gaussian wave packet, 29-30, 37

Gell-Mann-Goldberger result, 271

Generalized optical potential, 384, 388, 390

Generalized optical theorem, 301

Green's function, as matrix element of $G(z)$, 130

  bounds on, 211, 235

  channel, 375

  for regular solution, 201

  free, 170-172

  partial-wave, 189, 209

  see also Green's operator

Green's operator, 129-134

  analyticity of, 130-132

  channel, 362

  free, 131-133

  Lippmann-Schwinger equation for, 133

  standing-wave, 283

  see also Green's function

$\mathcal{H}_{as}$, defined, 334

  momentum basis of, 339-342

$\mathcal{H}_{cm}$ and $\mathcal{H}_{rel}$, 58

Hamiltonian, channel, 322-324

Hankel functions, Riccati, 183-184

  bounds on, 218-235

Hard sphere potential, 209-210

Heitler's damping equation, 282

Heitler's matrix, 281

  see also K matrix

Helicity, 119-127

  and angular-momentum states, 122-123

  and momentum eigenstates, 120-121

  defined, 119

  relation to S_3 states, 123

Hilbert space, 7

Holomorphic functions, 213

Identical particles, 434-462

  amplitudes involving several, 453, 462

  general form of amplitude for, 460-462

  general formalism of, 435-441

  multichannel scattering with, 448-454

  scattering of two, 441-448

  scattering probabilities for, 451-453

  total cross section for, 448

Impact parameter, 44-45, 47

Improper vectors, 8-9

In and out density matrices, 113-114

  for (spin ½) – (spin 0) scattering, 115

In asymptote, *see* Asymptotes

In spinors, 78-80

Incoming radial function, $\chi^-(r)$, 224

Inelasticity factor, 353

Interaction picture, 157

Intermediate states, 157, 160-161

Intertwining relations, 39-40

  multichannel, 342

Invariance, *see* Invariance under PT,
    Parity invariance, Rotational invariance,
    Time-reversal invariance, Translational
    invariance

Invariance under PT, 106, 357

Inverse of an operator, 12

Ionization, 371

Isometric operator, 14-16

  from $\mathcal{H}$ onto $\mathcal{H}'$, 333

  inverse and adjoint of, 16

Jost function, 204-207

  analyticity in l, 303

  analyticity in p, 217-223

  as enhancement factor, 428-430

  as $\lambda \to 0$, 206

  as $p \to \infty$, 206, 227

  as Wronskian of χ^+ and ϕ, 225

  bounds on, 206, 218

  cannot vanish for p real, 226

  integral for, 206

  power series for, 206

Jost function (*continued*)
  relation to S matrix, 205
  see also Zeros of Jost function
Jost matrix, 401-402
Jost solutions, $\chi^{\pm}(r)$, 224

K matrix, 280-283
  partial-wave, 230-231
Kohn variational method, 277-280
Kramers-Kronig relation, 289

$\mathcal{L}^2(\mathbb{R}^n)$, defined, 7
Lab frame, 64
λ, *see* Coupling constant λ
Left-hand cut, 287
  absence from full
    amplitude, 296-297
Left-right asymmetry, 117
Legendre function, bounds on, 312
Lehmann ellipse, 309-310
Levinson's theorem, 227-228, 233-234, 305
  (illustrated), 195-196
Lifetime of a resonance, 254-255
Linear operator, 10-11
  between different spaces, 333
Lippmann-Schwinger equation, for
    $G(z)$, 133
  for scattering states, 169
  for $T(z)$, 135
  multichannel, 362, 364
  partial-wave, 188-190
Low-energy behavior of f_l, 193-194,
    229-231, 234-235

Mandelstam representation, 297-302
  in relativistic scattering, 300-302
  subtracted, 299-300, 313
Mean life of a resonance, 254-255
Meromorphic functions, 220
Microreversibility, 354
  see also Time-reversal invariance
"Minus" state, $|\chi-\rangle$, 31
Mixed state, 110
Modulo π ambiguity in δ_l, 181, 192,
    198, 200-201
Møller operators, 30
  effect on Hamiltonian, 40
  in terms of $G(z)$ or $T(z)$, 135-137
  intertwining relations for, 39-40, 342
  isometry of, 34

Møller operators (*continued*)
  multichannel, 325, 334
Momentum eigenstates, 41
  multichannel, 339-342
  relation to angular-momentum
    states, 87, 184-185
Momentum-space wave functions, 9
Momentum transfer, 147
Mott scattering, 116
Multichannel scattering, 315-433
  defined, 1, 316
  see also Angular momentum,
    Asymptotes, Asymptotic
    completeness, etc.

N over D method, 290
N-state approximation, 382-383
Neumann function, Riccati, 183-184
  bounds on, 211, 235
Neutron-proton scattering, 235, 257
Nonforward dispersion relations, 294-296
Nonlocal potentials, 37
Norm of a vector, 7

Observables, 8
Off-shell T matrix, 43, 139-140
  see also T matrix
On-shell T matrix, 42-43
  relation to off-shell, 139
  see also T matrix
One-state approximation, 153, 382-383
Open channels, 319
Operators, anti-linear, 92-93
  anti-unitary, 92-93, 106
  between different spaces, 333
  isometric, 14-16, 333
  linear, 10-11
  projection, *see* Projector
  unitary, 5, 13-14
Optical potential, 384-388
  generalized, 384, 388, 390
Optical theorem, 53-55
  and forward dispersion relations, 293-294
  for particles with spin, 80
  for two identical particles, 459-460
  generalized, 301
  multichannel, 356
Orbit, 25
Orthogonal complement, 10
Orthogonality, 10

Orthogonality theorem, 32
  multichannel, 326-327
Orthonormal bases, 8-9
Out asymptote, *see* Asymptotes
Out spinors, 78-80
Outgoing radial function $X^{\pm}(r)$, 224

$\underline{p}$, set of momenta, 339
$\underline{\bar{p}}$, set of relative momenta, 344
Parity, defined, 90
  effect on helicity states, 126
  effect on spin states, 97
Parity invariance, 91
  and helicity states, 126
  for particles with spin, 104
  in (spin ½) – (spin ½) scattering, 107
  in (spin ½) – (spin 0) scattering,
    98-100, 102
Partial-wave amplitude, 88
  analyticity in l , 302-304
  analyticity in p or E, 220-223, 287-290
  as $l \to \infty$, 192-193, 305
  as $\lambda \to 0$, 192, 206
  as resonant part plus background, 256
  at high energy, 192, 206
  at low energy, 193-194, 229-231,
    234-235
  Born approximation for, 190
  Born series for, 190
  for square well, 210
  in helicity basis, 123
  in terms of full amplitude, 88
  in terms of radial wave function, 190
  multichannel, 352
  relation to radial function as $r \to \infty$, 187
  residue at poles, 236
  see also, Partial-wave S matrix
Partial-wave Born series, 190
  convergence at high energy, 208
  convergence for small λ, 207-208
Partial-wave cross section, 89
Partial-wave dispersion relation, 288-290
  in relativistic scattering, 290
Partial-wave Green's function, 189, 209
Partial-wave K matrix, 230-231
Partial-wave Lippmann-Schwinger
    equation, 188-190
Partial-wave S matrix, 86-87

Partial-wave S matrix (*continued*)
  analyticity in λ, 207-208
  analyticity in p, 220-222
  Born series for, 207-208
  for particles with spin, 103-105
  Hermiticity of, below lowest
    threshold, 404
  in (spin ½) – (spin 0) scattering, 105
  in terms of Jost function, 205
  multichannel, 351-353
  multichannel, analyticity of, 392-404
  poles of, and bound states, 223-226,
    405-407
  poles of, and resonances, 244
  relation to partial-wave amplitude, 88
  relation to phase shift, 87
  see also Partial-wave amplitude, S matrix
Partial-wave series, 88-90
  analyticity in $\cos \theta$, 308-310
  for coupled-channel equations, 389-390
  for (spin ½) – (spin 0) scattering, 105
  for two identical bosons, 460
  in helicity formalism, 124
  multichannel, 352
Partial-wave states, free, 86, 103
  multichannel, 351
  scattering, 185-188
Partial widths, 410, 412, 415-416
Pauli matrices, 101
Permutation operators, 436
Perturbation series for S matrix, 159
  see also Born series
Phase equation, 199-200
Phase shift, 87
  and radial wave function as $r \to \infty$, 187-188
  as $l \to \infty$, 192-193, 305
  as $\lambda \to 0$, 192, 200
  at high energy, 192, 200
  at low energy, 194, 231, 237
  at resonance, 240-242
  complex, 353
  Coulomb, 261
  derivative with respect to l, 304
  for Coulomb plus short-range
    potential, 268
  for square well, 195-196, 257
  limit on rate of decrease, 252
  modulo π ambiguity in, 181, 192,
    198, 200-201
  multichannel, 345

Phase shift (*continued*)
  relation to Jost function, 205
  sign of, for attractive and repulsive
      potentials, 200
  see also Partial-wave amplitude,
      Partial-wave S matrix
Phase space calculations, 357
$\phi_{l,p}(r)$, *see* Regular wave function
Photoelectric effect, 425-428
Physical region, 208, 215, 243
Physical sheet of E, 243
  multichannel, 395-396
Pion-nucleon scattering, 105, 432
Plane waves, 41
  expansion in terms of spherical waves, 185
"Plus" state, $|\phi+\rangle$, 31
Polarization analyzing power, 117
Polarization experiments, 116-119
Polarization vector, 112
Polarizing power, 118
Poles of f_l, residue at, 236
Poles of Green's operator, 130-131
Poles of multichannel S matrix, and
      bound states, 405-407
  and resonances, 407-413
Poles of s_l, and bound states, 223-226
  and resonances, 244
Post form of T matrix, 361, 365, 371
Potential, analytic, 220-221
  assumptions on, 26-28, 191-192
  channel, 360
  Coulomb, *see* Coulomb potential
  Coulomb plus short-range, 266-269
  exponential, 236
  exponentially bounded, 220, 229-230
  finite-range, 392
  hard sphere, 209-210
  nonlocal, 37
  $1/r^4$, 231
  optical, 384-388, 390
  separable, 37, 141
  singular, 27-28
  square well, *see* Square well
  Yukawa, *see* Yukawa potential
Potential matrix, 382, 389
Principal value integral, 289
Prior form of T matrix, 361, 365, 371
Projector, defined, 386
  onto channel subspace, 386
  symmetrizing, 437-438

Propagator, 161
Proper vectors, 8
$\psi_{l,p}(r)$, *see* Radial wave function
Pure state, 110

$\mathfrak{R}$, space of scattering states, 32-33, 329
$\mathfrak{R}_\pm$, defined, 32
$\mathfrak{R}_\pm^\alpha$, defined, 326
$\mathbb{R}^3$, defined, 5
$\mathfrak{R}_-^\alpha \neq \mathfrak{R}_+^\alpha$, 330
Radial Schrödinger equation, 186
  for coupled channels, 389-390, 399
  free, 182
Radial wave function, 185-188
  behavior as $r \to \infty$, 187-188
  Born series for, 207-208
  Coulomb, 261, 267
  for coupled channels, 389-390, 399-400
  free, 182-185
  incoming and outgoing, 224
  normalized, 186
  see also Regular wave function
Ramsauer-Townsend effect, 195
Range of an operator, 11
Reaction (or reactance) matrix, 281
  see also K matrix
Rearrangement collisions, 378-379
  Born approximation for, 455-457
Reciprocity theorem, 354
  see also Time-reversal invariance
Reduced mass, 59
Reflection in a plane, 99
Regge poles, 306-308
  and amplitude as $\cos\theta \to \infty$, 312-313
  and high energy limits, 313
Regge theory, 302-314
  in relativistic scattering, 313-314
Regge trajectories, 306-308
Regular functions, 213
Regular wave function, 197
  analyticity in p, 216-217
  bounds on, 203-204, 217
  by iteration, 201-204
  integral equation for, 201
  multichannel, 400-402
Relative coordinates, 58, 343-344
  for three particles, 356
Relative momenta, 59, 344
Residue of S matrix, at bound-state
      pole, 236, 406-407

Residue of S matrix (*continued*)
  at resonance pole, 409-412
Resolvent, *see* Green's operator
Resolvent equation, 133
Resonances, 195, 238-258
  and bound states, 245-249
  as nearly bound states, 254-255
  as poles of $s_l(p)$, 244
  as zeros of the Jost function, 240-244
  behavior of cross section at, 241-244,
    412-413
  behavior of δ_l at, 240-242
  Breit-Wigner, 242, 244, 247-248
  delay of scattered packet at, 251-253
  exponential decay of, 253-254
  (illustrated), 239, 242, 248
  multichannel, 407-417
  multichannel, and poles of S
    matrix, 407-413
  multichannel Breit-Wigner, 409-410,
    412-413
  multichannel, decay of, 413-416
  multichannel (illustrated), 411, 413
  multiple pole, 256, 258
  near threshold, 246-249
  of a square well, 257
  of subsystems, 431-432
  zero-energy, 235, 246
  see also Width of a resonance
Riccati-Bessel function, 182-184
  bounds on, 211, 235
  orthogonality of, 183
Riccati-Hankel function, 183-184
  bounds on, 218, 235
Riccati-Neumann function, 183-184
  bounds on, 211, 235
Riemann-Lebesgue lemma, 167
Riemann surface of E, 243
Right-hand cut, 287
Rotation matrix, 124-125
  orthogonality relation for, 125
Rotation of spin states, 96
Rotation operators, 83
Rotational invariance, and angular
    momentum conservation, 83-84, 86
  and partial-wave series, 85-90
  for particles with spin, 97-98, 101-104
  in multichannel scattering, 351-353
  in (spin ½) – (spin ½) scattering, 107
  in (spin ½) – (spin 0) scattering, 98-102

Rutherford cross section, 266
  see also Coulomb potential

S, *see* S operator
S^α, *see* Channel, subspaces
s_l, *see* Partial-wave S matrix
S matrix, 41
  in angular-momentum basis, 86-87
  in angular-momentum basis for particles
    with spin, 103-105
  in helicity and angular-momentum
    basis, 123
  in momentum-helicity basis, 122
  in terms of $T(z)$, 138-139
  multichannel, 341-345
  perturbation series for, 159
  symmetry of, 106, 282, 357
  see also Partial-wave S matrix, S operator
S operator, 34-37
  for particles with spin, 73
  for two particles, 62
  in interaction picture, 158
  multichannel, 335
  unitarity of, 36-37
Scattering, electron-helium, 155-156
  electron-hydrogen, 154-155, 294,
    368-370, 454-459
  neutron-proton, 235, 257
  of particles with spin, 72-73
  of two composite particles, 389
  of two identical particles, 441-448
  of two particles, 60-68
  off a composite target, 365-383
  off two potentials, 270-271
  pion-nucleon, 105, 432
  proton-carbon, 117-119
Scattering amplitude, 43-44
  as function of E and θ, 85
  behavior as $q^2 \to \infty$, 312-313
  behavior at low energy, 194
  direct, 435, 453, 455
  exchange, 435, 453-457
  for Coulomb potential, 265
  for helicity states, 122
  for particles with spin, 74
  for several identical particles, 453, 462
  for two-body processes, 349
  for two identical bosons, 444-446
  for two identical fermions, 447, 460-462
  for two particles, 63

Scattering amplitude (*continued*)
  in terms of stationary scattering states,
      169, 179
  matrix, 75
  relation to scattering states as $r \to \infty$,
      3-4, 172, 377, 379
  relation to T matrix, 43, 349
  spin-flip and spin-nonflip, 75
  see also Partial-wave amplitude,
      Partial-wave series, T matrix
Scattering length, 193-194, 230
  large, 235
  minimum principle for, 280
Scattering operator, *see* S operator
Scattering orbits, classical, 24
Scattering probability, 35
  for identical particles, 451-453
  multichannel, 325-326, 335
Scattering states, 31, 33
Schrödinger equation, coupled-channel,
      382-383, 398-399
  free radial, 182
  multichannel radial, 389-390, 399-400
  radial, 186
  time-dependent, 13
Schrödinger picture, 157
Schwartz inequality, 17
Schwartz reflection principle, 214, 288
Schwinger variational method, 274-277, 284
Self-adjoint operators, 8
Separable potentials, 37, 141
Single-channel scattering, 21-314
  defined, 1-2
Singular potentials, 27-28
Sommerfeld-Watson transform, 310-312
Spatial wave functions, 9
Spectral function, double, 298
Spherical Bessel functions, 182-183
Spherical harmonic addition theorem, 88
Spherical-wave basis functions, 86
Spin, 70-72
  of fragments in multichannel scattering,
      336, 342
  rotation of, 96
  sums and averages, 76-78
Spin-flip and nonflip amplitudes, 75
Spinors, 70-71
  in and out, 78-80
Spin-statistics, theorem, 437

Spiralling orbits, 24
Square well, amplitude for, 210
  analytic properties corresponding to, 220
  Born approximation for, 163
  phase shifts and cross sections, 195-196
  resonances of, 257
  s-wave phase shift for, 257
Standing-wave states, 283
State vector, 7
Static approximation, 153, 382-383
Stationary scattering states, 3-4, 165-168
  and asymptotic condition, 166-167
  as basis, 167-168
  asymptotic form of, 172
  Born series for, 170
  for Coulomb potential, 265
  Lippmann-Schwinger equation for, 169
  multichannel, 359
  multichannel, asymptotic form of,
      376-379, 389
  multichannel, Lippmann-Schwinger
      equation for, 362
  partial-wave, 185-188
  relation to actual wave packet, 173
  standing wave, 283
  target-state expansion of, 380-383, 389
Strength parameter λ, 191
  analyticity in, 207-208
  complex values of, 207
Stripping reactions, 421-424
Strong convergence, 19
Subspaces, 9-10
  channel, 323-324
  of bound states, 32, 326
  of scattering states, 32-33, 329
Subtracted dispersion relations, 299-300, 313
Sums over spins, 77-78
Symmetric wave functions, 436
Symmetrizing projectors, 437-438
Symmetry of S matrix, 106, 282, 357

T matrix, 42-43, 344
  for particles with spin, 73-74
  for two particles, 63
  in terms of stationary scattering states,
      169, 179, 361
  multichannel, 344, 361
  off-shell, 43, 139-140
  on-shell, 42-43

T matrix (*continued*)
  on-shell, in terms of off-shell, 139
  post and prior forms of, 361
  relation to amplitude, 43, 349
  see also Scattering amplitude, T operator
T operator, 134
  Lippmann-Schwinger equation for, 135
  multichannel, 363-364
$T(z)$, *see* T operator
Target-state expansion, 380-383, 389
Tensor product, 57-59
Threshold behavior of f_l, 193-194,
    229-231, 234-235
Thresholds, 318-319, 345
  singularities of S matrix at, 393-395
Time-delay at a resonance, 251-253
Time-dependent scattering theory,
    defined, 2
Time-development operator, *see* Evolution
    operator
Time-independent scattering theory,
    defined, 3, 128
Time reversal, 93
  effect on radial wave functions, 209
  effect on spin states, 97, 106
  effect on stationary scattering states, 178
Time-reversal invariance, 94-95
  for particles with spin, 104
  in multichannel scattering, 353-356
  in (spin ½) – (spin 0) scattering, 102-103,
    106-107
  in (spin ½) – (spin ½) scattering, 107
  in terms of helicity states, 127
Total cross section, 54
  for identical particles, 448
  multiparticle, 350
  see also Cross section
Translation operator, 82
Translational invariance, 82-83
Triangle inequality, 18
Triple scattering experiments, 119
Two-particle, scattering, 60-68
  relation to one-particle problem, 62
Two-particle states, 57-60
Two-potential formula, 271
Two-state approximation, 382

Unitarity, and the Born approximation,
    149-150, 190-191
  of multichannel S matrix, 341, 352-353

Unitarity (*continued*)
  of S and the optical theorem, 53-54
  of S and threshold singularities, 393-394
  of S operator, 36-37
Unitarity bound on partial cross sections, 89
Unitarity cut, 394
Unitary circle, 191, 353
Unitary operators, 13-14
  notation for, 5
Unpolarized beams and targets, 76
Unpolarized cross section, 77-79, 114-115
  for identical particles, 447
  for (spin ½) – (spin 0) scattering, 78, 100
Unstable states, *see* Resonances

Variable phase method, 197-201
Variational method, for ground
    states, 273-274
  Kohn, 277-280
  Schwinger, 274-277, 284
Vector-coupling coefficient, 71, 108
Virtual state, 246, 257
Volterra equations, 201

Watson transform, 310-312
Watson's theorem, 271
Wave operators, *see* Møller operators
Wave packet, free, 174-175
  Gaussian, 29-30, 37
  motion during collision, 175-176
Weak convergence, 19
Width of a resonance, 244, 250
  partial, 410, 412, 415-416
  related to mean life, 254-255
Wigner coefficient, 71, 108
Wigner's theorem, 91-92
Wigner-Wiesskopf model, 416
Wronskian, 225
  of two matrices, 417

$\underline{x}$, set of coordinates, 322

Yukawa potential, analyticity of f_l for,
    220-223, 287-290
  Born approximation for, 150-153, 178
  Mandelstam representation for, 297-300
  nonforward dispersion relation
    for, 294-297
  Regge trajectories for, 307-308
  Watson transform for, 309-312

Zero-energy bound states, 232
Zero-energy resonance, 235, 246
Zero of energy, choice of, 321
Zeros of det $\mathcal{F}$ and bound states, 405-406
Zeros of Jost function, and
    bound states, 223-226

Zeros of Jost function (*continued*)
  and resonances, 240-244
  are simple when Im $p > 0$,
    226, 236
  at threshold, 232-233
  motion near threshold, 245-249